AF769774

BIFURCATION AND CHAOS IN NONSMOOTH MECHANICAL SYSTEMS

WORLD SCIENTIFIC SERIES ON NONLINEAR SCIENCE

Editor: Leon O. Chua
University of California, Berkeley

Series A. MONOGRAPHS AND TREATISES

Volume 27: The Thermomechanics of Nonlinear Irreversible Behaviors — An Introduction
G. A. Maugin

Volume 28: Applied Nonlinear Dynamics & Chaos of Mechanical Systems with Discontinuities
Edited by M. Wiercigroch & B. de Kraker

Volume 29: Nonlinear & Parametric Phenomena*
V. Damgov

Volume 30: Quasi-Conservative Systems: Cycles, Resonances and Chaos
A. D. Morozov

Volume 31: CNN: A Paradigm for Complexity
L. O. Chua

Volume 32: From Order to Chaos II
L. P. Kadanoff

Volume 33: Lectures in Synergetics
V. I. Sugakov

Volume 34: Introduction to Nonlinear Dynamics*
L. Kocarev & M. P. Kennedy

Volume 35: Introduction to Control of Oscillations and Chaos
A. L. Fradkov & A. Yu. Pogromsky

Volume 36: Chaotic Mechanics in Systems with Impacts & Friction
B. Blazejczyk-Okolewska, K. Czolczynski, T. Kapitaniak & J. Wojewoda

Volume 37: Invariant Sets for Windows — Resonance Structures, Attractors, Fractals and Patterns
A. D. Morozov, T. N. Dragunov, S. A. Boykova & O. V. Malysheva

Volume 38: Nonlinear Noninteger Order Circuits & Systems — An Introduction
P. Arena, R. Caponetto, L. Fortuna & D. Porto

Volume 39: The Chaos Avant-Garde: Memories of the Early Days of Chaos Theory
Edited by Ralph Abraham & Yoshisuke Ueda

Volume 40: Advanced Topics in Nonlinear Control Systems
Edited by T. P. Leung & H. S. Qin

Volume 41: Synchronization in Coupled Chaotic Circuits and Systems
C. W. Wu

Volume 42: Chaotic Synchronization: Applications to Living Systems
E. Mosekilde, Y. Maistrenko & D. Postnov

Volume 43: Universality and Emergent Computation in Cellular Neural Networks
R. Dogaru

Volume 44: Bifurcations and Chaos in Piecewise-Smooth Dynamical Systems
Z. T. Zhusubaliyev & E. Mosekilde

*Forthcoming

WORLD SCIENTIFIC SERIES ON NONLINEAR SCIENCE Series A Vol. 45
Series Editor: Leon O. Chua

BIFURCATION AND CHAOS IN NONSMOOTH MECHANICAL SYSTEMS

Jan Awrejcewicz
Technical University of Łódź, Poland
Claude-Henri Lamarque
ENTPE, Vaulx-en-Velin, France

World Scientific
New Jersey • London • Singapore • Hong Kong

Published by

World Scientific Publishing Co. Pte. Ltd.

5 Toh Tuck Link, Singapore 596224

USA office: Suite 202, 1060 Main Street, River Edge, NJ 07661

UK office: 57 Shelton Street, Covent Garden, London WC2H 9HE

British Library Cataloguing-in-Publication Data
A catalogue record for this book is available from the British Library.

BIFURCATION AND CHAOS IN NONSMOOTH MECHANICAL SYSTEMS

Copyright © 2003 by World Scientific Publishing Co. Pte. Ltd.

All rights reserved. This book, or parts thereof, may not be reproduced in any form or by any means, electronic or mechanical, including photocopying, recording or any information storage and retrieval system now known or to be invented, without written permission from the Publisher.

For photocopying of material in this volume, please pay a copying fee through the Copyright Clearance Center, Inc., 222 Rosewood Drive, Danvers, MA 01923, USA. In this case permission to photocopy is not required from the publisher.

ISBN 981-238-459-6

Printed in Singapore.

Preface

The material of the book can be treated as interaction of the following key word fields: Oscillations, impacts, friction, modern theory of dynamical systems, applied mathematics, numerical techniques, piecewise linear and non-linear systems. In the standard terminology we deal with the lumped, mechanical systems governed by ordinary differential equations.

One may say that our attention is focused on old and classical problems on mechanics, i.e., impact and friction. Yes, this is true and this is a motivation to write the book. We principally are followers of original sources of mathematics and mechanics and vice versa. The applied mathematics took their many ideas from the natural sciences like mechanics, and mechanics looked carefully for development of mathematics to solve many of (still) unsolved problems. Both key problems roughly expressed by the terms "impact" and "friction" are still unsolved and require idea from mathematics for their solution. These ideas can be easily followed in our book. We try to attack old natural problems using more sophisticated approaches from the points of view of applied mathematics and computational physics (mechanics).

We use all recent developments in the field of modern dynamical systems (bifurcation and chaos theories, theory of differential inclusions, both modern analytical approaches to predict chaos and numerical algorithms to solve the "discontinuous" problems which are sometimes rigorously discussed, etc.) to look again for "impact" and "friction". The pendulum is taken as another paradigm tool that exhibits fascinating and complicated non-linear behavior. *The exhibition of simultaneous simplicity and complexity in nonlinear dynamical behaviour is one of the main goals of the reported research.*

This book is focused on mechanical (natural) insight than on overloading a reader with "heavy" mathematical treatment. However, we do not forget on mathematics addressing some questions in a more rigorous manner. We provide the reader with many illustrations and physical explanations of the system behaviour, and we try to use the most simplest objects to exhibit the most complicated discontinuous dynamics.

The book is intended to be used by physicists, applied mathematicians and engineers and it can also serve as a textbook for a class on nonlinear oscillations. Our simple and clear approaches yield many fascinating discoveries in the ocean of non-linear (piecewise linear) lumped dynamical system behaviour. Our examples taken from mechanics serve as a call to researchers: be patient, do not always believe in authorities, utilize mathematics since the fascinating non-linear hidden world awaits your discovery!

In Chapter 1 both Fillipov's and Aizerman's theories are described which are suitable to analyse piecewise continuous systems. Two simple examples introduce a reader (very often met in physical systems) to jumps of velocity and acceleration. Moreover, a boundary value problem is formulated to find analytically and numerically oscillations of one-degree-of-freedom non-autonomous oscillator with Coulomb friction.

Chapter 2 deals with the mathematical background for multivalued formulations. First the origin of both smooth and unsmooth nonlinearities is illustrated. Then two simple model examples with friction and impact are studied. The existence and uniqueness problems, maximal monotone operators and ill posed problems are rigorously discussed. Finally, some remarks concerning both deterministic and stochastic differential equations are given.

In Chapter 3 well-adapted numerical schemes to trace differential inclusions and differential equations are proposed. Since the book is intended to serve as a text for engineers and applied scientists, the methods and language heavily accessible to engineers are avoided, and hence a rigorous mathematical treatment is here omitted.

Chapter 4 is devoted to general questions of convergence of numerical schemes for relatively large classes of nonlinear problems governed either by differential inclusions or by ordinary differential equations and differential inclusions. Dynamical systems with friction (or elastoplastic terms) and impacts are studied. A special attention is paid to resolution methods, sticking motion and accumulation of impacts. More rigorous mathematical background including theorems and their proofs applied to numerical analysis of nonsmooth differential systems is included.

Bifurcation and chaos of a particular van der Pol-Duffing oscillator with Coulomb friction are discussed in Chapter 5. The so-called van der Pol averaging procedure is applied to trace "0"-type and complex bifurcations and their correspondence to chaos are discussed.

A systematic approach to trace stick-slip dynamics with friction using mechanical example with two-degrees-of-freedom is presented in Chapter 6. A special attention is paid to analyse transitions from the stick to slip, slip to stick and slip to slip states.

The shooting method is used to trace stability and bifurcations of equilibria and periodic orbits. Many interesting examples of bifurcations and chaotic behaviour of investigated slide-roll oscillator are reported and discussed. This chapter is based on the references [Awrejcewicz and Delfs (1990a); Awrejcewicz and Delfs (1990b)] written together with J. Delfs (the so-called stick-slip chaotic dynamics exhibited by our physical system has been reported for the first time in 1990).

Another two-degrees-of-freedom system with horizontally situated bodies and Coulomb friction is studied in section 6.3. Some standard numerical approaches to analyse nonlinear and discontinuous dynamics are reviewed and a validity of analytical approximation to friction is discussed. A variety of non-linear dynamical phenomena is reported.

The piecewise linear approximations are discussed in Chapter 7. First exact and approximated models are analysed, and then numerical results are given. This chapter is based on reference [Mesnier *et. al.* (1999)].

Chapter 8 is devoted to analysis of Chua circuit, and particularly to electromechanical or purely mechanical realization of Chua's circuit and Chua's unfolding circuit. In spite of some proposed mechanical devices governed by Chua's equations the rheological model of the generalized double scroll Chua's circuit is proposed, and existence and uniqueness of solutions of discontinuous Chua's model are addressed. This chapter used materials from reference [Lamarque *et. al.* (1999)].

The modal approaches to investigate one- and two-degrees-of- freedom mechanical systems with impacts are proposed in Chapter 9. A piecewise exact integration is used to study the periodic responses under sinusoidal excitation. Both weak and strong couplings are examined while studing two-degrees-of-freedom systems. The source of this chapter is [Janin (2001)] and [Lamarque and Janin (2000)].

An approximate solution of the differential inclusion model of a pendulum with friction is analysed in Chapter 10. Existence and uniqueness of results and numerical schemes convergence are studied. Special coexisting attractors generated by friction are investigated. The Melnikov's method is used to predict chaos. Chapter 10 is based on reference [Lamarque and Bastien (2000)].

In Chapter 11 several types of rheological models constituted of springs, St-Venant elements and dashpots are examined within a frame of a differential inclusion. Numerous simulations exhibiting hysteresis limit cycles are presented for different periodic forcings. Materials comes from the common work with J. Bastien and M. Schatzman [Bastien *et. al.* (2000)].

Seven-degrees-of-freedom mechanical system with impacts is studied in Chapter 12. It models a cable with its shaft and a gear level with its support. The numerical scheme, numerical results and comments are given.

In Chapter 13 dynamics of grazing periodic solutions with one impact at zero velocity per cycle are investigated. Stability of grazing periodic solutions are exhibited and illustrated, among others.

The triple pendulum with damping external forces and with impacts is investigated in Chapter 14. The theory of Aizerman and Gantmacher is applied to calculate the fundamental solution matrices in the system with discontinuities. The fundamental matrices are used during calculation of the Lyapunov exponents, during stability analysis of periodic solutions (Floquet multipliers) and in shooting procedure. Periodic, quasi-periodic and chaotic dynamics are reported and discussed.

For the first time in 1999 stick-slip chaotic dynamics has been predicted in a one-degree-of-freedom quasi-autonomous oscillator using the Melnikov's method [Awrejcewicz and Holicke (1996)]. Chapter 15 is based mainly on the mentioned reference. The critical chaotic threshold curves, where infinitely small external periodic perturbations may lead to chaos, are reported, among others.

In Chapter 16 the plane axially-symmetric problem dealing with thermoelastic contact of the rotating shaft with the rigid bush fixed elastically to the steady base by springs in conditions of frictional self-oscillations and wear is studied. This problem is rather complicated, since a shaft temperature, the contact pressure, wear of the bush, and the bush movement are coupled with each other creating a frictional joint. Steady state solutions with and without wear are analysed and some important conclusions are derived.

In the last Chapter 17, a control procedure for systems including elasto-plastic terms is described: It is based on classical Riccatti process obtained for linear systems. A linear system is associated with non linear initial one. Then control force is applied to the full non linear system.

This book project is a result of our seven years co-operation, which started in 1996 with the help of the Région Rhône-Alpes (France). In September 16-18, 1996 the International Conference on "Nonlinearity, Bifurcation and Chaos. The Doors to the Future" has been organised in the frame of co-operation between the Technical University of Łódź (Poland) and Ecole National des Travaux Publics de l'Etat of Vaulx-en-Velin (France). This conference has attracted many top researches from the field.

Then our common research has been supported by numerous grants from both Polish and French sides ("Tempra" Program in 1995-1996, and "Tempra" PECO Program in 2000-2001), including also common grants (like three years grant "Polonium" in 1999-2001).

Also many of our Ph. D. students and co-workers have been involved in the cooperation in the frame of Erasmus/Socrates Programs in the years 1996-2001.

Our cooperation has been continued and re-freshed by the every-two-year meetings during the series of international conferences "Dynamical Systems — Theory and Applications" sponsored by the Technical University of Łódź, Ministry of Education of Poland (KBN) and the Stefan Batory Foundation (Poland). Also the grant obtained from the Ministry of Education of France in 1999 is highly appreciated (J. A.)

Help from both our universities, the Technical University of Łódź and the National School of Public Works (ENTPE, especially the Geo-Materials Laboratory – LGM and the Building and Civil Engineering Department – DGCB, URA CNRS 1652) is appreciated.

The first author (J. A.) acknowledges the help and cooperation of his former Ph.D. students and co-workers: J. Delfs, M. Holicke, G. Kudra, P. Olejnik and Dr. Yu. Pyr'ev. The final realisation of the book would be impossible without important revisions of the draft and many valuable comments given by Dr. J. Nowakowski and M. Kaźmierczak.

This book is also the result of common works with several former students (Vincent Mesnier, Tristan Robert) or former Ph.D. students of Claude-Henri Lamarque: Jérôme Bastien, Olivier Janin, Stéphane Pernot. Without them many things would have not been possible. And even if they are not directly involved in this book, one of us (CHL) particularly remembers his friends and colleagues L. Jézéquel, V. Roberti, J.-M. Malasoma, J.-M. Cornet, F. Roberti. Especially thanks are addressed to Michelle Schatzman (Laboratory MAPLY, Lyon I University and UMR CNRS 5585) for common works and discussions. Her help in the mathematical (but not only) point of view is always invaluable.

The investigations included in this book have been supported by both the research program "Polonium" between Poland and France and by the Polish Scientific Research Committee (KBN) under the grant No. 5 T07A 019 23

Jan Awrejcewicz
Łódź (Poland)

Claude-Henri Lamarque
Vaulx-en-Velin (France)

Contents

Chapter 1

Introduction to Discontinuous ODEs

1.1 Introduction

Tracing research focused on friction it seems that the model introduced by Coulomb is still one of the most appropriate (perhaps with slight modifications) and therefore it will be further used in our considerations. It appears that friction causes not only problems in industrial equipments but also there are some not yet satisfactorily solved problems of mathematical nature. Although since the Coulomb's work published in 1781, a great amount of papers devoted to friction phenomena have appeared, the problem remains still open. We do not review the state of the art in this field but we only indicate some research directions oriented on friction. The oldest approach is based on an experiment and modelling of friction (see, for instance, works [Dowell and Schwartz (1983b)], [Tomlinson and Chen (1996)] and [Hess and Soom (1990)]). Some papers are devoted to numerical investigations of systems with friction ([Stewart (1996)], [Stewart and Trinkle (1997)]). Analytical and asymptotical approaches to analyse the friction phenomena are sometime also applied ([Capecchi and Vestroni (1995)] and [Altpeter *et. al.*(1998)]).

Stick-slip and chaotic phenomena caused by friction have been analysed by [Pratt and Williams (1981)], [Awrejcewicz and Delfs (1990a)], [Awrejcewicz and Delfs (1990b)], [Stelter (1992)], [Ionescu and Paumier (1993)], [Baumberger *et. al.* (1995)], [Shaw (1985)] and [Madan (1993)]. More theoretically oriented research has been carried out by [Deimling (1992)], [Deimling and Szilágyi (1994)] and [Fečkan (1997)]. For instance it is possible to generalize classical Conley index theory to nonsmooth system and use it to prove bifurcation results. Also the classical Kolmogorov-Arnold-Moser (KAM) theory can be used to prove the existence of periodic and quasiperiodic solutions of simple, one degree of freedom, conservative, nonsmooth system ([Kunze (2000)]). The mathematical results concerning existence and uniqueness of nonsmooth systems solutions have been reported by [Jean and Pratt (1985)], [Monteiro Marques (1994)], [Matrosov and Finogenko (1996a)], [Bastien (2000)]. The recently observed research involved with friction phenomena is oriented towards control. Here two paths of investigations can be distinguished. First, a mathematical treatment concerning stability of nonsmooth hybrid mechanical systems is recently developed (see [Brogliato (1996)]). Second, a real control of the systems with friction is complemented ([Armstrong-Helouvry (1991)], [Armstrong-Helouvry *et. al.* (1994)]).

Consider a simple one degree of freedom autonomous oscillator of the form:

$$\ddot{x} + c\dot{x} + \mu sgn\dot{x} + \alpha^2 x = 0, \tag{1.1}$$

which is dissipative due to occurrence of viscous damping and Coulomb type friction.

In the interval $[-\mu, \mu]$ for $\dot{x} = 0$ we have infinitely many solutions, i.e. every $(x, 0)$ with $x \in [-\mu, \mu]$ is an equilibrium, which is not represented by Eq. (1.1). The Eq. (1.1) governs a real behaviour only for $\dot{x} \neq 0$. If we define in a typical way sgn(0) = 0, it does not describe autonomous system (1.1). In the latter case a solution can possess a dead zone for $t \in [t_1, t_2]$, $t_2 > t_1$.

Based on the introduced examples it is clear that we need, at least from the mathematical point of view, to adapt mathematical approach in order to fit the described physical behaviour. This means that an extension of the differential equations concept is needed to describe discontinuous behaviour.

It corresponds to an earlier introduced theory related to the differential equations with their multivalued right-hand sides or so called differential inclusions [Filippov (1988)]:

$$\frac{dx}{dt} \in f(t, x), \tag{1.2}$$

where $f(x, t)$ is the multivalued function (a set) in R^n. However, if for all (x, t) a set includes only one point then (1.2) reduces to the ODEs. The theory of differential inclusions assumes that for each point (t_0, x_0) a set $f(t_0, x_0)$ is a convex, closed and bounded one. In addition, the following assumption holds: if $(t, x) \to (t_0, x_0)$ then a limit of $f(t, x) \subset \overline{f}(t_0, x_0)$.

We give some definitions below

Definition 1.1 *The set $\Omega \subset X$ is called convex if*

$$(x, y \in \Omega) \Rightarrow ([x, y] \subset \Omega).$$

The smallest convex set including a given set E is denoted as conv E. If E is composed of the points $x_1, x_2, \ldots, x_n$ then

$$conv\ E = \left\{ x : x = \sum_{i=1}^{n} \lambda_i x_i, \text{ where } \lambda_i \geq 0 \text{ and } \sum_{i=1}^{\lambda_i} = 1 \right\}.$$

This definition is illustrated in Fig. 1.1. Another equivalent definition of a convex set follows.

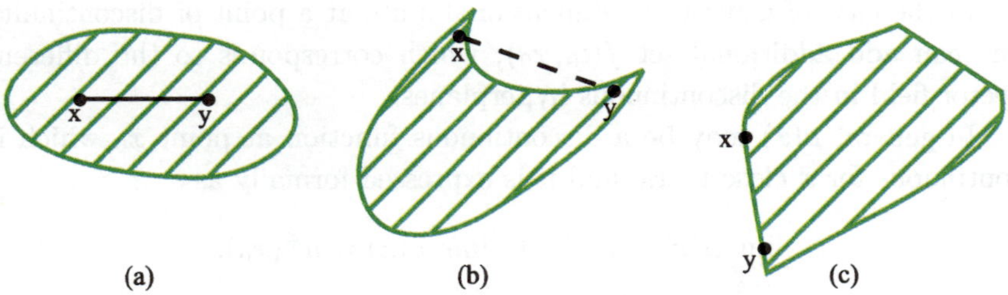

Fig. 1.1 A convex (a), non-convex (b) and *conv E* (c) sets.

Definition 1.2 *Let the series $\{x_n\}$, $\{y_n\}$,... belong to $E \subset X$. A set containing limits of all series $\{x_n\}$, $\{y_n\}$,... is called the closure of the set E and is denoted by $\overline{E}$. A set E is called closed if $E = \overline{E}$.*

Let X be a linear real space. A set $M \subset X$ is called convex if

$$(x, y \in M) \Rightarrow (\lambda x + (1 - \lambda)y) \in M \text{ for } \lambda \in [0, 1]).$$

Some examples of multivalued set represented by different friction approximations are given in Fig. 1.2.

For $\dot{x} = 0$ we have a multivalued static friction, which is often used in mechanical systems to describe either energy dissipation or self-excited oscillations. Therefore, one needs to add some other sets in the points of discontinuity which should satisfy earlier mentioned properties. Then, for an arbitrary (t_0, x_0) from a set of $f(x, t)$ and for t sufficiently close to t_0 a solution $x(t)$ exists which satisfies $x(t_0) = x_0$. In addition, all theorems related to solution extension and its continuous dependence on the parameter change hold.

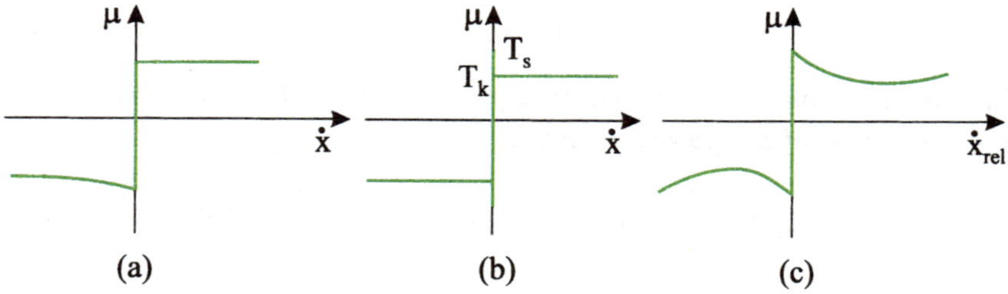

Fig. 1.2 Friction against velocities.

In the case of general n - dimensional state at a point of discontinuity one can add additional set $f(t_0, x_0)$, which corresponds to the different vector field in the discontinuous hyperplanes.

In general $\mu(x)$ may be a discontinuous function at point x_0 which is continuous for x close to x_0, and it is expressed formally as

$$\lim_{\substack{x \to x_0 \\ x < x_0}} \mu(x) = \mu^-(x_0), \ \lim_{\substack{x \to x_0 \\ x > x_0}} \mu(x) = \mu^+(x_0).$$

1.2 Filippov's Theory

Consider any dynamical system with a discontinuous vector field $f(x,t)$ on a certain smooth hypersurface in the space (t,x), which generates a continuous flow. We assume that when a point (t,x) approaches this hypersurface there are limiting values $f^+(t,x)$ and $f^-(t,x)$ depending on the direction of approach, i.e. from one or the other side of the hypersurface. The question appears how one can define a vector field in order to realize a sliding mode. According to Filippov we take two limiting values of vectors f^+ and f^- in the point x, and we link their ends (see Fig. 1.3).

The intersection between a triangle constructed by f^+ and f^- and the tangent hyperplane given by the vector f^0, which defines the vector field in the point x, i.e. gives a solution to the ODEs with discontinuous right hand side.

From a point of view of the earlier introduced description of the differential inclusions we have to add a multiple function on the discontinuous hyperplane, which should be bounded, convex and $f^{(\cdot)} \in (t,x)$, $(\cdot) = (+,-)$.

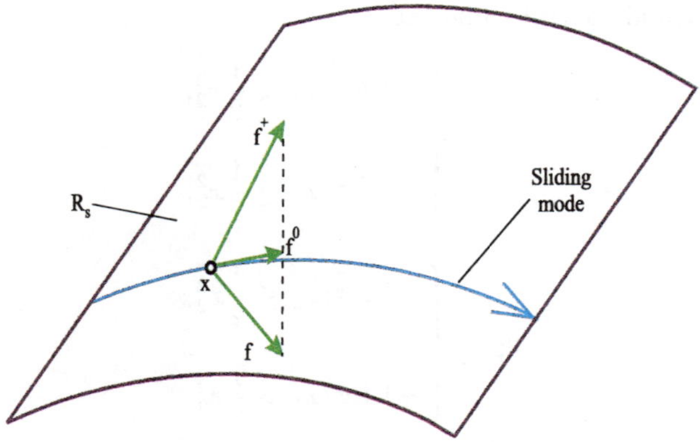

Fig. 1.3 Stable sliding mode.

The smallest set satisfying the mentioned requirements is defined by $\varrho = \{\lambda f^+(t,x) + (1-\lambda) f^-(t,x); \lambda \in [0,1]\}$. Therefore, we define $f(t,x) = \varrho$ for $x \in R_S$. In addition, a vector representing the velocity of the sliding mode must belong to a tangent hyperplane and this requirement leads to Filippov's results indicating that this is satisfied only by $f^0(t,x) \in \varrho$. It can be proved that Filippov's solution corresponds to a possible minimal set $f(t,x)$, but unfortunately there exist more than one solution satisfying the described requirements.

It should be emphasized that although the Filippov's theory has many advantages (mostly Filippov's choice results in solution uniqueness, and it provides an explanation for unstable branches of stick-slip vibrations), it is not a panaceum to solve all problems related to physically discontinuous systems behaviour.

In order to demonstrate it we consider the following example given in monograph [Filippov (1988)].

Consider the system

$$\begin{aligned} \frac{dx_1}{dt} &= x_2 u_1, \\ \frac{dx_2}{dt} &= u_2, \end{aligned} \tag{1.3}$$

where $|u_1| \leq 1$ and $|u_2| \leq 1$ are control variables, which should realize the quickest movement of an arbitrary point x to the origin. It can be shown that the optimal control values are

$$
u_1 = \begin{cases} +1 \text{ for } x_1 < \dfrac{1}{2}x_2^2, \\ \\ -1 \text{ for } x_1 \geq \dfrac{1}{2}x_2^2, \end{cases}
\qquad
u_2 = \begin{cases} -1 \text{ for } x_1 \leq \dfrac{1}{2}x_2^2, \\ \\ +1 \text{ for } x_1 > \dfrac{1}{2}x_2^2. \end{cases}
\tag{1.4}
$$

A particular optimal trajectory is defined by $2x_1 = x_2^2$, and it possesses the property $\dot{x}_1 = -x_2$, $\dot{x}_2 = -1$. Now we take the point $x = (x_1, x_2)$ lying on this trajectory (see Fig. 1.4).

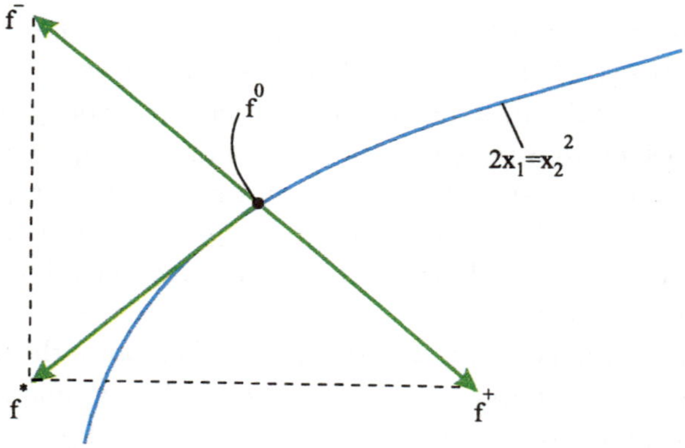

Fig. 1.4 Optimal and Filippov's sliding solutions.

For this case we have $f^-(x) = (-x_2, 1)$ and $f^+(x) = (x_2, -1)$. Because $f^-(x) = -f^+(x)$, therefore $f^0(x) = 0$. It means that the Filippov sliding mode is defined by $x(t) \equiv const$. However, another sliding solution corresponds to the optimal solution which is defined by $f^*(t, x) = (-x_2, -1)$.

1.3 Aizerman's Theory

Another approach to solve the problems with physically discontinuous systems is presented by Aizerman and his co-workers [Aizerman and Gantmacher (1958)], [Aizerman and Gantmacher (1963)] - see also recent comment on that in report [Leine *et. al.* (1999)].

Let us present a general approach related to a single discontinuity. We consider the differential inclusions (1.2), which are everywhere piecewise continuous in spite of $(t, x) \in R_S$, which is the hyperspace of the multiple valued function $f(t, x)$. The flow crosses the hyperplane R_S when

$$g(t, x) = 0, \tag{1.5}$$

and the normal n perpendicular to R_S is defined by

$$n(t, x(t)) = grad(g(t, x(t)). \tag{1.6}$$

For the crossing time instants t_c there are two derivatives f_c^- and f_c^+ with the magnitudes $n^T f_c^-$ nad $n^T f_c^+$. The flow crosses the hyperspace when

$$n^T f_c^- n^T f_c^+ > 0. \tag{1.7}$$

Now, in order to find a fundamental solution matrix we proceed in a typical way. We disturb by δx_0 the initial conditions and follow the corresponding changes on an orbit $x(t)$. The relations between the state changes in time t_0 and t are governed by equation

$$\delta x(t) = \phi(t, t_0)\delta x_0, \tag{1.8}$$

where $\phi(t, t_0)$ can be chosen as $\phi(t_0, t_0) = I$ (I is the identity matrix). In the latter case $\phi(t, t_0)$ is called the monodromy matrix, and the following equation holds:

$$\dot{\phi}(t, t_0) = J\phi(t, t_0), \ J = \frac{\partial f}{\partial x}, \tag{1.9}$$

and J is the Jacobian. Due to discontinuity exhibited in the time t_c one has two different monodromy matrices

$$\phi(t_c^{\mp}, t_0) = \lim_{t \to t_c^{\mp}} \phi(t, t_0), \tag{1.10}$$

and, generally, $\phi(t_c^+) \neq \phi(t_c^-)$. Because we have two different matrices, one can define a so called 'jump' or 'saltation' matrix in the form

$$\phi(t_c^+, t_0) = S\phi(t_c^-, t_0). \tag{1.11}$$

Due to the known properties of the evolution operator ϕ we multiply both sides of (1.11) by $\phi^{-1}(t_c^-, t_0)$ (note, that ϕ^{-1} exists for the time (t_0, t_c^-)) to get

$$\phi(t_c^+, t_0)\phi^{-1}(t_c^-, t_0) \equiv \phi(t_c^+, t_c^-) = S. \tag{1.12}$$

After intersecting of the discontinuous hyperplane, the fundamental matrix associated with the time t is defined by

$$\phi(t, t_0) = \phi(t, t_c^+)S\phi(t_c^-, t_0). \tag{1.13}$$

So, a state of the system after crossing the discontinuous hyperplane depends on the saltation matrix S.

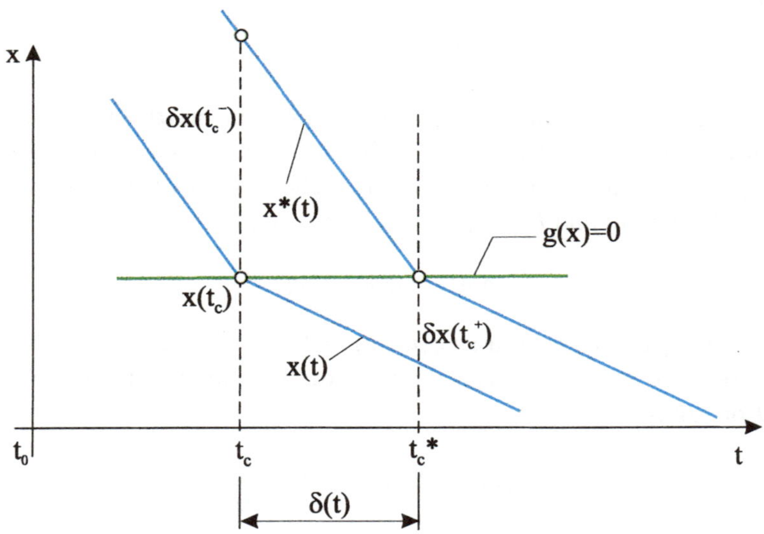

Fig. 1.5 Perturbed (x^*) and unperturbed (x) flows and the corresponding variations.

To define the saltation one can proceed in a classical way but taking into account two different values of f, i.e. f_c^+ and f_c^- at the time instant t_c. We consider the perturbation δx_0 to the initial state x_0, and in the space f^- the disturbed flow stays longer in $t_c^* - t_c$, where t_c^* denotes time of reaching the discontinuous hyperplane of the disturbed flow. The perturbated motion in the vicinity of t_c and t_c^* can be expressed in the following way (see Fig. 1.5):

$$x^*(t_c) = x(t_c) + \delta x(t_c^-), \tag{1.14}$$

$$x^*(t_c^*) = x(t_c^*) + \delta x(t_c^+), \tag{1.15}$$

where $t_c^* = t_c + \delta t$. Now we can apply a Taylor series expansion taking into account only linear terms

$$x(t_c^*) = x(t_c) + f^+\delta t,\ f^+ = \left.\frac{dx}{dt}\right|_{t=t_c^+}, \tag{1.16}$$

$$x^*(t_c^*) = x(t_c^*) + f^-\delta t,\ f^- = \left.\frac{dx}{dt}\right|_{t=t_c^-}. \tag{1.17}$$

From Eq. (1.15), taking into account (1.14) one obtains:

$$\begin{aligned}\delta x(t_c^+) &= x^*(t_c^*) - x(t_c^*) \cong \\ &\underbrace{x(t_c^*) - x(t_c)}_{\delta x(t_c^-)} + (f^- - f^+)\delta t,\end{aligned} \tag{1.18}$$

and finally

$$\delta x(t_c^+) \cong \delta x(t_c^-) + (f^- - f^+)\delta t. \tag{1.19}$$

This equation describes relation between perturbations acting onto two sides of the discontinuous hyperplane.

For $t = t_c^*$ (see Fig. 1.5) for the perturbed trajectory the following estimation holds:

$$g(t_c^*, x^*(t_c^*)) = 0. \tag{1.20}$$

The same holds for $t = t_c$.

Recalling that $x^*(t_c^*) \cong x(t_c)+\delta x(t_c^-)+f^-\delta t$, we can develop the implicit function in the neighbourhood of $(t_c, x(t_c))$:

$$\begin{aligned} g(t_c^*, x^*(t_c^*)) &\cong g(t_c^+ + \delta t, x(t_c) + \delta x(t_c^-) + f^-\delta t) = \\ &\cong g(t_c, x(t_c)) + \frac{\partial g}{\partial t}(t_c, x(t_c))\delta t + n^T(\delta x(t_c^-) + f^-\delta t). \end{aligned} \tag{1.21}$$

The time variation δt reads

$$\delta t = -\frac{n^T \delta x(t_c^-)}{n^T f^- + \dfrac{\partial g}{\partial t}(t_c, x(t_c))}. \tag{1.22}$$

The formula (1.21) can be simplified for an autonomous case since $\frac{\partial g}{\partial t}(t, x(t_c)) = 0$, i.e. the discontinuous hyperplane does not change in time.

From (1.19) and (1.22) one obtains

$$\delta x(t_c^+) = \delta x(t_c^-) + (f^+ - f^-)\frac{n^T \delta x(t_c^-)}{n^T f^- + \frac{\partial g}{\partial t}(t_c, x(t_c))}. \qquad (1.23)$$

However, the variations $\delta x(t_c^+)$ and $\delta x(t_c^-)$ are linked via the saltation matrix of the form

$$\delta x(t_c^+) = S\delta x(t_c^-), \qquad (1.24)$$

where $S = \phi(t_c^+, t_c^-)$.

Substituting (1.24) to (1.23) one gets

$$S = I + \frac{(f^+ - f^-)n^T}{n^T f^- + \frac{\partial g}{\partial t}(t_c, x(t_c))}, \qquad (1.25)$$

which defines the saltation matrix.

1.4 Examples

Two examples will be considered and attention will be focused on the velocity (example 1) and acceleration (example 2) jumps exhibited by piecewise linear systems.

1.4.1 *Example 1*

We consider the system governed by the loop presented in Fig. 1.6 with the piecewise linear term [Gessing *et. al.* (1981)]. The piecewise linear element P possesses in regions II and III saturation parts, whereas in the interval $(-a, a)$ we have $x = k_n u$. The other units are defined in the following way: $K(s) = \frac{1}{V}(1 + \frac{1}{sT_0} + sT)$, $s = \frac{d}{dt}$, $k_0 = 2s^{-1}$, $k_n = 1$, $a = 10$, $V = 0.1$, $T_0 = 10s$, $T = 2s$. The following input value is applied, $w(t) = w_0\mathbf{1}(t)$, $w_0 = 40$. We assume that for $t < 0$ all of the variables were equal to zero and we are going to find $y(t)$ as well as to discuss potential jumps of the variables.

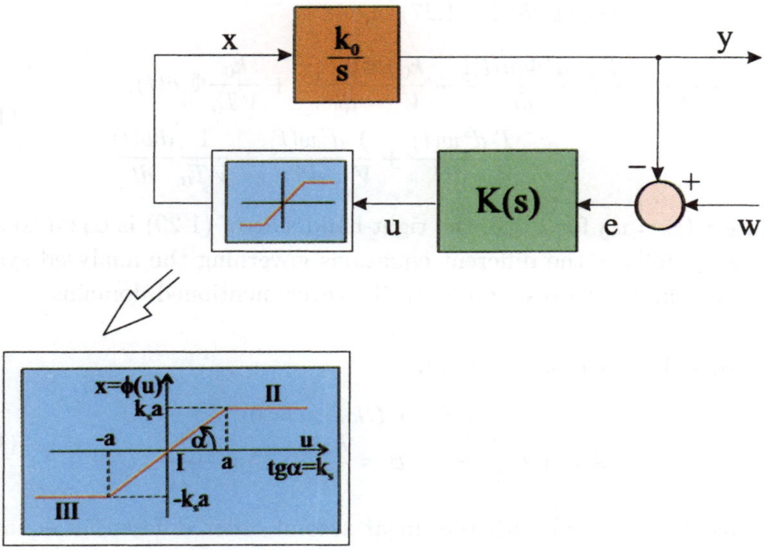

Fig. 1.6 The investigated dynamical system.

Applying standard Laplace operator transformations one finds

$$Y = \frac{k_o}{s}X,\ U = \frac{1}{V}[1 + \frac{1}{sT_0} + sT]E,\ E = W - Y. \tag{1.26}$$

After substraction of Y and E one gets

$$s^2 u = \frac{1}{V}(s^2 T + s + \frac{1}{T_0})(sW - k_0 X),$$

or in a time domain

$$\ddot{u} = \frac{T}{V}\dddot{w} + \frac{1}{VT_0}\dot{w} - \frac{T}{V}k_0\ddot{x} - \frac{k_0}{V}\dot{x} - \frac{1}{VT_0}k_0 x. \tag{1.27}$$

According to Fig. 1.6 the function $\Phi(u)$ is defined in the following way:

$$x \equiv \Phi(u) = \begin{cases} k_s u \text{ in the domain I, where} & |u| \leq a, \\ k_s u \text{ in the domain II, where} & u \geq a, \\ -k_s u \text{ in the domain III, where} & u \leq -a. \end{cases} \tag{1.28}$$

Taking into account (1.28) in (1.27) one gets

$$\frac{d^2u(t)}{dt^2} + \frac{T}{V}k_0\frac{d^2\Phi[u(t)]}{dt^2} + \frac{k_0}{V}\frac{d\Phi[u(t)]}{dt} + \frac{k_0}{VT_0}\Phi[u(t)] = \frac{T}{V}\frac{d^3w(t)}{dt^3} + \frac{1}{V}\frac{d^2w(t)}{dt^2} + \frac{1}{VT_0}\frac{dw(t)}{dt}. \tag{1.29}$$

Because $w(t) = w_0$ for $t > 0$, the right-hand side of (1.29) is equal to zero.

In what follows the different equations governing the analysed system will be formulated corresponding to the three mentioned domains.

Domain I. In this case we obtain

$$A\ddot{u} + B\dot{u} + Ck_0k_su = 0, \quad A = 1 + \frac{T}{V}k_0k_s,\ B = \frac{k_0k_s}{V},\ C = \frac{k_0k_s}{VT_0}. \tag{1.30}$$

It means that we deal with the linear second order differential equation, which generates the following characteristic equation

$$As^2 + Bs + C, \tag{1.31}$$

whith the following roots

$$s_{1,2} = -\frac{B}{2A} \pm \frac{\sqrt{B^2 - 4AC}}{2A}, \tag{1.32}$$

and for given $A = 41$, $B = 20s^{-1}$, $C = 2s^{-2}$ both of them are real and negative ($s_1 = -0.34,\ s_2 = -0.14$). It means that a singular point $(0,0)$ is the stable node, whose phase portrait is defined by

$$\frac{(\dot{u} - s_1u)^{s_1}}{(\dot{u} - s_2u)^{s_2}} = C, \tag{1.33}$$

where constant C depends on initial conditions.

Domains II and III. In this case remembering that $\ddot{u} = \frac{d\dot{u}}{du}\dot{u}$ we get

$$\dot{u}^2 = -\frac{2k_0k_sa}{VT_0}u + C \quad \text{(domain II)}, \tag{1.34}$$

and

$$\dot{u}^2 = \frac{2k_0k_sa}{VT_0}u + C \quad \text{(domain III)}. \tag{1.35}$$

In the phase plane $(\dot{u}, u)$ Eqs. (1.34) and (1.35) are represented by a family of parabolas.

The trajectories defined by Eqs. (1.33), (1.34) and (1.35) are shown in the Fig. 1.7.

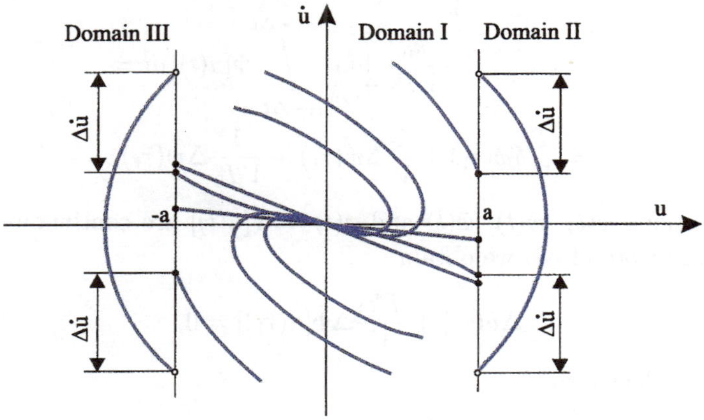

Fig. 1.7 Phase portrait of the investigated system with the marked switching lines.

However, a key attention should be focused on determination of the jumps between different families of solutions which occur on the switching vertical lines $u = \pm a$, which depend on the velocities $\dot{u}$.

Using the theory of distributions one can define the following operation

$$\Delta u(\tau) = u(\tau^+) - u(\tau^-), \tag{1.36}$$

where $u(\tau^+)$ and $u(\tau^-)$ correspond to the right- and left- hand side limits of the function $u(t)$ in the point τ. Note that in case of continuity at point τ we have $\Delta u(\tau) = 0$. A similar definition holds for the derivatives and integrals. The operation $u^{(j-i)}$ for $j-i > 0$ denotes the $(j-i)$th difference, whereas for $j - i < 0$ we have undefined integral of the function $u(t)$. For example, the ith integration of the functions $u^{(j)}(t)$ around the point $t = \tau$ is defined as follows

$$\begin{aligned}\lim_{\substack{\Delta t \to 0 \\ \Delta t > 0}} \int_{\tau-\Delta t}^{\tau+\Delta t} dt_1 \int_{\tau-\Delta t}^{t_1} dt_2 \ldots \int_{\tau-\Delta t}^{t_i-1} u^{(j)}(t_i)\, dt_i &= \\ = \Delta u^{(j-1)}(\tau^+) - u^{(j-i)}(\tau^-).\end{aligned} \tag{1.37}$$

Assume that τ_1 is the time instant when the piece-wise linear element achieves the saturation, i.e. for $\tau_1 + \varepsilon$ ($\varepsilon > 0$) we have $\Phi[u(\tau_1 + \varepsilon)] = \pm k_s a = const$, whereas for $\tau_1 - \varepsilon$ we have $|u(\tau_1 - \varepsilon)| < a$. Integrating Eq. (1.29) around the point $t = \tau_1$ and taking into account (1.37) we get

$$\begin{aligned} \Delta\dot{u}(\tau_1) + \frac{Tk_0}{V}\Delta\dot{\Phi}[u(\tau_1)] + \frac{k_0}{V}\Delta\Phi[u(\tau_1)] + \\ + \frac{k_0}{VT_0} \lim_{\substack{\Delta t \to 0 \\ \Delta t > 0}} \int\limits_{\tau_1 - \Delta t}^{\tau_1 + \Delta t} \Phi[u(t)]\,dt = \\ = \frac{T}{V}\delta\ddot{w}(\tau_1) + \frac{1}{V}\Delta\dot{w}(\tau_1) + \frac{1}{VT_0}\Delta w(\tau_1). \end{aligned} \tag{1.38}$$

The functions $w(t)$, $\dot{w}(t)$, $\ddot{w}(t)$ and $x(t) = \Phi[u(t)]$ are continuous for $t = \tau_1 > 0$ and from (1.38) we obtain

$$\Delta\dot{u}(\tau_1) + \frac{Tk_0}{V}\Delta\dot{\Phi}[u(\tau_1)] = 0, \tag{1.39}$$

or in equivalent form

$$\dot{u}(\tau_1^+) - \dot{u}(\tau_1^-) + \frac{Tk_0}{V}\left\{\dot{\Phi}[u(\tau_1^+)] - \dot{\Phi}[u(\tau_1^-)]\right\} = 0. \tag{1.40}$$

Remembering that the time instant τ_1 corresponds to a saturation achievement we have

$$\dot{\Phi}[u(\tau_1^-)] = \dot{x}(\tau_1^-) = k_s\dot{u}(\tau_1^-), \tag{1.41}$$

and

$$\dot{\Phi}[u(\tau_1^+)] = \dot{x}(\tau_1^+) = 0. \tag{1.42}$$

Taking into account (1.41) and (1.42) in (1.40) we obtain

$$\dot{u}(\tau_1^+) = (1 + \frac{T}{V}k_0k_s)\dot{u}(\tau_1^-). \tag{1.43}$$

One can proceed in a similar way by considering the time instant $t = \tau_2$ when we leave the saturation state. In this case, for $t = \tau_2 - \varepsilon$ ($\varepsilon > 0$), we have $\Phi[u_2(\tau_2 - u)] = \pm k_s a$, whereas for $t = \tau_2 + \varepsilon$ we have $|u(\tau_2 + \varepsilon)| < a$. It results in the equation

$$\dot{u}(\tau_2^+) = \frac{\dot{u}(\tau_2^-)}{1 + \frac{T}{V}k_0k_s}. \tag{1.44}$$

To conclude, Eqs. (1.43) and (1.44) define the jumps of the velocity $\dot{u}$, which occur when approaching or leaving the saturations.

1.4.2 *Example 2*

We consider oscillations of an autonomous oscillator damped by Coulomb friction (see Fig. 1.8).

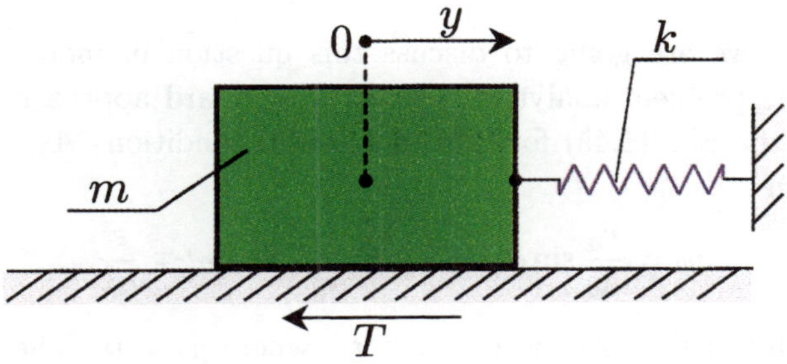

Fig. 1.8 One-degree of freedom oscillator damped by friction.

In the origin of the marked coordinate the spring k is unstretched. Then, the dynamics of the rigid body with mass m is governed by the following differential equation

$$m\ddot{y} + ky = T\mathrm{sgn}(-\dot{y}) \equiv \begin{cases} T \text{ for } \dot{y} < 0, \\ \text{not defined for } \dot{y} = 0, \\ -T \text{ for } \dot{y} > 0, \end{cases} \tag{1.45}$$

which can be transformed to the form

$$\ddot{y} + \alpha_0^2 y = \varepsilon \mathrm{Sgn}(-y), \tag{1.46}$$

where

$$\alpha_0^2 = \frac{k}{m}, \qquad \varepsilon = \frac{T}{m}. \tag{1.47}$$

Note, that for $\dot{y} = 0$ either the system is governed by the right-hand side $\mathrm{Sgn}(-\dot{y}) = \mathrm{sgn}(-\dot{y})$ (dynamics) or it is caught by friction force (statics) and we have so-called dead-zone. In the latter case we have infinitely many equillibrum positions, $y_o \in \frac{1}{k}[-T, T]$. Therefore

$$\mathrm{Sgn}(\dot{y} \equiv v) = \begin{cases} \{v/\,|v|\}: & v \neq 0, \\ \frac{T}{k}[-1, 1]: & v = 0, \end{cases} \tag{1.48}$$

which extends the standard definition of $\mathrm{sgn}(\dot{y})$, where $\mathrm{sgn}(0) = 0$. This replacement of typical sgn "=" by "$\in$" results in replacement of (1.45) by multi-valued extension, i.e. the standard theory of differential inclusions have been used (see [Deimling (1992)], [Filippov (1988)] and [Kunze (2000)]).

However, we are going to discuss this question in more detail later. We solve the problem analytically using a standard approach. A general solution to the Eq. (1.45) for arbitrary initial conditions A_0, $v_0 \equiv \dot{y}_0 > 0$ has the form

$$y_0 = \frac{v_0}{\alpha_0}\sin\alpha_0 t + (A \pm \frac{\varepsilon}{\alpha_0^2})\cos\alpha_0 t \mp \frac{\varepsilon}{\alpha_0^2}, \tag{1.49}$$

which is valid in the interval $0 < t < t_1$, where $\dot{y}_0 > 0$. The end of this interval can be found from the algebraic equation

$$\dot{y}_0 \equiv v_0 \cos\alpha_0 t_1 + (\alpha_0 A_0 - \frac{\varepsilon}{\alpha_0})\sin\alpha_0 t = 0, \tag{1.50}$$

which gives the value

$$t_1 = \frac{1}{\alpha_0}\arctan\left(-\frac{v_0}{\alpha_0 A - \dfrac{\varepsilon}{\alpha_0}}\right). \tag{1.51}$$

Taking into account (1.51) in (1.49) we define the initial conditions in the next time interval $t_1 < t < t_2$ for which $\dot{y}_0 < 0$:

$$\begin{aligned} A_1 &= y_0(t_1), \\ v_1 &= 0. \end{aligned} \tag{1.52}$$

A solution to the Eq. (1.46) for $\dot{y}_0 < 0$ and the initial conditions (1.52) has the form

$$y_1 = \left(A_1 \mp \frac{\varepsilon}{\alpha_0^2}\right)\cos\alpha_0(t - t_1) \pm \frac{\varepsilon}{\alpha_0^2}. \tag{1.53}$$

A condition of $\dot{y}_1(t_2) = 0$ leads to:

$$t_2 = \frac{\pi}{\alpha_0} + t_1. \tag{1.54}$$

The new initial conditions for $t_2 < t < t_3$, where $\dot{y}_2 > 0$, are as follows:

$$\begin{aligned} A_2 &= y_1(t_2), \\ v_2 &= 0. \end{aligned} \tag{1.55}$$

Proceeding further in a similar way for $t_n < t < t_{n+1}$ we find a general solution of the form

$$y_n = \left(A_n \pm \frac{\varepsilon}{\alpha^2}\right)\cos\alpha_0(t - t_n) \mp \frac{\varepsilon}{\alpha_0^2}, \tag{1.56}$$

where:

$$A_n \equiv y_{n-1}(t_n) = -A_{n-1} \mp 2\frac{\varepsilon}{\alpha_0^2} \equiv \pm A_1 \mp 2(n-1)\frac{\varepsilon}{\alpha_0^2}, \quad n = 1, 2, \ldots \tag{1.57}$$

The upper sign corresponds to odd n ($\dot{y_n} < 0$), whereas the lower sign to even n ($\dot{y_n} > 0$). The process of calculation ends when the following condition is satisfied

$$|A_n| \geq \frac{T}{k} \equiv \frac{\varepsilon}{\alpha_0^2}. \tag{1.58}$$

It means physically, that the mass is overcome by friction and it stops. Now we are going to explain a powerful idea of Coulomb introduced in 1781. We refer to the Fig. 1.6 and to Example 1 considering a limiting case of the straight line in interval $\dot{x} \in [-a, a]$ (see Fig. 1.9).

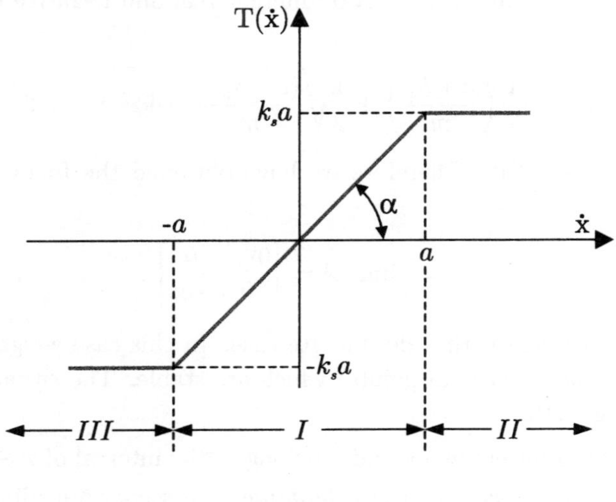

Fig. 1.9 Friction force when $\alpha \to \frac{\pi}{2}$.

We consider one degree of freedom free oscillations with viscous and

Coulomb damping of the form

$$m\ddot{x} + c\dot{x} + kx + T(\dot{x}) = 0, \tag{1.59}$$

where according to the introduced domains in Example 1 we have

$$T(\dot{x}) = \begin{cases} k_s\dot{x} \text{ in domain I } |\dot{x}| \leq a, \\ k_s a \text{ in domain II } \dot{x} \geq a, \\ -k_s a \text{ in domain III } \dot{x} \leq a. \end{cases} \tag{1.60}$$

Taking $x = x_1$ we transform (1.59) to two first order differential equations

$$\begin{Bmatrix} \dot{x}_1 \\ \dot{x}_2 \end{Bmatrix} = \begin{bmatrix} 0 & 1 \\ -\alpha_0^2 & -\dfrac{c+k_s}{m} \end{bmatrix} \begin{Bmatrix} x_1 \\ x_2 \end{Bmatrix}. \tag{1.61}$$

The characteristic equation related to (1.61) has the form

$$p(\lambda) = \lambda^2 - (trA)\lambda + \det A = 0, \tag{1.62}$$

where:

$$\Delta = (trA)^2 - 4\det A = \frac{(c+k_s)^2}{m} - 4\alpha_0^2.$$

Because we consider a limiting case then $k_s \equiv \tan\alpha \to +\infty$, and hence $\Delta > 0$. It means that we have two different real and negative eigenvalues of the form

$$\lambda_i = -\frac{1}{2}\left(\frac{c+k_s}{m}\right) \pm \frac{1}{2}\sqrt{\frac{(c+k_s)^2}{m} - 4\alpha^2}, i = 1, 2, \tag{1.63}$$

where: $\lambda_2 < \lambda_1 < 0$. Therefore we have obtained the following critical matrix:

$$A^* = \lim_{k_s \to \infty} A = \begin{bmatrix} 0^{(-)} & 0 \\ 0 & -\infty \end{bmatrix}, \tag{1.64}$$

and we have to deal with a degenerate case. In this case we get infinitely many equilibria (or critical points) which are stable. The situation is presented in Fig. 1.10.

Note, that infinitely many nodes belong to the interval of $x \in \dfrac{T_0}{k}[-1, 1]$, and in addition this case gives explanation to a scenario leading from dynamics to statics. Finally this transition is continuous. Now we apply the theory of distributions given in Example 1. Let $t = \tau_1$ be the time moment corresponding to $\dot{x}(\tau_1) = 0$. Thus Eq. (1.59) reads

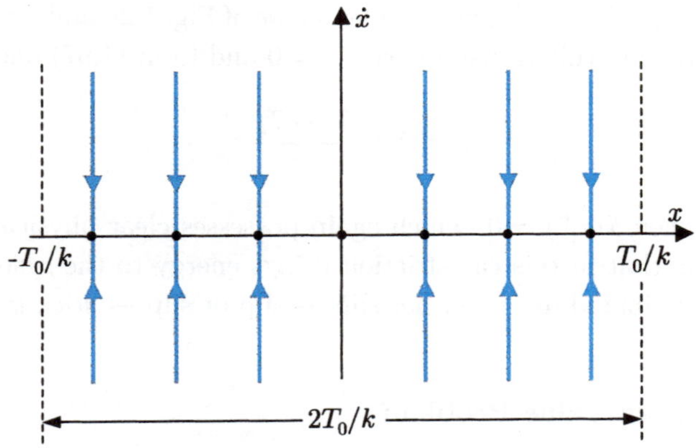

Fig. 1.10 Critical nodes for degenerated matrix $A^{\star}$.

$$m\Delta\ddot{x}(\tau_1) + c\Delta\dot{x}(\tau_1) + k\Delta x(\tau_1) + \Delta T[\dot{x}(\tau_1)] = 0. \tag{1.65}$$

Because $\dot{x}(\tau_1)$ and $x(\tau_1)$ are continuous therefore $\Delta\dot{x}(\tau_1) = 0$ as well as $\Delta x(\tau_1) = 0$, and from (1.65) one obtains

$$m[\ddot{x}(\tau_1^+) - \ddot{x}(\tau_1^-] + T[\dot{x}(\tau_1^+) - \dot{x}(\tau_1^-] = 0. \tag{1.66}$$

Finally we get the relation between the jumps of the accelerations of the form

$$\ddot{x}(\tau_1^+) = \frac{T[\dot{x}(\tau_1^-)] - T[\dot{x}(\tau_1^+)]}{m} + \ddot{x}(\tau_1^-). \tag{1.67}$$

It is clear, that jumps of accelerations depend only on friction (i.e. mass) and not on any other parameters of the system. It possesses also clear physical interpretation. The friction and inertial forces act on each other, but they always have the same direction and the first tries to stop the body faster. Knowing that we can also discuss the slip $\rightarrow$ slip, stick $\rightarrow$ slip and slip $\rightarrow$ stick phenomena.

The following hypothesis can be formulated. If

$$\left|\ddot{x}(\tau_1^-)\right| \leq \left|\frac{T[\dot{x}(\tau_1^-)] - T[\dot{x}(\tau_1^+)]}{m}\right|, \tag{1.68}$$

then $\ddot{x}(\tau_1^+) = 0$ (transition slip $\rightarrow$ stick). If (1.68) does not hold, then (1.67) holds and we have slip $\rightarrow$ slip transition with exactly defined jumps of accelerations. Finally, let us consider a case of Fig. 1.2b and a stick → slip transition. During this transition $\dot{x}(\tau_1^-) = 0$ and from (1.67) one gets

$$\ddot{x}(\tau_1^+) = \frac{T_s - T_k}{m}. \tag{1.69}$$

If $T_s = T_k$, then $\ddot{x}(\tau_1^+) = 0$, which again possesses clear physical interpretation. Note, that in this case friction brings energy to the system, which is however dissipated during either slip → slip or slip → stick transitions.

1.5 Boundary Value Problem

It has been already mentioned that systems with friction cause problems in numerical simulations. In this section we show, that in some cases for special solutions being sought the problem can be solved efficiently either numerically or analytically. We discuss the latter one. We consider the following nondimensional oscillator:

$$\ddot{x} + c\dot{x} + x + T\text{sign}\dot{x} = q\cos(\omega t + \beta + \theta), \tag{1.70}$$

where for $\dot{x} > 0$, $\text{sign}\dot{x} = 1$, $\beta = \arctan[c\omega/(1-\omega^2)]$, and θ is the phase between the exciting force and the system displacement.

Analyzing a periodic solution with the period $T = 2\pi/\omega$, which satisfies the following boundary conditions:

$$\begin{aligned} t = 0: \ & x(0) = -s, \ \dot{x}(0) = 0, \\ t = \frac{\pi}{\omega}: \ & x\left(\frac{\pi}{\omega}\right) = s, \ \dot{x}\left(\frac{\pi}{\omega}\right) = 0, \end{aligned} \tag{1.71}$$

where $s > 0$ is the value to be found. The following solution satisfies the boundary conditions (1.71):

$$x(t) = e^{-\frac{ct}{2}}(A\cos\lambda t + B\sin\lambda t) + F\cos(\omega t + \theta) - T, \tag{1.72}$$

where:

$$\dot{x}(t) > 0, \quad \lambda = \sqrt{1 - \frac{c^2}{4}}, \quad F = \frac{q}{\sqrt{(1-\omega^2)^2 + c^2\omega^2}}. \tag{1.73}$$

Taking into account (1.71) in (1.72) the problem is reduced to the solution of the following algebraic equations

$$\begin{aligned} A + s + F\cos\theta &= T, \\ B\lambda &= \frac{1}{2}Ac + F\omega\sin\theta, \\ e^{-\frac{c\pi}{2\omega}}(A\cos(\pi\lambda/\omega) + B\sin(\pi\lambda/\omega)) &= s + T + F\cos\theta, \\ (-Ac + 2B\lambda)\cos(\pi\lambda/\omega) &+ 2e^{\frac{c\pi}{2\omega}}F\omega\sin\theta + \\ -(Bc + 2A\lambda)\sin(\pi\lambda/\omega)) &= 0, \end{aligned} \tag{1.74}$$

which gives

$$A = \frac{e^{\frac{c\pi}{2\omega}}T(2e^{\frac{c\pi}{2\omega}}\lambda + 2\lambda\cos(\frac{\pi\lambda}{\omega}) - c\sin(\frac{\pi\lambda}{\omega}))}{\lambda(1 + e^{\frac{c\pi}{\omega}} + 2e^{\frac{c\pi}{2\omega}}\cos(\frac{\pi\lambda}{\omega})),},$$

$$B = \frac{e^{\frac{c\pi}{2\omega}}T(ce^{\frac{c\pi}{2\omega}}\lambda + c\cos(\frac{\pi\lambda}{\omega}) + 2\lambda\sin(\frac{\pi\lambda}{\omega}))}{\lambda(1 + e^{\frac{c\pi}{\omega}} + 2e^{\frac{c\pi}{2\omega}}\cos(\frac{\pi\lambda}{\omega}))},$$

$$\beta = \pm\arccos\left[\pm\sqrt{1 - e^{\frac{c\pi}{\omega}}\left(\frac{T(c^2 + 4\lambda^2)\sin(\pi\lambda/\omega)}{2F\lambda\omega(1 + e^{\frac{c\pi}{\omega}} + 2e^{\frac{c\pi}{2\omega}}\cos(\frac{\pi\lambda}{\omega}))}\right)^2}\,\right],$$

$$s = T\left(1 + \frac{ce^{\frac{c\pi}{2\omega}}\sin(\frac{\pi\lambda}{\omega}) - 2\lambda(e^{\frac{c\pi}{\omega}} + 2e^{\frac{c\pi}{2\omega}}\cos(\frac{\pi\lambda}{\omega}))}{\lambda(1 + e^{\frac{c\pi}{\omega}} + 2e^{\frac{c\pi}{2\omega}}\cos(\frac{\pi\lambda}{\omega}))}\right) \mp$$

$$\mp F\sqrt{1 - e^{\frac{c\pi}{\omega}}\left(\frac{T(c^2 + 4\lambda^2)\sin(\pi\lambda/\omega)}{2F\lambda\omega(1 + e^{\frac{c\pi}{\omega}} + 2e^{\frac{c\pi}{2\omega}}\cos(\frac{\pi\lambda}{\omega}))}\right)^2}.$$

The exact analytical and numerical solutions obtained for $F = 0.1$, $c = 0.1$, $q = 1.0$, $\omega = 0.5$, for initial conditions $x(0) = \dot{x}(0) = 1.0$ and using the Runge-Kutta method with the step $2\pi/400$ are undistinguishable, which is shown in Fig. 1.11. A similar like boundary value problem can be formulated even if the left hand side of Eq. (1.70) is more complicated but

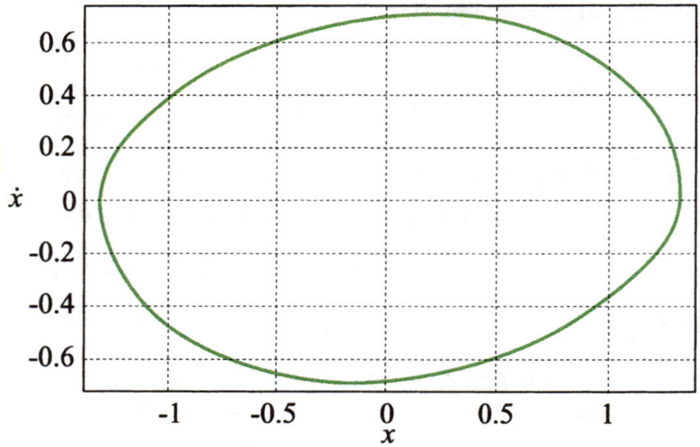

Fig. 1.11 Analytical and numerical solutions in the phase plane $(\dot{x}, x)$.

its analytical periodic solution is known between the boundaries related to $T/2$.

Chapter 2

Mathematical Background for Multivalued Formulations

2.1 Origin of Nonlinearities

First, we recall the origins of nonlinearities in the frame of mechanics. It is well known that one can the origin of nonlinearities a follows. First, there is the geometrical origin of nonlinearities. The simplest example of such a nonlinear behaviour is provided by large oscillations of a pendulum. Geometrical nonlinearities arise in the field of continuum mechanics due to the nonlinear quadratic terms that occur in the relation between the displacement and the strain tensors [Eringen *et. al.* (1975)], [Sedov (1971)].

Second, nonlinearities may occur from constitutive laws: nonlinear elasticity for example [Antman (1995)] or any other nonlinear constitutive laws [Palmov (1998)] such as plasticity, viscoelastoplasticity, etc. It is well known that mathematical expressions of nonlinear constitutive laws do not always use smooth mathematical expressions but on the contrary must often involve nonsmooth terms [Alber (1998)].

The third origin of nonlinearities is due to links or joints that are needed to describe physical systems: for example, two shells can be linked together using screws; locally if screws are locked not enough, friction occurs. Clearly, even if the behaviour of each shell is linear, this joint may be highly nonlinear; or a beam clamped at one extremity can impact on an obstacle at the other one. In some sense the beam is linked to an external system consisting of the obstacle.

2.2 Smooth and Nonsmooth Nonlinearities

We distinguish between smooth and nonsmooth nonlinearities.

Definition 2.1 A nonlinearity is smooth *if it can be formulated using a state variable $X(t)$ that is a function of time t and a smooth nonlinear function f so that the contribution of nonlinearity to the fundamental law of dynamics is*

$$f(t, X(t)), \tag{2.1}$$

where:

$$\begin{cases} \mathbb{R} \times E \longrightarrow & E, \\ (t, X) \mapsto & f(t, X), \end{cases} \tag{2.2}$$

is a C^∞ function of (t, X) on $\mathbb{R} \times E$. E is a vector space with norm $\|\cdot\|$. E is a finite or infinite vector space.

It is clear that any other kind of nonlinearity is called **nonsmooth**. But one can see at once that there are different ways to obtain nonsmooth nonlinearities. For example

$$\begin{cases} f : \mathbb{R} \times \mathbb{R} \longrightarrow & \mathbb{R}, \\ \quad (t, X) \mapsto & |X|, \end{cases} \tag{2.3}$$

provides an nonsmooth nonlinear term with a C^0 function. But the first reason for the nonsmoothness of a nonlinear term is a worse one. One can imagine that f is no longer a function. For example, we can consider:

$$\begin{cases} f : \mathbb{R} \times \mathbb{R} \longrightarrow & \mathbf{P}(\mathbb{R}), \\ & \{1\} \quad \text{if} \quad X > 0, \\ (t, X) \mapsto & \{-1\} \quad \text{if} \quad X < 0, \\ & [-1, 1] \quad \text{if} \quad X = 0. \end{cases} \tag{2.4}$$

In order to simplify notations, we shall write $f(X)$ instead of $f(t, X)$ if f does not depend on t.

Why do we decide to introduce such a partition between nonlinearities? There are several explanations. The first one is linked to mathematics. Classical results [Crouzeix and Mignot (1984)] provide (at least locally) existence and uniqueness of the solution to smooth nonlinear ordinary differential equations under reasonable assumptions (Lipschitz-continuity of functions). In the case of nonsmooth nonlinear terms, mathematical tools that have to be used in order to establish the existence and the uniqueness may become more sophisticated as we shall see later.

The second reason is related to applications, especially in the field of mechanics: one is often interested in periodic solutions of dynamical equilibrium equations. If the considered nonlinearity is nonsmooth, many analytical methods or perturbation methods (such as center manifold theory [Carr (1981)], normal forms methods [Arnold (1980)], [Brjuno (1989)], [Iooss and Adelmayer (1992)], [Nayfeh (1993)], Lie series [Giacaglia (1972)], multiple scales method [Nayfeh and Mook (1979)], [Van Horssen (1999)], method of averaging [Roseau (1986)], [Verhulst (1996)], etc.) can not be useful. In the case of nonsmooth nonlinear terms these methods either fail or their applications do not lead to a general method but only cover special examples.

2.3 Examples and Dynamical Equilibria

Let us consider two simple basic examples.

2.3.1 *Models with friction*

First, we will examine a one degree of freedom model consisting of one mass m fixed to a vertical support via one spring with stiffness k. The mass moves on an horizontal support; an external excitation is exerted on the mass and friction occurs between the mass and the horizontal support. Usually, one can think that this mechanical system is governed by the equation:

$$\begin{cases} x(t_0), \dot{x}(t_0) \text{ given}, \forall t \in [t_0, t_0 + T], T > 0, \\ m\ddot{x}(t) + kx(t) + c\dot{x} + \text{ friction } = f(t), \end{cases} \tag{2.5}$$

where x is displacement of mass m from initial equilibrium position, $c\dot{x}(t)$ is a classical dissipation term due to viscous damping c. Friction term can be chosen so that

$$\begin{cases} \text{friction } = \Phi(t) = F_n \mu_D \text{sgn}(\dot{x}(t)) \quad \text{if} \quad \dot{x}(t) \neq 0, \\ \text{friction } = \Phi(t) \text{ with } \mid \Phi(t) \mid \leq F_n \mu_S \quad \text{if} \quad \dot{x}(t) = 0, \end{cases} \tag{2.6}$$

where F_n is the normal (to the support) force exerted on the mass m, μ_D (μ_S) is respectively the dynamic (static) friction coefficient, with $0 \leq \mu_D \leq \mu_S$. The case $\mu_S = \mu_D = \mu$ corresponds to the classical Coulomb friction. One can think that μ could be a function of t, but we shall see later that this choice can lead to nonuniqueness problems. Sophisticated models [Bothe (1999)], [Hinrichs *et. al.* (1998)] may include functions $\mu(\dot{x}(t))$ or $\mu(x(t))$.

Here we intend to explain via this first simple example why the Eq. (2.5) should be written again like a differential inclusion in order to verify at each time t in $[t_0, t_0 + T]$ fundamental equation of dynamics. So let us examine a particular behaviour of the mass m. Let us assume that at t_0 we have $x(t_0) = 0$, $\dot{x}(t_0) = 0$ and a positive friction and no external force ($f(t_0) = 0$). Then within $[t_0, t_0 + T]$, with $T > 0$, $f(t)$ is assumed to be weak but strictly increasing so that the mass m does not move.

Physically this is a reasonable situation. We have:

$$\forall t \in [t_0, t_0 + T], \quad x(t) = 0, \quad \dot{x}(t) = 0.$$

Now if we write friction term as a function of x, (2.5) becomes:

$$\forall t \in]t_0, t_0 + T], m\ddot{x}(t) + kx(t) + c\dot{x}(t) + \text{friction}(x(t)) = f(t), \tag{2.7}$$

or clearly

$$\forall t \in [t_0, t_0 + T], 0 + 0 + 0 + \text{friction}(0) = f(t).$$

So it is clear that *friction*(0) can not be simply a unique real value, but has to provide a *reaction* force in order to obtain equilibrium versus $f(t)$. According to (2.6), one can see that a better modelling of the friction problem is given by

$$\begin{cases} x(t_0), \dot{x}(t_0) \text{ given}, \forall t \in [t_0, t_0 + T], \\ m\ddot{x}(t) + c\dot{x}(t) + kx(t) + G(\dot{x}(t)) \ni f(t), \end{cases} \tag{2.8}$$

with $G(\dot{x}(t))$ which is a graph verifying in the case of the Coulomb friction:

$$G(\dot{x}(t)) = \mu F_n \text{sgn}(\dot{x}(t)), \tag{2.9}$$

in other words

$$\begin{cases} G(\dot{x}(t)) = \{\mu F_n\} \quad \text{if} \quad \dot{x}(t) > 0, \\ G(\dot{x}(t)) = \{-\mu F_n\} \quad \text{if} \quad \dot{x}(t) < 0, \\ G(\dot{x}(t)) = [-\mu F_n, \mu F_n] \quad \text{if} \quad \dot{x}(t) = 0. \end{cases} \tag{2.10}$$

So it is clear that (2.8) can be detailed in the form:

$$\forall t \in [t_0, t_0{+}T], f(t) \in \{m\ddot{x}(t){+}c\dot{x}(t){+}kx(t){+}\mu F_n\} \quad \text{if} \quad \dot{x}(t) > 0, \tag{2.11}$$

that is obviously

$$f(t) = m\ddot{x}(t) + c\dot{x}(t) + kx(t) + \mu F_n, \tag{2.12}$$

or:

$$\forall t \in [t_0, t_0{+}T], f(t) \in \{m\ddot{x}(t){+}c\dot{x}(t){+}kx(t){+}\mu F_n\} \quad \text{if} \quad \dot{x}(t) < 0, \tag{2.13}$$

that is

$$f(t) = m\ddot{x}(t) + c\dot{x}(t) + kx(t) + \mu F_n, \tag{2.14}$$

or:

$$\begin{cases} \forall t \in [t_0, t_0 + T], \quad \text{if} \quad \dot{x}(t) = 0, \\ f(t) \in [m\ddot{x}(t) + c\dot{x}(t) + kx(t) + \mu F_n, m\ddot{x}(t) + c\dot{x}(t) + kx(t) + \mu F_n]. \end{cases} \tag{2.15}$$

The last case provides origin of using graphs and differential inclusions in order to verify everywhere dynamical equilibrium.

Many other examples including friction (or plastic terms) will be introduced elsewhere.

2.3.2 *Another simple example with impact*

Now let us consider a material point with mass m fixed to a support via a spring k, submitted to external forcing $F(t)$: this mass can impact against a fixed wall at x_{max}. The model is simply written as:

$$m\ddot{x}(t) + kx(t) = F(t) \quad \text{if} \quad x(t) < x_{max}, \tag{2.16}$$

and at $x(t) = x_{max}$, impact occurs and a restitution law is chosen as follows:

$$\text{if} \quad x(t) = x_{max}, \quad \dot{x}(t^+) = -e\dot{x}(t^-), \quad e \in [0, 1]. \tag{2.17}$$

$F(t)$ denotes external solicitation.

Indeed, the fundamental equation of dynamics is not verified at every time if we consider the model (2.16), (2.17). It could seem better to introduce reaction R of the wall (depending on x) and to write

$$m\ddot{x}(t) + kx(t) + R(x(t)) = F(t), \tag{2.18}$$

but again a special configuration with impact at t_1 with

$$\begin{cases} x(t_1) = x_{max}, \quad \dot{x}(t_1) = 0, \\ \dot{x}(t_1^+) = -e\dot{x}(t_1) = 0, \end{cases} \tag{2.19}$$

and $F(t)$ (strictly increasing for example) so that for $t > t_1$, the mass m is constrained against the wall; this leads to

$$\begin{cases} \forall t > t_1, m\ddot{x}(t) + kx(t) + R(x(t)) - F(t) = 0 \\ \Longleftrightarrow \forall t > t_1, kx_{max} + R(x_{max}) - F(t) = 0, \end{cases} \tag{2.20}$$

and this last expression can not be equal to zero for every t!

Again we have to introduce a differential inclusion in order to write in a proper way the dynamical equilibrium at each time t and in order to take into account the correct reactions at impacts.

It is convenient to write

$$\begin{cases} x(t_0), \quad \dot{x}(t_0) \quad \text{given}, \\ m\ddot{x}(t) + kx(t) + \partial\Psi(x(t)) \ni F(t), \\ \dot{x}(t^+) = -ex(t^-) \quad \text{if} \quad x(t) \in \partial K, \end{cases} \tag{2.21}$$

where K is a convex set (here $K = [-\infty, x_{max}]$) with boundary ∂K ($= \{x_{max}\}$) and Ψ_K is function of K defined (in the frame of convex analysis) by:

$$\begin{cases} \Psi_K(u) = 0 \quad \text{if} \quad u \in K, \\ \Psi_K(u) = +\infty \quad \text{if} \quad u \notin K. \end{cases} \tag{2.22}$$

Clearly Ψ_K is defined in the domain:

$$\text{dom}(\Psi_K) = \{u/\Psi_K(u) < +\infty\} = K. \tag{2.23}$$

The subdifferential $\partial\Psi_K$ of Ψ_K is defined by

$$\partial\Psi_K = \begin{cases} 0 \quad \text{if} \quad u \in \text{Int}(K), \\ N_K(u) \quad \text{if} \quad u \in \partial K, \end{cases} \tag{2.24}$$

where $N_K(u)$ is the normal cone at u, defined by

$$N_K(u) = \Big(T_K(u)\Big)^{\perp} = \{y \in \mathbb{R}, \forall z \in T_K(u), (y,z) \leq 0\}, \tag{2.25}$$

where $T_K(u)$ is the tangential cone at u defined by

$$T_K(u) = \Big(\overline{\cup_{\lambda>0}\lambda(K-u)}\Big)^{\perp}, \tag{2.26}$$

and $(\cdot,\cdot)$ is a scalar product in $\mathbb{R}$. Clearly, in our simple case we have:

$$\partial\Psi_K(u) = 0 \text{ if } u \in [-\infty, x_{max}], \tag{2.27}$$

$$\partial\Psi_K(x_{max}) = \Big(\overline{\cup_{\lambda>0}\lambda([-\infty, x_{max}] - x_{max})}\Big)^{\perp} = (\mathbb{R}^-)^{\perp} = [0, +\infty]. \tag{2.28}$$

Now clearly, we can chose any positive value for "reaction" in order to realize dynamical equilibrium of our special previous case:

$$kx_{max} + r - F(t) = 0, \tag{2.29}$$

because $r =$ "$R(x_{max})$" $\in [0, +\infty]$ is chosen so that the equality (2.29) occurs.

2.4 Existence and Uniqueness

2.4.1 *The frame of maximal monotone operators*

2.4.1.1 *General case*

Let us consider a Hilbert space H with scalar product $(\cdot,\cdot)$ and corresponding norm $\|\cdot\|$.

Definition 2.2

i) *A is a multivalued operator on H if and only if (iff) A maps H on* **P(H)** *where* **P(H)** *denotes the set of all the subsets of H.*

ii) *Domain of A -denoted $D(A)$- is defined by:*

$$D(A) = \{x \in H / Ax \neq \emptyset\}, \tag{2.30}$$

where Ax abbreviates $A(x)$.

iii) *Any operator on H is identified to its graph in $H \times H$ defined by*

$$\{(x,y)/x \in H, \quad y \in Ax\}. \tag{2.31}$$

Definition 2.3 *A is a monotone operator on H iff*

$$\forall(x_1, x_2) \in D(A) \times D(A), (Ax_1 - Ax_2, x_1 - x_2) \geq 0. \tag{2.32}$$

Let us notice that if A is single valued operator, monotone simply means that A is "increasing" function. So a simple example of monotonous operator on $\mathbb{R}$ is given by

$$\tilde{f} : \begin{cases} \mathbb{R} \longrightarrow \mathbf{P}(\mathbb{R}), \\ x \mapsto [f(x^-), f(x^+)] \cap \mathbb{R}, \end{cases}$$

with $f : \mathbb{R} \longrightarrow \mathbb{R}$ is an increasing function.

Definition 2.4 *An operator A on H is a maximal monotone operator iff*
i) *A is monotone on H,*
ii) *A is maximal in the set of all the monotone operators on H.*

Let us notice that "maximal" is associated to inclusion of graphs;

$$A \subset B \quad \text{iff} \quad \forall x \in D(A), \quad x \in D(B) \text{ and } Ax \subset Bx. \tag{2.33}$$

Now let us consider a maximal monotone operator A on H.

Definition 2.5 *The section of Ax for $x \in H$, denoted A^0x, is defined by*

$$A^0x = proj_{Ax}(0). \tag{2.34}$$

This is the projection on Ax in the sense of the theorem of the projection on a convex set.

Details, examples and properties of maximal monotone operators could be found in [Brezis (1973)] or [Brezis (1992)]. Now we intend to claim the main results without details or proofs. The basic result is due to Brézis [Brezis (1973)]. It explains how the solution of differential inclusions evolves, and especially how the choice of an element among the multiple values of (x, Ax) helps eliminating the difficulty due to the multivalued term.

Theorem 2.1 *Let A be a maximal monotone operator defined on $D(A)$. Then we have $\forall x_0 \in D(A), \exists! x : [t_0, +\infty] \longrightarrow H$ so that:*
i) $\forall t > t_0, x(t) \in D(A)$,

ii) x is Lipschitz-continuous on $[t_0, +\infty]$ and

$$\| \frac{dx}{dt} \|_{L^\infty} ([t_0, +\infty], H) \leq \| A^0 x \|, \tag{2.35}$$

iii) x verifies:

$$\forall t > t_0, \quad \frac{dx}{dt} + Ax(t) \ni 0, \tag{2.36}$$

iv) x starts from x_0:

$$x(t_0) = x_0, \tag{2.37}$$

v) the right-hand side derivative of x at t, $\frac{du^+}{dt}(t)$ exists and is given by

$$\forall t \in [t_0, +\infty[, \frac{du^+}{dt}(t) + A^0 x(t) = 0. \tag{2.38}$$

Property (v) is very important because it holds everywhere whereas the relationship (3) holds almost everywhere.

Theorem 2.2 *Let A be a maximal monotonous operator with domain $D(A)$. Then for every $f \in L^1([0,T], H)$ and every $u_0 \in \overline{D(A)}$, there exists a unique weak solution u of the equation $\frac{du}{dt} + Au \ni f$ verifying $u(0) = u_0$.*

In order to obtain a result similar to the property (5) of the Theorem 9.2, some assumptions have to be added. We have:

Theorem 2.3 *Let A be a maximal monotonous operator of H, $f \in L^1([0,T], H)$ and let $u \in C([0,T], H)$ be a weak solution of the equation $\frac{du}{dt} + Au \ni f$. Let $t_0 \in [0,T]$ be a right Lebesgue point of f (respectively $t_0 \in [0,T]$ a Lebesgue point of f). Let us set $f(t_0+0) = \lim_{h\to 0} \frac{1}{h} \int_{t_0}^{t_0+h} f(s)ds$. Then the following properties are equivalent:*

(1) $u(t_0) \in D(A)$,

(2) $\liminf_{h\to 0} \frac{1}{h} |u(t_0 + h) - u(t_0)| < +\infty$

(respectively $\liminf_{\substack{h \longrightarrow 0 \\ h \neq 0}} \frac{1}{h} |u(t_0 + h) - u(t_0)| < +\infty$*)*

(3) u has a right hand side derivative at t_0
and $\frac{d^+u}{dt}(t_0) = f(t_0+0) - Proj_{Au(t_0)} f(t_0+0)$.

The next result leads to a strong solution under convenient assumptions.

Theorem 2.4 *Let* H *be a finite dimensional Hilbert space. Let* A *be a maximal monotonous operator so that* $Int(D(A)) \neq \emptyset$. *Let* f *belong to* $L^1([0,T],H)$. *Then every weak solution of the equation* $\frac{du}{dt} + Au \ni f$ *is a strong solution.*

Indeed a strong solution is a function $u \in C([0,T],H)$ absolutely continuous on every compact subset of $[0,T]$ verifying:

$$u(t) \in D(A) \quad \text{and almost everywhere in} \quad [0,T] \quad \frac{du}{dt}(t) + Au(t) \ni f.$$

A function $u \in C([0,T],H)$ is a weak solution if and only if there exists a series $f_n \in L^1([0,T],H)$ and $u_n \in C([0,T],H)$ so that u_n is a strong solution of the equation $\frac{du_n}{dt}(t) + Au_n(t) \ni f_n(t)$, with $\lim_{n\longrightarrow+\infty} f_n = f \in L^1([0,T],H)$ and uniformly on $[0,T]$: $\lim_{n\longrightarrow+\infty} u_n = u$.

Hence it can be seen that if $f \in L^1([0,T])$, u is a strong solution if u is a weak solution and u is absolutely continuous on every compact subset of $[0,T]$.

Theorem 2.5 *([Brezis (1973)] page 105): Let* A *be a maximal monotonous operator on* H. *Let us chose* $\omega > 0$, $f \in L^1([0,T],H)$ *and* $u_0 \in \overline{D(A)}$. *Then there is a unique weak solution of the equation:*

$$\begin{cases} \frac{du}{dt}(t) + Au(t) - \omega u(t) \ni f, \\ \\ u(0) = u_0. \end{cases}$$

One can derive from the Theorem 2.4 and from the proof of the Theorem 2.5 that if $dim(H) < +\infty$ and $Int(D(A)) \neq \emptyset$ then the weak solution of the Theorem 2.5 is a strong solution.

This basic result could be extended to the equation:

$$\frac{dx}{dt} + F(t,x) + Ax \ni 0, \tag{2.39}$$

with F Lipschitz-continuous function with respect to x and bounded with respect to t. In this book we consider only special cases of operators A - those that derives from a particular potential - so we give the accurate result hereafter in this convenient frame.

2.4.1.2 *Convex particular potentials*

Let us consider a function $\phi : \mathbb{R}^p \longrightarrow [-\infty, +\infty]$.

Definition 2.6 *ϕ is lower-semi-continuous (l.s.c.) function on $\mathbb{R}^p$ iff*

$$\liminf_{y \longrightarrow x} \phi(y) \geq \phi(x). \tag{2.40}$$

Definition 2.7 *ϕ is convex iff*

$$\forall (x, y) \in \mathbb{R}^p \times \mathbb{R}^p, \forall \lambda \in [0, 1], \phi(\lambda x + (1-\lambda)y) \leq \lambda \phi(x) + (1-\lambda)\phi(y). \tag{2.41}$$

Definition 2.8 *ϕ is said to be proper iff*

$$\exists x \in \mathbb{R}^p, \phi(x) \neq +\infty. \tag{2.42}$$

Now we can define the subdifferential of the l.s.c. proper convex function ϕ.

Definition 2.9 *For l.s.c. proper convex function $\phi : \mathbb{R}^p \longrightarrow [-\infty, +\infty]$, we introduce the subdifferential of ϕ at $x \in \mathbb{R}^p$, denoted as $\partial\phi(x)$ by*

$$y \in \partial\phi(x) \Longleftrightarrow \forall h \in \mathbb{R}^p, \quad \phi(x+h) - \phi(x) \geq < y, h >, \tag{2.43}$$

where $< \cdot, \cdot >$ is the canonical scalar product on $\mathbb{R}^p$, and

$$D(\partial\phi) = \{x \in \mathbb{R}^p / \partial\phi(x) \neq \emptyset\}. \tag{2.44}$$

Theorem 2.6 *For any ϕ l.s.c. proper convex function on $\mathbb{R}^p$, $\partial\phi$ is a maximal monotone operator on $\mathbb{R}^p \times \mathbb{R}^p$ (or $D(\partial\phi) \times \mathbb{R}^p$).*

Proof is given in [Brezis (1973)], [Brezis (1992)].

Examples:

i) If σ is the graph of sgn function defined by:

$$\begin{cases} \sigma : \mathbb{R} \longrightarrow & \mathbf{P}(\mathbb{R}), \\ \quad x \mapsto & \begin{cases} 1 & \text{if } x > 0, \\ -1 & \text{if } x < 0, \\ [-1,1] & \text{if } x = 0, \end{cases} \end{cases} \tag{2.45}$$

one can see easily that $\sigma = \partial\phi$, where:

$$\begin{cases} \phi : \mathbb{R} \longrightarrow [-\infty, +\infty], \\ \quad x \mapsto \phi(x) = \mid x \mid . \end{cases} \tag{2.46}$$

ii) For $\beta = \sigma^{-1}$,

$$\begin{cases} \beta : \mathbb{R} \longrightarrow & \mathbf{P}(\mathbb{R}), \\ \quad x \mapsto & \begin{cases} \emptyset & \text{if } \mid x \mid > 1, \\ [0,+\infty] & \text{if } x = 1, \\ [-\infty,0] & \text{if } x = -1, \\ 0 & \text{if } \mid x \mid < 1, \end{cases} \end{cases} \tag{2.47}$$

we have

$$\beta = \partial\phi, \tag{2.48}$$

with $\phi = \psi_{[-1,1]}$ defined by

$$\begin{aligned} \psi_{[-1,1]} : \mathbb{R} \longrightarrow & \quad [-\infty, +\infty], \\ x \mapsto & \begin{cases} 0 & \text{if } x \in [-1,1], \\ +\infty & \text{if } \mid x \mid > 1. \end{cases} \end{aligned} \tag{2.49}$$

2.4.1.3 *Changing scalar product*

Let us change the canonical scalar product $< \cdot, \cdot >$ on $\mathbb{R}^p$ into $< \cdot, \cdot >_M$ defined by

$$\begin{aligned} < \cdot, \cdot >_M \mathbb{R}^p \times \mathbb{R}^p \longrightarrow & \quad \mathbb{R} \\ (x, y) \mapsto & < x, y >_M = {}^t x M^{-1} y, \end{aligned} \tag{2.50}$$

where M is a symmetric definite positive $p \times p$ matrix.

Let us consider a function ϕ l.s.c. proper convex with subdifferential $\partial\phi$ for scalar product $< \cdot, \cdot >$ and $\partial\phi_M$ for scalar product $< \cdot, \cdot >_M$. Clearly one can prove

$$\partial\phi_M(x) = M\partial\phi(x). \tag{2.51}$$

2.4.1.4 *General result for discrete systems*

Let us consider the following general problem that can describe problems with impacts or friction or elastoplastic behaviour:

$$\begin{cases} \text{For almost every (a.e.)} \quad t \in [t_0, t_0 + T], \\ \quad \dot{X}(t) + M\partial\phi(X(t)) \ni G(X(t), t), \\ \\ \qquad X(t_0) = \xi \in \mathbb{R}^p, \end{cases} \tag{2.52}$$

with $T > 0$, $G : [y_0, t_0 + T] \times \mathbb{R}^p \longrightarrow \mathbb{R}^p$ Lipschitz-continuous i.e.,

$$\begin{cases} \exists \omega_G, \forall t \in [t_0, t_0 + T], \forall (X_1, X_2) \in \mathbb{R}^p \times \mathbb{R}^p, \\ \\ \parallel G(t, X_1) - G(t, X_2) \parallel \leq \omega_G \parallel X_1 - X_2 \parallel, \end{cases} \tag{2.53}$$

and

$$\forall Y \in \mathbb{R}^p, G(., Y) \in L^\infty\Big([t_0, t_0 + T], \mathbb{R}^p\Big), \tag{2.54}$$

where M is a positively defined symmetric $p \times p$ matrix, ϕ is a convex, proper, l.s.c. function on $\mathbb{R}^p$.

Such a frame is easily obtain from fundamental relation of dynamics and M is then derived from mass matrix. We have the following result [Bastien (2000)]:

Theorem 2.7

$$\forall \xi \in D(\partial\phi), \exists! X \in W^{1,1}\Big([t_0, t_0 + T], \mathbb{R}^p\Big) \textit{ so that}$$

$$\begin{cases} \textit{for almost every } t \in [t_0, t_0 + T], \dot{X}(t) + M\partial\phi(X(t)) \ni G(X(t), t), \\ \\ \qquad X(t_0) = \xi. \end{cases} \tag{2.55}$$

2.4.2 *Ill-posed problems*

2.4.2.1 *First example of Ill-posed problems*

This subsection is based on reference [Schatzman *et. al.* (1999)]. We consider the one-dimensional motion of a material point on a plane support moving at velocity $v(t)$. This point is submitted to a normal force orthogonal to the plane support which creates a longitudinal friction force. Moreover, the material point is submitted to given dynamics. The mathematical formulation of the problem is the following:

$$\ddot{x}(t) = F(t, x(t), \dot{x}(t)) - \mu_D F_n \mathrm{sgn}(\dot{x}(t) - v(t)), \text{ if } \dot{x}(t) - v(t) \neq 0, \quad (2.56)$$

$$\ddot{x}(t) = F(t, x(t), \dot{x}(t)) + F_l, \text{ with } F_l \in [-\mu_S, \mu_S], \text{ if } \dot{x}(t) - v(t) = 0, \quad (2.57)$$

where μ_D and μ_S are the dynamical and static friction coefficients, $\dot{x}(t)$ denotes velocity relative to the support, F describes the longitudinal dynamics. F is assumed to be Lipschitz-continuous with respect to $(x, \dot{x})$.

Let us write again this model in the form:

$$\ddot{x}(t) = F(t, x(t), \dot{x}(t)) + \phi(t), \quad (2.58)$$

$$x(0) = x_0, \quad (2.59)$$

$$\dot{x}(0) = \dot{x}_0, \quad (2.60)$$

where $\phi(t)$ denotes friction force. Let T be a nonnegative real. Let w be a continuous function on $[0, T]$ and let

$$b(t, w) = \begin{cases} t, & \text{if } w(t) \neq 0, \\ \sup\{s \in [0, t[\text{ such that } w(s) \neq 0\}, & \text{if } w(t) = 0. \end{cases} \quad (2.61)$$

We assume indeed that if $w(0) = 0$, a nonpositive $b(0, w)$ is given. Now let us consider two different phases: static ones (the relative velocity is null), and dynamic ones (the relative velocity is strictly positive). Two coefficients are introduced: a dynamic coefficient μ_D (strictly positive constant), and a static friction coefficient $\mu_S(t)$ (an increasing positive function, bounded on every static phase). So we have:

$$\begin{cases} \phi(t) = -\mu_D \mathrm{sgn}(\dot{x}(t) - v(t)), & \text{if } \quad \dot{x}(t) - v(t) \neq 0, \\ \phi(t) \in [-\mu_S(t), \mu_S(t)], & \text{if } \quad \dot{x}(t) - v(t) = 0, \end{cases} \quad (2.62)$$

with

$$\mu_S(t) = h(t - b(t, \dot{x} - v)), \quad (2.63)$$

where h is an increasing bounded positive function from $[0, +\infty]$ to $[0, +\infty]$. The Eqs. (2.62) and (2.63) can be simplified by setting

$$h(0) = \mu_D, \tag{2.64}$$

and assumption $\mu_D \leq \mu_S$. Introducing the maximal monotone graph σ defined by

$$\sigma(x) = \begin{cases} -1, & \text{if } x < 0, \\ 1, & \text{if } x > 0, \\ [-1, 1], & \text{if } x = 0, \end{cases} \tag{2.65}$$

the Eqs. (2.62) and (2.63) are equivalent to the following one:

$$\phi(t) \in -\sigma(\dot{x}(t) - v(t))h(t - b(t, \dot{x} - v)). \tag{2.66}$$

Thus, for given F, v, b, h, x_0 and $\dot{x}_0$, we seek functions x and ϕ from $[0, T]$ to $\mathbb{R}$ such that the Eqs. (2.58) and (2.66) are verified with the initial conditions (2.59) and (2.60). Now, ϕ can be eliminated from (2.58) and (2.66) and the following equation is obtained:

$$\ddot{x}(t) + \sigma(\dot{x}(t) - v(t))h(t - b(t, \dot{x} - v)) \in F(t, x(t), \dot{x}(t)). \tag{2.67}$$

If h is a positive constant $\mu = \mu_D$, (2.66) is equivalent to the particular case of Coulomb model.

If velocity $\dot{x}$ is denoted by y, (2.59), (2.60) and (2.67) are equivalent to:

$$\dot{y}(t) + \sigma(y(t) - v(t))h(t - b(t, y - v)) \in F(t, x_0 + \int_0^t y(s)ds, y(t)), \tag{2.68}$$

where y verifies the initial condition

$$y(0) = \dot{x}_0. \tag{2.69}$$

F is continuous with respect to all its arguments and Lipschitz-continuous with respect to its last two arguments x and $\dot{x}$. Then there exists a solution to (2.59), (2.60) and (2.67). In the particular case of Coulomb's friction, x is unique. In the general case uniqueness is not true. One can exhibit an example with nonuniqueness for the problem (2.68) and (2.69). Now, let us assume:

$$y(0) = 0, \quad v = 0, \tag{2.70}$$

$$h \text{ is increasing, continuous and strictly positive on } \mathbb{R}_+, \tag{2.71}$$

$$F(t,a,b) = h(t). \tag{2.72}$$

A family of distinct functions $\{y_u\}_{u\in[0,T]}$ is defined by

$$\forall t \in [0,u], y_u(t) = 0, \text{ and } \forall t \in [u,T], y_u(t) = \int_u^t h(s)ds - h(0)(t-u). \tag{2.73}$$

It can be verified readily that y_u solves (2.68) and (2.69) and that $y_u \neq y_v$ when $u \neq v$. In the reference [Schatzman *et. al.* (1999)], the same ideas are used to prove the sensitivity of the solutions to data. In the same reference, it is proved that the natural numerical scheme of Euler type (that can be build from Eqs. (2.68) and (2.69)) provides several solutions.

2.4.2.2 *Second example of Ill-posed models*

We consider first the connection with two springs (k_1, k_2) and two St-Venant elements (α_1, α_2), as shown in Fig. 2.1.

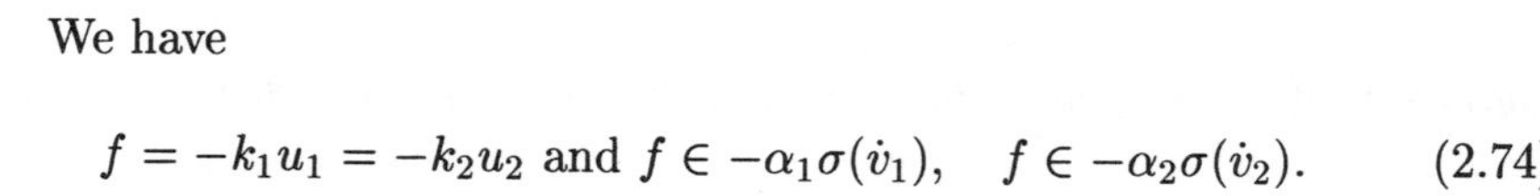

Fig. 2.1 Two springs and two St-Venant elements.

We have

$$f = -k_1 u_1 = -k_2 u_2 \text{ and } f \in -\alpha_1\sigma(\dot{v}_1), \quad f \in -\alpha_2\sigma(\dot{v}_2). \tag{2.74}$$

If $\alpha_1 < \alpha_2$, we have according to (2.74),

$$|f| \leq \alpha_1 < \alpha_2, \tag{2.75}$$

and

$$\dot{v}_2 = 0. \tag{2.76}$$

Hence, the St-Venant element 2 is always locked and does not alter the mechanical system. If $\alpha_1 = \alpha_2$, this system is undetermined from a mechanical point of view. Thus, if

$$|f| < \alpha_1 = \alpha_2, \tag{2.77}$$

both St-Venant elements are locked, whereas if

$$|f| = \alpha_1 = \alpha_2, \tag{2.78}$$

we cannot determine which St-Venant element oscillates. Therefore this ill-posed model is not of mechanical interest.

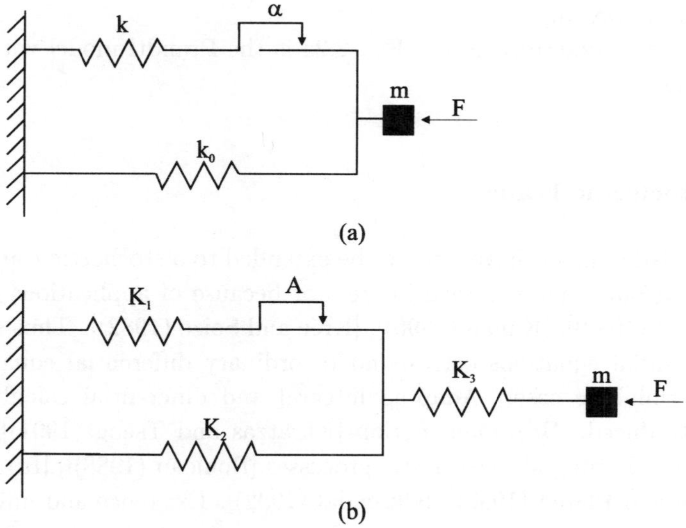

Fig. 2.2 The Prandtl model with linear hardening (a), and model with three springs and one St-Venant element (b).

Let us show now an equivalence between the two models given in the Figs. 2.2a and 2.2b.

We can observe that the model of the Fig. 2.2a is governed by the system (with $\eta = \alpha/k$):

$$\begin{cases} \dot{x} = y, \\ \dot{y} = (F - k_0 x - ku)/m, \\ \dot{u} + \beta(u/\eta) \ni y, \end{cases} \tag{2.79}$$

with the initial data

$$x(0) = x_0, \quad y(0) = y_0, \quad u(0) = u_0 \in [-\eta, \eta]. \tag{2.80}$$

The model of the Fig. 2.2b is governed by the same system by setting:

$$k_0 = \frac{K_2 K_3}{K_2 + K_3}, \quad k = \frac{K_1 {K_3}^2}{(K_1 + K_2 + K_3)(K_2 + K_3)} \text{ and } \alpha = A \frac{K_3}{K_2 + K_3}. \tag{2.81}$$

In this case, the problem is not mathematically ill-posed, but its mathematical formulation is not the simplest one: in some sense, the problem is physically ill-posed.

The model described in the Fig. 2.2a is the Prandtl model with linear hardening.

2.5 Stochastic Frame

Deterministic models have often to be extended to a stochastic frame work because of pure mathematical interest or because of applications [Boussa and Labbé (1989)], [Kahan (1996)], [Krée and Soize (1982)]. Thus stochastic differential equations correspond to ordinary differential equations of the deterministic case. Adapted integral and differential calculus have been introduced: Brownian motion [Karatzas and Tseng (1991)], Itô or Stratonovitch integrals, stochastic processes [Bouleau (1988)], [Itô (1951)], [Kloeden and Platen (1992)], [Oksendal (1992)]. Existence and uniqueness results similar to those obtained in the deterministic case have been obtained (see in the references [Kloeden and Platen (1992)], [Oksendal (1992)], [Yamada and Watanabe (1971)]). Numerical methods have been developed taking into account the stochastic term (see in the references [Kloeden and Platen (1992)], [Pardoux and Talay (1985)]).

Obviously, to the deterministic multivalued differential equations correspond the multivalued stochastic differential equations. These have been introduced and studied by several authors: Krée [Kree (1982)], Lépingle and Marois [Lépingle and Marois (1987)], and recently by Cépa [Cépa (1995a)], [Cépa (1995b)]. This last reference proposes well adapted numerical schemes.

Here we do not intend to recall all the mathematical background that is necessary to claim the existence and uniqueness results and and their proofs. We only intend to set the models and recall briefly some typical existence and uniqueness results and then to build adapted numerical schemes similar to those that have been given in the deterministic cases.

Let us consider $\mathbf{E} = (\Omega, \mathbf{F}, \{\mathbf{F}_t\}, \mathbb{P})$. $W = \{W_t, \mathbf{F}_t, t \in [0,T]\}$ is the standard n dimensional Brownian motion defined on $(\Omega, \mathbf{F}, \mathbb{P})$ with initial condition $W_0 = 0$.

Let us choose $|\cdot|$ the usual norm on $\mathbb{R}^n$, and let $T > 0$. Nonsmooth problems can be extended to the stochastic frame in the form:

$$dX_t + A(X_t)dt \ni b(t, X_t)dt + \sigma(t, X_t)dW_t, \tag{2.82}$$

where X_t is the unknown stochastic process, W_t is the Brownian motion, A is a maximal monotone operator on $\mathbb{R}^n$, b and σ (the diffusion coefficient) are Lipschitz-continuous functions; this means that,

$$\begin{cases} \exists C > 0, \forall t \in [0,T], \forall (x,y) \in \mathbb{R}^n \times \mathbb{R}^n, \\ \mid b(t,x) - b(t,y) \mid + \mid \sigma(t,x) - \sigma(t,y) \mid \leq C \mid x - y \mid, \end{cases} \tag{2.83}$$

with linear growth written as

$$\exists K > 0, \forall t \in [0,T], \forall x \in \mathbb{R}^n, \mid b(t,x) \mid + \mid \sigma(t,x) \mid \leq K(1+ \mid x \mid), \tag{2.84}$$

and initial condition ξ_0 that is a given random variable defined on $(\Omega, \mathbf{F}, \mathbb{P})$.

Under previous assumptions, [Kree (1982)] or [Cépa (1995a)] prove some existence and uniqueness results by using rather sophisticated mathematical frame and methods (see in the quoted references for details). They provide a correct mathematical sense to the words "solution", "existence" (weak and strong existence of solutions) and "uniqueness" (in a probability sense). We do not intend to go inside further mathematical developments. We shall simply provide the corresponding numerical scheme in the next chapter.

Chapter 3

Numerical Schemes and Analytical Methods

3.1 Numerical Schemes

In this section we intend to build numerical schemes that are well-adapted to dynamical equilibrium and that can be described by differential inclusions (or by ordinary differential equations and differential inclusions in the case of determined or stochastic case). Let us assume that existence and uniqueness results are obtained for all the mathematical problems considered in this chapter. First we deal with practical formulations of these numerical schemes without rigorous applied mathematical background. We often give the references where these mathematical questions are resolved. Many other numerical schemes can be built: based on classical numerical methods (see e.g. section 4.2) they are not adapted to very general cases as the next ones. But they are also commonly used and they can be efficient.

3.1.1 *Deterministic cases*

3.1.1.1 *First model with differential inclusion*

Let us consider a dynamical system described by the following differential inclusion of first order with initial conditions :

$$\begin{cases} \dot{X} + F(t, X) + A(X) \ni 0, & t \in]0, T] \\ X(0) = \tilde{X}_0, \end{cases} \tag{3.1}$$

with $T > 0$, $X : [0, T] \longrightarrow E$ the (un)known exact unique vector function solution defined from interval $[0, T]$ to (finite or infinite dimensional) vector space E; F is a smooth function of t and X. A is a maximal monotone graph (or operator) on E for a given scalar product $(\cdot, \cdot)$. $\tilde{X}_0$ is a given vector in E (initial condition).

In order to build numerical schemes, we introduce a discretization of $[0, T]$ using time $t_0 = 0, t_1, t_2, \ldots, t_n, \ldots$ with $\forall i, t_i \in [0, T]$, and note by X_n the approximated value of the exact value $X(t_n)$ that is expected from the numerical scheme. We introduce the time step $h_n = t_n - t_{n-1}$ for $n \in \{1, 2, \ldots\}$. It is often very convenient to choose a constant time step $h_n = h$. Here in order to simplify the presentation of the different numerical schemes, we choose a constant time step.

Proposition 3.1 (**Full Implicit Euler Scheme**) *The scheme is described by the following recursive relations:*

$$\left[\begin{array}{c} X_0 = \tilde{X}_0, \\ \\ \dfrac{X_{n+1} - X_n}{h} + F(t_{n+1}, X_{n+1}) + A(X_{n+1}) = 0. \end{array}\right. \tag{3.2}$$

Clearly this scheme is not very convenient, even if $A(X) = 0$, when F is a nonlinear term depending strongly on X. In such a case one must solve a nonlinear problem. This is why, it is much more convenient to use the following numerical scheme.

Proposition 3.2 (**Partial Implicit Euler Scheme**) *The scheme is described by the following recursion:*

$$\left[\begin{array}{ll} & X_0 = \tilde{X}_0, \\ \\ n \geq 0, & \dfrac{X_{n+1} - X_n}{h} + F(t_n, X_n) + A(X_{n+1}) = 0. \end{array}\right. \tag{3.3}$$

Here, one must simply compute the inverse graph $(I + hA)^{-1}$. If this calculation is simple the (PIES) scheme can be written in the form:

$$\left[\begin{array}{ll} & X_0 = \tilde{X}_0, \\ \\ n \geq 0, & X_{n+1} = (I + hA)^{-1}\Big(X_n + hF(t_n, X_n),\Big) \end{array}\right. \tag{3.4}$$

where the term $X_n + hF(t_n, X_n)$ is evaluated at each iteration.

The key point for building these numerical schemes is to make the implicit choice $A(X_{n+1})$: the properties of A imply that $(I + hA)$ is invertible.

3.1.1.2 *Second model with differential inclusion*

Let us consider a dynamical system described by the following differential inclusion and ordinary differential system of the first order with initial conditions:

$$\begin{cases} \dot{X} + F(t, X, U) \ni 0, \quad t \in [0, T], \\ \dot{U} + A(U) + G(t, X) \ni 0, \quad t \in [0, T], \\ X(0) = \tilde{X}_0, U(0) = \tilde{U}_0 \in D(A), \end{cases} \tag{3.5}$$

with $T > 0$, $X : [0, T] \longrightarrow E_1$ and $U : [0, T] \longrightarrow E_2$ the (un)known exact unique vector function (X, U) solution defined from interval $[0, T]$ to (finite or infinite dimensional) vector space $E_1 \times E_2$, F, G are smooth functions of t, X and U. A is a maximal monotone graph (or operator) on E_2 for a given scalar product $(\cdot, \cdot)_{E_2}$. $(\tilde{X}_0, \tilde{U}_0)$ is a given vector in $E_1 \times D(A) \subset E_1 \times E_2$ (initial conditions).

Again we introduce discretization of $[0, T]$ using time $t_0 = 0$, $t_1, t_2, \ldots, t_n, \ldots$ with $\forall i, t_i \in [0, T]$, and let us note X_n, U_n the approximated values of the exact values $X(t_n), U(t_n)$ that are expected from the numerical scheme. Let us set again $h_n = t_n - t_{n-1}$ for $n \in \{1, 2, \ldots\}$. To simplify, let us consider a constant time step $h_n = h, \forall n \in \mathbb{N}$.

Again, a fully implicit Euler scheme is not very convenient for computations. So, we can build different partial implicit Euler schemes.

Proposition 3.3 (First Partial Implicit Euler Scheme) *The scheme is described by the following recursion:*

$$\left[\begin{array}{l} X_0 = \tilde{X}_0, \quad U_0 = \tilde{U}_0, \\ n \geq 0, \quad \dfrac{U_{n+1} - U_n}{h} + A(U_{n+1}) + G(t_n, X_n) = 0, \\ \dfrac{X_{n+1} - X_n}{h} + F(t_{n+1}, X_{n+1}, U_{n+1}) = 0. \end{array} \right. \tag{3.6}$$

Clearly this scheme is not very convenient, even if $A(U) = 0$, when F is a nonlinear term depending strongly on X. In such a case one must solve a nonlinear problem. This is why, it is much more convenient to use the following numerical scheme.

Proposition 3.4 (Second Partial Implicit Euler Scheme) *The scheme is described by the following recursive relations*

$$\left[\begin{array}{ll} & X_0 = \tilde{X}_0, \quad U_0 = \tilde{U}_0, \\ n \geq 0, & \dfrac{U_{n+1} - U_n}{h} + A(U_{n+1}) + G(t_n, X_n) = 0, \\ & \dfrac{X_{n+1} - X_n}{h} + F(t_n, X_n, U_n) = 0. \end{array}\right. \tag{3.7}$$

Here, one must simply compute the inverse graph $(I + hA)^{-1}$. If this calculation is simple the (SPIES) scheme can be written in the form:

$$\left[\begin{array}{ll} & X_0 = \tilde{X}_0, \quad U_0 = \tilde{U}_0, \\ n \geq 0, & U_{n+1} = (I + hA)^{-1}\Big(U_n + hG(t_n, X_n),\Big) \\ & X_{n+1} = X_n + hF(t_n, X_n, U_n) = 0. \end{array}\right. \tag{3.8}$$

where the term $X_n + hF(t_n, X_n, U_n)$ is evaluated at each iteration.

Again the key point for building these numerical schemes is the choice of $A(U_{n+1})$.

Indeed, one can imagine **for numerical experiments implemented from "physical intuitive reasons"** some other reasonable numerical scheme of the form

$$\left[\begin{array}{ll} & X_0 = \tilde{X}_0, \quad U_0 = \tilde{U}_0, \\ n \geq 0, & U_{n+1} = (I + hA)^{-1}\Big(U_n + hG(t_n, X_n),\Big) \\ & X_{n+1} = g(h, t_n, X_n, U_n, U_{n+1}). \end{array}\right. \tag{3.9}$$

One has simply to pay attention to the order of the computations! It is not interesting to strongly improve the accuracy for the approximations of X since the approximations of U are quite rough.

3.1.2 *Stochastic case*

In the stochastic smooth context (without A), first order numerical scheme of Euler-Maruyama type are built (see [Kloeden and Platen (1992)], [Pardoux and Talay (1985)]) in the form:

$$\begin{cases} \forall p \in \{0,\dots,N-1\}, \forall k \in \{0,\dots,n\}, \\ X_k^{p+1} = hb_k(t_p, X^p) + \sum_{j=1}^{n} \sigma_{kj}(t_p, X^p)\Delta W_j^p, \\ X^0 = X_0, \end{cases} \tag{3.10}$$

where: $N \in \mathbb{N}^\star$, $h = T/N$, $\forall p \in \{0,\dots,N\}$, $t_p = ph$ and $\Delta W_j^p = W_{t_p} - W_{t_{p+1}}$ corresponds to the increment of the real Brownian motion between t_p and t_{p+1}.

Under the assumptions (2.83) and (2.84), it could be proved that this numerical scheme is convergent and of order 1/2 in the following sense:

$$\left[\mathbb{E}(\sup_{0\leq t\leq T} \mid X_t - X_t^h \mid^2)\right]^{1/2} \leq C\sqrt{h}, \tag{3.11}$$

where X^h is the function built on $\mathbb{R}^n$, the kth component of which is a piecewise linear function interpolating the $(X_k^p)_{0\leq p\leq N}$ and $\mathbb{E}$ denotes the expectation with respect to the probability measure $\mathbb{P}$. See [Kloeden and Platen (1992)] and [Pardoux and Talay (1985)] for details.

In the stochastic non smooth case, Cépa [Cépa (1995a)] proposed to extend the classical scheme by using the Skorohod map. He built an Euler-Peano scheme. In such a case it seems that it is difficult to prove convergence and to give an order of convergence.

Here we intend to propose (see [Bernardin (2000)]) a new heuristic numerical scheme based on the method developed for the deterministic case. We have not proved the convergence result and obviously no results for the order of convergence. But in practice (see [Bernardin (2000)]) the numerical behaviour sounds correct.

The proposed numerical scheme is again based on the univoque character of $(I + \lambda A)^{-1}$ for any $\lambda > 0$. It reads as follows:

$$\begin{cases} \forall p \in \{0,\dots,N-1\}, X^{p+1} - X^p + hA(X^{p+1}) \ni hb(t_p, X^p) + \sigma(t_p, X^p)\Delta W^p, \\ X^0 = x \in \overline{D(A)}, \end{cases} \tag{3.12}$$

(with $D(A)$ the domain of A) that is equivalent to

$$\begin{cases} \forall p \in \{0,\dots,N-1\}, (I+hA)(X^{p+1}) \ni X^p + hb(t_p,X^p) + \sigma(t_p,X^p)\Delta W^p, \\ X^0 = x \in \overline{D(A)}, \end{cases} \tag{3.13}$$

or

$$\begin{cases} \forall p \in \{0,\dots,N-1\}, X^{p+1} = (I+hA)^{-1}(X^p + hb(t_p,X^p) + \sigma(t_p,X^p)\Delta W^p), \\ X^0 = x \in \overline{D(A)}, \end{cases} \tag{3.14}$$

because of the recalled property of $(I+\lambda A)^{-1}$.

To our knowledge, there is no proof for the convergence order of numerical schemes in the stochastic frame, even if one can feel that a result similar to the deterministic one should exist, as given in [Bastien (2000)], [Bastien and Schatzman (1999)] or [Bastien and Schatzman (2000)].

3.2 Analytical Methods

3.2.1 *Simple case*

Let us consider a partition of $\mathbb{R}^n$:

$$\mathbb{R}^n = R_1 \cup R_2 \cup \cdots \cup R_m, \tag{3.15}$$

with:

$$\forall i \in \{1,\dots,m\}, \forall j \in \{1,\dots,m\}, i \neq j \Longrightarrow R_i \cap R_j = \emptyset. \tag{3.16}$$

Let us assume that we are able to solve exactly problems defined by:

$$\left.\begin{cases} \dfrac{dX}{dt} + F_j(t,X) = 0, \\ X(t_0) = X_0 \in \mathbb{R}^n \end{cases}\right] (P_j)\text{problem}, j \in \{1,\dots,\}, \tag{3.17}$$

so that

$$X_j(t) = \phi(t_0, X_0, t), \tag{3.18}$$

is the exact solution of (3.17). Then it is clear that one can solve the following piecewise defined problem:

$$\begin{cases} \frac{dX}{dt}(t) + F_j(t, X(t)) = 0, \quad X(t) \in R_j, \quad j \in \{1, \ldots, \}, \\ \\ X(t_0) = X_0 \in \mathbb{R}^n, \end{cases} \tag{3.19}$$

with left or right continuity conditions at the boundaries between two different R_i and R_j. For example let us consider $X_i(t)$ which is solution of P_i on $[0, t_i)$ with $\forall t \in [0, t_i[, \quad X_i(t) \in R_i$ and t_i is the unique time so that either $X_i(t_i) \in R_i$ and $X_i(t) \in R_j, \quad t > t_i$ or $X_i(t_i) \in R_j$. Then one can solve

$$\frac{dX}{dt} + F_j(t, X) = 0,$$

for $t > t_i$ with initial condition $X_i(t_i)$ given by $\lim_{t \longrightarrow t_i} X_i(t)$. So generally, one can join the exact solutions that are computed piecewisely using continuity properties at each boundary crossing.

3.2.2 *Second case*

Very often we have to consider problems a little bit more complex. They occur when starting from second order problems, when the continuity condition can not be used. This is the case when the state variables are separated into displacements and velocities. Displacements are governed by continuous functions, but velocities have to be found at each transition. In this case, the solution is very similar to the case of the previous subsection except that physical or mathematical reasons have to be added in order to find the correct initial condition at each boundary.

Chapter 4

Properties of Numerical Schemes

In this chapter properties of some results will be claimed. Some of them will solve general questions of convergence of numerical schemes for rather large classes of nonlinear problems described by differential inclusions or by ordinary differential equations and differential inclusions. Some others are only particular results that can illustrate the complex behaviour of numerical schemes in the case of special nonlinear behaviours: even if we do not give proofs for general cases, it is interesting to pay attention to the behaviours of the numerical schemes that can occur from simple examples because it seems that the same kind of behaviour may arise for general situations.

4.1 Dynamics of Systems with Friction or Elastoplastic Terms

Properties of numerical schemes for discrete (and even more general) systems including friction or elastoplastic terms have been studied in [Bastien (2000)], [Bastien and Schatzman (1999)] and [Bastien and Schatzman (2000)]. Let us recall the main results, adapted from these two last references. Let H be a separable Hilbert space with the scalar product $(\cdot, \cdot)$ and norm $||\cdot||$. Let us consider the problem:

$$\begin{cases} \dot{X}(t) + BX(t) + AX(t) \ni f(t, X(t)), a.e.\ t \in [t_0, t_0 + T], \\ X(t_0) = \xi \in H, \end{cases} \tag{4.1}$$

with

$$X : \begin{cases} [t_0, t_0 + T] \longrightarrow \quad H, \\ \\ \quad\quad t \quad\quad \mapsto X(t), \end{cases} \tag{4.2}$$

where B is a Lipschitz-continuous operator and, H is an elliptic one, i.e.:

$$\exists \omega_B, \forall (x_1, x_2) \in H \times H, \| Bx_1 - Bx_2 \| \leq \omega_B \| x_1 - x_2 \|, \tag{4.3}$$

and

$$\exists \alpha_B, \forall (x_1, x_2) \in H \times H, (Bx_1 - Bx_2, x_1 - x_2) \geq \alpha_B \| x_1 - x_2 \|^2 . \tag{4.4}$$

A is a maximal monotone operator on H with bounded section A^0, and

$$f : \begin{cases} [t_0, t_0 + T] \times H \longrightarrow \quad H, \\ \\ \quad\quad (t, x) \quad\quad \mapsto f(t, x), \end{cases} \tag{4.5}$$

is Lipschitz-continuous on H, i.e.

$$\begin{gathered} \exists \omega_f, \forall t \in [t_0, t_0 + T], \forall (x_1, x_2) \in H \times H, \\ \| f(t, x_1) - f(t, x_2) \| \leq \omega_f \| x_1 - x_2 \| \end{gathered} \tag{4.6}$$

and

$$\forall x \in H, f(., x) \in L^2([t_0, t_0 + T], H). \tag{4.7}$$

The map

$$\begin{cases} H \longrightarrow L^2([t_0, t_0 + T], H), \\ \\ x \;\mapsto\; \quad \dfrac{\partial f}{\partial t}(., x), \end{cases} \tag{4.8}$$

is bounded in $L^2([t_0, t_0 + T], H)$ on every compact subset of H. Then, existence and uniqueness results have been obtained in [Brezis (1973)] and the following numerical scheme [Bastien and Schatzman (2000)] with constant time step $h > 0$ can be used in order to approximate the exact solution $X(t)$ of (4.1):

$$\begin{cases} \dfrac{X_{n+1} - X_n}{h} + BX_{n+1} + AX_{n+1} \ni f(nh, X_n), \\ \\ X_0 = \xi, \; h = \dfrac{T}{N+1}, \; N \in \mathbb{N}^\star, \; n = 0, 1, 2, \ldots, N. \end{cases} \tag{4.9}$$

It is proved in [Bastien (2000)] or [Bastien and Schatzman (1999)] (and an idea of proof is given in [Bastien and Schatzman (2000)]) that this numerical scheme is of order 1/2, i.e., it verifies:

$$\max\Big(\| X(t) - \tilde{X}_h(t) \|^2 + \alpha_B \int_{t_0}^{t} \| X(s) - \tilde{X}_h(s) \|^2 \, ds\Big)^{\frac{1}{2}} = o(\sqrt(h)),$$
$$t \in [t_0, t_0 + T], \tag{4.10}$$

where $\tilde{X}_h$ is the function defined by linear interpolation of X_{nh}.

In many physical situations as we already explained, A derives from a particular potential, i.e., a function ϕ lower semi-continuous convex proper exists so that

$$A = \partial\phi. \tag{4.11}$$

In such a case the order is improved [Bastien (2000)], [Bastien and Schatzman (2000)]:

$$\max\Big(\| X(t) - \tilde{X}_h(t) \|^2 + \alpha_B \int_{t_0}^{t} \| X(s) - \tilde{X}_h(s) \|^2 \, ds\Big)^{\frac{1}{2}} = o(h),$$
$$t \in [t_0, t_0 + T]. \tag{4.12}$$

So the order of the numerical scheme is one.

4.2 Systems with Impacts

4.2.1 *Introduction*

Nonlinear dynamical systems exhibiting impacts are often encountered in practice, for instance in the models of hammering in engines [Kunert and Pfeiffer (1989b)], [Pfeiffer and Prestl (1994)], rotor-casing dynamical systems [Li and Paidoussis (1994)], collisions of solids, ships moored at dockside, hand-held percussion machine [Babitsky (1998a)], [Babitsky (1999)], rigid blocks [Iyengar and Roy (1996)], control of joint's loose [Panet *et. al.* (1995)], [Paoli (1993)], [Paoli *et. al.* (1992)]. Many applications are given in [Babitsky (1998a)], [Brogliato (1996)]. Such mechanical systems have been mainly investigated in the single degree of freedom or the two degree of freedom case, in order to study the behaviour of the response: periodic responses [Babitsky (1998a)], [Babitsky and Krupenin (2001)], [Van De Vorst *et. al.* (1996)] including sticking motions [Budd and Dux (1994)], [Toulemonde and Gontier (1998)] in the undamped case and chaotic responses [Shaw and Shaw (1989)] have been found, bifurcations, transitions and global behaviours have been examined [Budd and Dux (1996)], [Peterka (1999a)], [Peterka and Vacik (1992)], [Whiston (1987)], [Yim and Lin (1991)] including grazing bifurcations [Foale and Bishop (1994)], [Frederiksson and Nordmark (1997)], [Nordmark (1991)] and nonsmooth phenomena [Lamba and Budd (1994)], [Lenci and Rega (1999)], [Whiston (1992)]. Some authors have studied more complicated models including both impacts and friction [Cone and Zadoks (1995)], [Peterka (1999b)], [Pfeiffer (1999)], or clearance variations [Budd *et. al.* (1995)]. Theoretical and experimental analysis of elastic systems undergoing vibro-impact oscillations were carried out in [Emaci *et. al.* (1997)], [Azeez and Vakakis (1999)], [Azeez and Vakakis (2001)], whereas methods for studying exactly vibro-impact oscillations and their bifurcations were outlined in [Mikhlin *et. al.* (1998)], [Azeez *et. al.* (1999)]. Some authors also considered the issue of controlling systems with impacts [Awrejcewicz *et. al.* (1996b)], [Brogliato (1996)], [Lenci and Rega (1998b)], [Lenci and Rega (1998c)], [Peterka and Vacik (1992)]. In almost every case, mechanical systems are modelled by using linear or integrable differential equations, with implicit equations governing impact times which are numerically solved at any accuracy. In some cases, authors use classical schemes like Runge-Kutta schemes. Nevertheless, performances of such numerical schemes have not been investigated yet and mathematical results are valid and known only for smooth systems (see [Crouzeix and Mignot (1984)], [Dekker and Verwer (1984)], [TNAG (1997)], [Hairer *et. al.* (1987)], [Hairer and Wanner (1991)] for example). On the other hand, particular numerical methods have already been introduced for nonsmooth dynamical systems from a practical point of view (see [Van Der Spek *et. al.* (1994)] for example). As impacts problems can be understood as unilateral constraints, Glocker and Pfeiffer developed efficient algorithms in [Glocker and Pfeiffer (1992)], [Pfeiffer (1988b)], [Pfeiffer and Glocker (1996)], [Stiegelmeyr and Pfeiffer (1999)]. Finally, starting from the mathematical point of view a new numerical scheme has been introduced in [Panet *et. al.* (1995)], [Paoli (1993)], [Paoli and Schatzman (1993b)][Paoli and Schatzman (1994)], [Paoli *et. al.* (1992)].

Finally, an interesting analytical method based on the use on nonsmooth functions that result from vibro-impact oscillations are outlined in the earlier works by Pilipchuk and then further developed by Pilipchuk and Vakakis [Pilipchuk *et. al.* (1997)].

Therefore, except in the last works, authors do not address the problem of the choice between the existing numerical schemes adapted to the occurrence of impacts. This is why in this chapter we intend to test some numerical methods applied to a simple single degree of freedom example with two different kinds of responses.

This section is outlined as follows: we introduce the considered mechanical system and we present several ways to obtain its solutions, first in an analytical way and then in a numerical way with several methods for localizing phase transitions. Then, we prove some properties about the responses with infinite sequences of impacts that can occur, even in periodic solutions. We investigate the behaviour of the different numerical schemes already defined from a theoretical point of view, and we prove some consistency results. Numerical experiments are performed in order to illustrate and complete the theoretical results. We present and analyse the results of computations, and the behaviour of the numerical methods (order, accuracy, computational cost) is investigated. Finally, we conclude about the relevance of each category of considered numerical methods.

4.2.2 *Dynamical system and resolution methods*

Model

We consider a simple model of a single degree of freedom oscillator consisting of a mass which is subjected to an external forcing and which can impact against a plane fixed rigid stop (see the Fig. 4.1).

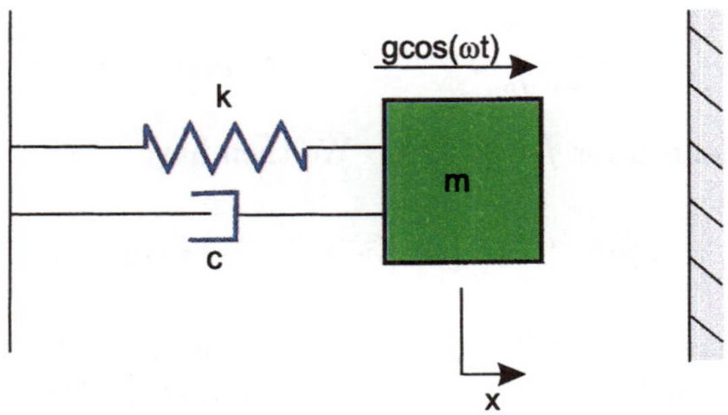

Fig. 4.1 Single degree of freedom system with $k = \omega_1^2 m$, $c = am$, $g = fm$.

The free flight of the system is governed by the equation:

$$m\ddot{x}(t) + c\dot{x}(t) + kx(t) = F(t), \tag{4.13}$$

which, setting $\omega_1 = \sqrt{\dfrac{k}{m}}$, $a = \dfrac{c}{2m}$ and $F(t) = mf(t)$, can be written in the form:

$$\ddot{x}(t) + 2a\dot{x}(t) + \omega_1^2 x(t) = f(t). \tag{4.14}$$

We will assume that $f \in \mathbf{L}^1_{loc}(\mathbb{R})$, namely that f is locally integrable on $\mathbb{R}^+$. An elastoplastic impact law is used to account for impacts against the stop when $x = x_{max}$. This law is characterized by the restitution coefficient $e \in [0, 1]$:

$$x(t) = x_{max} \Rightarrow \dot{x}(t^+) = -e\dot{x}(t^-). \tag{4.15}$$

Therefore the velocity of the mass is discontinuous at every impact time, except at a zero velocity impact.

Analytical Resolution

The simplicity of the previously introduced model enables to solve analytically the obtained system (4.14), (4.15). Indeed, between two consecutive impacts the differential Eq. (4.14) can be solved for any initial conditions:

$$x(t) = \mathrm{e}^{-at/2}[A\cos(\tilde{\omega_1}t) + B\sin(\tilde{\omega_1}t)] + x_p(t), \tag{4.16}$$

where $\tilde{\omega_1} = \sqrt{\omega_1^2 - a^2}$, x_p denotes a particular solution of (4.14) and A and B are determined by the initial conditions of the system. Let us set $\eta = \dfrac{a}{\tilde{\omega_1}}$. We will choose as a particular solution of (4.14):

$$x_p(t) = \frac{1}{\tilde{\omega_1}} \int_0^t f(u)\mathrm{e}^{-a(t-u)} \sin[\tilde{\omega_1}(t-u)]\mathrm{d}u,$$

which is well defined for $f \in \mathbf{L}^1_{loc}(\mathbb{R})$. We then have

$$\dot{x_p}(t) = \int_0^t f(u)\mathrm{e}^{-a(t-u)}\{\cos[\tilde{\omega_1}(t-u)] - \eta\sin[\tilde{\omega_1}(t-u)]\}\mathrm{d}u.$$

Moreover, the impact law (4.15) provides at each impact new initial conditions that enable to get an analytical form of the trajectory until the next impact:

$$x(t) = \mathrm{e}^{-at}[A_n\cos(\tilde{\omega_1}t) + B_n\sin(\tilde{\omega_1}t)] + x_p(t) \text{ on } [t_{n-1}, t_n], \tag{4.17}$$

where $t_0 = 0$, t_n verifies $x(t_n) = x_{max}$, $A_1 = x_0$, $B_1 = \dot{x_0}$ and

$$\left\{\begin{aligned} A_{n+1} &= A_n + \frac{1+e}{\tilde{\omega_1}} e^{at_n} \sin(\tilde{\omega_1} t_n)\dot{x}(t_n^-), \\ B_{n+1} &= B_n - \frac{1+e}{\tilde{\omega_1}} e^{at_n} \cos(\tilde{\omega_1} t_n)\dot{x}(t_n^-). \end{aligned}\right. \tag{4.18}$$

The recursive relations (4.17), (4.18) describe the motion of the system provided that the mass does not stick to the stop. This phenomenon occurs after a zero velocity impact when the external forcing constrains the mass on the stop. The equation of the system during a sticking phase is then:

$$\ddot{x} + 2a\dot{x} + \omega_1^2 x = f(t) + R(t), \tag{4.19}$$

where $R(t)$ denotes the reaction of the stop. Yet as long as the mass sticks to the stop we have $x(t) = x_{max}$ and $\dot{x}(t) = \ddot{x}(t) = 0$. Thus (4.19) becomes:

$$R(t) + f(t) = \omega_1^2 x_{max}. \tag{4.20}$$

This equation holds as long as the reaction of the stop is negative, namely as long as:

$$f(t) \geq \omega_1^2 x_{max}. \tag{4.21}$$

For the contact to cease, the function $f(t) - \omega_1^2 x_{max}$ must become zero and changes sign.

The recursive relations (4.17), (4.18), coupled with the system (4.20), (4.21) when sticking occurs, give an analytical solution to the system (4.14), (4.15).

Remark 4.1 *The relation (4.21) implies that sticking can not occur if the system's parameters are such that*

$$f(t) < \omega_1^2 x_{max} \ \forall t \in \mathbb{R}^+. \tag{4.22}$$

Remark 4.2 *The trajectory is not known in a fully analytical way: the impact times can only be obtained by a numerical solution of the nonlinear equation* $x(t) = x_{max}$.

Numerical Resolution

The discontinuity in the velocity, due to the impact law, requires the use of an adapted numerical scheme. In the whole chapter we will only consider numerical methods with constant time step.

The Schatzman-Paoli Method

We assume here that f is continuous. A numerical method valid for mechanical systems with unilateral constraint can be found in [Paoli (1993)]. It is a two step method of Euler type which can be written in our case as follows:

$$\begin{cases} y_0 = x_0, \\ y_1 = x_0 + \dot{x}_0 h + \dfrac{h^2}{2}(f(0) - 2a\dot{x}_0 - \omega_1^2 x_0), \\ x_n = \dfrac{h^2 f(t_n) + (2 - h^2\omega_1^2)y_n - [(1-e) - (1+e)ah]\, y_{n+1}}{1+ah}, \\ y_{n+1} = -e y_{n-1} + \min\{x_n, (1+e)x_{max}\}. \end{cases} \tag{4.23}$$

This method does not require any explicit approximation of impact times.

Some Classical Methods

In addition to the preceding scheme, we are going to consider some classical numerical methods often used to solve smooth ordinary differential equations: Newmark's method, the Runge-Kutta RK24 method and the Runge-Kutta method of Dormand and Prince DOPRI5 (see [Hairer *et. al.* (1987)]). We need to modify these methods in order to adapt them to the resolution of (4.14), (4.15) by adding an impact time approximation procedure. If f is continuous, then between two consecutive impact times the solution is of class C^2, hence Newmark's method is of order 2 and converges towards the exact solution. If we make further assumptions on f, namely if we suppose that f is of class C^2, then the solution is piecewise of class C^4, so that the RK24 method converges and is of order 4 on each time interval without impacts. The DOPRI5 method is a fifth order one assuming f is of class C^5.

Hence the key point of those numerical methods lies on precise location of the times of impacts. Two main different methods can be used to deal with impacts: the simplest method is to try to estimate at most one impact time per step. This method obviously leads to a loss of information if more than one impact occurs on the same time step, but its main advantage is that it requires fewer computations. The second method consists in trying to find as many impact times as possible on each time step, until two successive impact times are closer than a chosen precision.

Several ways to approximate the time of impact can be considered. First of all, an impact is deemed to occur when the value of (y_{n+1}, z_{n+1}) given by the scheme at the $n+1^{\text{th}}$ step verifies $y_{n+1} > x_{max}$: at least one impact then happens between t_n and t_{n+1}. The simplest way to approximate the impact time consists in linearly interpolating the approximation of the solution given by the scheme, that is to say to determine the linear mapping which maps t_n to y_n and t_{n+1} to y_{n+1}, and to infer the time which is mapped to x_{max}. This method will be referred to as (IM1). Another way to determine impact times is similar to the previous one, but uses an interpolation by a second order polynomial which maps t_n to y_n, t_{n+1} to y_{n+1} and whose derivative at t_n is z_n: this method will be referred to as (IM2). We will test a third method for localizing impact times which is based on a dichotomy between t_n and t_{n+1} until the chosen precision is reached, and we will refer to this method as (IM3). The precision required in the impact time approximation will be set to h^2 for Newmark's method, and to h^4 in the RK24 and DOPRI5 method. Finally, we will also investigate the behaviour of Newton's method for solving nonlinear equations, which will be referred to as (IM4). An obvious drawback of the interpolations (IM3) and (IM4) lies in the precision that has to be set to h^p where p is the order of the initial numerical method: from the numerical point of view the computer precision is quickly reached when h is relatively small (for instance if $h = 10^{-4}$ in the RK24 or DOPRI5 case). Once the impact time t_* has been approximated by one of the four preceding methods, it remains to compute the actual value (y_{n+1}, z_{n+1}) that the numerical method will provide at the $(n+1)$-th step. To achieve this, we first start to compute the approximation z_*^- of the velocity just before impact, by applying the chosen numerical method between t_n and t_*. The impact law then yields $z_*^+ = -ez_*^-$, and we apply the numerical scheme starting from (x_{max}, z_*^+) with time step $t_{n+1} - t_*$ in order to obtain (y_{n+1}, z_{n+1}).

As we do not allow the approximation of the solution to lie inside the non-admissible area, we must eventually check that the obtained value (y_{n+1}, z_{n+1}) does not verify $y_{n+1} > x_{max}$. If it does, we consider that the system is in a sticking phase at time t_{n+1} so we set y_{n+1} to x_{max} and z_{n+1} to 0. Finally, if sticking is occurring at time t_n, namely if $(y_n, z_n) = (x_{max}, 0)$, then we only need to check the sign of the acceleration: if it is negative, then the system is still sticking at t_{n+1} so we set $(y_{n+1}, z_{n+1}) = (x_{max}, 0)$. If it is positive, the algorithm depends on the chosen point of view. In the case of a scheme with at most one impact localization per step, we simply apply the procedure introduced above in order to get the value of the displacement and velocity at t_{n+1}. In the case of a scheme with computation of several impact times per step, we use the same interpolation method as we previously used to estimate impact times in order to approximate the time $\bar{t}$ when sticking ends, which verifies $f(\bar{t}) - \omega_1^2 x_{max} = 0$. Then the usual procedure is applied on $[\bar{t}, t_{n+1}]$.

Let us recall the expressions of the different numerical schemes that we previously mentioned. The Newmark method with time step h, associated with (β, γ), is given by:

$$\left\{\begin{array}{l} y_{n+1} = \dfrac{1}{1+2ah\gamma+\omega_1^2h^2\beta} \times \{y_n\,[1+2ah\gamma+ \\ \qquad -\omega_1^2h^2(0.5-\beta)+a\omega_1^2h^3\,(2\beta-\gamma)]+ \\ \qquad +z_nh\,[1+ah(2\gamma-1)+2a^2h^2(2\beta-\gamma)]+ \\ \qquad +h^2\{\beta f(t_{n+1})+[0.5-\beta-ah\,(2\beta-\gamma)]\,f(t_n)\}\}, \\ \\ z_{n+1} = \dfrac{1}{1+2ah\gamma+\omega_1^2h^2\beta} \times \{-y_nh\omega_1^2\left[1+\omega_1^2h^2\left(\beta-\dfrac{\gamma}{2}\right)\right]+ \\ \qquad +z_n\,[1-2ah(1-\gamma)+\omega_1^2h^2(\beta-\gamma)-a\omega_1^2h^3\,(2\beta-\gamma)]+ \\ \qquad +h\{\gamma f(t_{n+1})+\left[1-\gamma+\omega_1^2h^2\left(\beta-\dfrac{\gamma}{2}\right)\right]f(t_n)\}\}. \end{array}\right. \tag{4.24}$$

The Runge-Kutta RK24 method with the time step h is given by:

$$\left\{\begin{array}{l} k_1^n(1) = z_n, \\ k_1^n(2) = -\omega_1^2y_n - 2ak_1^n(1) + f(t_n), \\ k_2^n(1) = z_n + \dfrac{h}{2}k_1^n(2), \\ k_2^n(2) = -\omega_1^2\left[y_n+\dfrac{h}{2}k_1^n(1)\right] - 2ak_2^n(1) + f\left(t_n+\dfrac{h}{2}\right), \\ k_3^n(1) = z_n + \dfrac{h}{2}k_2^n(2), \\ k_3^n(2) = -\omega_1^2\left[y_n+\dfrac{h}{2}k_2^n(1)\right] - 2ak_3^n(1) + f\left(t_n+\dfrac{h}{2}\right), \\ k_4^n(1) = z_n + hk_3^n(2), \\ k_4^n(2) = -\omega_1^2\,[y_n+hk_3^n(1)] - 2ak_4^n(1) + f(t_{n+1}), \\ y_{n+1} = y_n + \dfrac{h}{6}[k_1^n(1)+2k_2^n(1)+2k_3^n(1)+k_4^n(1)], \\ z_{n+1} = z_n + \dfrac{h}{6}[k_1^n(2)+2k_2^n(2)+2k_3^n(2)+k_4^n(2)]. \end{array}\right. \tag{4.25}$$

The Runge-Kutta method of Dormand and Prince is a method that can provide dense output: it gives a fourth order approximation of the solution at every point between t_n and t_{n+1} ($\theta \in [0,1]$), and a fifth order approximation at t_{n+1} ($\theta = 1$) (see [Hairer *et. al.* (1987)]).

Table 4.1 The Runge-Kutta method of Dormand and Prince.

0							
$\frac{1}{5}$	$\frac{1}{5}$						
$\frac{3}{10}$	$\frac{553}{40}$	$\frac{9}{40}$					
$\frac{4}{5}$	$\frac{44}{45}$	$\frac{-56}{15}$	$\frac{32}{9}$				
$\frac{8}{9}$	$\frac{19372}{6561}$	$\frac{-25360}{2187}$	$\frac{64448}{6561}$	$\frac{-212}{729}$			
1	$\frac{9017}{3168}$	$\frac{-355}{33}$	$\frac{46732}{5247}$	$\frac{49}{176}$	$\frac{-5103}{18656}$		
1	$\frac{35}{384}$	0	$\frac{500}{1113}$	$\frac{125}{192}$	$\frac{-2187}{6784}$	$\frac{11}{84}$	
	$b_1(\theta)$	$b_2(\theta)$	$b_3(\theta)$	$b_4(\theta)$	$b_5(\theta)$	$b_6(\theta)$	$b_7(\theta)$

The Table 4.1 shows the coefficients associated with this Runge-Kutta method, where:

$$\begin{aligned}
b_1(\theta) &= \theta(1 + \theta(-1337/480 + \theta(1039/360 + \theta(-1163/1152)))), \\
b_2(\theta) &= 0, \\
b_3(\theta) &= 100\theta^2(1054/9275 + \theta(-4682/27825 + \theta(379/5565)))/3, \\
b_4(\theta) &= -5\theta^2(27/40 + \theta(-9/5 + \theta(83/96)))/2, \\
b_5(\theta) &= 18225\theta^2(-3/250 + \theta(22/375 + \theta(-37/600)))/848, \\
b_6(\theta) &= -22\theta^2(-3/10 + \theta(29/30 + \theta(-17/24)))/7, \\
b_7(\theta) &= 0.
\end{aligned} \tag{4.26}$$

The four impact approximation procedures are given by:

- (IM1):

$$t_* = t_n + h\frac{x_{max} - y_n}{y_{n+1} - y_n}, \tag{4.27}$$

- (IM2):

$$X = \frac{y_{n+1} - y_n - hz_n}{h^2},$$
$$t_* = t_n + \frac{-z_n + \sqrt{z_n^2 + 4X(x_{max} - y_n)}}{2X}, \tag{4.28}$$

- (IM3) and (IM4) are defined by an iterative procedure where Scheme $(t_1, X_1, Y_1, h, X_2, Y_2)$ returns in (X_2, Y_2) the value obtained after one iteration of the initial numerical method with time step h starting from (X_1, Y_1) at t_1:

  - (IM3):

$$\left[\begin{array}{c} \text{inf } = t_n \\ \text{sup } = t_{n+1} \\ \text{REPEAT} \\ \hat{t} = \dfrac{inf + sup}{2} \\ \text{Scheme } (t_n, y_n, z_n, \hat{t} - t_n, X_2, Y_2) \\ \text{IF } X_2 < x_{max} \\ \text{inf } = \hat{t} \\ \text{ELSE} \\ \text{sup } = \hat{t} \\ \text{ENDIF} \\ \text{UNTIL sup-inf } \leq \text{ Precision} \\ t_* = \dfrac{inf + sup}{2} \end{array}\right. \tag{4.29}$$

– (IM4):

$$\left[\begin{array}{c} \hat{t} = t_n + h/2 \\ \text{REPEAT} \\ \text{Scheme } (t_n, y_n, z_n, \hat{t} - t_n, X_2, Y_2) \\ \Delta = \dfrac{x_{max} - X_2}{Y_2} \\ \hat{t} = \hat{t} + \Delta \\ \text{UNTIL } (\mid \Delta \mid \leq \text{ Precision) OR } (\hat{t} < t_n - h) \text{ OR } (\hat{t} > t_n + 2h) \\ \text{IF } (\hat{t} > t_n + 2h) \text{ OR } (\hat{t} < t_n - h) \\ t_* = t_n \\ \text{ELSE} \\ t_* = \hat{t} \\ \text{ENDIF} \end{array}\right. \tag{4.30}$$

4.2.3 *Sticking motion and accumulation of impacts*

When the impact law is a purely plastic one, namely when $e = 0$, a sticking phase may occur after any impact where the acceleration of the system is positive. On the other hand, when dealing with an elastic impact law, we note that sticking happens after an accumulation of impact times.

We are going to introduce in this part some results about accumulations of impact times. Let us start with a lemma that will be useful afterwards.

Lemma 4.1 *Let us consider a sequence of impact times $(t_n)_{n\geq 1}$ such that $\lim_{n\to+\infty} t_n = t_\infty$. We assume moreover that $e < 1$. Then $\dot{x}$ is bounded on $[t_1, t_\infty]$.*

Proof *The result given in this lemma can be seen as a consequence of the proofs given in [Paoli (1993)]. In our case, the simplicity of the system can lead a direct proof of this result.*

We saw in (4.17) the analytical writing of x on $[t_{n-1}, t_n]$:

$$\begin{aligned} \dot{x}(t) = {} & -\tilde{\omega_1} \mathrm{e}^{-at} \left\{A_n[\sin(\tilde{\omega_1} t) + \eta \cos(\tilde{\omega_1} t)]\right\} + \\ & + B_n[-\cos(\tilde{\omega_1} t) + \eta \sin(\tilde{\omega_1} t)] + \dot{x_p}(t). \end{aligned} \tag{4.31}$$

Therefore, in order to show that $\dot{x}$ is bounded, it is sufficient to show that $(A_n)_{n\geq 1}$, $(B_n)_{n\geq 1}$ and $\dot{x_p}$ are bounded. First of all, we have:

$$\dot{x_p}(t) = \int_0^t f(u) \mathrm{e}^{-a(t-u)} \{\cos[\tilde{\omega_1}(t-u)] - \eta \sin[\tilde{\omega_1}(t-u)]\} \mathrm{d}u,$$

hence, because $f \in \mathbf{L}^1_{loc}(\mathbb{R})$, $\dot{x_p}$ *is clearly bounded on* $[t_1, t_\infty]$.

Let us set $U_n = \begin{pmatrix} A_n \\ B_n \end{pmatrix}$, $c_n = \cos(\tilde{\omega_1} t_n)$ *and* $s_n = \sin(\tilde{\omega_1} t_n)$. *According to (4.18), we have:*

$$U_{n+1} = M_n U_n + N_n, \tag{4.32}$$

where

$$M_n = \begin{pmatrix} 1-(1+e)s_n(s_n+\eta c_n) & -(1+e)s_n(-c_n+\eta s_n) \\ (1+e)c_n(s_n+\eta c_n) & 1+(1+e)c_n(-c_n+\eta s_n) \end{pmatrix}, \tag{4.33}$$

and

$$N_n = \frac{1+e}{\tilde{\omega_1}} \mathrm{e}^{at_n} \dot{x_p}(t_n) \begin{pmatrix} s_n \\ -c_n \end{pmatrix}. \tag{4.34}$$

The eigenvalues of M_n *are* $-e$ *and* 1 *and setting*

$$P_n = \begin{pmatrix} c_n & s_n \\ s_n+\eta c_n & -c_n+\eta s_n \end{pmatrix}$$

we obtain:

$$M_n = P_n \Lambda P_n^{-1},$$

where $\Lambda = \begin{pmatrix} 1 & 0 \\ 0 & -e \end{pmatrix}$. *Let us set* $V_n = P_n U_n$ *and* $K_n = P_n N_n$. *Then* V_n *is given by the following recursion:*

$$V_{n+1} = \Lambda V_n + K_n, \tag{4.35}$$

where $K_n = \dfrac{1+e}{\tilde{\omega_1}} \mathrm{e}^{at_n} \dot{x_p}(t_n) \begin{pmatrix} 0 \\ 1 \end{pmatrix}$.

Since we have seen that $\dot{x_p}$ *was bounded, there exists* $\alpha \in \mathbb{R}^+$ *such that* $\forall n \geq 1$, $\| K_n \| \leq \alpha$. *Moreover, according to (4.35), we have*

$$V_n = \Lambda^{n-1} V_1 + K_{n-1} + \Lambda K_{n-2} + \ldots + \Lambda^{n-2} K_1.$$

But $\forall n \geq 1$, $\| \Lambda^{n-1} \| \leq \| \Lambda \|^{n-1} < 1$ *because* $e \in [0,1]$.
Moreover, $\forall i \in \{1, \ldots, n-2\}$, *that*

$$\Lambda^i K_{n-1-i} = \frac{1+e}{\tilde{\omega_1}} \mathrm{e}^{at_{n-1-i}} \dot{x_p}(t_{n-1-i}) \begin{pmatrix} 0 \\ (-e)^i \end{pmatrix},$$

then $\| \Lambda^i K_{n-1-i} \| \leq e^i \alpha$.

Thus, $\| V_n \| \leq \| V_1 \| + \alpha + e\alpha + \ldots + e^{n-2}\alpha$. *Yet* $\alpha + e\alpha + \ldots + e^{n-2}\alpha = \alpha\frac{e^{n-1}-1}{e-1} \leq \frac{\alpha}{1-e}$ *since* $e < 1$. *Hence if we set* $M = \| V_1 \| + \frac{\alpha}{1-e}$, *we obtain* $\| V_n \| \leq M \; \forall n \geq 1$.
Then $U_n = P_n^{-1} V_n$, *and, as there clearly exists* $\beta \in \mathbb{R}^+$ *such that* $\| P_n^{-1} \| \leq \beta \; \forall n \geq 1$, *we obtain* $\| U_n \| \leq \beta M \; \forall n \geq 1$. *Therefore* $(A_n)_{n\geq 1}$ *and* $(B_n)_{n\geq 1}$ *are bounded which implies, because* $\dot{x}_p$ *is also bounded, that* $\dot{x}$ *is bounded on* $[t_1, t_\infty]$.

Thanks to the preceding lemma we can prove the following result:

Proposition 4.1 *Let us assume that* $e < 1$ *and that* f *is continuous on* $\mathbb{R}^+$. *Let* $(t_n)_{n\geq 1}$ *be a time sequence such that:*

$$\begin{cases} x(t_n) = x_{max} & \forall n \geq 1, \\ \lim\limits_{n\to+\infty} t_n = t_\infty, \\ f(t_\infty) \neq \omega_1^2 x_{max}. \end{cases}$$

Then a sticking motion starts at t_∞. *Furthermore, the sequence verifies:*

$$\lim_{n\to+\infty} \frac{t_{n+1} - t_n}{t_n - t_{n-1}} = \lim_{n\to+\infty} \frac{\dot{x}(t_{n+1}^-)}{\dot{x}(t_n^-)} = e. \tag{4.36}$$

Proof *On* $[t_{n-1}, t_n]$, *the solution* x *is of class* C^2. *A second order Taylor expansion between* t_{n-1}^+ *and* t_n^- *yields:*

$$x(t_{n-1}^+) = x(t_n^-) + (t_{n-1} - t_n)\dot{x}(t_n^-) + \frac{1}{2}(t_{n-1} - t_n)^2 \ddot{x}(\theta_n), \tag{4.37}$$

where $\theta_n \in [t_{n-1}, t_n]$.
Hence, because x *is continuous on* $\mathbb{R}^+$ *and* $x(t_{n-1}) = x(t_n) = x_{max}$ *by assumptions, we get:*

$$\dot{x}(t_n^-) = \frac{t_n - t_{n-1}}{2}\ddot{x}(\theta_n), \tag{4.38}$$

$$= \frac{t_n - t_{n-1}}{2}[f(\theta_n) - 2a\dot{x}(\theta_n) - \omega_1^2 x(\theta_n)]. \tag{4.39}$$

But $t \mapsto f(t) - \omega_1^2 x(t)$ *is continuous on* $[t_1, t_\infty]$ *so it is bounded: there exists* $M \in \mathbb{R}^+$ *such that* $| f(\theta_n) - \omega_1^2 x(\theta_n) | \leq M \; \forall n \geq 1$. *Moreover, according to the Lemma 4.1,* $\dot{x}$ *is bounded on* $[t_1, t_\infty]$. *Then* $\ddot{x}$ *is bounded as well, which implies:*

$$\lim_{n\to+\infty} \dot{x}(t_n^-) = 0,$$

and then:

$$\begin{cases} x(t_\infty) = x_{max}, \\ \dot{x}(t_\infty^-) = 0. \end{cases} \tag{4.40}$$

Therefore a zero velocity impact occurs at t_∞. *Furthermore, according to (4.38), we have:*

$$\ddot{x}(\theta_n) = \frac{2\dot{x}(t_n^-)}{t_n - t_{n-1}}, \tag{4.41}$$

where $\theta_n \in [t_{n-1}, t_n]$. *But* $\dot{x}(t_n^-) \geq 0$ *because impacts occur with a positive velocity, and* $t_n - t_{n-1} > 0$. *We consequently have* $\ddot{x}(\theta_n) \geq 0 \ \forall n \geq 1$, *so passing to the limit we obtain* $\ddot{x}(t_\infty^-) \geq 0$. *The response thus verifies at* t_∞*:*

$$\begin{cases} x(t_\infty) = x_{max}, \\ \dot{x}(t_\infty^-) = 0, \\ \ddot{x}(t_\infty^-) \geq 0. \end{cases} \tag{4.42}$$

This means that a sticking motion starts at t_∞ *because* $\ddot{x}(t_\infty) = f(t_\infty) - \omega_1^2 x_{max}$ *is nonzero by assumptions.*

By using a second order Taylor expansion between t_n^+ *and* t_{n+1}^-, *and because* $x(t_n) = x(t_{n+1}) = x_{max}$, *we obtain* $\dot{x}(t_n^+) + \dfrac{t_{n+1} - t_n}{2}\ddot{x}(\gamma_n) = 0$ *where* $\gamma_n \in [t_n, t_{n+1}]$, *and then:*

$$t_{n+1} - t_n = \frac{-2\dot{x}(t_n^+)}{\ddot{x}(\gamma_n)}. \tag{4.43}$$

Moreover, a first order Taylor expansion of $\dot{x}$ *yields* $\dot{x}(t_{n+1}^-) = \dot{x}(t_n^+) + \ddot{x}(\beta_n)(t_{n+1} - t_n)$ *with* $\beta_n \in [t_n, t_{n+1}]$. *According to (4.43), we obtain:*

$$\dot{x}(t_{n+1}^-) = \dot{x}(t_n^+)\left[1 - 2\frac{\ddot{x}(\beta_n)}{\ddot{x}(\gamma_n)}\right]. \tag{4.44}$$

Since $\ddot{x}(\beta_n) = f(\beta_n) - a\dot{x}(\beta_n) - \omega_1^2 x(\beta_n)$, $\lim\limits_{n\to+\infty} x(\beta_n) = x(t_\infty) = x_{max}$ *and* $\lim\limits_{n\to+\infty} \dot{x}(\beta_n) = \dot{x}(t_\infty^-) = 0$, *and because the limits are identical for* γ_n, *we have*
$\lim\limits_{n\to+\infty} \dfrac{\ddot{x}(\beta_n)}{\ddot{x}(\gamma_n)} = 1$, *from which we infer* $\lim\limits_{n\to+\infty} \dfrac{\dot{x}(t_{n+1}^-)}{\dot{x}(t_n^+)} = -1$. *Thus, because of the impact law (4.15), we obtain:*

$$\lim_{n\to+\infty} \frac{\dot{x}(t_{n+1}^-)}{\dot{x}(t_n^-)} = e. \tag{4.45}$$

Finally, according to (4.43) and (4.38), we have:

$$\frac{t_{n+1}-t_n}{t_n-t_{n-1}} = \frac{-2\dot{x}(t_n^+)}{\ddot{x}(\gamma_n)}\frac{\ddot{x}(\theta_n)}{2\dot{x}(t_n^-)} = e\frac{\ddot{x}(\theta_n)}{\ddot{x}(\gamma_n)}, \tag{4.46}$$

with $\gamma_n \in [t_n, t_{n+1}]$ *and* $\theta_n \in [t_{n-1}, t_n]$. *As before, it can be shown that* $\lim_{n\to+\infty} \frac{\ddot{x}(\theta_n)}{\ddot{x}(\gamma_n)} = 1$, *hence*

$$\lim_{n\to+\infty} \frac{t_{n+1}-t_n}{t_n-t_{n-1}} = e.$$

4.2.4 *Behaviour of the numerical methods*

4.2.4.1 *Convergence and order of one step numerical methods for smooth differential systems*

Let us recall a few definitions for one step numerical methods which can be found in [Crouzeix and Mignot (1984)]. Let us deal with a scheme of the following form:

$$y_{n+1} = y_n + h_n\phi(t_n, y_n, h_n) \ \forall n \geq 0, \tag{4.47}$$

where $t_0 = 0$ and $t_{n+1} = t_n + h_n$, used to approximate the solution of the following differential equation on $\mathbb{R}^n$:

$$\begin{cases} \dot{y}(t) = f(t, y(t)), & t \in [0, T], \\ y(0) = y_0. \end{cases} \tag{4.48}$$

Then the following definitions hold:

Definition 4.1 *The numerical method (4.47) is said to be consistent for the differential Eq. (4.48) if for any solution y of (4.48) the quantity*

$$\sum_{n=0}^{N-1} \| y(t_{n+1}) - y(t_n) - h_n\phi(t_n, y(t_n), h_n) \|,$$

tends to 0 when $h = \max_{0\leq n<N} h_n$ *tends to 0.*

Definition 4.2 *The numerical method (4.47) is said to be stable if there exists a constant* M, *not depending on* h, *such that for all sequences*

$(y_n)_{0\leq n\leq N}$, $(z_n)_{0\leq n\leq N}$ *and* $(\varepsilon_n)_{0\leq n\leq N}$ *verifying*

$$\begin{cases} y_{n+1} = y_n + h_n\phi(t_n, y_n, h_n), \\ z_{n+1} = z_n + h_n\phi(t_n, z_n, h_n) + \varepsilon_n, \end{cases} \tag{4.49}$$

one has

$$\max_{0\leq n\leq N} \| z_n - y_n \| \leq M\Big[\| y_0 - z_0 \| + \sum_{n<N} \| \varepsilon_n \| \Big]. \tag{4.50}$$

Definition 4.3 *The numerical method (4.47) is said to be convergent if*

$$\lim_{h\to 0} \max_{0\leq n\leq N} \| y(t_n) - y_n \| = 0. \tag{4.51}$$

Then the following classical theorem gives a sufficient condition for the scheme to be convergent:

Theorem 4.1 *If the numerical method (4.47) is consistent and stable, then it is convergent.*

The order of a one step numerical method is defined by:

Definition 4.4 *The numerical method (4.47) is said to be of order* p *(*$p > 0$*) if there exists a constant* K*, depending only on* y *and* ϕ*, such that:*

$$\sum_{n=0}^{N-1} \| y(t_{n+1}) - y(t_n) - h_n\phi(t_n, y(t_n), h_n) \| \leq Kh^p, \tag{4.52}$$

for any solution y *of (4.48) such that* $y \in \mathrm{C}^{p+1}([0,T])$.

The difference between the exact solution and the approximation provided by the scheme can then be estimated by means of the order of the numerical method with the following theorem:

Theorem 4.2 *If the numerical method (4.47) is stable and of order* p*, and if* $f \in \mathrm{C}^p([0,T] \times \mathbb{R})$*, one has*

$$\forall n \leq N,\ \| y(t_n) - y_n \| \leq MKh^p. \tag{4.53}$$

4.2.4.2 *Convergence and order of one step numerical methods applied to nonsmooth differential systems*

The Definition 4.4 and Theorem 4.2 cannot directly apply to the system (4.14) since the solution is not of class C^1 (except of course if there are no impacts, in which case the system is a usual linear oscillator). Nevertheless, the nature of the solution enables to adapt the Definition 4.4, considering the solutions y of (4.48) such that y is piecewise of class C^{p+1} on $[0, T]$. Then the Theorem 4.2 still holds, provided we modify in the same way the assumption on the regularity of f.

First of all, the Schatzman-Paoli method is convergent when f is continuous (see [Paoli (1993)]). Nevertheless, no general result has been proved about its order, but according to [Paoli (1993)] the order is expected to be at most 1 for the case that we consider. In the following sections, we are going to prove results of consistency about numerical methods associated with the impact localization methods defined in 4.2.2: we are going to deal with at most one impact detection at each time step. These results will then apply to the Newmark and RK24 methods.

Systems with Finite Number of Impacts

In this subsection, we are dealing with responses with a finite number of impacts on any finite time interval. The following results only hold in that particular case.

Theorem 4.3 *Let us consider a one step numerical method convergent and at least of order 2 for any ordinary differential equation sufficiently smooth, to which we associate (IM1) in order to obtain an approximation of the solution of Eq. (4.14) with impact law (4.15) and where f is assumed to be differentiable with a bounded derivative on $\mathbb{R}^+$.*

Then the resulting numerical method is consistent and of order 2.

Proof *Let $[0, T]$ be an interval of $\mathbb{R}$. (NM) will refer to the numerical method at least of order 2 when applied to a smooth system, and (NMIM1) to the scheme obtained by adding the method (IM1) of impact localization. Let x be the solution of (4.14), (4.15) starting from the initial conditions $(x_0, \dot{x}_0)$. On any time interval where the system does not impact against the stop, the scheme is convergent and the consistency error is of order 2 since the solution of (4.14) is of class C^2. Moreover, as by assumptions the number of impacts on $[0, T]$ is finite, we can choose a step h sufficiently small so that between two successive steps at most one impact occurs.*

Let $n \in \mathbb{N}$ be such that an impact occurs at t_1 with $t_n < t_1 < t_{n+1}$, the *velocity just before impact being nonzero (otherwise the solution is tangent to the stop at t_1 and therefore it is of class C^2 in the neighbourhood of t_1). We are going to estimate the consistency error $\varepsilon_n = (\varepsilon_n^1, \varepsilon_n^2)$ between t_n and t_{n+1}.*

We must first of all make sure that the scheme (NM) provides an approximation y_{n+1} at t_{n+1} which verifies $y_{n+1} > x_{max}$ for h sufficiently small. According to the assumptions on (NM), we obtain an approximation at t_{n+1} at least with an accuracy of order 3 of the solution y of $\ddot{y} + a\dot{y} + \omega_1^2 y = f(t)$ with initial conditions $(y(t_n), \dot{y}(t_n)) = (x(t_n), \dot{x}(t_n))$. Moreover, as $t_1 \in [t_n, t_{n+1}]$, a second order Taylor expansion of y yields:

$$\begin{aligned} y_{n+1} &= y(t_{n+1}) + \mathbf{O}(h^3), \\ &= y(t_1) + \dot{y}(t_1)(t_{n+1} - t_1) + \mathbf{O}(h^2), \\ &= x_{max} + \dot{x}(t_1^-)(t_n - t_1 + h) + \mathbf{O}(h^2), \end{aligned} \tag{4.54}$$

because $0 < t_{n+1} - t_1 < h$. Thus $y_{n+1} - x_{max}$ has the same sign as $\dot{x}(t_1^-)(t_{n+1} - t_1)$ when h is sufficiently small. The velocity just before an impact being strictly positive (because we are dealing with a nonzero velocity impact), one has $y_{n+1} > x_{max}$ for h sufficiently small. Hence the impact localizing procedure (IM1) has to be applied in order to obtain the approximation of x at t_{n+1}.

The first step consists of determining the quality of the approximation to t_1 by the procedure (IM1). According to (4.27), the approximation is given by

$$t_1^* = t_n + h\frac{x_{max} - x(t_n)}{y_{n+1} - x(t_n)}, \tag{4.55}$$

where y_{n+1} is the value provided by (NM) at t_{n+1}. The Taylor expansion of x at t_n is given by:

$$x(t_n) = x_{max} + \dot{x}(t_1^-)(t_n - t_1) + \mathbf{O}(h^2), \tag{4.56}$$

because $0 < t_1 - t_n < h$. According to (4.54) we get:

$$\begin{aligned} \frac{h}{y_{n+1} - x(t_n)} &= \frac{1}{\dot{x}(t_1^-) + \mathbf{O}(h)}, \\ &= \frac{1}{\dot{x}(t_1^-)}[1 + \mathbf{O}(h)], \end{aligned} \tag{4.57}$$

and then:

$$t_1^* = t_n - (t_n - t_1)[1 + \mathbf{O}(h)], \\ = t_1 + \mathbf{O}(h^2). \tag{4.58}$$

The impact time t_1 is consequently approximated to the second order in h by the method (IM1).

Once t_1^* has been computed, the numerical method (NM) has to be applied on $[t_n, t_1^*]$ in order to get an approximation v^- of the velocity just before impact. By the assumptions on (NM), it provides an approximation at least of order 3 of the solution y of (4.14) at t_1^*, starting from $(x(t_n), \dot{x}(t_n))$ at t_n. Thus we have:

$$v^- = \dot{y}(t_1^*) + \mathbf{O}((t_1^* - t_n)^3) = \dot{y}(t_1) + \mathbf{O}(t_1^* - t_1).$$

Because $\dot{y}(t_1) = \dot{x}(t_1^-)$ and because of (4.58) we get:

$$v^- = \dot{x}(t_1^-) + \mathbf{O}(h^2). \tag{4.59}$$

It finally remains to apply the scheme (NM) on $[t_1^*, t_{n+1}]$ to obtain the actual value of $(x_{n+1}, \dot{x}_{n+1})$ provided by (NMIM1) at t_{n+1}. The numerical method (NM) gives an approximation at least of order 3 to the solution y of (4.14) at t_{n+1}, starting from $(x_{max}, -ev^-)$ at t_1^*. Therefore, we have:

$$\begin{cases} x_{n+1} = y(t_1^*) + \dot{y}(t_1^*)(t_{n+1} - t_1^*) + \mathbf{O}(t_{n+1} - t_1^*)^2, \\ \dot{x}_{n+1} = \dot{y}(t_1^*) + \ddot{y}(t_1^*)(t_{n+1} - t_1^*) + \mathbf{O}(t_{n+1} - t_1^*)^2. \end{cases} \tag{4.60}$$

But

$$t_{n+1} - t_1^* = t_n - t_1 + h + \mathbf{O}(h^2),$$

and

$$\ddot{y}(t_1^*) = f(t_1^*) - a\dot{y}(t_1^*) - \omega_1^2 y(t_1^*), \\ = f(t_1) + ae\dot{x}(t_1^-) - \omega_1^2 x_{max} + \mathbf{O}(h), \\ = \ddot{x}(t_1^+) + \mathbf{O}(h),$$

since f is differentiable. Hence:

$$x_{n+1} = x_{max} + \dot{x}(t_1^+)(t_n - t_1 + h) + \mathbf{O}(h^2), \\ = x(t_{n+1}) + \mathbf{O}(h^2), \tag{4.61}$$

and:

$$\begin{aligned}\dot{x}_{n+1} &= \dot{x}(t_1^+) + \ddot{x}(t_1^+)(t_n - t_1 + h) + \mathbf{O}(h^2),\\ &= \dot{x}(t_{n+1}) + \mathbf{O}(h^2).\end{aligned}$$

We have finally obtained for the consistency error:

$$\begin{cases}\varepsilon_n^1 = \mathbf{O}(h^2),\\ \varepsilon_n^2 = \mathbf{O}(h^2).\end{cases} \tag{4.62}$$

We have therefore just showed that the consistency error was of second order in h on a time step where an impact occurs:

$$\| \varepsilon_n \| \leq Ch^2, \tag{4.63}$$

where C denotes a constant. The scheme being at least of order 2 on every interval without impacts, the overall consistency error on $[0,T]$ is then given by:

$$\begin{aligned}\Upsilon &= \sum_{\text{impacts}} |\varepsilon_n| + \sum_{\text{no impacts}} |\varepsilon_n|,\\ &\leq Ch^2 N_{\text{impacts}} + Kh^3(N - N_{\text{impacts}}),\\ &\leq C' h^2 + KTh^2,\\ &\leq \tilde{M}h^2,\end{aligned} \tag{4.64}$$

where N_{impacts} denotes the number of impacts on $[0,T]$. The numerical method (NMIM1) is then consistent and of second order.

In the same way, with the impact localization method (IM2), the numerical method obtained is of order 3 provided the starting scheme is at least of order 3.

Theorem 4.4 *Let us consider a one step numerical method convergent and at least of order 3 for any ordinary differential equation sufficiently smooth, to which we associate (IM2) in order to obtain an approximation of the solution of Eq. (4.14), with the impact law (4.15) and where f is assumed to be twice differentiable with a bounded second derivative on $\mathbb{R}^+$. Then the resulting numerical method is consistent and of order 3.*

Proof *Let $[0,T]$ be an interval of $\mathbb{R}$. (NM) will refer to the numerical method at least of order 3 for a smooth differential equation, and (NMIM2) to the scheme obtained by adding the impact localization procedure (IM2). Let x be the solution of (4.14), (4.15) starting from the initial conditions $(x_0, \dot{x}_0)$. On every time interval where the system does not impact, the* *numerical method (NMIM2) is identical to (NM), hence it is convergent and the consistency error is of order 3 since the solution of (4.14) is of class* C^3. *Moreover, as by assumption the number of impacts on* $[0, T]$ *is finite, the time step* h *can be chosen sufficiently small so that between two successive steps at most one impact occurs.*

Like in the preceding proof, it is sufficient to deal with the time steps $n \in \mathbb{N}$ *such that a nonzero velocity impact occurs at* t_1 *with* $t_n < t_1 < t_{n+1}$. *We are going to estimate the consistency error* $\varepsilon_n = (\varepsilon_n^1, \varepsilon_n^2)$ *between* t_n *and* t_{n+1}.

According to the proof of the Theorem 4.3, for h *small enough the numerical method (NM) gives at* t_{n+1} *an approximation* $y_{n+1} > x_{max}$. *The impact localization procedure (IM2) must then be applied. The first step consists in estimating the closeness of the approximation to* t_1 *by means of (IM2). According to (4.28), the approximation is given by*

$$t_1^* = t_n + \frac{-\dot{x}(t_n) + \sqrt{\dot{x}(t_n)^2 + 4X(x_{max} - x(t_n))}}{2X}, \tag{4.65}$$

where $X = \dfrac{y_{n+1} - x(t_n) - h\dot{x}(t_n)}{h^2}$ *and* y_{n+1} *is provided by (NM) at* t_{n+1}. *But (NM) gives an approximation at least of order 4 of the solution* y *of (4.14) at* t_{n+1}. *Moreover, because* $t_1 \in [t_n, t_{n+1}]$, *a third order Taylor expansion of* y *yields:*

$$\begin{aligned} y(t_{n+1}) = y(t_1) + \dot{y}(t_1)(t_{n+1} - t_1) + \frac{1}{2}\ddot{y}(t_1)(t_{n+1} - t_1)^2 + \\ + \frac{1}{6}\dddot{y}(t_1)(t_{n+1} - t_1)^3 + \mathbf{O}(h^4). \end{aligned} \tag{4.66}$$

Hence we have:

$$\begin{aligned} y_{n+1} = x_{max} + \dot{x}(t_1^-)(t_n - t_1 + h) + \frac{1}{2}\ddot{x}(t_1^-)(t_n - t_1 + h)^2 + \\ + \frac{1}{6}\dddot{x}(t_1^-)(t_n - t_1 + h)^3 + \mathbf{O}(h^4). \end{aligned} \tag{4.67}$$

In the same way, Taylor expansions at t_n *yield:*

$$\begin{aligned} x(t_n) &= x_{max} + \dot{x}(t_1^-)(t_n - t_1) + \frac{1}{2}\ddot{x}(t_1^-)(t_n - t_1)^2 + \\ &\quad + \frac{1}{2}\ddot{x}(t_1^-)(t_n - t_1)^2 + \mathbf{O}(h^4), \\ \dot{x}(t_n) &= \dot{x}(t_1^-) + \ddot{x}(t_1^-)(t_n - t_1) + \frac{1}{2}\dddot{x}(t_1^-)(t_n - t_1)^2 + \mathbf{O}(h^3), \end{aligned} \tag{4.68}$$

since $t_n - t_1 < h$. *Then we obtain:*

$$X = \frac{1}{2}\ddot{x}(t_1^-) + \frac{1}{2}\dddot{x}(t_1^-)\left(t_n - t_1 + \frac{h}{3}\right) + \mathbf{O}(h^2), \tag{4.69}$$

and hence:

$$\dot{x}(t_n)^2 + 4X(x_{max} - x(t_n)) =$$
$$= \dot{x}(t_1^-)^2 - \dot{x}(t_1^-)\dddot{x}(t_1^-)(t_n - t_1)\left(t_n - t_1 + \frac{2h}{3}\right) + \mathbf{O}(h^3).$$

Therefore

$$2X(t_1^* - t_n) = -\ddot{x}(t_1^-)(t_n - t_1) - \dddot{x}(t_1^-)(t_n - t_1)\left(t_n - t_1 + \frac{h}{3}\right) + \mathbf{O}(h^3),$$

and finally:

$$t_1^* = t_1 + \mathbf{O}(h^3). \tag{4.70}$$

The impact time t_1 *is thus approximated to the order 3 by the procedure (IM2).*

By applying the same procedure as in the proof of the Theorem 4.3, it is then easy to show that $v^- = \dot{x}(t_1^-) + \mathbf{O}(h^3)$, *and to infer that for any time step where an impact occurs the consistency error is of order 3 in h:*

$$\| \varepsilon_n \| = \mathbf{O}(h^3). \tag{4.71}$$

We can then conclude that the numerical method (NMIM2) is consistent and of order 3.

Finally, with the impact localization method (IM3), the numerical method obtained is of order 4 provided the starting scheme is at least of order 4 and the precision chosen in the impact localization procedure is h^4:

Theorem 4.5 *Let us consider a one step numerical method convergent and at least of order 4 for any ordinary differential equation sufficiently smooth, to which we associate (IM3) in order to obtain an approximation of the solution of Eq. (4.14), with the impact law (4.15) and where* f *is assumed to be three times differentiable with a bounded third derivative on* $\mathbb{R}^+$.

Then the resulting numerical method is consistent and of order 4.

Proof *Let* $[0, T]$ *be an interval of* $\mathbb{R}$. *(NM) will refer to the numerical method at least of order 4 for a smooth differential equation, and (NMIM3) to the scheme obtained by adding the impact localization procedure (IM3). Let* x *be the solution of (4.14), (4.15) starting from the initial conditions*

$(x_0, \dot{x}_0)$. *On every time interval where the system does not impact, the numerical method (NMIM3) is identical to (NM), hence it is convergent and the consistency error is of order 4 since the solution of (4.14) is of class* C^4. *Moreover, as by assumptions the number of impacts on* $[0, T]$ *is finite, the time step* h *can be chosen sufficiently small so that between two successive steps at most one impact occurs.*

Like in the preceding proofs, it is sufficient to deal with the time steps $n \in \mathbb{N}$ *such that a nonzero velocity impact occurs at* t_1 *with* $t_n < t_1 < t_{n+1}$. *We are going to estimate the consistency error* $\varepsilon_n = (\varepsilon_n^1, \varepsilon_n^2)$ *between* t_n *and* t_{n+1}.

According to the proof of the Theorem 4.3, for h *small enough the numerical method (NM) gives at* t_{n+1} *an approximation* $y_{n+1} > x_{max}$. *The impact localization procedure (IM3) must then be applied. The first step consists in estimating the closeness of the approximation to* t_1 *by means of (IM3). Let us define the function* $X : \left| \begin{array}{l} [t_n, t_{n+1}] \to \mathbb{R}^2 \\ \zeta \mapsto X(\zeta) = (X_1(\zeta), X_2(\zeta)) \end{array} \right.$ *where* $X(\zeta)$ *is the value given by (NM) starting from* $(t_n, x(t_n), \dot{x}(t_n))$, *with time step* $\zeta - t_n$. *The dichotomy method generates two sequences* $(r_k)_{k \geq 0}$ *and* $(s_k)_{k \geq 0}$ *defined by* $r_0 = t_n$, $s_0 = t_{n+1}$ *and*

$$\begin{aligned} X_1\left(\frac{r_k + s_k}{2}\right) \geq x_{max} &\Rightarrow \begin{cases} r_{k+1} = r_k, \\ s_{k+1} = \dfrac{r_k + s_k}{2}, \end{cases} \\ X_1\left(\frac{r_k + s_k}{2}\right) < x_{max} &\Rightarrow \begin{cases} r_{k+1} = \dfrac{r_k + s_k}{2}, \\ s_{k+1} = s_k. \end{cases} \end{aligned} \tag{4.72}$$

Both sequences converge to λ *such that* $X_1(\lambda) = x_{max}$ *with the estimate*

$$| r_k - \lambda | \leq s_k - r_k = h2^{-k}. \tag{4.73}$$

But if h *is sufficiently small, since (NM) gives an approximation of order 4 to the solution* y *of (4.14) with initial conditions* $(y(t_n), \dot{y}(t_n)) = (x(t_n), \dot{x}(t_n))$, *we have* $y(\lambda) = X_1(\lambda) + \mathbf{O}(h^5) = x_{max} + \mathbf{O}(h^5) = y(t_1) + \mathbf{O}(h^5)$, *and then we infer*

$$\lambda = t_1 + \mathbf{O}(h^5). \tag{4.74}$$

The dichotomy process is stopped at the kth *step such that* k *is the smallest integer verifying* $h2^{-k} \leq h^4$, *and the approximation to the impact time is then given by* $t_1^* = \dfrac{r_k + s_k}{2}$. *Hence, from (4.73) and (4.74) we get* $t_1^* = t_1 + \mathbf{O}(h^4)$.

As we did in the proof of the Theorem 4.3, we can then easily show that $v^- = \dot{x}(t_1^-) + \mathbf{O}(h^4)$ *and finally that* $\| \varepsilon_n \| = \mathbf{O}(h^4)$. *The numerical method (NMIM3) is then consistent and of order 4.*

Remark 4.3 *It can easily be shown that in all preceding Taylor expansions, the terms* $\mathbf{O}(h^k)$ *stand for* $h^k K(h)$ *where* K *is a bounded function of* h *uniformly in* n.

Systems with Infinite Number of Impacts

Let us now deal with systems for which accumulation of impact times – followed by a sticking phase – can occur.

Proposition 4.2 *Let us consider a one step numerical method convergent and at least of order 2 for any smooth ordinary differential equation, to which we associate (IM1) in order to obtain an approximation of the solution of the equation (4.14) on* $[0,T]$, *with the impact law (4.15) and where* f *is assumed to be differentiable with locally Lipschitz-continuous derivative. We assume that there is one accumulation of impact times on* $[0,T]$. *Let* $(t_n = nh)_{0 \le n \le N}$ *be the sequence of time steps on* $[0,T]$ *and* $\varepsilon_n = (\varepsilon_n^1, \varepsilon_n^2)$ *the consistency error between* t_n *and* t_{n+1}. *Then we have:*

(1) if there is no impact between t_n *and* t_{n+1}, *then* $\varepsilon_n^1 \le C_1 h^3$ *and* $\varepsilon_n^2 \le C_2 h^3$ *where* C_1 *and* C_2 *are constants,*
(2) if there is at least one impact between t_n *and* t_{n+1}, *then* $\varepsilon_n^1 \le C_3 h^2$ *and* $\varepsilon_n^2 \le C_4 h$ *where* C_3 *and* C_4 *are constants,*
(3) if t_n *is in a sticking phase between, then* $\varepsilon_n^1 \le C_5 h^3$ *and* $\varepsilon_n^2 =\le C_6 h$ *where* C_5 *and* C_6 *are constants.*

Proof *Let* $(\bar{t}_k)_{k \ge 0}$ *be the sequence of impact times such that* $\lim_{k \to +\infty} \bar{t}_k = t_\infty$. *According to the Proposition 4.1, the system sticks to the stop from* t_∞, *until the first time* $t^* > t_\infty$ *such that* $f(t) - \omega_1^2 x_{max}$ *becomes zero and changes sign.*

Let us first of all deal with the time steps n *where there is no impact. The solution being three times differentiable on* $[t_n, t_{n+1}]$, *the numerical method is of order 2 on this interval and we can infer that the consistency error verifies:*

$$\begin{cases} \varepsilon_n^1 \le K_1 h^3, \\ \varepsilon_n^2 \le K_2 h^3. \end{cases} \tag{4.75}$$

Let us now deal with a time step where the system admits one or several impacts. Let h be a fixed time step. Since the sequence $(\bar{t}_k)_{k\geq 0}$ converges to t_∞, for k large enough the impact time $\bar{t}_k$ satisfies the relation,

$$\bar{t}_{k+1} - \bar{t}_k \leq h. \tag{4.76}$$

But we saw in the proof of the Proposition 4.1 that the velocities just before impacts of the infinite sequence of impacts verify, according to (4.38), $\dot{x}(\bar{t}_k^-) = (\bar{t}_{k-1} - \bar{t}_k)M(k)$ where $k \mapsto M(k)$ is bounded when k describes $\mathbb{N}^*$. We then have $\dot{x}(\bar{t}_k^-) = \mathbf{O}(\bar{t}_{k-1} - \bar{t}_k)$, and if $\bar{t}_{k+1} - \bar{t}_k \leq h$ we obtain $\dot{x}(\bar{t}_k^-) = \mathbf{O}(h)$.

For the time steps with one impact and where the impact time does not verify (4.76), we can apply the Theorem 4.3 and obtain a consistency error of order 2. On the other hand, if there is one impact at $\bar{t}_k$ verifying (4.76), the proof of the Theorem 4.3 no longer holds since the velocity just before impact is of order 1. Let us consider such a case: let h be a time step and $\bar{t}_k$ an impact time such that $\dot{x}(\bar{t}_k^-) = \mathbf{O}(h)$. There exists an integer n such that $t_n = nh \leq \bar{t}_k < t_{n+1}$, so we are going to consider the consistency error on that n^{th} step. Since we can no longer be sure that the numerical method (NMIM1) considered in this proposition actually detects an impact between t_n and t_{n+1}, we have to distinguish between two cases. Firstly, if we suppose that (NMIM1) does detect an impact at $t_k^* \in [t_n, t_{n+1}]$ then we have $\mid t_k^* - \bar{t}_k \mid< h$, and hence $t_k^* = \bar{t}_k + \mathbf{O}(h)$. With this estimate we can infer that $v^- = \dot{x}(\bar{t}_k^-) + \mathbf{O}(h) = \mathbf{O}(h)$ (where v^- is the approximation of the velocity at t_k^* given by (NMIM1)). Finally we use two Taylor expansions to estimate the consistency error:

$$\begin{aligned}
x(t_{n+1}) &= x_{max} - (t_n + h - \bar{t}_k)e\dot{x}(\bar{t}_k^-) + \mathbf{O}(h^2),\\
&= x_{max} + \mathbf{O}(h^2),\\
x_{n+1} &= x_{max} - (t_n + h - \bar{t}_k)ev^- + \mathbf{O}(h^2),\\
&= x_{max} + \mathbf{O}(h^2),
\end{aligned}$$

which leads to $\varepsilon_n^1 = \mathbf{O}(h^2)$. Similar calculations give for the velocity $\varepsilon_n^2 = \mathbf{O}(h)$. We now have to consider the case when the numerical method (NMIM1) does not detect an impact on $[t_n, t_{n+1}]$. We can still write:

$$\begin{aligned}
x(t_{n+1}) &= x_{max} + (t_n + h - \bar{t}_k)\dot{x}(\bar{t}_k^+) + \mathbf{O}(h^2),\\
&= x_{max} + \mathbf{O}(h^2).
\end{aligned}$$

Moreover, the value given by (NMIM1) at t_{n+1} *verifies* $x_{n+1} = x(t_n) + h\dot{x}(t_n) + \mathbf{O}(h^2)$ *because the scheme without impacts is assumed to be of second order. By using again Taylor expansions of* x *and* $\dot{x}$*, we obtain* $x(t_n) = x(\bar{t}_k) + (t_n - \bar{t}_k)\dot{x}(\bar{t}_k^-) + \mathbf{O}(h^2) = x_{max} + \mathbf{O}(h^2)$ *and* $\dot{x}(t_n) = \dot{x}(\bar{t}_k^-) + \mathbf{O}(h) = \mathbf{O}(h)$*, which yields* $x_{n+1} = x_{max} + \mathbf{O}(h^2)$*. Finally we obtain the same estimate of the consistency error as in the preceding case, namely* $\varepsilon_n^1 = \mathbf{O}(h^2)$*, and similarly* $\varepsilon_n^2 = \mathbf{O}(h)$*. Therefore, the consistency error on a step with one impact verifies in the general case:*

$$\begin{cases} \varepsilon_n^1 \leq K_3 h^2, \\ \varepsilon_n^2 \leq K_4 h. \end{cases} \tag{4.77}$$

Let us now consider a step n such that the system admits K impacts at $\bar{t}_1, \bar{t}_2, \ldots \bar{t}_K$*, where K is finite and strictly greater than 1. We cannot show anymore, as we did in the proof of the Theorem 4.3, that the numerical method (NMIM1) detects an impact between* t_n *and* t_{n+1}*. If it does, then we get at* t_{n+1}*:*

$$\begin{cases} x_{n+1} = x_{max} + \dot{x}(\bar{t}_1^+)(t_{n+1} - \bar{t}_1) + \mathbf{O}(h^2) \text{ if } x_{n+1} \leq x_{max}, \\ \qquad x_{n+1} = x_{max} \qquad\qquad\qquad\qquad \text{otherwise} \end{cases}$$

and

$$\begin{cases} \dot{x}_{n+1} = \dot{x}(\bar{t}_1^+) + \ddot{x}(\bar{t}_1^+)(t_{n+1} - \bar{t}_1) + \mathbf{O}(h^2) \text{ if } x_{n+1} \leq x_{max}, \\ \qquad \dot{x}_{n+1} = 0 \qquad\qquad\qquad\qquad \text{otherwise} \end{cases}$$

and if it does not then we get at t_{n+1}*:*

$$\begin{cases} x_{n+1} = y(t_{n+1}) + \mathbf{O}(h^3) = x_{max} + \dot{x}(\bar{t}_1^-)(t_{n+1} - \bar{t}_1) + \mathbf{O}(h^2), \\ \dot{x}_{n+1} = \dot{y}(t_{n+1}) + \mathbf{O}(h^3) = \dot{x}(\bar{t}_1^-) + \mathbf{O}(h). \end{cases} \tag{4.78}$$

Let us deal with the first case. A first order Taylor expansion gives:

$$\begin{cases} x(t_{n+1}) = x_{max} + \dot{x}(\bar{t}_K^+)(t_{n+1} - \bar{t}_K) + \mathbf{O}((t_{n+1} - \bar{t}_K)^2), \\ \dot{x}(t_{n+1}) = \dot{x}(\bar{t}_K^+) + \ddot{x}(\bar{t}_K^+)(t_{n+1} - \bar{t}_K) + \mathbf{O}((t_{n+1} - \bar{t}_K)^2). \end{cases} \tag{4.79}$$

Thus the consistency error is given by:

$$\begin{cases} \varepsilon_n^1 = \dot{x}(\bar{t}_K^+)(t_{n+1} - \bar{t}_K) - \dot{x}(\bar{t}_1^+)(t_{n+1} - \bar{t}_1) + \mathbf{O}(h^2), \\ \varepsilon_n^2 = \dot{x}(\bar{t}_K^+) + \ddot{x}(\bar{t}_K^+)(t_{n+1} - \bar{t}_K) - \dot{x}(\bar{t}_1^+) - \ddot{x}(\bar{t}_1^+)(t_{n+1} - \bar{t}_1) + \mathbf{O}(h^2). \end{cases} \tag{4.80}$$

Moreover, we have here $0 < \bar{t}_{k+1} - \bar{t}_k \leq h$ *for all* $k \in \{1, \ldots, K-1\}$*, so we can infer* $\dot{x}(\bar{t}_1^+) = \mathbf{O}(h)$ *and* $\dot{x}(\bar{t}_K^+) = \mathbf{O}(h)$ *as we did in the case* $K = 1$*.*

The consistency error is then given by:

$$\begin{cases} \varepsilon_n^1 = \mathbf{O}(h^2), \\ \varepsilon_n^2 = \mathbf{O}(h). \end{cases}$$

If (NMIM1) provides at the $(n+1)^{\text{th}}$ *step the approximation* $(x_{n+1}, \dot{x}_{n+1}) = (x_{max}, 0)$*, the result is the same. Indeed, according to (4.80), we have:*

$$\begin{cases} \varepsilon_n^1 = \dot{x}(\bar{t}_K^+)(t_{n+1} - \bar{t}_K) + \mathbf{O}(h^2), \\ \varepsilon_n^2 = \dot{x}(\bar{t}_K^+) + \ddot{x}(\bar{t}_K^+)(t_{n+1} - \bar{t}_K) + \mathbf{O}(h^2), \end{cases}$$

so that we obtain the same estimation for the consistency error:

$$\begin{cases} \varepsilon_n^1 = \mathbf{O}(h^2), \\ \varepsilon_n^2 = \mathbf{O}(h). \end{cases}$$

Finally, in the third case the consistency error is given by:

$$\begin{cases} \varepsilon_n^1 = -(1+e)\dot{x}(t_1^-)(t_{n+1} - t_1 + h) + \mathbf{O}(h^2), \\ \varepsilon_n^2 = -(1+e)\dot{x}(t_1^-) + \mathbf{O}(h), \end{cases}$$

and hence

$$\begin{cases} \varepsilon_n^1 \leq K_5 h^2, \\ \varepsilon_n^2 \leq K_6 h. \end{cases}$$

Let us now consider the step where the system admits an infinite number of impacts: it is the nth *step where* n *is such that* $t_n \leq t_\infty < t_{n+1}$*. If* h *is sufficiently small,* t_{n+1} *is in a sticking phase, which means that* $x(t_{n+1}) = x_{max}$ *and* $\dot{x}(t_{n+1}) = 0$*. Let* $\bar{t}_k$ *be the first impact time after* t_n*. As we did before, we have to examine two cases depending on whether (NMIM1) leads to sticking at* t_{n+1} *or not. If it does then* $(x_{n+1}, \dot{x}_{n+1}) = (x_{max}, 0)$*, and so the consistency error is zero. In the other case, the numerical method (NMIM1) gives an approximation at least of order 2 of the solution* y *of (4.14) at* t_{n+1}*, starting from* $(x(t_n), \dot{x}(t_n))$ *at* t_n *and impacting against the stop at* $\bar{t}_k$*, according to the proof of the Theorem 4.3. One has then according to (4.61) and (4.62):*

$$\begin{cases} x_{n+1} = x_{max} + \dot{y}(\bar{t}_k^+)(t_n - \bar{t}_k + h) + \mathbf{O}(h^2), \\ \dot{x}_{n+1} = \dot{y}(\bar{t}_k^+) + \mathbf{O}(h), \end{cases}$$

and hence:

$$\begin{cases} \varepsilon_n^1 = -\dot{x}(\bar{t}_k^+)(t_n - t_k + h) + \mathbf{O}(h^2), \\ \varepsilon_n^2 = -\dot{x}(\bar{t}_k^+) + \mathbf{O}(h). \end{cases}$$

Because $t_n < \bar{t}_k < \bar{t}_{k+1} < t_{n+1}$, as we saw before, we have $\dot{x}(\bar{t}_k^+) = \mathbf{O}(h)$. The consistency error is therefore given by the following estimation:

$$\begin{cases} \varepsilon_n^1 \le K_7 h^2, \\ \varepsilon_n^2 \le K_8 h, \end{cases}$$

where K_7 and K_8 are constants.

Let us finally deal with the time steps n for which the system is sticking to the stop. One has then $x(t_n) = x_{max}$ and $\dot{x}(t_n) = 0$. The scheme (NM) gives an approximation at least of order 2 of the solution y of (4.14) at t_{n+1}, starting from $(x_{max}, 0)$ at t_n:

$$y_{n+1} = y(t_n) + \dot{y}(t_n)h + \frac{1}{2}\ddot{y}(t_n)h^2 + \mathbf{O}(h^3).$$

Hence

$$y_{n+1} = x_{max} + \frac{1}{2}\ddot{x}(t_n)h^2 + \mathbf{O}(h^3). \tag{4.81}$$

Thus $y_{n+1} - x_{max}$ has the same sign as $\ddot{x}(t_n)$ for h sufficiently small, and this quantity is positive since at t_n the system is sticking to the stop. The numerical method (NMIM1) hence detects an impact at

$$t_1^* = t_n + h\frac{x_{max} - x(t_n)}{y_{n+1} - x(t_n)} = t_n. \tag{4.82}$$

Once t_1^* has been computed, the numerical method (NM) must be applied on $[t_n, t_1^*]$, which is here a zero length interval. The approximation v^- is hence equal to $\dot{x}(t_n)$, namely 0, and it remains to apply (NM) on $[t_1^*, t_{n+1}] = [t_n, t_{n+1}]$. Consequently, we find again (4.81) which is greater than x_{max} for h sufficiently small. The numerical method (NMIM1) therefore provides at the $n+1^{\text{th}}$ step the value $(x_{max}, 0)$.

Two cases must be examined, depending on whether the system is still sticking to the stop at t_{n+1} or not. If it is, then $x(t_{n+1}) = x_{max}$ and $\dot{x}(t_{n+1}) = 0$ so that the consistency error at the nth step is zero. Otherwise we have at order 2:

$$x(t_{n+1}) = x(\bar{t}) + (t_{n+1} - \bar{t})\dot{x}(\bar{t}) + \frac{1}{2}(t_{n+1} - \bar{t})^2\ddot{x}(\bar{t}) + \mathbf{O}((t_{n+1} - \bar{t})^3),$$

$$\dot{x}(t_{n+1}) = \dot{x}(\bar{t}) + (t_{n+1} - \bar{t})\ddot{x}(\bar{t}) + \frac{1}{2}(t_{n+1} - \bar{t})^2\dddot{x}(\bar{t}) + \mathbf{O}((t_{n+1} - \bar{t})^3),$$

where $\bar{t}$ is the time when sticking ends. But because of the definition of $\bar{t}$, one has $x(\bar{t}) = x_{max}$ and $\dot{x}(\bar{t}) = \ddot{x}(\bar{t}) = 0$. Hence we have:

$$\begin{cases} x(t_{n+1}) = x_{max} + \mathbf{O}(h^3), \\ \dot{x}(t_{n+1}) = \frac{1}{2}(t_{n+1} - \bar{t})^2 \dot{f}(\bar{t}) + \mathbf{O}(h^3). \end{cases} \tag{4.83}$$

Thus

$$\begin{cases} \varepsilon_n^1 \leq K_9 h^3, \\ \varepsilon_n^2 \leq K_{10} h^2, \end{cases} \tag{4.84}$$

where K_9 and K_{10} are constants.

Remark 4.4 *The preceding result has been stated for a numerical method associated with (IM1), but obviously it would still hold for the methods (IM2), (IM3) and (IM4), since a better estimation of impact times cannot improve the consistency error.*

Remark 4.5 *As in the case of a finite number of impacts, it can easily be shown that in all the preceding Taylor expansions, the terms $\mathbf{O}(h^k)$ stand for $h^k K(h)$ where K is a bounded function of h uniformly in n, thanks to the results of the section 4.2.3.*

The Proposition 4.2 gives an estimate of the consistency error on one step for different kinds of behaviour of the system. In order to get a global consistency error and an idea of the order of the scheme, we need in particular to know an estimation of the number of time steps where the system admits at least one impact. The Proposition 4.1 describes the asymptotic behaviour of the impact times sequence: this sequence behaves asymptotically like a geometrical sequence whose ratio is e. In this theoretical part, we are only to consider an exact geometrical sequence: The general case will only be examined by numerical tests.

Proposition 4.3 *Let $(\bar{t}_k)_{k \geq 1}$ be a sequence defined by the recursive relation*

$$\frac{\bar{t}_{k+1} - \bar{t}_k}{\bar{t}_k - \bar{t}_{k-1}} = e$$

with $0 < \bar{t}_1 < \bar{t}_2$ and $0 \leq e < 1$, and $(t_n)_{n \geq 0}$ defined by $t_n = nh$. Let us define $t_\infty = \lim_{k \to +\infty} \bar{t}_k$. Let us denote $k(h)$ the number of time steps that contain at least one element of the sequence $(\bar{t})_{k \geq 0}$. Then we have $k(h) = \mathbf{O}_0 \ln h$.

Proof *Let us introduce* $\bar{t}_\infty = \lim_{k \to +\infty} \bar{t}_k$*:* $< +\infty$ *since the ratio* e *verifies* $0 < e < 1$*. Let the time step* h *be fixed. We can determine the smallest integer* $a(h)$ *so that* $h > e^{a(h)}(\bar{t}_2 - \bar{t}_1)$*: it is*

$$\lfloor 1 + \frac{\ln h - \ln(\bar{t}_2 - \bar{t}_1)}{\ln e} \rfloor,$$

where $\lfloor x \rfloor$ *denotes the greatest integer smaller or equal to* x*. Since the sequence* $(\bar{t}_k)_{k \geq 1}$ *is a geometrical one with the ratio* e*, we have*

$$\bar{t}_{k+1} - \bar{t}_k = e^{k-1}(\bar{t}_2 - \bar{t}_1).$$

Thus, if $k < a(h)$ *we have* $h \leq e^k(\bar{t}_2 - \bar{t}_1)$ *and then* $\bar{t}_{k+1} - \bar{t}_k \geq h$*. So the elements of the sequence* $(\bar{t}_k)_{1 \leq k \leq a(h)}$ *are all located on distinct time steps (except perhaps* $\bar{t}_1$ *and* $\bar{t}_2$*). For the other elements of the sequence* $(\bar{t}_k)_{k \geq 1}$*, we know*

$$\bar{t}_\infty - \bar{t}_{a(h)+1} = \frac{e^{a(h)}(\bar{t}_2 - \bar{t}_1)}{1 - e},$$

that implies by definition of $a(h)$*:*

$$\bar{t}_\infty - \bar{t}_{a(h)+1} \leq \frac{h}{1 - e}.$$

The interval $[\bar{t}_{a(h)+1}, \bar{t}_\infty]$ *contains at most* $1 + \lfloor \frac{1}{1-e} \rfloor$ *time steps. From the previous properties we obtain:*

$$k(h) \leq 1 + \lfloor \frac{1}{1-e} \rfloor + a(h).$$

Using the expression $a(h)$ *given here above we can deduce that* $k(h) \leq A + B \ln h$ *(*A *and* B *are constants) and so:*

$$k(h) = \mathbf{O}(\ln h).$$

As for the actual impact time sequence, it is not possible to imitate the preceding proof. Nevertheless, we observed that the numerical behaviour of $k(h)$ when h decreases to 0 is similar to the ideal case of an exact geometrical sequence. Thus if we assume that the result still holds for the impact time sequence, we can obtain an estimation of the global consistency error on any interval $[0, T]$ containing a unique or a finite number of accumulation of impact times.

Let N be the number of steps on $[0,T]$ ($N = \mathrm{E}(T/h)$). Let us deal with the infinite sequence of impact times. If we consider the sticking phase, the proof of the Proposition 4.2 shows that the consistency error is zero except on the last step where the sticking phase ends. In that case, we have

$$\begin{cases} \varepsilon_n^1 \le C_1 h^3, \\ \varepsilon_n^2 \le C_2 h^2. \end{cases} \tag{4.85}$$

Moreover, the number of steps during the sticking phase, where the consistency error is zero, is given by $s(h) \le \dfrac{\bar{t} - t_\infty}{h}$ where $\bar{t}$ is the time when sticking ends.

As for the steps containing a finite number of impacts, the consistency error is

$$\begin{cases} \varepsilon_n^1 \le C_3 h^2, \\ \varepsilon_n^2 \le C_4 h, \end{cases} \tag{4.86}$$

and the number of such steps is the integer $k(h)$ previously defined. The consistency error is the same for the step with the infinite number of impacts. Lastly, the steps without any impacts lead to a consistency error:

$$\begin{cases} \varepsilon_n^1 \le C_5 h^3, \\ \varepsilon_n^2 \le C_6 h^3, \end{cases} \tag{4.87}$$

and the number of such steps is given by $n(h) = N - k(h) - 2 - s(h) \le N \le \dfrac{T}{h}$. The global consistency error can then be estimated by:

$$\begin{cases} \varepsilon^1 = \displaystyle\sum_{n=0}^{N} \varepsilon_n^1 \le M_1 h^2 \ln(h), \\ \varepsilon^2 = \displaystyle\sum_{n=0}^{N} \varepsilon_n^2 \le M_2 h \ln(h), \end{cases} \tag{4.88}$$

where M_1 and M_2 are constants. The overall consistency error is clearly the same if we use (IM2), (IM3) or (IM4) instead of (IM1) in order estimate impact times, since the steps containing a finite number of impacts will still behave like h^2 for the displacement and h for the velocity.

4.2.5 *Numerical results*

In the case of a smooth problem, when the numerical method verifies some Lipschitz condition, the consistency and order p is equivalent to

$$\max_{0\leq n\leq N} \mid y(t_n) - y_n \mid \leq Ch^p. \tag{4.89}$$

As the problem (4.14), (4.15) is not smooth, we cannot use such a result to estimate the order of convergence as long as stability has not been proved. However, it is interesting to numerically investigate (4.89) rather than (4.52), because from a practical point of view we want the solution provided by the numerical method to be as close as possible to the exact solution. We are then going to estimate the order of the numerical methods by plotting $-\log\left(\max_{0\leq n\leq N} \mid y(t_n) - y_n \mid\right)$ versus $-\log(h)$: the slope of the set of points obtained gives an idea of the numerical order of the scheme. The theoretical results that we obtained in the section 4.2.4 in terms of consistency of the numerical schemes are summarized in the Table 4.2.

In this section we apply the numerical methods that we previously studied from a mathematical point of view to the resolution of a differential equation similar to (4.14), where the forcing f is a periodic one and is given by $f(t) = f\cos(\omega t)$. The behaviour of the schemes is investigated with two examples of responses that the system may exhibit. We first choose the following values for parameters: $e = 0.9$, $f = 20$, $\omega_1 = 2.5$, $a = 0.05$, $\omega = 2.5$, $x_{max} = 14$, $x_0 = 11.36263$ and $\dot{x_0} = 31.40358$. For these values, the response is $3T$-periodic with two impacts per cycle, where $T = \dfrac{2\pi}{\omega}$ (see Fig. 4.2).

Table 4.2 Theoretical consistency orders of the numerical methods.

Numerical method	Interpolation method	Finite no. of impacts		Infinite no. of impacts	
		Displacem.	Velocity	Displacem.	Velocity
Newmark	IM1	2	2	2	1
	IM2	2	2	2	1
	IM3	2	2	2	1
RK24	IM1	2	2	2	1
	IM2	3	3	2	1
	IM3	4	4	2	1
DOPRI5	IM3*	4	4	2	1

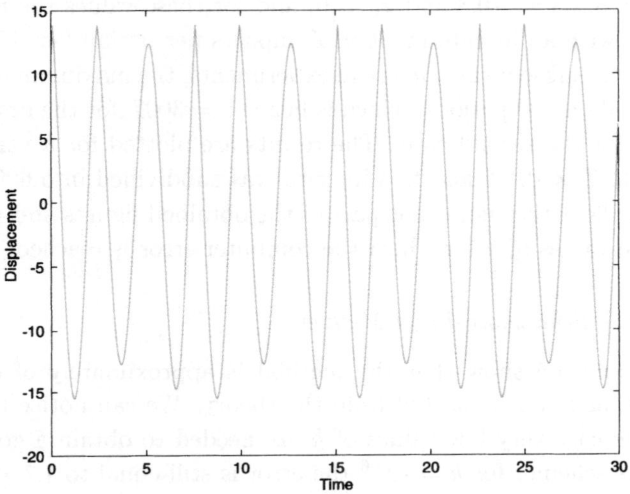

Fig. 4.2 3T-periodic solution with two impacts per period.

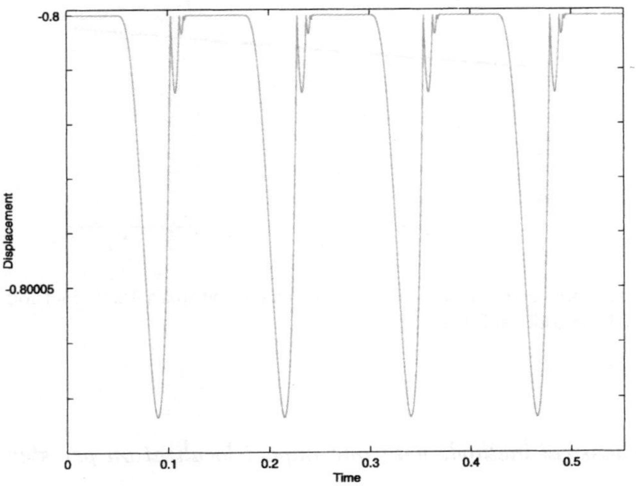

Fig. 4.3 T-periodic solution with accumulation of impacts.

The second set of parameters is $e = 0.5$, $f = 1$, $\omega_1 = 1$, $a = 1$, $\omega = 50$, $x_{max} = -0.8$, $x_0 = -0.8$ and $\dot{x_0} = 0$, and for these values the response is T-periodic with an infinite number of impacts per period (see Fig. 4.3).

For all the subsequent numerical experiments, the maximum of the error is considered on 100 periods, which is hence $\tau = 300T$ for the first case and $\tau = 100T$ for the second case. The results are plotted for h varying from $T \times 10^{-1}$ to $T \times 10^{-6}$, and this interval was subdivided into 600 values of h. We only kept the significant part of the obtained figures and we did not include the values of h for which the computer error is reached.

4.2.5.1 *The Schatzman-Paoli Method*

Figs. 4.4 and 4.5 show that the method is approximately of order 1 in both cases, as it was expected from the theory. We can notice that in the $3T$-periodic case very low values of h are needed to obtain a good overall error of the scheme: for $h = 10^{-6}$ the error is still equal to 1.7×10^{-4}.

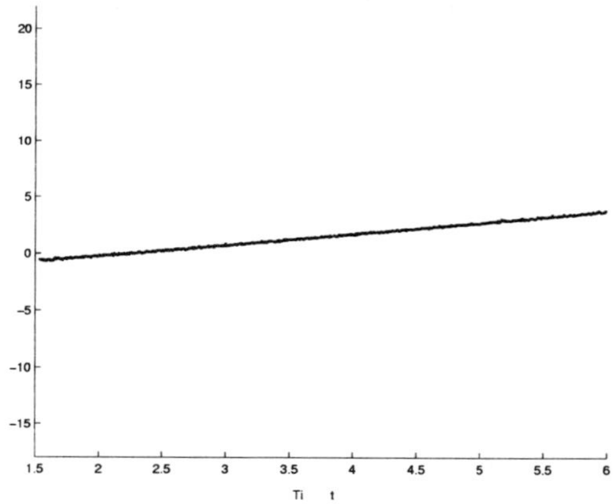

Fig. 4.4 Maximum error in displacement of the Schatzman-Paoli method for different time steps, $3T$-periodic solution.

4.2.5.2 *Classical methods with one impact localization per step*

The Newmark's Method

We have proved in the section 4.2.4 the consistency and the order 2 of Newmark's method with (IM1), for a solution without accumulation point of impact times. Moreover, the proofs of the Theorems 4.4 and 4.5 can

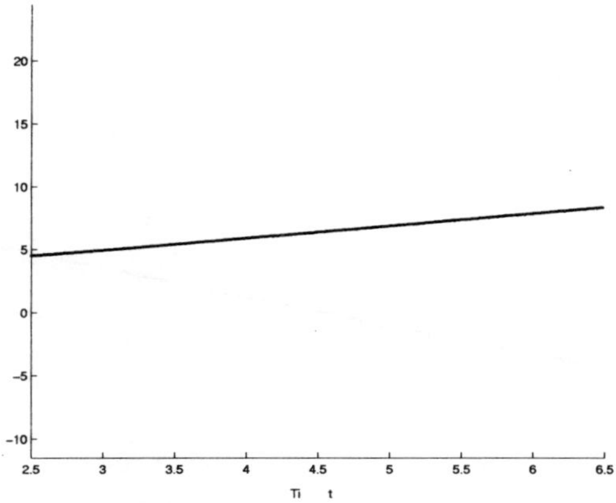

Fig. 4.5 Maximum error in displacement of the Schatzman-Paoli method for different time steps, T-periodic solution with infinite number of impacts.

easily be adapted to a numerical method of order 2 so that we can conclude that Newmark's method associated with (IM2) or (IM3) is consistent and of second order.

Numerical results show that whatever the method we choose among (IM1), (IM2) and (IM3) to localize impacts, the corresponding numerical method seems to be convergent, and the expected order 2 is found (see Fig. 4.6 for (IM1)). Thus it is not worth refining the approximation of impact times since a simple linear interpolation leads to an order 2 for this method. This is not surprising since the order cannot obviously be higher than 2 for any impact localization method because, between two successive impacts, Newmark's method is of order 2. Moreover, we can notice that for relatively high values of h, the velocity of the response is not accurately approximated, and the actual slope 2 in the velocity error of Fig. 4.6 only starts below some value of the time step h. This behaviour can be easily explained by the fact that if h is too big, the approximation to the exact solution by the numerical method is too rough, and therefore some impacts are localized on wrong time steps. The error obtained is thus due to the jumps in the velocity at impact times.

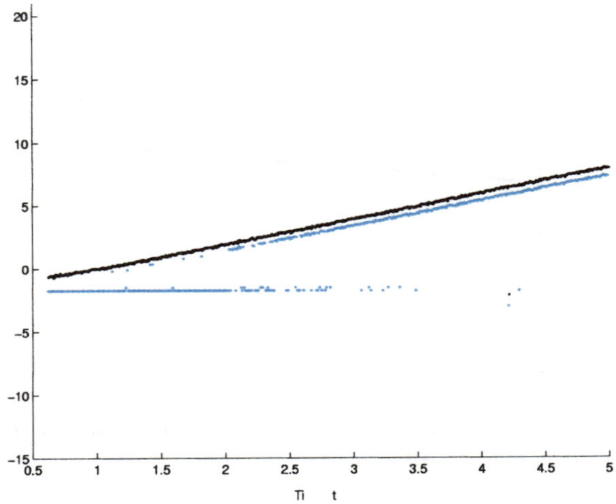

Fig. 4.6 Maximum error of Newmark's method with (IM1) for different time steps, $3T$-periodic solution. Black: displacement, blue: velocity.

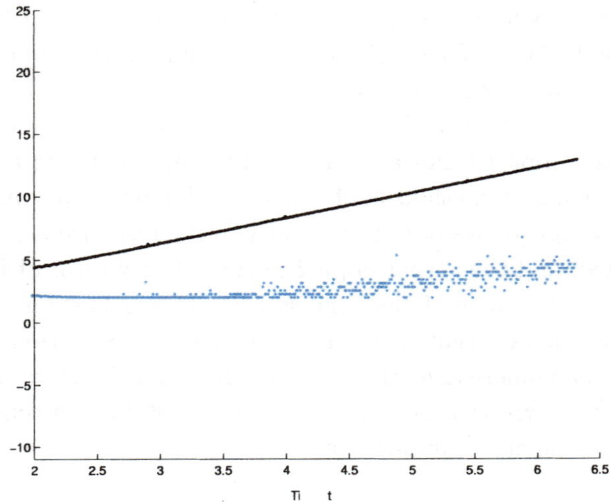

Fig. 4.7 Maximum error of Newmark's method with (IM1) for different time steps, T-periodic solution with infinite number of impacts. Black: displacement, blue: velocity.

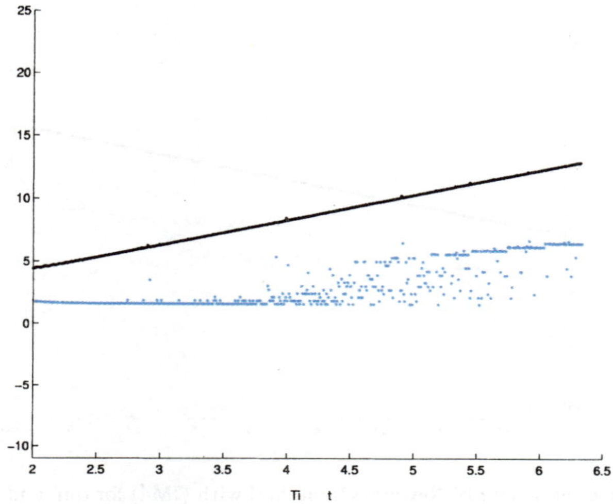

Fig. 4.8 Maximum error of Newmark's method with (IM2) for different time steps, T-periodic solution with infinite number of impacts. Black: displacement, blue: velocity.

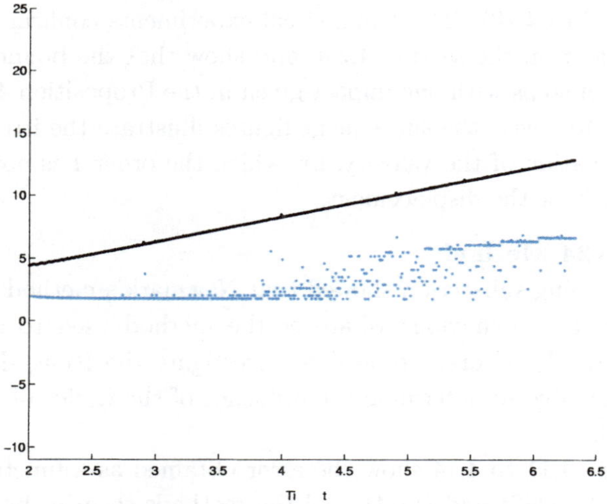

Fig. 4.9 Maximum error of Newmark's method with (IM3) for different time steps, T-periodic solution with infinite number of impacts. Black: displacement, blue: velocity.

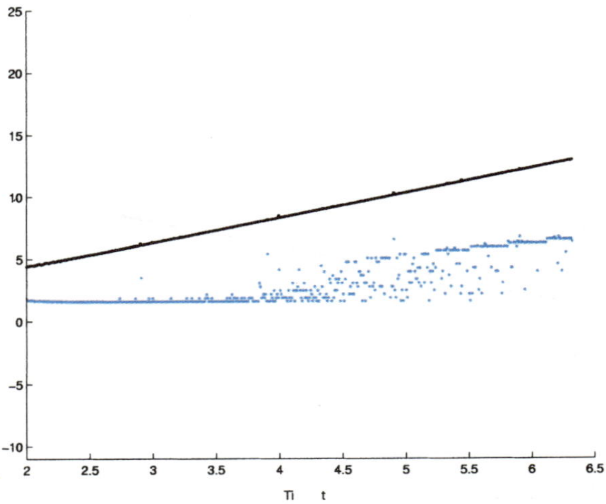

Fig. 4.10 Maximum error of Newmark's method with (IM4) for different time steps, T-periodic solution with infinite number of impacts. Black: displacement, blue: velocity.

Moreover, for solutions with an infinite number of impacts, we numerically observe that all the methods lead to a convergent scheme, and the order obtained seems to be 2 for the displacement and 1 for the velocity (see Figs. 4.7 to 4.10). These numerical experiments confirm the theoretical results seen in the section 4.2.4, and show that the bound in $\ln(h)$ of the number of steps with one impact given in the Proposition 4.3 was overestimated. Moreover, the subsequent figures illustrate the irregularities in the approximation of the velocity, for which the order 1 is not as obvious as the order 2 for the displacement.

The RK24 Method

The preceding subsection showed that Newmark's method has approximately the same behaviour whatever the method used to approximate impact times. Therefore, we need to investigate the Runge-Kutta RK24 method to be able to determine the influence of the choice of the localization method.

The Figs. 4.11 to 4.14 show the error obtained as a function of h, for (IM1), (IM2), (IM3) and (IM4). All the methods seem to be convergent, and the numerical order, which can be estimated from the figures, confirms the theory seen in the section 4.2.4: we get an order 2 for (IM1), 3 for (IM2) and 4 for (IM3) and (IM4).

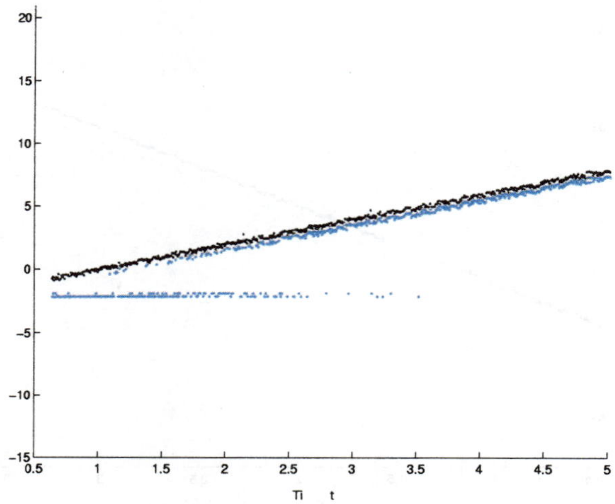

Fig. 4.11 Maximum error of the RK24 method with (IM1) for different time steps, $3T$-periodic solution. Black: displacement, blue: velocity.

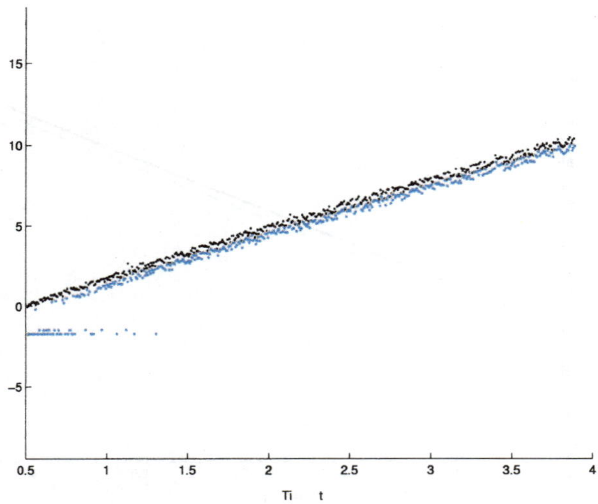

Fig. 4.12 Maximum error of the RK24 method with (IM2) for different time steps, $3T$-periodic solution. Black: displacement, blue: velocity.

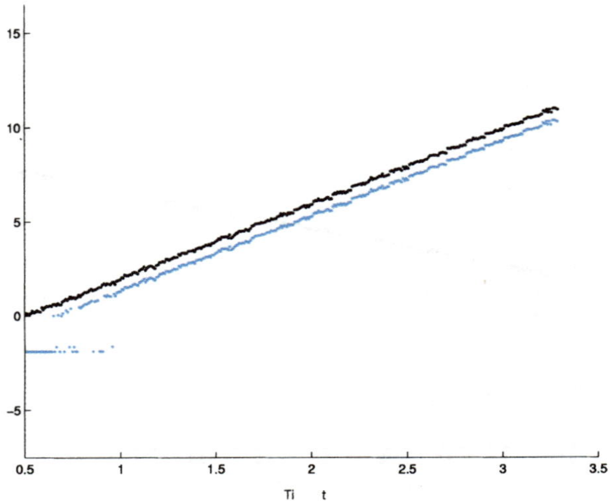

Fig. 4.13 Maximum error of the RK24 method with (IM3) for different time steps, $3T$-periodic solution. Black: displacement, blue: velocity.

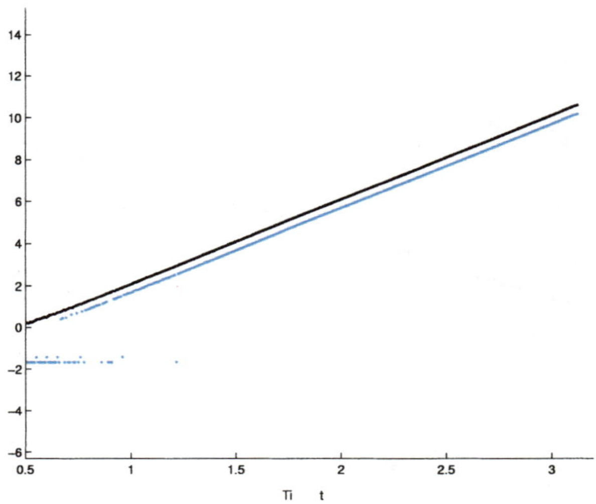

Fig. 4.14 Maximum error of the RK24 method with (IM4) for different time steps, $3T$-periodic solution. Black: displacement, blue: velocity.

As for the solutions with infinite sequences of impact times, we obtain the same results as Newmark's scheme: The impact approximation method does not seem to affect the order of the RK24 method as we can see in Figs. 4.15 to 4.18. In such a case, an accurate impact time approximation turns out to be useless since the order obtained is always 2 for the displacement and approximately 1 for the velocity.

If we take a closer look at the consistency error obtained numerically, we understand where these orders 2 and 1 come from. In the Fig. 4.19 we plotted the maximum of the consistency error for (IM3), depending on the behaviour of the solution on each time steps: we distinguished between steps with K impacts where $K > 1$, steps where sticking ends and all the other ones. For the second type of steps, we obtain an error in h^3 for the displacement and h^2 for the velocity as the theory predicted, but if we include an approximation of the time $\bar{t}$ when sticking ends as we mentioned in the section 4.2.2, the consistency error on this step then behaves like h^4. Therefore, the steps which restrict the overall order are the ones where K impacts occur where $K > 1$. The number n_h of such steps on one period cannot be easily analytically calculated: we numerically observe that when h varies, this number varies in the interval $\{0, 1\}$.

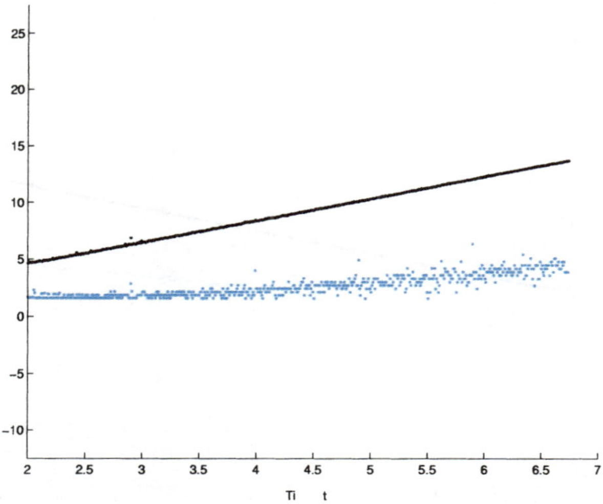

Fig. 4.15 Maximum error of the RK24 method with (IM1) for different time steps, T-periodic solution with infinite number of impacts. Black: displacement, blue: velocity.

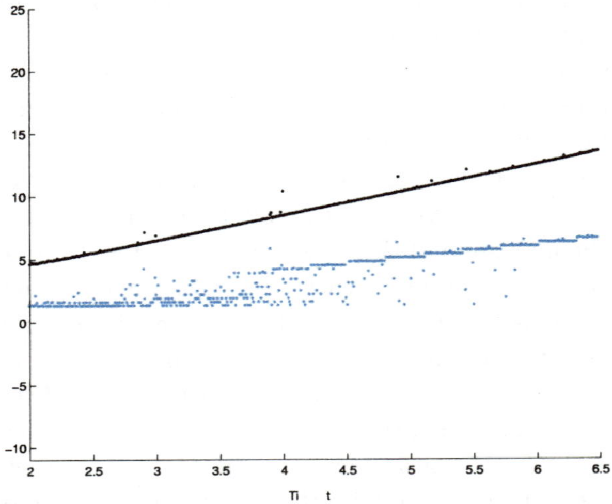

Fig. 4.16 Maximum error of the RK24 method with (IM2) for different time steps, T-periodic solution with infinite number of impacts. Black: displacement, blue: velocity.

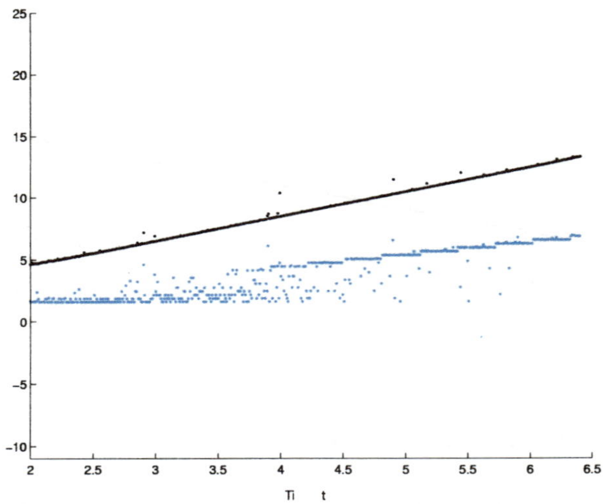

Fig. 4.17 Maximum error of the RK24 method with (IM3) for different time steps, T-periodic solution with infinite number of impacts. Black: displacement, blue: velocity.

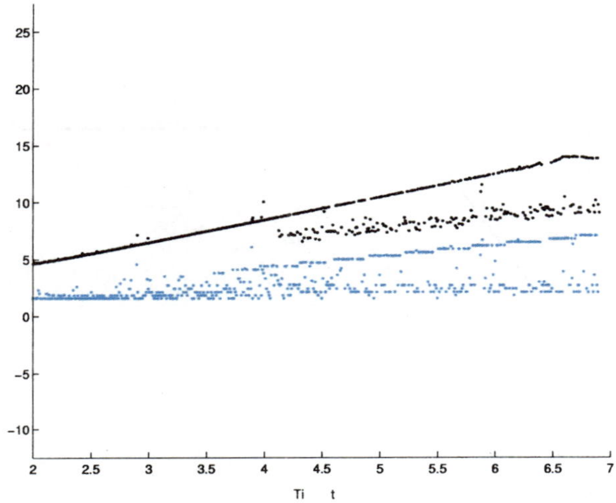

Fig. 4.18 Maximum error of the RK24 method with (IM4) for different time steps, T-periodic solution with infinite number of impacts. Black: displacement, blue: velocity.

The computation of times when sticking ends is consequently not very interesting in that case since it does not improve the order of the numerical method. Nevertheless, numerical experiments show that the computation of these times of sticking end can be useful from a practical point of view: for the RK24 method with (IM2), (IM3) or (IM4), it results in a better behaviour of the scheme mainly for the approximation of the velocity which is much more regular than in the previous cases. Moreover, the method (IMi) with localization of sticking ends, which will be referred to as (IM$i\#$) where $i \in 1, 2, 3, 4$, leads to a smaller constant K_i in the estimation of the error $\max_{0 \leq n \leq N} \mid y(t_n) - y_n \mid \leq K_i h^2$ for $i \in 2, 3, 4$, and then to a smaller final error in displacement (see the Figs. 4.20 and 4.21). This numerical property no longer holds for Newmark's method or for the RK24 method with (IM1). Therefore we can ultimately conclude that the detection of times when sticking ends is interesting only in the case of a RK24 method used with (IM2), (IM3) or (IM4).

The preceding results show that if we wish to improve the scheme for solutions with accumulations of impacts, we need to try to approximate as many impact times as possible.

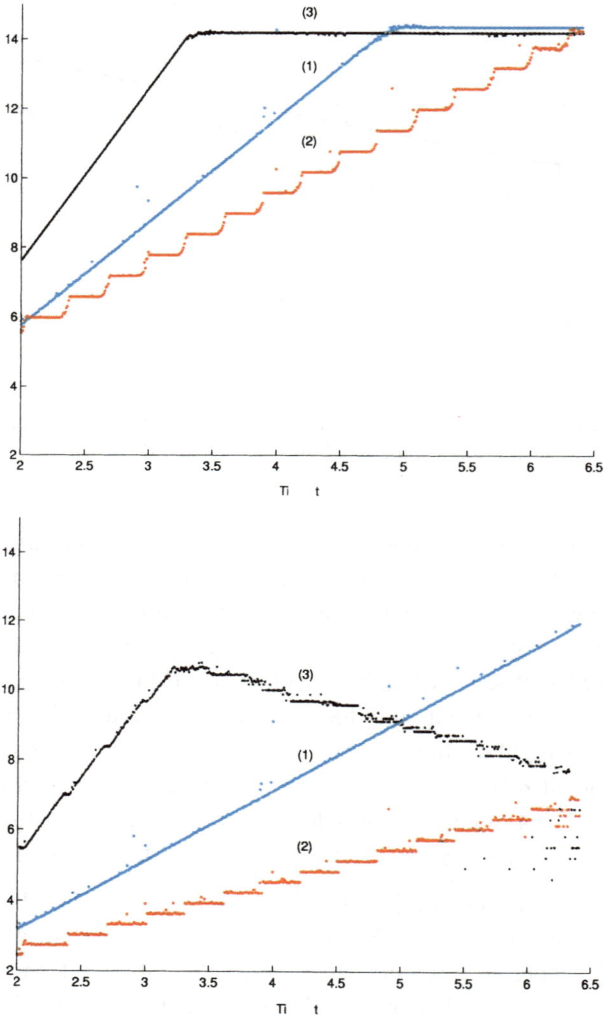

Fig. 4.19 Consistency error of the RK24 method with (IM3), T-periodic solution with infinite number of impacts. (1): step with a time of sticking end, (2): step with more than one impact, (3): other steps.

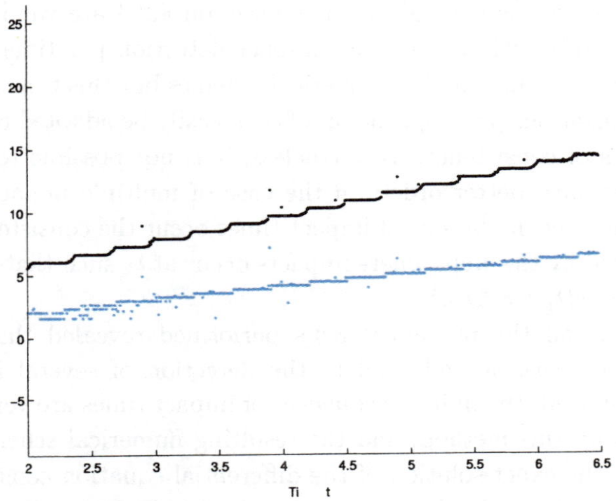

Fig. 4.20 Maximum error of the RK24 method with (IM2#) for different time steps, T-periodic solution with infinite number of impacts. Black: displacement, blue: velocity.

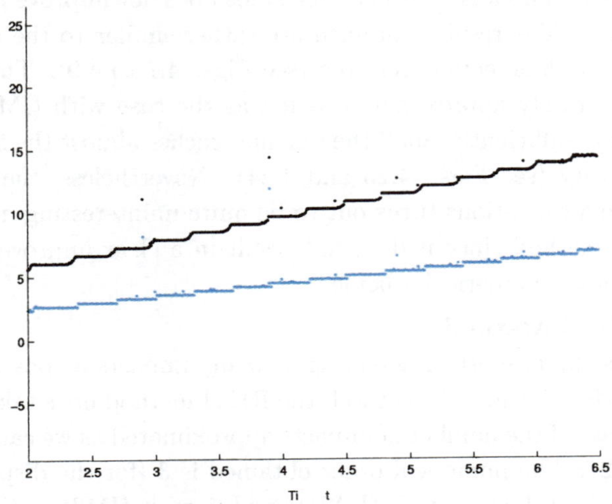

Fig. 4.21 Maximum error of the RK24 method with (IM3#) for different time steps, T-periodic solution with infinite number of impacts. Black: displacement, blue: velocity.

4.2.5.3 *Classical methods with several impact localizations per step*

The results of consistency given in the section 4.2.4 are valid for the numerical methods with at most one impact detection per time step. If we now consider the same kind of numerical schemes but this time with several impact localizations per step, the proofs can easily be adapted to obtain the same results of consistency. Nevertheless, it is not possible to prove that the schemes have better orders in the case of multiple impact detection, because when accumulations of impact times occur the consistency order is still restricted by the steps where impacts occur at t_k such that the velocity at t_k verifies $\dot{x}(t_k^-) = \mathbf{O}(h)$.

Moreover, all the numerical tests performed revealed that Newton's method (IM4) was not adapted to the detection of several impacts per time step. Indeed, the infinite sequences of impact times are very badly approximated by this method, and the resulting numerical scheme does not converge to the exact solution of the differential equation considered. The results of the classical schemes associated with (IM4*) will consequently not be represented in the subsequent study.

The Newmark's Method

Figs. 4.22 to 4.24 show the error obtained with Newmark's method associated with (IM1*), (IM2*) and (IM3*). We can notice that the localization of many impacts on every time steps does not improve the precision of the scheme: The results obtained are rather similar to the case with at most one impact detection per step (see Figs. 4.7 to 4.9). The velocity is still not accurately approximated as it was the case with (IMi) methods, but when h is sufficiently small the scheme reaches almost the second order for the velocity (see Figs. 4.23 and 4.24). Nevertheless, the addition of impact time localizations turns out to be quite uninteresting in the case of Newmark's method since it does not result in a clear improvement of the behaviour of the numerical scheme.

The RK24 Method

Whereas the interest of detecting as many impacts as possible was not clearly seen for Newmark's method, the RK24 method does take advantage of the increase of the number of impacts approximated as we can see in Figs. 4.26 and 4.27: The numerical order obtained is 3 (for the displacement as well as for the velocity) with (IM2*), and 4 with (IM3*). On the other hand, the method using linear interpolation (IM1*) leads to results very similar to (IM1) (compare Figs. 4.25 and 4.15).

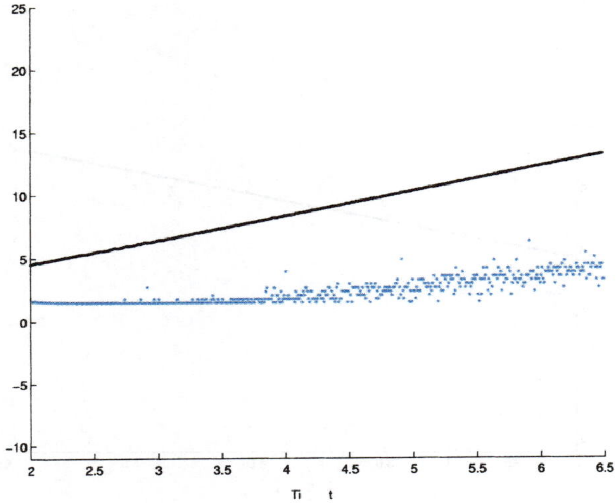

Fig. 4.22 Maximum error of Newmark's method with (IM1*) for different time steps, T-periodic solution with infinite number of impacts. Black: displacement, blue: velocity.

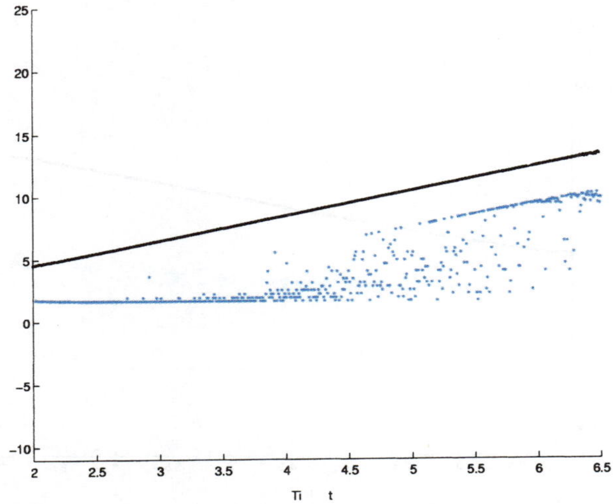

Fig. 4.23 Maximum error of Newmark's method with (IM2*) for different time steps, T-periodic solution with infinite number of impacts. Black: displacement, blue: velocity.

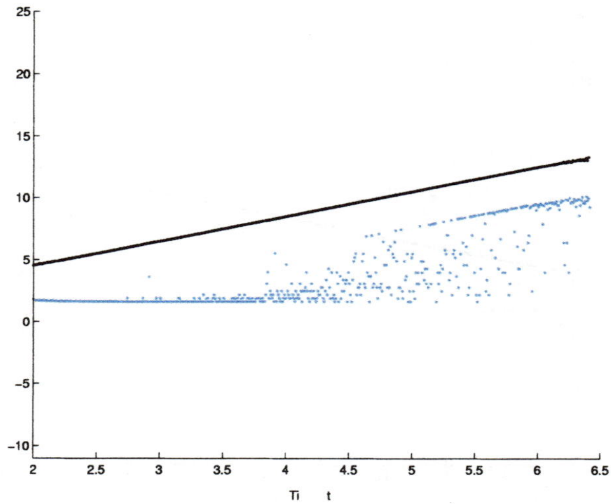

Fig. 4.24 Maximum error of Newmark's method with (IM3*) for different time steps, T-periodic solution with infinite number of impacts. Black: displacement, blue: velocity.

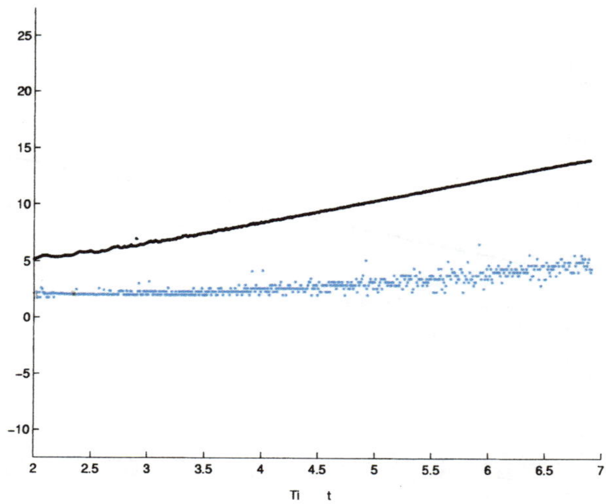

Fig. 4.25 Maximum error of the RK24 method with (IM1*) for different time steps, T-periodic solution with infinite number of impacts. Black: displacement, blue: velocity.

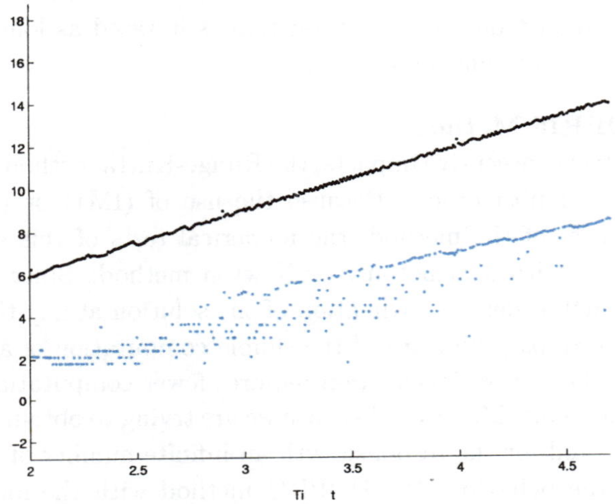

Fig. 4.26 Maximum error of the RK24 method with (IM2*) for different time steps, T-periodic solution with infinite number of impacts. Black: displacement, blue: velocity.

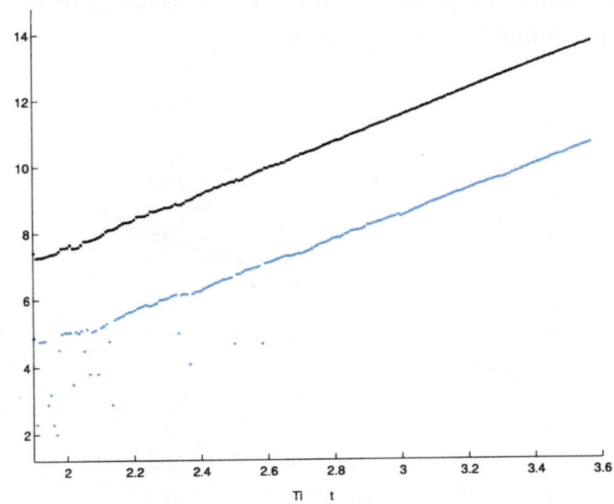

Fig. 4.27 Maximum error of the RK24 method with (IM3*) for different time steps, T-periodic solution with infinite number of impacts. Black: displacement, blue: velocity.

Thus the difficulties that we met with infinite sequences of impact times are overcome with the use of a high order numerical scheme coupled with a high order impact detection method that is iterated as long as impacts are found on a same time step.

The DOPRI5 Method

Between two consecutive impacts, the Runge-Kutta method of Dormand and Prince is of fifth order. Because the use of (IM1) or (IM2) would reduce the order of the method, the numerical tests of this scheme were only performed with a dichotomy or Newton method: Since the method provides a fourth order approximation of the solution at any time between two successive steps (by means of the simple computation of a third order polynomial), the impact localization requires fewer computations than the usual RK24 method. Moreover, because we are trying to obtain a high order method also valid for the response with an infinite number of impacts, we investigated the behaviour the DOPRI5 method with the localization of all the impacts occurring on a time step (to the given precision h^4). The numerical results are shown in Figs. 4.28, 4.29 and 4.30 for the interpolation methods (IM3*) and (IM4*) (only in the finite number of impacts case for (IM4*))). The corresponding methods occur to be at least of fourth order in displacement and velocity even in the case of infinite sequences of impact, with the surprising exception of (IM4*) which seems to be of fifth order in the case of finite number of impacts on $[0, T]$.

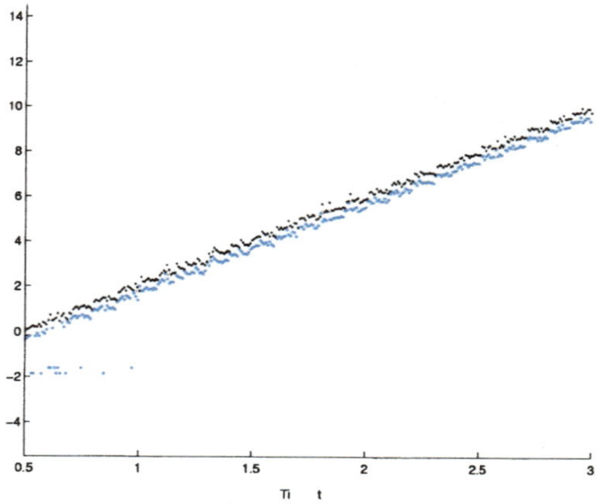

Fig. 4.28 Maximum error of the DOPRI5 method with (IM3*) for different time steps, $3T$-periodic solution. Black: displacement, blue: velocity.

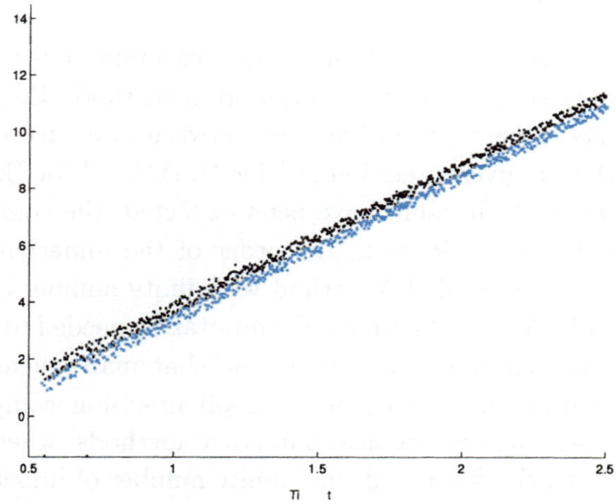

Fig. 4.29 Maximum error of the DOPRI5 method with (IM4*) for different time steps, T-periodic solution with infinite number of impacts. Black: displacement, blue: velocity.

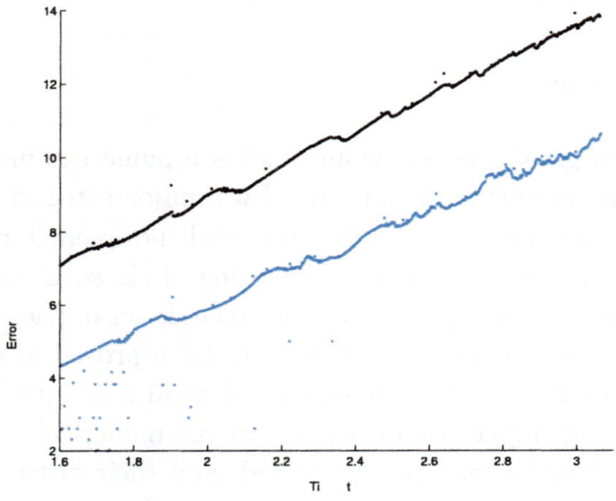

Fig. 4.30 Maximum error of the DOPRI5 method with (IM3*) for different time steps, T-periodic solution with infinite number of impacts. Black: displacement, blue: velocity.

4.2.6 *Computing times*

The computing times for a fixed time step are summarized in the Table 4.3, where (IMi*) stands for the interpolation method (IMi) iterated as long as it detects impacts, the difference between two successive impacts being greater than a given precision (h^2 for (IM1*), h^3 for (IM2*) and h^4 for the other ones). As it could have been expected, the computing times are approximately proportional to the order of the underlying numerical method, except for the DOPRI5 method with finite number of impacts.

Tables 4.4 and 4.5 show the times of computation needed to approximate the solution to a given precision. The Paoli-Schatzman scheme appears to need huge computing times to reach a good precision compared to the other methods. Among the classical numerical methods, when considering an average between the finite and the infinite number of impacts case, several methods lead to similar results in terms of computing efficiency: We can conclude from these experiments that the RK24 method with (IM2#), (IM3#) and (IM2*) and the DOPRI5 method with (IM3*) are the fastest numerical schemes for getting approximations of the solutions of the mechanical system with impacts considered.

4.3 Conclusion

We have investigated the behaviour of several numerical methods adapted to mechanical systems with impacts. Two major categories of numerical schemes were considered: The first one with no explicit computation of impact times, and the second one consisting of classical methods of Newmark or Runge-Kutta type to which we include an impact approximation procedure. Partial theoretical results have been proved in the latter case, showing the difficulties to establish general results in particular when infinite sequences of impact times occur. All the numerical methods defined have been thoroughly numerically tested, and their order has been computed. The different sources of numerical errors have been identified in the case of accumulations of impact times and has led to the definition of a new algorithm iterating the procedure of impact approximation as long as impacts occur on every time step.

Table 4.3 Computing times for the different numerical methods, $h = T \times 10^{-4}$

Numerical method		Periodic solution (3, 2)	Periodic solution $(1, \infty)$
Schatzman-Paoli		100	100
Newmark	IM1	285	203
	IM2	285	204
	IM3	286	210
	IM4	285	206
	IM1*	291	207
RK24	IM1	538	349
	IM2	537	350
	IM3	543	383
	IM4	539	354
	IM3*	546	455
	IM4*	543	/
DOPRI5	IM3*	849	531
	IM4*	851	/
Analytical method		57	214

Table 4.4 Computing times for the different numerical methods, $\max\limits_{0 \leq n \leq N} \mid x(t_n) - x_n \mid = 10^{-4}$, $3T$-periodic solution, 2 impacts per cycle.

Numerical method		Computing time
Schatzman-Paoli		3,062,372
Newmark	IM1	4,094
RK24	IM1	4,273
	IM2	365
	IM3	199
	IM4	154
DOPRI5	IM3*	289
	IM4*	100
Analytical method		791

Table 4.5 Computing times for the different numerical methods, $\max_{0 \le n \le N} | x(t_n) - x_n | = 10^{-7}$, T-periodic solution, infinite number of impacts per cycle.

Numerical method		Computing time
Schatzman-Paoli		8,155
Newmark	IM1	257
	IM2	257
	IM3	344
	IM4	280
	IM1*	272
RK24	IM1	352
	IM2	304
	IM3	608
	IM2*	100
	IM3*	603
	IM2#	107
	IM3#	288
DOPRI5	IM3*	222
Analytical method		23,908

From the mathematical point of view, the Paoli-Schatzman scheme is the best choice because it has been proved to be convergent, provided the equations of the physical model that have to be solved can be written under a form that satisfies the mathematical conditions required to define the scheme. Its main numerical interest is that it does not explicitly compute impact times, but the low order of this method makes it very slow when a relatively high precision is needed. On the other hand, the classical methods associated with impact detection have shown good numerical properties, provided that the algorithms are designed to compute as many impacts as possible. In terms of speed of the different schemes, the tests that we performed showed that the Runge-Kutta method of Dormand and Prince associated with an impact approximation by dichotomy, and the RK24 method with second order interpolation are the best compromise between order and computing time for any kind of response.

All this study has been carried out in a simple single degree of freedom case where analytical calculations were possible, in order to investigate the qualitative properties of some numerical methods, and understand the reason why some schemes lead to better results than the other ones. In the general case of multi degree of freedom systems, we can expect that most of the results given in this chapter still hold for the schemes with impact localizations. However, in such a case the classical methods with explicit computation of all the impacts can become less convenient than the Paoli-Schatzman scheme, because when impacts involve several degrees of freedom the impact localization procedure can become very penalizing for the numerical method.

Chapter 5

Bifurcations of a Particular van der Pol-Duffing Oscillator

Interest in "0" type and Hopf bifurcation phenomena has grown recently because of the possibility of transition from regular to chaotic motion in physical systems via consecutive or sudden bifurcations of previously stationary regular solutions. Such bifurcations have been described and analysed extensively by others [Aichara and Matsumoto (1986)], [Chow and Hale (1982)], [Dold and Eckmann (1984)], [Gilmore (1981)], [Guckenheimer and Holmes (1983)], [Hassard *et. al.* (1981)], [Iooss (1979)], [Iooss and Joseph (1980)], [Marsden and McCracken (1976)], [Ruelle and Takens (1971)], [Thompson and Ghaffari (1982)], [Tomita (1986)].

In addition there has been a recent tendency to seek chaotic motion in the simplest physical systems [Awrejcewicz (1986); Awrejcewicz (1989a); Awrejcewicz (1990a)], [Dmitriev *et. al.* (1983)], [Kriukov (1984)], [Tomita (1986)], [Ueda (1979)], [Ueda and Akamatsu (1981)], [Ueda *et. al.* (1978)], [Ueda and Nanahara (1978)]. The study of nonlinear dynamics in these simpler systems provides the advantage of less complex analysis of the behaviour of strange attractors that can also appear in higher dimensional analogues.

Extensive work in this field has been done by Ueda and co-workers [Ueda (1979)], [Ueda and Akamatsu (1981)], [Ueda *et. al.* (1978)], [Ueda and Nanahara (1978)]. He has analysed, for instance, a particular Duffing oscillator governed by the equation

$$\ddot{x} + k\dot{x} + x^3 = B\cos t. \tag{5.1}$$

For $(k, B) = (0.1, 12.0)$ the solution of Eq. (5.1) has chaotic behaviour [Ueda (1979)]. The van der Pol-Duffing oscillator of the type

$$\ddot{x} + \varepsilon(1 - x^2)\dot{x} + x^3 = B\cos\nu t, \tag{5.2}$$

for $(\varepsilon, B, \nu) = (0.2, 17.0, 4.0)$ possesses a strange attractor as well [Ueda and Akamatsu (1981)].

Chaos in the Duffing and van der Pol-Duffing oscillators has also been found by Kriukov and Dmitriev with co-workers [Dmitriev *et. al.* (1983)], [Kriukov (1984)].

Ueda, Doumoto and Suehiro [Ueda *et. al.* (1978)] have found the chaotic solutions in numerical simulations of the nonlinear Mathieu equation, and Ueda and Nanahara have found chaos in a system which models a phase-locked feedback loop with time-delays [Ueda and Nanahara (1978)].

Chaotic motion in one, two and more degrees of freedom systems and friction has also been investigated by one of the authors [Awrejcewicz (1986); Awrejcewicz (1988)].

Some examples of physical systems in which chaos has been found have been described in the work of Holmes and Moon [Holmes and Moon (1983)].

5.1 The Analysed System and the Averaged Equations

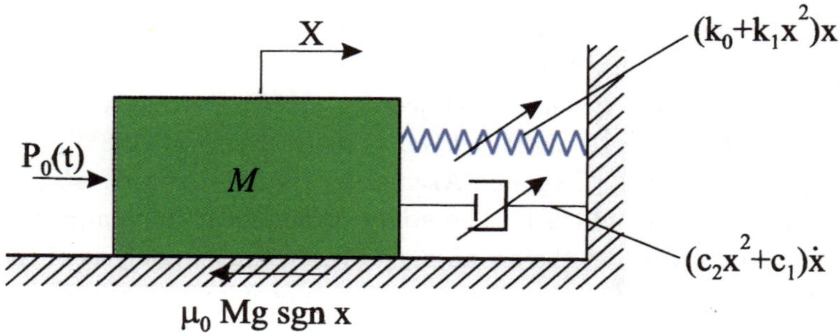

Fig. 5.1 A schematic of the analysed system.

The equation of motion for the system analysed, shown in Fig. 5.1 is

$$M\ddot{x} + (c_2x^2 - c_1)\dot{x} + k_0x + k_1x^3 + \mu_0 Mgsgn\dot{x} = P_0(t), \tag{5.3}$$

where $P_0(t) = a\mu\nu^2 \cos \nu t$.

The excitation force $P_0(t)$ originates from the rotating engine rotor with the mass m and the unbalance μ, where a is the amplification coefficient.

From (5.3) one obtains

$$\dot{y} = z, \qquad \dot{z} = \varepsilon(1 - y^2)\dot{y} - \delta y - \gamma y^3 - \alpha sgn\dot{y} + p_0\nu^2 \cos \nu t, \tag{5.4}$$

where:

$$\begin{gathered} y = \sqrt{(c_2/c_1)x}, \quad \varepsilon = c_1/M, \quad \delta = k_0/M, \\ \gamma = k_1 c_1/M c_2, \quad \alpha = \mu_0 g\sqrt{c_2/c_1}, \quad p_0 = a\mu m/M(c_2/c_1)^{-1/2}. \end{gathered} \tag{5.5}$$

An approximate analytical van der Pol method is used to solve the system of (5.4). It is assumed that

$$y = u(t)\cos \nu t - v(t)\sin \nu t, \quad z = -\nu[u(t)\sin \nu t + v(t)\cos \nu t], \tag{5.6}$$

where $u(t)$ and $v(t)$ are slowly changing functions of t. After substituting (5.6) in (5.4) one obtains

$$\dot{u} = N\sin \nu t, \quad \dot{v} = N\cos \nu t, \tag{5.7}$$

where:

$$\begin{aligned} N = {} & (\delta - \nu^2)\nu^{-1}(u\cos \nu t - v\sin \nu t) + \varepsilon\nu^{-1}(u\cos \nu t - v\sin \nu t)^3 \\ & +\varepsilon(1 - (u\cos \nu t - v\sin \nu t)^2)(u\sin \nu t + v\cos \nu t) \\ & +(\alpha/\nu)sign(-v(u\sin \nu t + v\cos \nu t)) - p_0\nu\cos \nu t. \end{aligned} \tag{5.8}$$

The right hand sides of (5.7) being periodic functions of t with period $2\pi/\nu$ are expanded into Fourier series. Upon taking into consideration both that u and v are slowly changing functions of t and that only the first terms of the expansions are significant the following averaged equations system is obtained:

$$\begin{aligned} \dot{u} &= -\frac{\delta - \nu^2}{2\nu}v + \frac{\varepsilon}{2}u - \frac{\varepsilon}{2}u(u^2 + v^2) - \frac{3\gamma}{8\nu}v(u^2 + v^2) - \frac{2\alpha}{\pi\nu}\frac{u}{\sqrt{u^2 + v^2}}, \\ \dot{v} &= \frac{\delta - \nu^2}{2\nu}u + \frac{\varepsilon}{2}v - \frac{\varepsilon}{8}v(u^2 + v^2) + \frac{3\gamma}{8\nu}u(u^2 + v^2) + \\ & \quad -\frac{2\alpha}{\pi\nu}\frac{v}{\sqrt{u^2 + v^2}} - \frac{p_0\nu}{2}. \end{aligned} \tag{5.9}$$

Assuming that

$$\omega = \frac{\delta - \nu^2}{\nu}, \quad \varepsilon = 8, \quad \frac{3\gamma}{8\nu} = 1, \quad \frac{2\alpha}{\pi\nu} = 1, \quad \frac{\nu}{2} = P, \quad p_0 = 1, \tag{5.10}$$

one obtains

$$\dot{u} = -\omega v + 4u - u(u^2+v^2) - v(u^2+v^2) - u/\sqrt{u^2+v^2},$$
$$\dot{v} = \omega u + 4v - v(u^2+v^2) + u(u^2+v^2) - v/\sqrt{u^2+v^2} - P. \quad (5.11)$$

Using the polar coordinates (r, Θ) one has:

$$u = r\cos\Theta, \quad v = r\sin\Theta, \quad (5.12)$$

and the system of Eqs. (5.11) acquires the form

$$\dot{r} = -r^3 + 4r - 1 - P\sin\Theta, \quad \dot{\Theta} = r^2 + \omega - (P/r)\cos\Theta. \quad (5.13)$$

At singular points $\dot{r} = \dot{\Theta} = 0$ and from Eqs. (5.13) one obtains

$$r^6 + (\omega - 4)r^4 + r^3 + \frac{\omega^2+16}{2}r^2 - 4r + \frac{1-P^2}{2} = 0. \quad (5.14)$$

The use of an approximation method, known as van der Pol method is very convenient, particularly in order to obtain the bifurcations of an averaged system of forced oscillators. Only one assumption must be taken into account, that the amplitudes $u(t)$ and $v(t)$ change slowly in time t. The same method was used, for instance, by Arrowsmith and Taha [Arrowsmith and Taha (1983)], where the local and global bifurcations of the averaged two parameter system were investigated (see also reference [Hale (1969)]).

5.2 "0" Type Bifurcations

The "0" type bifurcations occur when singular points overlap. The necessary condition for the existence of multiple roots of the Eq. (5.14) has the form

$$a_0^{10} \begin{vmatrix} s_0 & s_1 & s_2 & s_3 & s_4 & s_5 \\ s_1 & s_2 & s_3 & s_4 & s_5 & s_6 \\ s_2 & s_3 & s_4 & s_5 & s_6 & s_7 \\ s_3 & s_4 & s_5 & s_6 & s_7 & s_8 \\ s_4 & s_5 & s_6 & s_7 & s_8 & s_9 \\ s_5 & s_6 & s_7 & s_8 & s_9 & s_{10} \end{vmatrix} = 0, \quad (5.15)$$

where

$$s_k = (-1/a_0)^k \begin{vmatrix} a_1 & a_0 & 0 & \dots & 0 \\ 2a_2 & a_1 & a_0 & \dots & 0 \\ 3a_3 & a_2 & a_1 & \dots & 0 \\ \vdots & \vdots & \vdots & & \\ ka_k & ak_1 & & & a_1 \end{vmatrix},$$

$$\begin{aligned} a_0 = 1, a_4 = (\omega^2 + 16)/2, a_8 = 0, a_1 = 0, a_5 = -4, a_9 = 0, \\ a_2 = \omega - 4, a_6 = (1 - P^2)/2, a_{10} = 0, a_3 = 1, a_7 = 0, s_0 = 6. \end{aligned} \quad (5.16)$$

From Eq. (5.16) one obtains

$$x^5 + b_4x^4 + b_3x^3 + b_2x^2 + b_1x + b_0 = 0, \quad (5.17)$$

where

$$\begin{aligned} x = P^2, b_4 = 0.7\omega^3 + 16\omega^2 - 71.1\omega - 1.1, \\ b_3 = 0.2\omega^6 + 21\omega^5 - 28\omega^4 - 360.1\omega^3 + 704.2\omega^2 + 5244.2\omega - 2412.4, \\ b_2 = -0.2\omega^9 + 3\omega^8 + 18.5\omega^7 - 336.4\omega^6 - 1440\omega^5 + 36330\omega^4 \\ +1388.5\omega^3 + 7608.7\omega^2 - 19960\omega + 62152, \\ b_1 = 0.001\omega^{12} + 0.6\omega^{11} - 0.07\omega^{10} - 125.2\omega^9 - 623.4\omega^8 \\ +9400.9\omega^7 + 30163\omega^6 - 26595\omega^5 + 427720\omega^4 + 93038\omega^3 \\ +333350\omega^2 + 3347200\omega + 4174400, \\ b_0 = -0.8\omega^{13} - 14.5\omega^{12} - 95.5\omega^{11} - 625.1\omega^{10} - 25695.5\omega^9 \\ -41611.7\omega^8 + 1058200\omega^7 - 419820\omega^6 - 5740700\omega^5 - 378800\omega^4 \\ +5201600\omega^3 + 1165600\omega^2 + 5370900\omega + 977300. \end{aligned} \quad (5.18)$$

In view of the character of Eq. (5.17), the dependence $P(\omega)$ has been determined numerically by means of substituting the value ω (a real number) and then solving the algebraic equation in terms of x by using the secant method.

5.3 Complex Bifurcations

In order to determine the equation for complex bifurcations, the terms of (5.9) containing the square root, are expanded into Taylor series around the temporarily unknown point of equilibrium (u_0, v_0). The expansion is limited to the linear terms of u and v. As a result, one obtains:

$$\begin{aligned}\dot{u} = -\omega v + (4 - (u^2 + v^2))u - (u^2 + v^2)v - \frac{u_0}{\sqrt{u_0^2 + v_0^2}} - \\ \frac{v_0^2}{(u_0^2 + v_0^2)^{3/2}}(u - u_0) + \frac{u_0 v_0}{(u_0^2 + v_0^2)^{3/2}}(v - v_0), \\ \dot{v} = \omega u + (4 - (u^2 + v^2))v + (u^2 + v^2)u - \frac{v_0}{\sqrt{u_0^2 + v_0^2}} - \\ \frac{u_0^2}{(u_0^2 + v_0^2)^{3/2}}(v - v_0) + \frac{u_0 v_0}{(u_0^2 + v_0^2)^{3/2}}(u - u_0) - P.\end{aligned} \tag{5.19}$$

One can now proceed to a new coordinate system (u', v') whose origin is the singular point (u_0, v_0). From Eqs. (5.19), after linearization one obtains:

$$\begin{aligned}\dot{u}' = \left(4 - 3u_0^2 - v_0^2) - 2u_0 v_0 - \frac{v_0^2}{(u_0^2 + v_0^2)^{3/2}}\right) u' + \\ (-\omega - u_0 - 3v_0^2 - 2u_0 v_0 + \frac{u_0 v_0}{(u_0^2 + v_0^2)^{3/2}})v', \\ \dot{v}' = (\omega + 3u_0^2 + v_0^2 - 2u_0 v_0 + \frac{u_0 v_0}{(u_0^2 + v_0^2)^{3/2}})u' + \\ (4 - u_0^2 - 3v_0^2 + 2u_0 v_0 + \frac{u_0^2}{(u_0^2 + v_0^2)^{3/2}})v'.\end{aligned} \tag{5.20}$$

The characteristic equation of the system Eqs. (5.20) has the form

$$\delta^2 - (A + D)\delta + AD - BC = 0, \tag{5.21}$$

where

$$\begin{aligned}A &= 4 - 3u_0^2 - v_0^2 - 2u_0 v_0 - v_0^2/(u_0^2 + v_0^2)^{3/2}, \\ B &= -\omega - u_0^2 - 3v_0^2 - 2u_0 v_0 + u_0 v_0/(u_0^2 + v_0^2)^{3/2}, \\ C &= \omega + 3u_0^2 + v_0^2 - 2u_0 v_0 + u_0 v_0/(u_0^2 + v_0^2)^{3/2}, \\ D &= 4 - u_0^2 - 3v_0^2 + 2u_0 v_0 - u_0^2/(u_0^2 + v_0^2)^{3/2}.\end{aligned} \tag{5.22}$$

In order for Hopf bifurcation to exist, it is necessary for the roots of the characteristic Eq. (5.21) to be strictly imaginary. This leads to the condition that

$$A + D = 0, \quad AD - BC > 0. \tag{5.23}$$

From the first equation of the system (5.23) and from the Eq. (5.11) one obtains the set of bifurcation equations:

$$\begin{aligned} \frac{1}{4\sqrt{u_0^2+v_0^2}} &= 2-(u_0^2+v_0^2), \\ -\omega v_0 + (4-(u_0^2+v_0^2))u_0 - (u_0^2+v_0^2)v_0 - \frac{u_0}{\sqrt{u_0^2+v_0^2}} &= 0, \\ \omega u_0 + (4-(u_0^2+v_0^2))v_0 + (u_0^2+v_0^2)u_0 - \frac{v_0}{\sqrt{u_0^2+v_0^2}} - P &= 0. \end{aligned} \tag{5.24}$$

This equation set, after replacing the Cartesian coordinates (u_0, v_0) by polar ones (r_0, Θ_0) acquires the following form:

$$\begin{aligned} 4r_0(2-r_0^2) &= 1, \\ -\omega r_0 \sin\Theta_0 + (4-r_0^2)r_0\cos\Theta_0 - r_0^3\sin\Theta_0 - \cos\Theta_0 &= 0, \\ \omega r_0\cos\Theta_0 + (4-r_0^2)r_0\sin\Theta_0 + r_0^3\cos\Theta_0 - \sin\Theta_0 - P &= 0. \end{aligned} \tag{5.25}$$

With regard to the symmetry of the equations in relation to P, the figure only presents the half plane $P \geq 0$. From the first of Eqs. (5.25) three values of r_0 are obtained: $r_{01} = 1.347$, $r_{02} = 0.126$ and $r_{03} = -1.473$. After substituting these values into the other two of Eqs. (5.25), three necessary conditions for the Hopf bifurcations are obtained:

$$\begin{aligned} P^2 &= 1.814\omega^2 + 6.584\omega + 2.274, \\ P^2 &= 0.016\omega^2 + 0.001\omega + 2.01, \\ P^2 &= 2.17\omega^2 + 9.415\omega - 34.409. \end{aligned} \tag{5.26}$$

Fig. 5.2 presents the complex (Hopf) bifurcation curves along with the diagram of the "0" type bifurcation curves.

5.4 Observations of Strange Attractors Using Numerical Simulations

As many authors connect the occurrence of chaos with bifurcation, it seems appropriate to seek irregular motion on the plane with parameters P, ω near the bifurcation curves (Fig. 5.2). For this purpose the differential Eq. (5.4) was numerically solved, after taking into account dependencies (5.10). Numerical calculations were made by using the variable-order, variable-step Gear method. The unknown solution was obtained by interpolation on

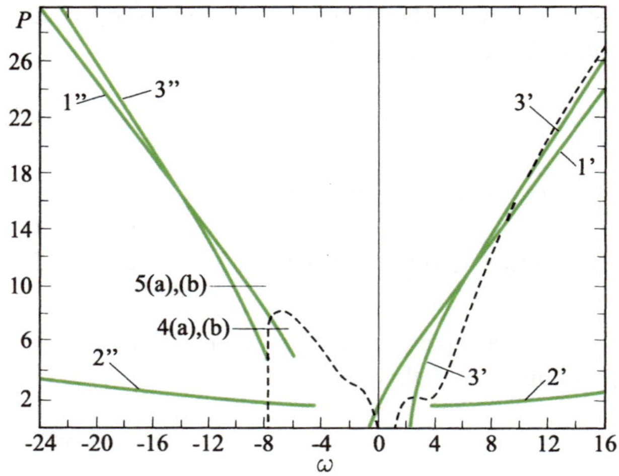

Fig. 5.2 Hopf (green) and the "0" type (- - -) bifurcation curves.

solution values produced by the Gear method. The accuracy of the integration and the interpolation is controlled by two parameters. The first is a calculation step with which integrations are made, while the latter determines the type of error control. At each step in the numerical solution an estimate of the local error is made. If the appropriate condition is not satisfied then the step-size is reduced and the solution is recomputed on the current step. The Gear method was used in order to avoid problems with numerical accuracy, when dealing with such an ill-behaved nonlinearity as dry friction. A description of Gear methods and their practical implementation is given in reference [Hall and Watt (1976)]. The introduction of the connections (5.10) between parameters of the nonlinear oscillator (5.4) makes it possible to observe the behaviour of this oscillator on the two parameter plane when bifurcation appears (see Fig. 5.2). Generally chaotic motion has been found for the parameters lying near the bifurcation curves. In addition, with increasing ω for the established values of P, a tendency towards ordering of the motion is observed. Previously Grebogi and Ott [Grebogi and Ott (1983)] have discussed the occurrence of sudden qualitative changes of chaotic dynamics as a single parameter is varied. Sudden changes in the size of chaotic attractors, sudden appearances of chaotic attractors and sudden destruction of chaotic attractors have been connected with the collision crises of an unstable periodic orbit and coexisting chaotic attractor. In our example, we encounter a similar situation in the sense that, for ω lying sufficiently left of the bifurcation curves, we have discovered regular orbits. With increasing ω, for the parameters near the bifurcation points the orbits fail to be closed curves and strange chaotic attractors were detected. Earlier, Curry and Yorke [Curry and Yorke (1978)] have shown using the example of particular maps in R^2 that at certain parameter values Hopf bifurcations occur, and as parameters change, the attracting invariant circle grows and starts to warp, eventually becoming a strange attractor. We should underline here the benefit of using the approximate method described earlier. Thanks to the average Eqs. (5.7) obtained by this method the Hopf bifurcation is related to the bifurcations of the amplitudes u and v of the periodic orbits in Eqs. (5.4). This means that the conditions for Hopf bifurcations are related to the bifurcation of periodic orbits in the starting Eqs. (5.4). With the bifurcation of periodic orbits either the new subharmonic periodic motion appears, or a quasi-periodic torus is born (see reference [Arnold (1983)]) or finally a chaotic orbit can appear.

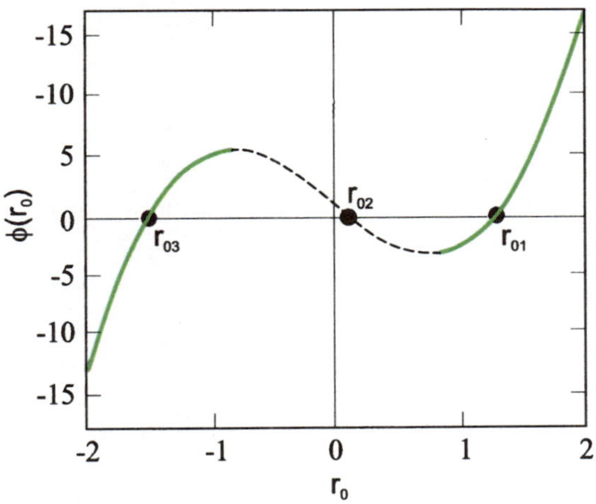

Fig. 5.3 The equilibrium points corresponding to the Hopf bifurcation curves presented in Fig. 5.2.

Some of the examples of chaotic motion detected near the bifurcation curves are given in the reference [Awrejcewicz and Mrozowski (1989)]. During computer experiments, chaos has not been found near the bifurcation curves 2' and 2". Fig. 5.3 presents the diagram of dependencies $f(r_0)$, where the roots of the first equation of system (5.24) are marked. It is evident that the equillibrum point r_{02} is unstable. For this reason, the bifurcation curves marked in Fig. 5.2 as 2' and 2" correspond to the unstable equilibrum point of the averaged equations. For the parameters near these bifurcation curves we have detected periodic and quasi-periodic attractors. Furthermore, the numerical investigations have not proven the existence of any relationship between the occurrence of "0" type bifurcations and the occurrence of chaos on the considered oscillator.

Chapter 6

Stick-Slip Oscillator with Two Degrees of Freedom

6.1 Introduction

In this chapter we present a systematic approach used as a tool for considering nonlinear dynamics from the applications point of view. This approach consist of several parts. We underline that technical problems often need to be considered under realistic friction conditions which unfortunately usually cause numerical difficulties in the case of simulation. We establish a physical model including geometric and friction-nonlinearities. The model is designed to describe the technically important phenomenon of self-excited oscillations. The nonlinear model considered, with two degrees of freedom then serves as an example to present a systematic approach of investigating such types of problems. Firstly we resolve the equation of motion into a nondimensional form in order to reduce the parameter space. By this the system defines its own (physically important) measures. We have successfully used an analytical approach to investigate the local behaviour of the phase flow of this four dimensional system and we have also obtained the parameter portrait of the trivial solutions. We used the numerical technique which allows us not only to find the solutions for the freely chosen parameter sets but also to obtain their stability.

Secondly, we investigate in detail the transitions from the stick to slip, slip to stick and slip to slip states. A possible physical interpretation of the behaviour of the system is also given. We concentrate on the systematic numerical approach to investigate the behaviour of the periodic orbits. The shooting method enables us to use the original differential equations and, after solving the boundary value problem we obtain full information about the investigated system with high accuracy. Using this technique we can also calculate the bifurcation points, new bifurcated solutions and the critical parameter sets, for which chaotic orbits appear. Then chaotic behaviour is analysed using the same tools, based on solving the initial value problem.

Self-excited vibrations caused by friction are widely described in the technical literature. The first example was given by Stoker [Stoker (1950)] (see also [Babakov (1968)]), where a simple one degree of freedom oscillator was considered. The necessary condition for oscillations to appear was the existence of a decreasing slope of the friction dependence on the relative velocity, between the tape and the mass lying on it. This oscillator was reconsidered a few years ago by Hassard, Kazarinoff and Wan [Hassard *et. al.* (1981)] and in their approach the bifurcation theory of dynamical systems was used to explain this phenomenon.

The friction function against relative velocity of two sliding bodies (more exactly its decreasing part, see Fig. 6.1b) also causes oscillations in the first of two examples of two degree of freedom autonomous mechanical systems. Moreover, the possible transition from sliding to sticking state for our oscillator plays an essential role on its dynamics as the dimension of the phase space switches between 4 and 2. There are many examples of vibrations in mechanical engineering caused by friction [Awrejcewicz (1981); Awrejcewicz (1985); Awrejcewicz (1991f); Awrejcewicz *et. al.* (1998)], [Awrejcewicz and Someya (1992)].

In the first analysed model, equilibria and their behaviour with the change of some parameters will be examined in detail analytically and then numerically. The numerical approach based on Newton's method allows us to obtain the stability of equilibria. Investigated further, a trivial solution point (y_{s1}, ψ_{s1}) becomes unstable for some parameter configurations and a periodic orbit is born due to Hopf bifurcation. In our numerical technique we solve the boundary value problem by the use of shooting and Newton's method. This approach is a continuation of earlier works by Brommundt [Brommundt (1975); Brommundt (1977)], where the Urabe-Reiter [Urabe and Reiter (1966)] technique was used to obtain the global bifurcation portrait of the system analysed. For more information the reader is referenced also to Seydel [Seydel (1988)] and Kreuzer [Kreuzer (1987)].

The eigenvectors in the Hopf point allow one to find the periodic orbit in the phase space. Then a continuation technique with linear prediction is used to trace the further behaviour of the observed periodic orbit. The numerical calculations yield a Floquet matrix whose eigenvalues (characteristic multipliers) decide about stability in the Lyapunov sense. Following Arnold [Arnold (1983)] three general bifurcation cases should be considered. When one of the multipliers crosses the unit circle of the complex plane through -1 the period doubling bifurcation takes place. The previously stable orbit becomes unstable and a new periodic orbit is born which is twice the original one. The successive periodic doubling bifurcation can also lead to chaos ([Guckenheimer and Holmes (1983)], [Holmes and Moon (1983)]). Further, it will be shown, that this period doubling scenario can be found also in our system. The second possibility of bifurcation arises when a pair of complex conjugate eigenvalues crosses the unit circle with nonzero velocity. Either a ultrasubharmonic resonance appears or the new quasi-periodic solution is born. In the third case, one of the multipliers goes through the unit circle at $+1$. In this case there are three possibilities. The investigated periodic orbit becomes either equilibrium or the new ultraharmonic solution appears. It is also possible that all of multipliers (shortly before one of them reaches $+1$) are real. Then corresponding stable and unstable periodic orbits intersects each other, which leads to the form of strange chaotic attractors. This scenario will be discussed in detail.

The above mentioned approach allows us to study systematically the considered dynamical system governed by ordinary autonomous nonlinear differential equations. It is possible to obtain accurate bifurcation points and also to find the critical parameter set for which chaotic orbits appear. Then, in order to testify strange chaotic attractors and to investigate their behaviour, some standard method based on solving the initial value problems can be used (time histories, phase portraits, Poincaré maps, Fourier spectra).

In a second model we consider two masses lying on a belt moving with a constant velocity. In this case in addition to the mentioned standard methods the bifurcational diagrams as well as the Lyapunov exponents will support our considerations. Also we discuss a possibility of approximation a discontinuity introduced by friction approximation with the sign term by applying arctan instead. Many interesting discontinuous behaviours of nonlinear dynamics of coupled oscillators are reported using numerical simulation.

6.2 Disc - Flexible Arm Oscillator

6.2.1 *Equations of motion and phase flow*

The investigated system is presented in Fig. 6.1a. It is composed of a disc of mass m and radius R sitting on a tangentially uniformly moving tape

(v_B) and arm of length l to which it is coupled via elasticity and damping (k_2, c_2, rotational). One end of the arm is supporting the disc in its centre while the entire system is fixed to the environment by the arm's other end. The disc/tape system in connection with the friction characteristics is meant to model typical slip-stick situations (here rolling and sliding); the arm being elastic and damped in its longitudinal direction realizes a nonconstant pressure of the disc against the tape. Moreover it introduces (due to the geometry) a saddle point-like behaviour.

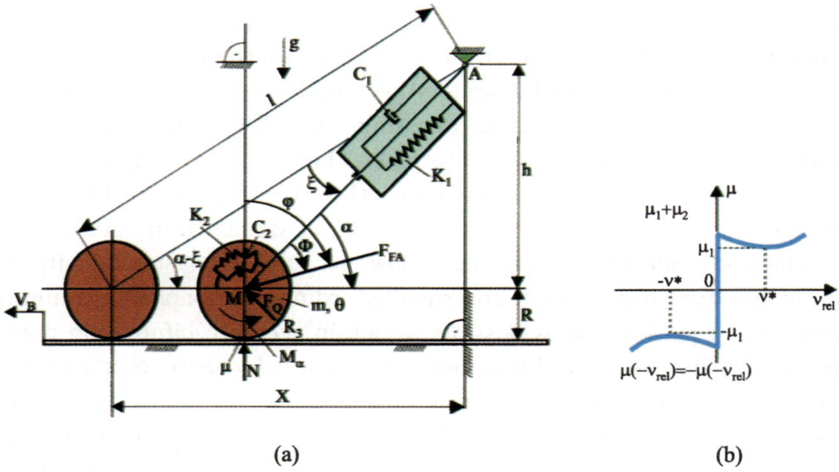

Fig. 6.1 Roll-slide oscillator with two degrees of freedom (a), and the friction coefficient against relative velocity (b). $\varphi(\Phi)$ is the absolute (relative) angle of rotation, $M_K = c_2\dot{\Phi} + K_2\Phi$.

The friction characteristics $\mu(v_{rel})$ (illustrated) was chosen to model, at least qualitatively, the realistic effects of self-excitement as well as damping depending on the relative velocity between the two surfaces in contact. Based on the notation given in Fig. 6.1 we have

$$\begin{cases} l(x) = \sqrt{h^2 + x^2}, \\ \dot{\Phi} = \dot{\varphi} - \dot{\xi}(x), \\ \xi(x) = \arcsin \dfrac{h}{l} + \alpha, \end{cases} \tag{6.1}$$

where α denotes the angle between the arm and the horizontal in relaxed state of spring K_1. In this case we define the length of the arm as $l = l_0$ (derivatives with respect to time t are marked with a dot).

The relative velocity between the disc and the tape is

$$v_{rel} = v_B - \dot{\Phi}R - \left(1 + \frac{Rh}{x^2 + h^2}\right)\dot{x}. \tag{6.2}$$

These relationships allow us to obtain the following equations of motion in the slip state

$$\mathbf{M}\begin{pmatrix}\ddot{x}\\ \ddot{\Phi}\end{pmatrix} + \mathbf{B}\begin{pmatrix}\dot{x}\\ \dot{\Phi}\end{pmatrix} + \mathbf{K}\begin{pmatrix}x\\ \Phi\end{pmatrix} = \mu\mathbf{N}\begin{pmatrix}1\\ 1\end{pmatrix}, \tag{6.3}$$

where

$$\mathbf{M} = \begin{pmatrix} m & 0 \\ \Theta\dfrac{h}{Rl^2} & \dfrac{\Theta}{R} \end{pmatrix},$$

$$\mathbf{B} = \begin{pmatrix} \dfrac{x^2}{l^2}c_1 & \dfrac{-h}{l^2}c_2 \\ -2\Theta\dfrac{hx\dot{x}}{Rl^4} & \dfrac{c_2}{R} \end{pmatrix},$$

$$\mathbf{K} = \begin{pmatrix} \left(1 - \dfrac{l_0}{l}\right)K_1 & -\dfrac{h}{l^2}K_2 \\ 0 & \dfrac{K_2}{R} \end{pmatrix},$$

$$N = mg - \left\{\left(1 - \frac{l_0}{l}\right)K_1 + \frac{i}{l}c_1 + \frac{K_2}{l^2}\Phi + \frac{c_2}{l^2}\dot{\Phi}\right\}h,$$

$$\mu = \begin{cases} \operatorname{sign}(v_{rel})\left(\mu_1 + \mu_2\left(1 - \left|\dfrac{v_{rel}}{v^*}\right|\right)^2\right), \quad v_{rel} \neq 0, \\ \qquad\qquad \operatorname{sign}\left(\dfrac{d^i v_{rel}}{dt^i}\right)(\mu_1 + \mu_2) \\ \dfrac{d^k v_{rel}}{dt^k} = 0 \text{ for } k = 0 \ldots (i-1) \text{ and } v_{rel} \not\equiv 0, \\ \text{not defined for } v_{rel} \not\equiv 0. \end{cases} \tag{6.4}$$

In the stick state we have

$$\left(\frac{\Theta}{R^2} + m\right)\ddot{x} + \left(1 - \left(\frac{h}{l}\right)^2\right)c_1\dot{x} + \left(1 - \frac{l_0}{l}\right)K_1 x =$$
$$= \left(\frac{1}{R} + \frac{h}{l^2}\right)(c_2\dot{\Phi} + K_2\Phi), \tag{6.5}$$

where:

$$\begin{cases} \dot{\Phi} = \dfrac{v_B}{R} - \left\{ \dfrac{1}{R} + \dfrac{h}{x^2 + h^2} \right\} \dot{x}, \\ \Phi = \Phi_H + \dfrac{v_B}{R}(t - t_H) - \arctan\left(\dfrac{x}{h}\right) + \arctan\left(\dfrac{x_H}{h}\right), \end{cases} \tag{6.6}$$

and we mark the initial values of t, x, Φ for the beginning of the stick with "H". By the use of transformations

$$\begin{cases} y_1(\equiv y) = \dfrac{x}{l_0}, \quad y_2 = y_1', \quad y_3(\equiv \phi) = \vartheta\Phi, \quad y_4 = y_3', \\ \tau = \sqrt{\dfrac{K_1}{m}} t \quad \text{and} \quad \dfrac{d(\ )}{d\tau} =: (\ )', \end{cases} \tag{6.7}$$

Eqs. (6.3) with fourteen parameters l_0, h, R, K_1, K_2, m, Θ, g, c_1, c_2, v^*, v_B, μ_1, μ_2, are reduced to dimensionless differential equations with the eleven parameters χ, ρ, κ, γ, β_1, β_2, ϑ, V^*, V_B, μ_1, μ_2 of the following (first order) form at the slip state:

$$\begin{aligned} y_1' &= y_2, \\ y_2' &= -\left[\left(1 - \frac{1}{\lambda}\right) + \frac{y_1 y_2}{\lambda^2}\beta_1\right](y_1 + \mu\chi) + \\ &\quad -\frac{1}{\vartheta\rho^2}(\beta_2 y_4 + \kappa y_3)\left(\frac{\mu y_1 - \chi}{\lambda^2}\right) + \mu\gamma, \\ y_3' &= y_4, \\ y_4' &= -\vartheta\frac{\chi}{\lambda^2}\left(y_2' - 2\frac{y_2^2 y_1}{\lambda^2}\right) - \frac{1}{\vartheta}\left(1 + \frac{\mu}{\rho}\frac{y_1}{\lambda^2}\right)(\beta_2 y_4 + \kappa y_3) - \\ &\qquad \mu\rho\left[\left(1 - \frac{1}{\lambda}\right)\chi + \frac{y_1 y_2}{\lambda^2}\beta_1\chi - \gamma\right], \end{aligned} \tag{6.8}$$

where

$$\begin{aligned} &\lambda = \frac{l}{l_0}, \\ &\chi = \frac{h}{l_0}, \quad \rho = \frac{l_0}{R}, \quad \kappa = \frac{1}{R^2}\frac{K_2}{K_1}, \quad \gamma = \frac{mg}{K_1 l_0}, \quad \vartheta = \frac{\Theta}{mR^2}, \\ &\beta_1 = \frac{c_1}{\sqrt{K_1 m}}, \quad \beta_2 = \frac{1}{R^2}\frac{c_2}{\sqrt{K_1 m}}, \\ &V^* = \frac{v^*}{l_0\sqrt{K_1/m}}, \quad V_B = \frac{v_B}{l_0\sqrt{K_1/m}}. \end{aligned} \tag{6.9}$$

For the stick state we have

$$
\begin{aligned}
y_1' &= y_2, \\
y_2' &= -\frac{1}{(\vartheta+1)}\left[\left(1-\left(\frac{\chi}{\lambda}\right)^2\right)\beta_1 y_2 + \left(1-\frac{1}{\lambda}\right) y_1 \right. \\
&\qquad \left. -\frac{1}{\rho\vartheta}\left(1+\frac{\chi}{\rho}\frac{1}{\lambda^2}\right)(\beta_2 y_4 + \kappa y_3)\right], \\
y_5' &= 1,
\end{aligned}
\tag{6.10}
$$

where

$$
\begin{aligned}
y_3 &= y_{3H} + \vartheta\rho\left\{V_B(\tau-\tau_H) - (y_1 - y_{1H}) + \right. \\
&\qquad \left. -\frac{1}{\rho}\arctan\left(\frac{y_1}{\chi}\right) + \frac{1}{\rho}\arctan\left(\frac{y_{1H}}{\chi}\right)\right\}, \\
y_4 &= \vartheta\rho\left\{V_B - \left(1+(\frac{\chi}{\rho}\frac{1}{\lambda^2}\right) y_2\right\}, \\
y_5 &= \tau.
\end{aligned}
$$

The trace of the Jacobian of (6.8) (divergence of the vector field, see for instance Guckenheimer and Holmes [Guckenheimer and Holmes (1983)]) for the slide state is:

$$
\left\{
\begin{aligned}
& T_{sl} = \left[-\frac{y_1^2}{\lambda^2}\beta_1 - \left(\frac{1}{\vartheta}\left(\frac{\chi}{\lambda}\right)^2\frac{1}{\lambda^4}\right)\beta_2\right]_1 + \left[2\frac{\mu_2}{V^*}(1+ \right. \\
& \left. -\left|\frac{V_{rel}}{V^*}\right|\right)\frac{\vartheta+1}{\vartheta}N^*\Bigg]_2 + \left[-\mu\frac{y_1}{\lambda^2}\left(\chi\beta_1 + \left(\frac{1}{\rho\chi} - \frac{\chi}{\rho^2\lambda^2}\right)\beta_2\right)\right]_3, \\
& N^* = \gamma - \left[\left(1-\frac{1}{\lambda}\right) + \frac{y_1 y_2}{\lambda^2}\beta_1\right]\chi - \frac{1}{\vartheta\rho^2}\frac{y_1}{\lambda^2}(\beta_2 y_4 + \kappa y_3), \\
& V_{rel} = V_B - \frac{1}{\rho\vartheta}y_4 - \left(1+\frac{\chi}{\rho}\frac{1}{\lambda^2}\right) y_2.
\end{aligned}
\right.
\tag{6.11}
$$

The expression in the first brackets of (6.11) is always negative. The term in the second brackets of (6.11) is positive if $|V_{rel}/V^*| < 1$ (since $N^* \geq 0$) (see Fig. 6.1b). The last expression shows that, for a constant friction coefficient, self-excited oscillations are also possible if $y_1 < 0$ and damping β_1 is increased.

Generally there exist parameters for which the investigated dynamic system can be locally conservative, dissipative or with positive divergence of the phase flow.

For the stick state the divergence of the vector field is always negative,

$$T_{sl} = -\frac{1}{\vartheta+1}\left[\left(\frac{y_1}{\lambda}\right)^2\beta_1 + \left(1+\frac{\chi}{\rho}\frac{1}{\lambda^2}\right)^2\beta_2\right], \tag{6.12}$$

which is important, since this shows that no self-excited oscillations are possible in pure stick state.

6.2.2 *Trivial solutions - analytical investigations*

Dropping the transient terms, from (6.8) with $y = y_1$ and $\psi = y_3$ we have

$$\begin{aligned} &\left(1-\frac{1}{\lambda_s}\right)(y_s+\mu_s\chi) + \frac{\kappa}{\vartheta\rho^2}\left(\frac{\mu_s y_s - \chi}{\lambda_s^2}\right)\psi_s - \mu_s\gamma = 0,\\ &\frac{\kappa}{\vartheta}\left(1+\frac{\mu_s}{\rho}\frac{y_s}{\lambda_s^2}\right)\psi_s + \mu_s\left[\left(1-\frac{1}{\lambda_s}\right)\chi - \gamma\right] = 0. \end{aligned} \tag{6.13}$$

For $\rho\lambda_s^2 + \chi \neq 0$ we obtain an uncoupled version

$$\begin{aligned} &\frac{\kappa}{\vartheta}\frac{\rho\lambda_s^2+\chi}{\rho\lambda_s^2}\psi_s - \left(1-\frac{1}{\lambda_s}\right)\rho y_s = 0,\\ &\frac{\rho\lambda_s^2+\mu_s y_s}{\rho\lambda_s^2+\chi}\left(1-\frac{1}{\lambda_s}\right)y_s + \mu_s\left[\left(1-\frac{1}{\lambda_s}\right)\chi - \gamma\right] = 0. \end{aligned} \tag{6.14}$$

In order to get a general overview about the possible solutions only the physically meaningful solution domains are considered. Since $N^* > 0$ and we can choose (without loss of generality) $V_B > 0$, $(\mu_s(V_B) > 0)$ we find that $\psi_s = (\vartheta/\chi)\rho\mu_s N^*$ is always positive. Furthermore we get

$$\pm\psi_s^* = \pm\frac{\kappa}{\vartheta}\frac{\psi_s}{\lambda_s^2\rho^2} \leq \pm\left[\gamma - \left(\frac{1-1}{\lambda_s}\right)\chi\right]\frac{1}{y_s}, \text{ for } \operatorname{sign}(y_s) = \pm 1. \tag{6.15}$$

On the other hand the first Eq. of (6.14) implies

$$\pm\frac{\kappa}{\vartheta}\frac{\psi_s}{\lambda_s^2\rho^2} = \pm\frac{(1-(1/\lambda_s))}{\rho\lambda_s^2+\chi}y_s. \tag{6.16}$$

From (6.15) and (6.16) we obtain

$$\gamma \geq (1+\rho\chi)\left(1-\frac{1}{\lambda_s}\right)\frac{\lambda_s^2}{\rho\lambda_s^2+\chi} = f^*(y_s, \text{geometry}), \tag{6.17}$$

where

$$\lim_{y\to\pm\infty} f^* = \chi + \frac{1}{\rho}. \tag{6.18}$$

Hence, the physically meaningful solution domains are bounded by the conditions

$$\psi_s^* \geq 0 \text{ and } \gamma \geq f^*.$$

6.2.3 *Discussion of the analytical results*

Results obtained in the previous section allow us to discuss in detail all of the eight cases presented in Fig. 6.2, including their physical interpretation. The eight principally different solution cases classified by the geometry (χ, ρ) are characterized by the different amount of possible steady states as well as the location of the solution domains. Cases a_1, b_1, b_2 and b_3 show an interesting phenomenon. It is remarkable that here solutions exist even for arbitrary low γ-values since it comes to a self-locking situation where the disc is additionally pressed against the band. Due to this phenomenon, the angle of rotation ψ_s at this point can get arbitrarily large while y_s remains bounded. The case d will be further analysed.

Here, for large γ two solution domains can be spotted: $\sqrt{1-\chi^2} < y_s < 0$ and $y_s > \sqrt{1-\chi^2}$. With the decrease of γ below the value $\chi + (1/\rho)$, the domain of positive solutions becomes bounded, until it entirely disappears at $\gamma = 0$. Only negative solutions remain possible for reversed gravity $(\gamma < 0)$ due to self-locking effects. However, getting below a critical value of γ all real solutions vanish (the disc drops from the band).

The second of (6.14) is an implicit equation in y_s so that y_s can only be determined iteratively. However, in order to investigate modes of motion of the system it is much more interesting to know *how many* steady states (solutions) there can be, and to consider under what parameter conditions solutions come into being or vanish. For this reason the second of (6.14) should be considered from a different point of view. Given a certain type of geometry (χ, ρ) we can ask ourselves, for example, which parameters (μ_s, γ) will lead to a steady state solution at a given position y_s. This sort of consideration has the advantage that we are able to solve explicitly for the parameter functions $\mu_s(y_s, \gamma)$ or $\gamma(y_s, \mu_s)$. Dropping subscript "s" we actually obtain:

$$\mu = y\left(\frac{\rho}{1+\chi\rho}\right)\frac{f^*}{\gamma \;-f^*}, \tag{6.19}$$

$$\gamma = \left[1 + \frac{\rho\gamma}{\mu(1+\chi\rho)}\right] f^*. \tag{6.20}$$

We can interpret the meaning of function f^* by inspection of (6.20). We find,

$$f^* = \lim_{\mu \to \infty} \gamma(y, \mu). \tag{6.21}$$

Local extremes of the parameter functions are important since the solution curves do not reach points lying opposite to these extremes. Local ex-

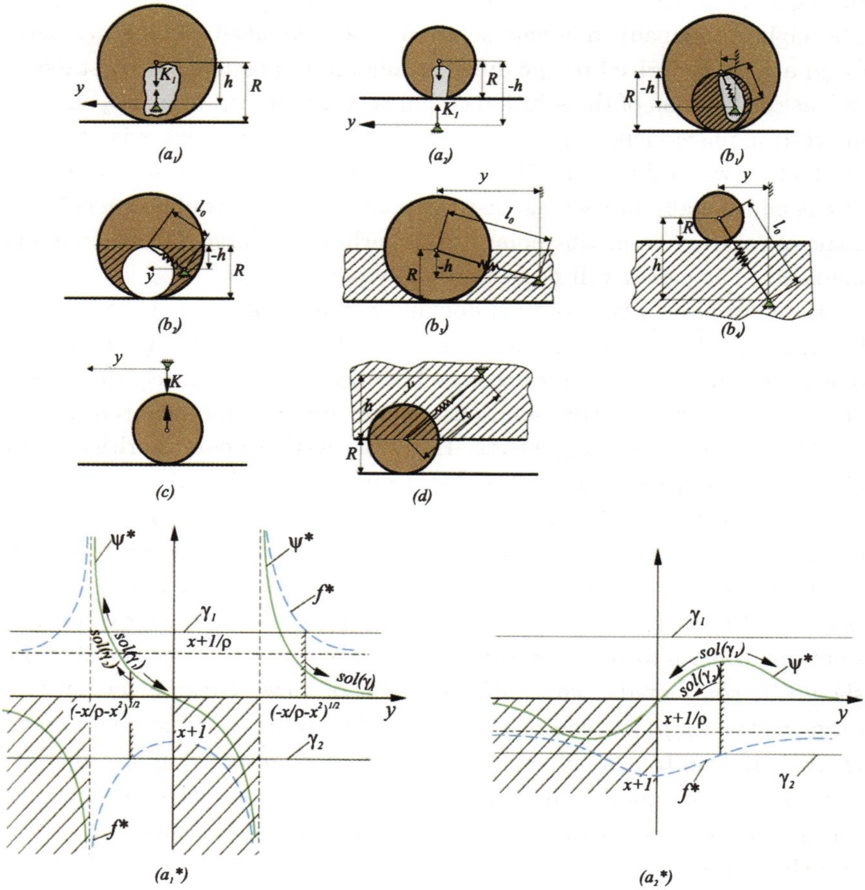

Fig. 6.2 The scheme of the static configuration of the considered system and the corresponding solutions (marked with asterisks): (a_1) $-\chi > 1$, $-\chi < 1/\rho$; (a_2) $-\chi > 1$, $-\chi > 1/\rho$; (b_1) $0 < -\chi < 1$, $-\chi > \rho$; (b_2) $0 < -\chi < 1$, $-\chi < \rho$, $\rho < 1$; (b_3) $0 < -\chi < 1$, $-\chi < 1/\rho$, $\rho > 1$; (b_4) $0 < -\chi < 1$, $-\chi > 1/\rho$; (c)$\chi > 1$; (d)$0 < \chi < 1$. Directions of arrows indicate the possible solutions sol(γ_i).

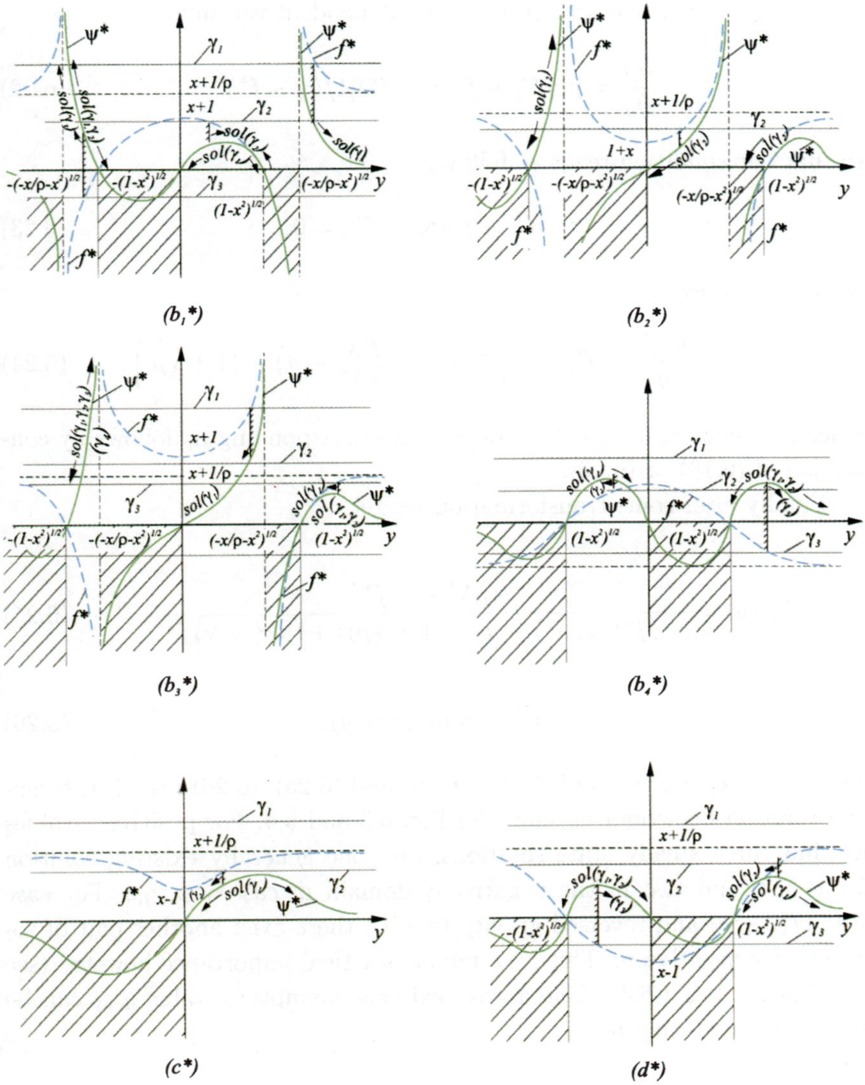

Fig. 6.2: Continued.

tremes of the parameter functions represent the locations where solutions are born or disappear. The corresponding parameters are called "critical parameters". We look for these extremes using the necessary condition of a vanishing derivative with respect to independent variables:

$$\frac{\partial \mu}{\partial y} = [\gamma y f^*_{,y} + f^*(\gamma - f^*)]1/(\gamma - f^*)^2, \tag{6.22}$$

so that our condition reads as follows

$$\gamma_c y f^*_{,y} + f^*(\gamma_c - f^*) = 0, \tag{6.23}$$

with the abbreviations

$$(\ )_{,y} := \frac{\partial(\ )}{\partial y}, \quad f^*_{,y} = \frac{y}{\rho\lambda^2 + \chi}\left[f^*\left(\frac{\chi}{\lambda^2} - \rho\right) + (1 + \chi\rho)\right]. \tag{6.24}$$

Since the relationship $\mu(\gamma)$ is unique, the corresponding μ_c follows by conversion of (6.19) or (6.20).

Finally after some transformation we get

$$\gamma_c(y) = \frac{(\rho\lambda^2 + \chi)f^{*2}}{y^2(f^*((\chi/\lambda^2) - \rho) + (1 + \chi\rho)) + (\rho\lambda^2 + \chi)f^*}, \tag{6.25}$$

$$\mu_c(y) = \mu(\gamma_c(y), y). \tag{6.26}$$

An analytical discussion of (6.19), (6.20) and (6.25), (6.26) reveals the general solution behaviour as shown in Fig. 6.3 and 6.4. For positive γ-values we find three steady state solutions, i.e. one generally existing solution for $y > 0$ and two in the negative γ domain in case $\gamma > \gamma_c$. For case $\chi < 1/(2 + \rho)$ and reversed gravity ($\gamma < 0$) there exist another pair of solutions for negative γ. This is of minor practical importance however (see also Figs. 6.3 and 6.4). The always existing asymptotic value y_{cr1} can be calculated according to

$$y_{1cr} : \rho\lambda^5_{1cr} + 3\chi\lambda^3_{1cr} - (2\chi + \chi^2\rho)\lambda^2_{1cr} - 2\chi^3\lambda_{1cr} + \chi^3 = 0 (\lambda_{1cr} = \sqrt{y^2_{1cr} + \chi^2}).$$

Fig. 6.5 allows for geometric interpretation of the configuration of the roll-slide oscillator for three different types of equilibria. The change in location of points for increasing friction is indicated by arrows below the characters. y_{s2} and y_{s3} move toward each other, eventually becoming identical at $y = y_{cr}(\gamma_c, \mu_c)$ and vanish.

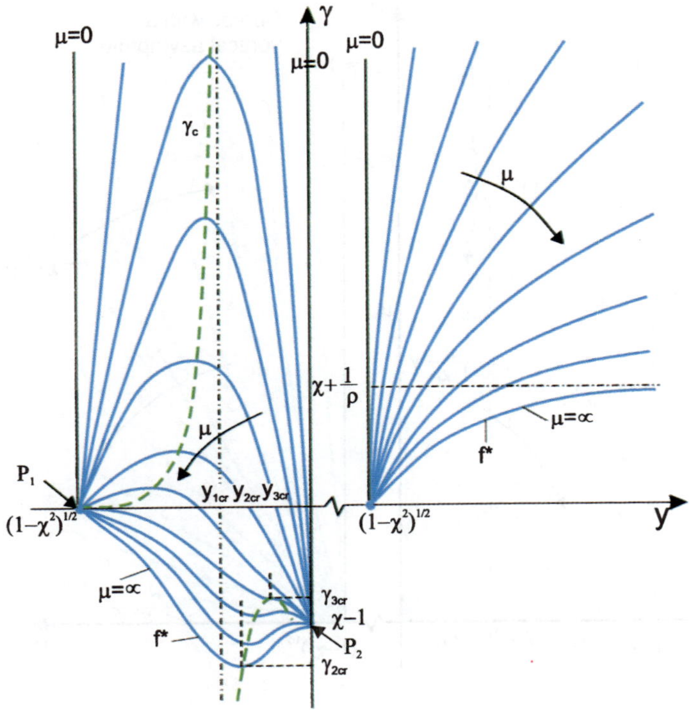

Fig. 6.3 Constant solutions against γ (local extrema for $\gamma < 0$ are possible only for $\chi < 1/(\rho+2)$).

6.2.4 *Stability of equilibria. Numerical investigations*

Thanks to the analytical approach the number and positions of equilibria have been obtained. Now we locally pertubate these solutions in order to investigate their stability in the Lyapunov sense. This leads us to solve the eigenvalue problem

$$(J_{ij} - \nu_k \delta_{ij}) u_j^k = 0, \quad i, j, k = 1, \ldots, 4, \tag{6.27}$$

where the Jacobi matrix includes the investigated solutions $(y_s, \psi_s(y_s))$ and

$$J_{ij} = \left. \frac{\partial y_i'}{\partial y_j} \right|_s =: \left(y_{i,j}' \big|_s \right). \tag{6.28}$$

Our attention will now be focused on the critical cases, when new equlibria are born or disappear due to the change of freely chosen parameters.

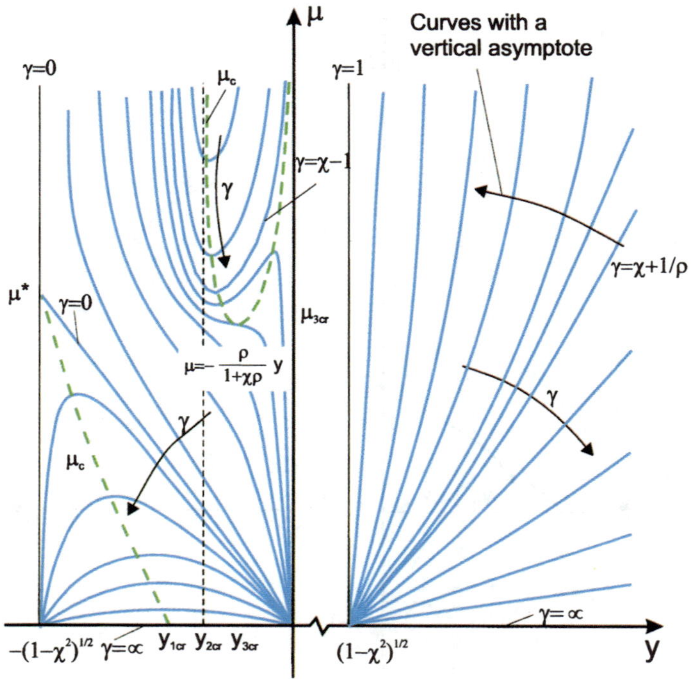

Fig. 6.4 Constant solutions against μ, where $\mu^* = (1-\chi^2)^{1/2}\rho/(1+\chi\rho)$. $\gamma = \chi - 1$ is the seperatrice between the amount of solutions (two or three) in $-(1-\chi^2)^(1/2) \leq y \leq 0$. Local extrema are possible for $\gamma < 0$ only if $\chi < 1/(\rho+2)$.

Consider the case for the following fixed parameters: $\chi = 0.6, \rho = 2.5, V_B = 0.5, V^* = 1.0, \kappa = 1.0, \beta_1 = \beta_2 = 0.05, \mu_1 = 0.05, \mu_2 = 0.2, \vartheta = 0.5, \gamma = (0.0-3.0)$ (Fig. 6.6). Because $\chi > 1/(2+\rho)$ it is possible to have either three or one equilibria. For the positive equilibrium y_{s1} (see Fig. 6.6) we have two pairs of complex conjugate eigenvalues. For this reason the equilibrium will be called a "sink-sink" type. For $\gamma = 0.12$ this equilibrium is unstable for the first time (the lower pair).

The real part of the second pair of eigenvalues becomes positive at $\gamma = 0.41$ (Fig. 6.7a). Increasing γ causes the lower pair of complex eigenvalues to wander in the positive direction of the real axis, whereas the other pair goes firstly in the same direction, and after $\gamma = 1.6$ is reached, it turns back. The map in Fig. 6.7a shows also that the upper pair of eigenvalues possesses a limit. Consider now the two other equilibria (Fig. 6.7b and 6.7c).

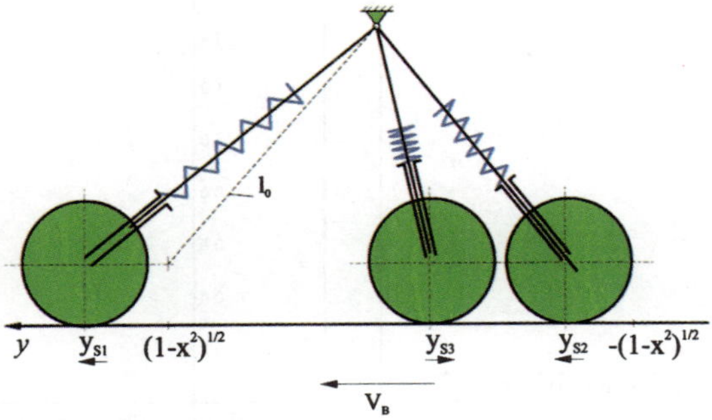

Fig. 6.5 Geometric interpretation of the constant solutions.

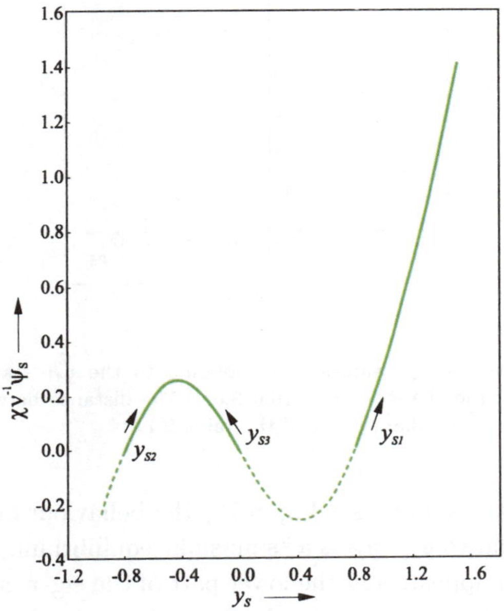

Fig. 6.6 Angle against displacement for $0 < \chi < 1$; $\rho > 0$; $\chi = 0.6$; $\rho = 2.5$ (see first Eq. 6.14).

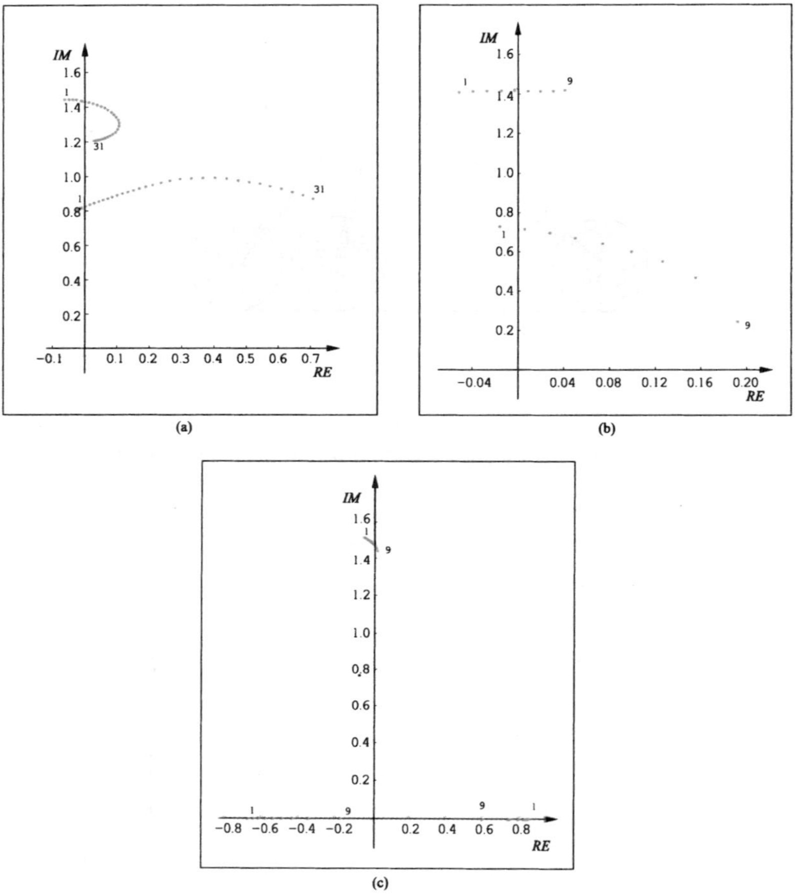

Fig. 6.7 The travel of eigenvalues corresponding to the y_{s1} stationary point from Fig. 6.6, caused by the change of $\gamma \in (0.0, 3.0)$. The distance between two successive points corresponds to the change in γ of the value 0.1.

For very small values of γ (see Fig. 6.7b) the behaviour of the equilibrium point y_{s2} is similar to y_{s1}. y_{s2} is a "sink-sink" equilibrium. For $\gamma \approx 0.07$ the Hopf bifurcation appears and the lower part of the eigenvalues moves more and more in the direction of real axis. The other pair of eigenvalues crosses the imaginary axis at $\gamma \approx 0.4$. The equilibrium y_{s3} is of the type "sink-saddle" (Fig. 6.7c). When γ increases, equilibria y_{s2} and y_{s3} approach each other and then for $\gamma_c = 0.81817$ both of them disappear. This situation is in principle clearly shown by the use of the projection $y'(y)$ (Fig. 6.8).

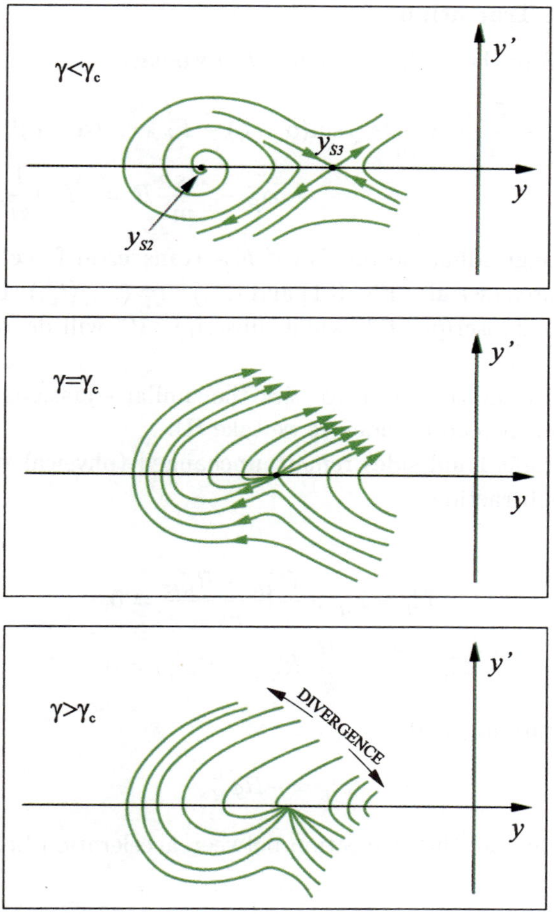

Fig. 6.8 The qualitative illustration of the changes of equilibria y_{s2} and y_{s3} "sink-sink" and "sink-saddle" type respectively with the increase of γ. Note, that even for $\gamma > \gamma_c$ the divergence of the phase flow remains extremely large.

6.2.5 *The integration of the equations of motion*

Dynamic friction phenomena are characterized by (quasi-) non-smoothness (see the $\mu(v_{rel})$ characteristic). Since numerical integrations across such points of non-smoothness fail, an additional physical statement has to be introduced to overcome this difficulty (i.e transition of those points). Here we furnish transition conditions without having to manipulate the friction characteristic in order to enable numeric calculation.

(i) Slip-Stick Transition

Shortly before the stick state (time t_{tr}^-) we have:

$$\ddot{x}_R^- - \frac{R_{SR}}{m} = \frac{1}{m}[F_Q \sin(\alpha - \zeta) - F_{FA}\cos(\alpha - \zeta)], \qquad (6.29)$$

$$\ddot{\phi}_R^- - \frac{R_{SR}}{m} R = -F_Q l \frac{1}{\Theta},$$

where F_{FA}- longitudinal spring force, F_Q- transversal force, R_S- stick or slide friction force (see also Fig. 6.1) and $(\ldots)^{\pm} := (\ldots)(t_{tr}^{\pm})$. For stick state we will use the subscript "H", while subscript "R" will denote slide state throughout.

Shortly after the transition to stick the similar equations as (6.29) are valid (only now instead of index R we take H).

Since the right hand sides remain unchanged (physical statement) we obtain upon subtraction

$$\ddot{x}_H^+ - \ddot{x}_R^- + \frac{R_{SR}^- - R_{SH}^+}{m} = 0, \qquad (6.30)$$

$$\ddot{\phi}_H^+ - \ddot{\phi}_R^- + \frac{R}{\Theta}(R_{SR}^- - R_{SH}^+) = 0.$$

At the stick state $v_{rel} \equiv 0$,

$$\ddot{x}_H^+ = -R\ddot{\phi}_H^+, \qquad (6.31)$$

and we can calculate the connection between acceleration before and after the transition

$$\ddot{x}_H^+ = \frac{1}{1+\vartheta}\ddot{x}_R^- - \frac{\vartheta}{1+\vartheta}R\ddot{\phi}_R^-, \quad \vartheta = \frac{\Theta}{mR^2}. \qquad (6.32)$$

The acceleration jumps during the slip-stick transition. The sudden change of velocity is in this case expected.

(ii) Slip-Slip Transition

We use the same physical statement as in the slip-stick transition and we have

$$\ddot{x}_R^+ = \frac{1}{m}(R_{SR}^+ - R_{SR}^-) + \ddot{x}_R^-,$$

$$\ddot{\phi}_R^+ = \frac{R}{\Theta}(R_{SR}^+ - R_{SR}^-) + \ddot{\phi}_R^-, \qquad (6.33)$$

where:

$$R_{SR}^{\pm} = \pm(\mu_1 + \mu_2) N \text{sign}(R_{SH}^{f}). \tag{6.34}$$

R_{SH}^{f} denotes a fictive resultant stick force, which we would obtain if we assumed a real slip-stick transition to be possible at this point (R_{SH}^{f} can easily be determined from (6.31) with $R_{SH}^{f} = R_{SH}^{+}$ which will obviously exceed the possible stick force limit $|(\mu_1 + \mu_2)N|$). The relative acceleration is

$$\dot{v}_{rel} = -R\ddot{\phi}_R - \ddot{x}_R. \tag{6.35}$$

Taking into account (6.35) and (6.33) one obtains

$$\dot{v}_{rel}^{+} = \dot{v}_{rel}^{-} - 2(\mu_1 + \mu_2)\left(\frac{R^2}{\Theta} + \frac{1}{m}\right) \text{sign}(R_{SH}^{f})N. \tag{6.36}$$

Eq. (6.36) shows that when the system "shoots" through the stick area, the acceleration jumps.

(iii) Stick-Slip Transition

The integration of the equations of motion can be problematic in this case. In this state $v_{rel} \equiv 0$ or $v_{rel} = 0$ and either a further stick or slip is possible. If we assume the transition to the slip state, the friction coefficient μ has to pass continuously through one of the spikes of the function $\mu(V_{rel})$. So in this case the acceleration changes continuously.

From (6.31) we obtain

$$\begin{aligned} \Theta(\dddot{\phi}_R^{+} - \dddot{\phi}_H^{-}) - R(\dot{R}_{SR}^{+} - \dot{R}_{SH}^{-}) &= 0, \\ m(\dddot{x}_R^{+} - \dddot{x}_H^{-} - (\dot{R}_{SR}^{+} - \dot{R}_{SH}^{-}) &= 0. \end{aligned} \tag{6.37}$$

For t_{tr}^{-}, v_{rel} it is still identically equal to zero, and therefore

$$-R\dddot{\phi}_H^{-} - \dddot{x}_H^{-} = 0. \tag{6.38}$$

But for t_{tr}^{+} we have

$$\ddot{v}_{rel}^{+} = \left(\frac{R^2}{\Theta} + \frac{1}{m}\right)(\dot{R}_{SH}^{-} - \dot{R}_{SR}^{+}), \tag{6.39}$$

where:

$$\dot{R}_{SH} = \left(\frac{\vartheta}{\vartheta+1}\right)\left[F_F A\frac{x}{l} + F_Q\left(-\frac{h}{l} + \frac{l}{R}\frac{1}{\vartheta}\right)\right], \qquad (6.40)$$
$$\dot{R}_{SR} = [\mu N].$$

The stick-slip transition accompanies the jump change of the derivative of the acceleration. This case can also be looked at as the jump passage from $v_{rel} \equiv 0$ or $v_{rel} = 0$. One can expect a C^1-smooth change of velocities for this case. All of the three possible transitions are presented in Fig. 6.9 as a calculated example.

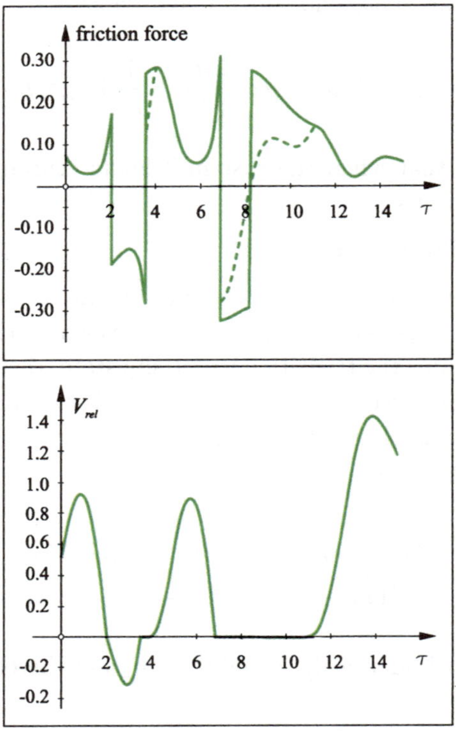

Fig. 6.9 Comparison among the maximum transferable stick friction force (H——) and the resulting friction force (R- - -). Both curves are the same for the slip situation. V_{rel} against time is presented below. The parameters and initial conditions are: $\chi = 0.6$, $\rho = 2.5$, $V_B = 0.5$, $V^* = 1.0$, $\kappa = 1.0$, $\beta_1 = \beta_2 = 0.05$, $\mu_1 = 0.05$, $\mu_2 = 0.2$, $\vartheta = 0.5$, $\gamma = 0.82$, $y_0 = -1.5$, $y_0' = 0.0$, $\psi_0 = 1.0$, $\psi_0' = 0.0$.

(iv) Numerical Integrations

The numerical integration of our system divides into two alternating phases: The integration of either slip or stick state and the transition phases. The first was carried out by a (implicit) Backward-Differentiation Method which is less accurate but provides good numerical stability, which is much more important in our case. The proper determination of the transition however is a rather delicate problem. It is difficult to determine the type of transition in those cases where the slipping system "jumps" onto the spikes of the $\mu(v_{rel})$ - characteristic or generally where the stick-slip transition takes place as mentioned earlier. Therefore we have to find the tendency of our test-functions (distance to stick force limit in the stick state or v_{rel} in the slip state) as accurately as possible. A multistep BD-Method would obviously be incorrect, so a simple Euler-Method for a very small period of time is used. We get considerably less round off error (fewer operations) without any difficulties in convergence.

There is however a lower limit for the integration step Δt in the Euler-Method due to the calculation precision ε of a computer. If we take the value of v_{rel} as a criterion for a decision, its magnitude must not be smaller than ε. To determine Δt for the stick-slip transition, we represent $v_{rel}(t \geq_{tr}^{+}$) by its simplified Taylor expansion

$$v_{rel} \approx v_{rel}(t_{tr}^{+}) + \left.\frac{dv_{rel}}{dt}\right|_{t_{tr}^{+}} \Delta t + \frac{1}{2}\left.\frac{d^2 v_{rel}}{dt^2}\right|_{t_{tr}^{+}} \Delta t^2. \tag{6.41}$$

Knowing that the first two terms on the right hand side vanish, we get

$$|v_{rel}| \approx \left|\frac{1}{2}\left.\frac{d^2 v_{rel}}{dt^2}\right|_{t_{tr}^{+}} \Delta t^2\right| \geq \varepsilon, \tag{6.42}$$

or

$$\Delta t \geq \sqrt{2\varepsilon / \left|\ddot{v}_{rel}\right|_{t_{tr}^{+}}}, \tag{6.43}$$

where $\ddot{v}_{rel}|_{t_t r^+}$ is obtainable from (6.39). This estimation cannot be done for $d^2 v_{rel}/dt^2\big|_{t_{tr}^{+}} = 0$. This shows the general problem of the transitions. Theoretically situations where the first n derivatives are all equal to zero are possible. If n is high enough, a further integration does not make sense due to round off errors. New physical statements would then have to be introduced, possibly by using a new dynamical model. Practical integrations show however that this is usually not necessary.

6.2.6 *Calculations of periodic orbits and their stability*

We take an approximate fixed point $y_F^{(k)}$ near the unknown "true" one and a numerical integration over an estimated period $T^{(k)}$ is carried out. Hence a point mapping $G(y_F^{(k)}) = G_F^{(k)}$ is defined. The error $E = y_F^{(k)} - G_F^{(k)}$ shows the accuracy of the estimation (k). Thanks to a Newton-Raphson procedure we can look for zeros in the error function E.

The problem of an examination of stability is reduced to the analysis of linear differential equations with periodic coefficients

$$\Delta p' = J(\tau)\Delta p = \left.\frac{\partial y'}{\partial y}\right|_{y_p(\tau)} \Delta p, \tag{6.44}$$

where $J(\tau + T) = J(\tau)$ and Δp is a perturbation vector of the periodic solution $y_p(\tau)$. The general solution of 6.44 is

$$\Delta p = \Phi(\tau)\Delta p(0), \tag{6.45}$$

where $\Phi(\tau) = \Phi(\tau + T)$ is the fundamental matrix.

The characteristic equation

$$\chi(\nu) = \det H_F - \nu 1) = 0, \quad H_F = \Phi(T) = \frac{\partial G_F}{\partial y_F}, \tag{6.46}$$

yields the characteristic multipliers where H_F is already known from the above mentioned iterations as a Jacobi matrix of the point mapping at the fixed point (for details see for instance [Jakubovich and Starzinskii (1972)]).

If we want to find an analogy between the stability criteria for the trivial solutions and the periodic orbits we can express the earlier in the form of point mappings:

$$\nu_k = e^{\omega_k T} = e^{\omega_k (2\pi/\omega)},$$

where ω denotes the angular frequency of the considered periodic solution, or vice versa:

$$\omega_k = \frac{\omega}{2\pi}\left[\ln|\nu_k| + i \arctan\left(\frac{\nu_{ki}}{\nu_{kr}}\right)\right], \quad \nu_k = \nu_{kr} + i\nu_{ki}.$$

We notice that we get real ω_k for $\nu_{ki} = 0$, independently of the sign of ν_{kr}. Through the real part of ω_k we can only learn about the rate of growth of the magnitude of the eigenvectors of the fundamental matrix H_F but not about any change in their orientation (in the case $\nu < 0$). In the case the corresponding eigenvectors moves on the Möbius band along the limit cycle ([Thompson and Stewart (1986)]). The crossing of the unit circle at +1 usually results in a complicated phase flow structure.

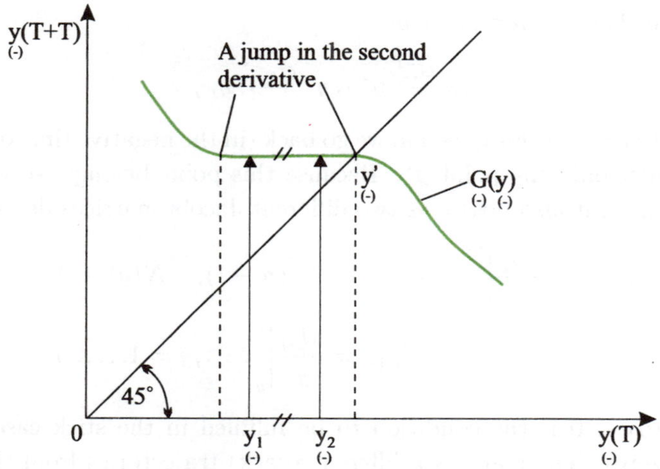

Fig. 6.10 A one dimensional point mapping, similar to a point mapping which is defined by a periodic orbit with slip-stick transition.

Our oscillator possesses another interesting property when slip transition is observed. More exactly, in this case we always get a zero eigenvalue. The question arises: why one of the characteristic multipliers must be equal to zero during the slip-stick transition. The point mapping shown in Fig. 6.10 serves as an illustration of this for the very simple one dimensional case. Suppose that the fixed point y^* is the limit point of the slip-stick transition and that it is surrounded by a very small space $\Delta U(y^*)$ with $\Delta U_{(k)} = \Delta U_{(k)}(y^*, \Delta U_i)$, $i, k = 1, ..., n, i \neq k$ (k-fixed) where the stick is possible and that another fixed point y_F of the limit cycle lies near y^* but just before the transition. Then by numerical integration over the period of the limit cycle (in the positive time direction) we have a map defined as

$$y_F = G(y_f). \tag{6.47}$$

This map can be regarded as consisting of two parts

$$G(y_f) = G_R(y^*) = G_R G_{tr}(y_F)), \tag{6.48}$$

where

$$y_F \overset{G_{tr}}{\longmapsto} y^* \wedge y^* \overset{G_R}{\longmapsto} y_F. \tag{6.49}$$

G_{tr} denotes the passage to the stick, whereas G_R represents the rest of the limit cycle. Is the inverted chain

$$y_F \overset{G_{tr}^{-1}}{\longmapsto} y^* \wedge y^* \overset{G_R{}^{-1}}{\longmapsto} y_F, \tag{6.50}$$

unique? To explain this question we go back (in the negative time direction) and consider only the point y^*. Because this point belongs to both stick and slip cases, it also possesses two different Jacobi matrices defined as:

$$J_H|_* = \left.\frac{\partial y_k'}{y_l}\right|_y^*, \quad k,l = 1,\ldots,(n-1), \quad N(y) \equiv 0,$$

$$J_R|_* := \left.\frac{\partial y_i'}{y_j}\right|_y^*, \quad i,j = 1,\ldots,n, \tag{6.51}$$

where $N(y) \equiv 0$ is the condition to be fulfilled in the stick case (v_{rel} is equal to zero). There are two different (back) trajectories from this point and also two possible return maps. That is G_{tr}^{-1} and so G^{-1} is not unique. Moreover G cannot contain any jumps. So its Jacobian H_F is also not uniquely invertible. This means, that (at least) one of the eigenvalues is equal to zero.

6.2.7 *Evolution of periodic orbits*

Our numerical calculations were made for the following parameters: $\chi = 0.6$, $\rho = 2.5$, $V_B = 0.5$, $V* = 1.0$, $\kappa = 1.0$, $\beta_1 = \beta_2 = 0.05$, $\mu_1 = 0.05$, $\mu_2 = 0.2$, $\vartheta = 0.5$, $\gamma = (0.0 - 3.0)$ and the investigations were limited to the change of the parameter γ. The main reason for choosing γ is that the change of this parameter changes not only the mass of the disc and the inertia of the system but also the friction force. The increase of γ accompanies the increase of γ_{s1}. The corresponding angle $\psi_{s1}(y_{s1})$ increases and potential energy is brought into the system. The considered interval of γ is (0.18-3.0) and, as will be shown further, there exist six different types of periodic motion within this interval. In order to clarify the evolution of orbits schematically Fig. 6.11 shows the sequence of observed orbits represented by their y'/y projections in dependence of the bifurcation parameter γ.

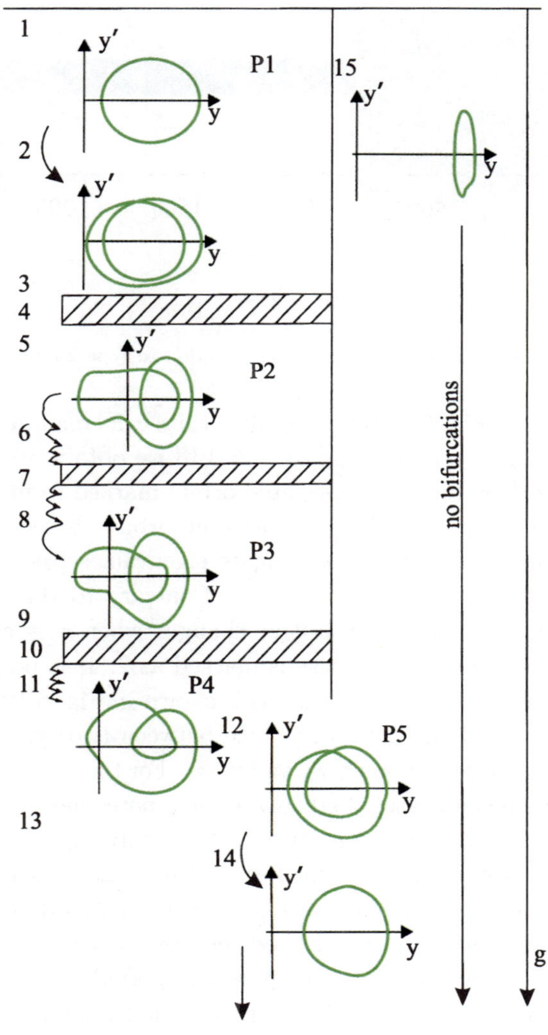

Fig. 6.11 Evolution of periodic orbits in dependence of the bifurcation parameter γ. (1) sudden birth of P1 (because of condition $N > 0$) from stationary point (y_{s1}, ψ_{s1}) to type 1-motion; (2) one period doubling; (3) saddle point; (4) chaos; (5) saddle point; (6) successive period doublings; (7) chaos; (8) successive period halvings; (9) Hopf point; (10) chaos; (11) successive period halvings; (12) saddle point; (13) saddle point; (14) one period doubling; (15) Hopf-bifurcation from stationary point (y_{s1}, ψ_{s1}) to type 2-motion.

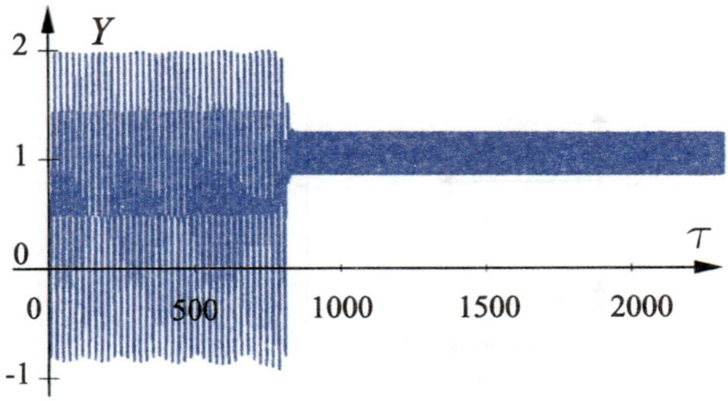

Fig. 6.12 A sudden transition from the type 1-motion to type 2-motion for $\gamma = 1.815$.

It was shown in Fig. 6.7a that periodic orbits can emanate from a trivial solution point (y_{s1}, ψ_{s1}). Starting at $\gamma = 0.18$ we obtain orbit "P1". This orbit, and the other further investigated orbits marked from P2 to P5, are referred as type-1 motion. Another periodic orbit marked as P6 is born when the other pair of complex conjugate eigenvalues (see also Fig. 6.7a) cross the imaginary axis for $\gamma = 0.415$. Contrary to the previous case, this orbit undergoes only quantitative changes when γ is changed. The qualitative motion type in this case is marked as 2. It is possible to jump from one type of motion to the other as is shown in Fig. 6.12. As is shown in Fig. 6.13 and 6.14 the main difference between two types of motion is clearly visible in the projection γ'/γ and ψ'/ψ. For the first case $E_{kt}/E_{k\omega}$ is greater than the second case(where E_{kt} is the kinetic energy of translation, and $E_{k\omega}$ the kinetic energy of rotation). Generally, the motion in case 1 has two main qualitatively different forms. The first situation takes place when the disc has no possibility of reaching negative values of y and the second when the kinetic energy is high enough to let the disc cross the point $y = 0$. Periodic orbit P1 undergoes one period doubling bifurcation for $\gamma = 0.60964$ (see Fig. 6.13a). Further period doubling is impossible, because for $\gamma = 0.6695$ the observed limit cycle disappears at a saddle point for $\gamma = 0.836071$ and in this case the mass passes through point $y = 0$. The sink-saddle point (y_{s3}, ψ_{s3}) has vanished for $\gamma = 0.81817$, but the phase flow retains a very large divergence close to this spot. We should emphasize the similar structure of the behaviour of fixed points (now periodic orbits), when it is compared with the behaviour of the equilibria y_{s2} and y_{s3} with the change of γ (see Fig. 6.6 and 6.8).

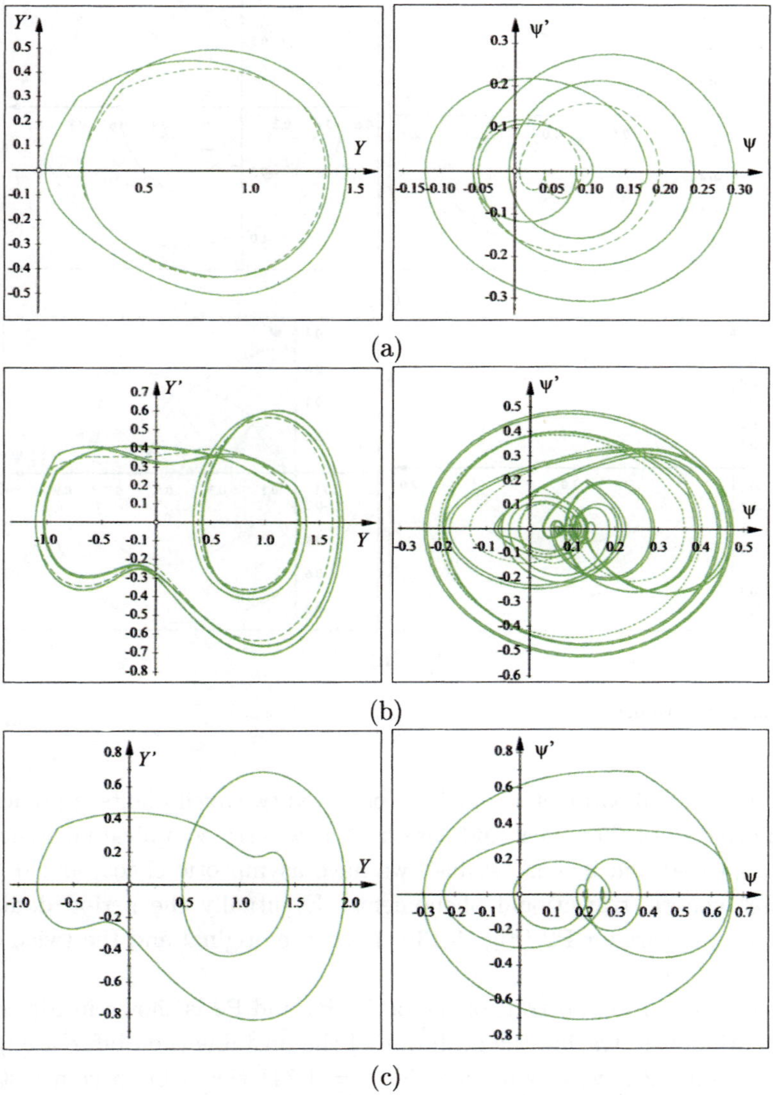

Fig. 6.13 Two projections of the periodic orbits for the type-1 motion with the change of γ: (a) 0.5 (original P1-orbit, broken), 0.65 (P1 after period doubling, full); (b) 0.85 (original P2-orbit, broken), 1.1 (P2 after two periods doublings, full); (c) 1.75(P3); (d) 2.0 (P4); (e) 2.6 (doubled version of P5, full), 3.0 (P5 nonbifurcated, broken).

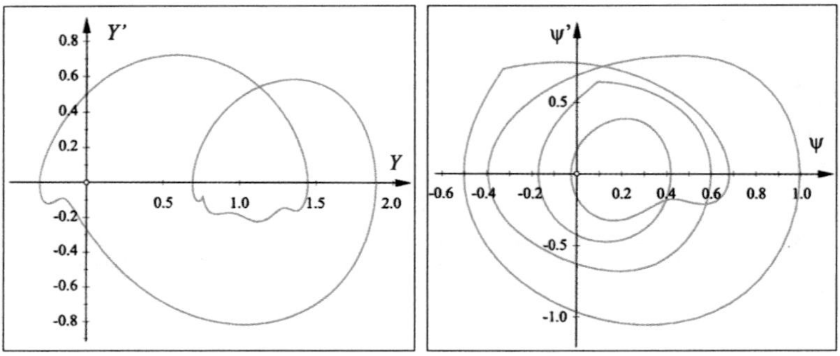

(d)

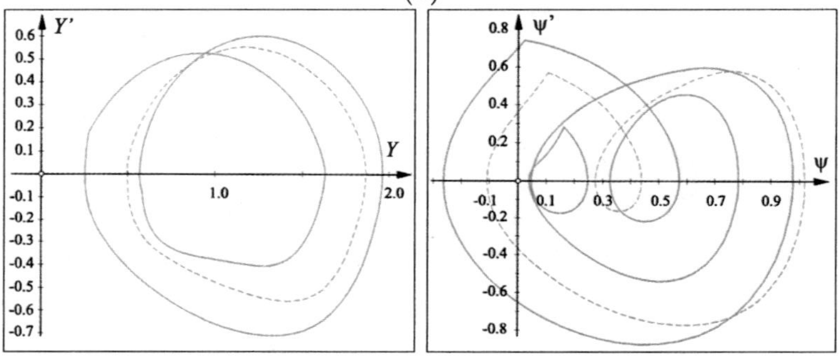

(e)

Fig. 6.13: Continued

Due to a critical parameter γ we have observed two fixed points approaching each other until they meet and vanish. A new, very complicated, structure has appeared and if it is "stable" we have asymptotic chaos, and if "unstable" chaotic transitional phenomena. Eventually the period doubling scenario appears for P2 (Fig. 6.13b shows the original and the twice doubled limit cycle).

The bifurcation portrait of the orbits P1 and P2 is shown in Fig. 6.15. By continuously tracing the evolution of the unstable (non-bifurcated) periodic orbit P2 it was found that for $\gamma = 1.741$ this orbit becomes stable again. Shortly before this value of γ one of the multipliers of P3 crosses the unit circle at -1. When the phase portrait of the orbit P2 in its original form (Fig. 6.13b) and P3 are compared (Fig. 6.13c) it is easy to check that both have a very similar form. We suppose that, with increase of γ, the orbit P2 goes through the successive period doubling bifurcation and, after crossing the critical value, successive period halving appears. Periodic orbit P4 has the form presented in Fig. 6.13d which is stable in the interval $1.895 < \gamma < 2.2545$. With the decrease of γ two successive period doubling bifurcations have been observed for $\gamma = 1.895$ and $\gamma = 1.83$. As can be seen from the phase portrait, the mass is given an acceleration in the negative y direction for negative y. Further increase of γ leads to the disappearance of this orbit at a saddle point for $\gamma = 2.2545$.

It should be pointed out that in the interval $2.007 < \gamma < 2.2545$ there exist together two periodic stable orbits of the type-1 motion: P4 and P5.

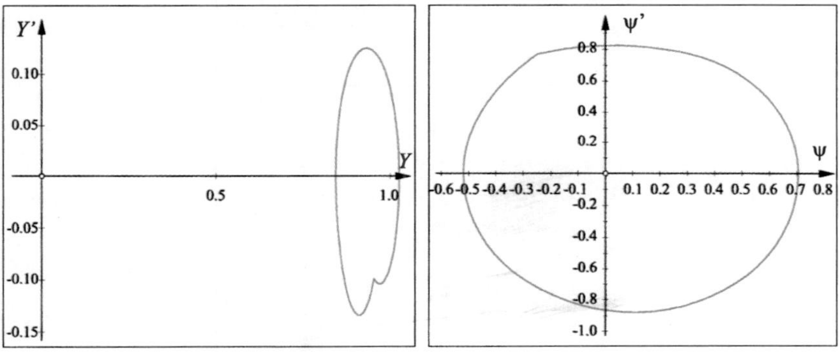

Fig. 6.14 Two projections of the periodic orbits for the type-2 motion for $\gamma = 0.82$.

The latter orbit P5 (see Fig. 6.13e) bifurcates for decreasing γ to another stable periodic orbit with twice the period. This one finally disappears at $\gamma = 2.007$ when one of the multipliers goes through $+1$ of the unit circle.

Finally, we discuss the periodic orbit P6 (Fig. 6.14) which belongs to the second type of the motion. Contrary to the previous cases, the rotation (see Fig. 6.7) predominates over the translation of the disc. Because of the small translational amplitudes of oscillations of the disc, geometric nonlilnearities do not have substantial influence on the motion. In the considered interval of γ, bifurcations of the periodic orbit were not found.

6.2.8 *Observations of chaos*

Generally, we have found chaotic orbits for the following intervals of γ: (a) transition from P1 to P2 ($0.6695 < \gamma < 0.836071$), (b) transition from P2 to P3 ($1.15 < \gamma < 1.7$), (c) transition from P3 to P4 ($1.81 < \gamma < 1.83$). These observations lead to the conclusion that chaos has been found where a qualitative change of motion has appeared.

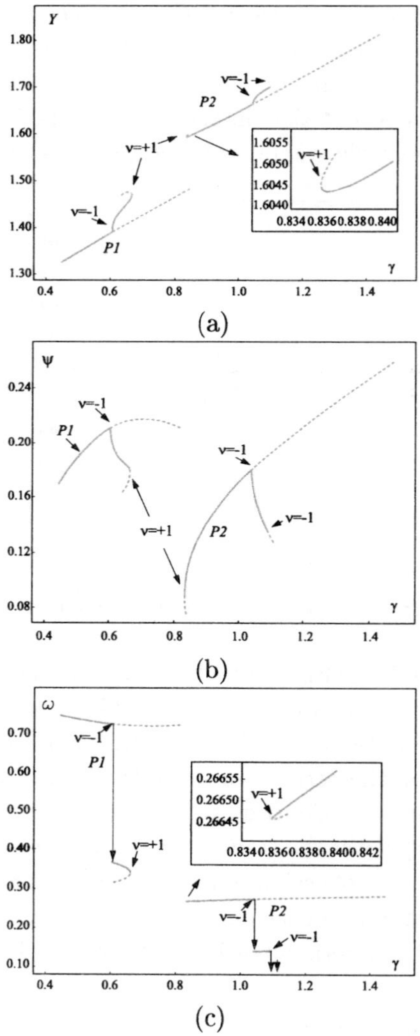

Fig. 6.15 Bifurcation diagram of the investigated P1 and P2 orbits with marked period doubling and saddle points (a) $y(\gamma)$; (b) $\psi(\gamma)$; (c)$\omega(\gamma)$.

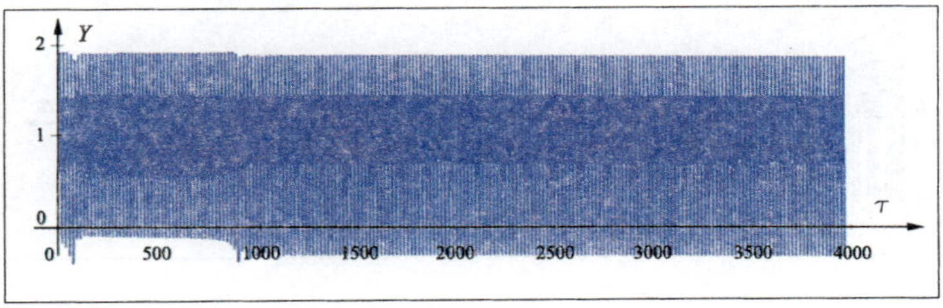

Fig. 6.16 Intermittency transition from P5 to P4; $\chi = 0.6$, $\rho = 2.5$, $V_B = 0.5$, $V^* = 1.0$, $\kappa = 1.0$, $\beta_1 = \beta_2 = 0.05$, $\mu_1 = 0.05$, $\mu_2 = 0.2$, $\vartheta = 0.5$, $\gamma = 2.007$, $y_0 = -0.129$, $y_0' = -0.066$, $\psi_0 = 0.024$, $\psi_0' = -0.101$ (time history $y(\tau)$).

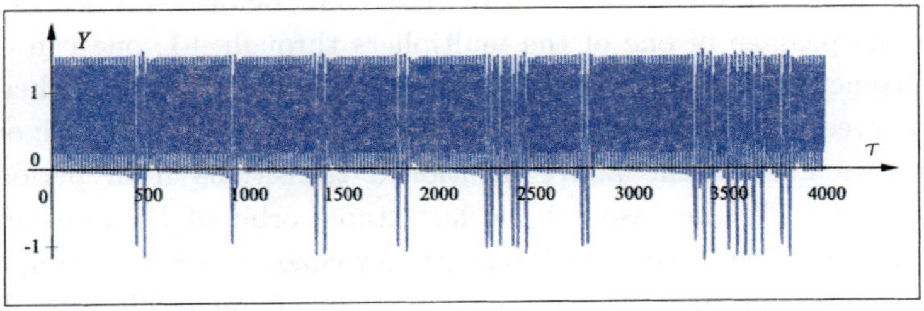

Fig. 6.17 Intermittency chaos in the neighbourhood of P1 for $\gamma = 0.67$ and $y_0 = 0.258$, $y_0' = -0.177$, $\psi_0 = 0.194$, $\psi_0' = 0.036$ (time history).

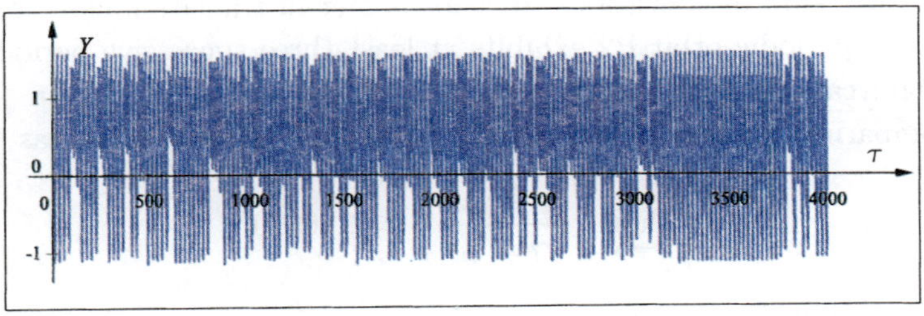

Fig. 6.18 Intermittency chaos in the neighbourhood of P2 for $\gamma = 0.836$ and $y_0 = 0.409$, $y_0' = -0.110$, $\psi_0 = 0.095$, $\psi_0' = 0.416$ (time history).

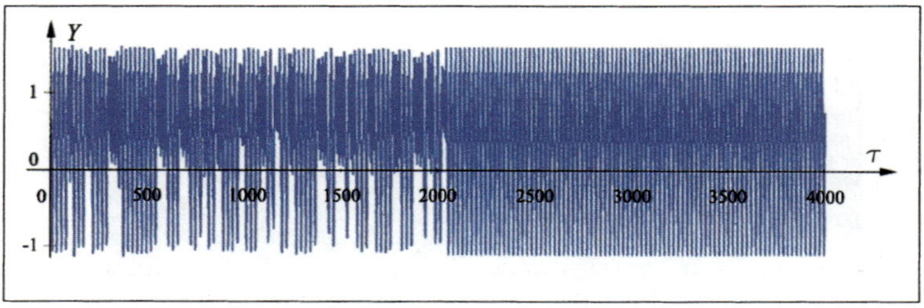

Fig. 6.19 Intermittency transition to P2 for $\gamma = 0.837$ and $y_0 = 1.290$, $y'_0 = -0.127$, $\psi_0 = 0.336$, $\psi'_0 = -0.201$ (time history $y(\tau)$).

An exception is the change from P4 to P5 since they coexists in a stable form for $2.007 < \gamma < 2.2545$. Fig. 6.16 shows a situation for which P5 has just disappeared in a saddle (the trajectory had started close to the old P5). Taking into consideration the fact that the orbit P5 has vanished during the passage of one of the multipliers through +1, one can expect intermittency phenomena. Contrary to the classical examples, where this was observed, periodic orbit P4 was reached instead of the chaotic one.

Now we analyse the above mentioned transitions from periodic to chaotic orbits. In the case (a) the last stable orbit of P1 vanishes in a saddle at $\gamma = 0.6995$ leading to intermittency chaos at $\gamma = 0.67$ (Fig. 6.17). Shortly before entering the parameter domain of the stable orbit P2 we get intermittency chaos at $\gamma = 0.836$ (Fig. 6.18). The time-evolution for y shows the shift from the first predominant motion of form P1 to form P2. A transitional chaotic phenomenon could be observed for $\gamma = 0.837$ (see Fig. 6.19) showing an irregular motion similar to that for $\gamma = 0.836$ for a long period of time until suddenly transitionless point P2 is reached.

Another kind of transitions to chaos is met in transition case (b). For $\gamma \geq 1.044$ periodic orbit P2 exhibits at least three successive period doubling bifurcations.

Comparing our results with the Feigenbaum constant defined as

$$\delta = \lim_{i \to \infty} \left(\frac{\gamma_i - \gamma_{i-1}}{\gamma_{i+1} \gamma_i} \right) = 4.66920... \,,$$

we have obtained

$$\delta_1 = \frac{(1.0966 - 1.044)}{(1.1075 - 1.0966)} = 4.826.$$

as a first estimation.

The periodic orbit P3 is a product of period halvings as γ is increased. Figs. 6.20 and 6.23 show chaotic motions close to P2 at $\gamma = 1.155$ and close to P3 at $\gamma = 1.7$, respectively. We have found band-type attractors showing the general shape of the corresponding periodic orbits (see y'/y projections). Comparing the Poincaré maps ψ'/ψ with the maps made for the so called Rössler band attractor we have found similar folded line-type structures. Due to the Thompson and Stewart [Thompson and Stewart (1986)] this folded band is the simplest structure for chaotic attractors. It is remarkable that as γ is increased from $\gamma = 1.155$ to $\gamma = 1.158$ (Fig. 6.21) or decreased from $\gamma = 1.7$ to $\gamma = 1.65$ (Fig. 6.22) we obtain the same phenomenon in either case.

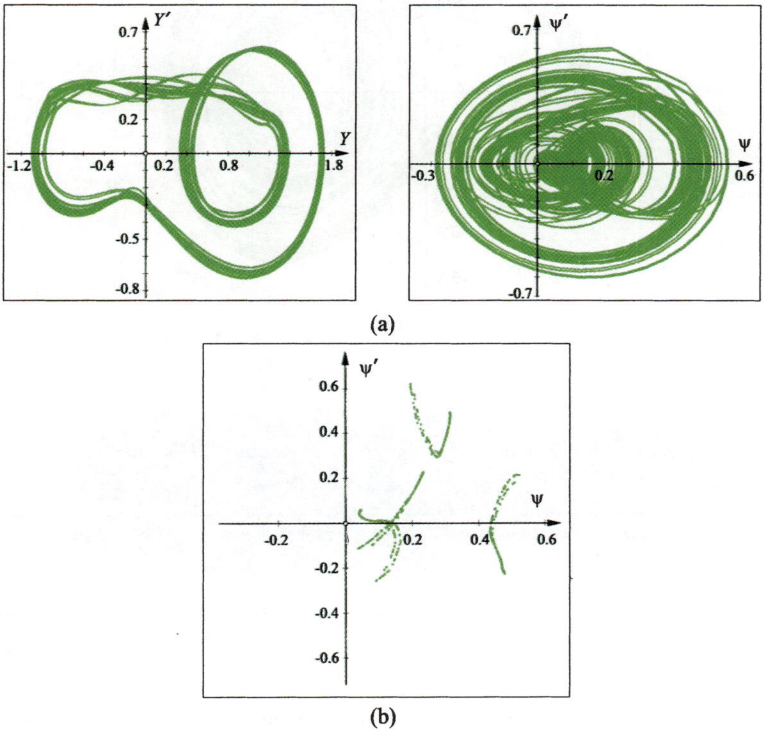

Fig. 6.20 Transition to chaos via period doubling bifurcations for $\gamma = 1.155$ and $y_0 = 1.393$, $y'_0 = -0.004$, $\psi_0 = 0.448$, $\psi'_0 = -0.013$; (a) phase portraits; (b) Poincaré map $\psi'/\psi(y' = 0)$.

The general structure of the earlier band-type attractors are still visible (see projections y'/y), but from time to time single trajectories escape from the old attractors performing a completely different motion. This observed behaviour can be interpreted as bifurcations of the previous band-attractors through an intermittency leading to a more complicated motion. The strong intermittency tendency, obviously caused by the geometric nonlinearity, seems to perturbe a smooth evolution of the band-attractors.

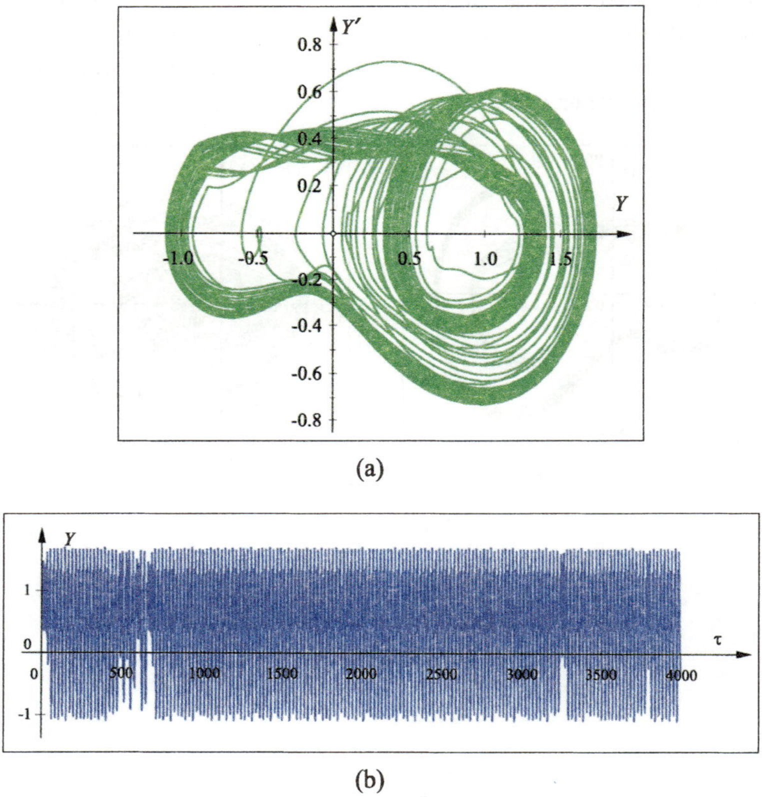

Fig. 6.21 Chaos after bifurcation of the attractor from Fig. 6.20 through an intermittency phenomenon for $\gamma = 1.158$ and $y_0 = 1.150$, $y_0' = -0.048$, $\psi_0 = 0.600$, $\psi_0' = -0.320$; (a) phase portrait y'/y; (b) time history $y(\tau)$.

Finally in the case (c), the transition from P3 to P4 is again different. Starting the system for $\gamma = 1.812$ close to the old P3 orbit (which had vanished in a Hopf bifurcation), we obtain a motion which is, qualitatively, completely different when compared with the previously considered cases. However, the transition phase to this motion is a sort of nonstationary quasi-periodic motion. The chaotic attractor presented in Fig. 6.24 is a result of mixed orbit P4 and orbit P5 form of motion (compare Fig. 6.24 with Figs. 6.13d, e). A horizontal line (made up of densely spaced points) which is clearly visible on the ψ'/ψ Poincaré maps indicates states of stick. It is easy to obtain from the dimensionless form of 6.8 that $\psi'(y' = 0) = \vartheta\rho V_B = 0.625$.

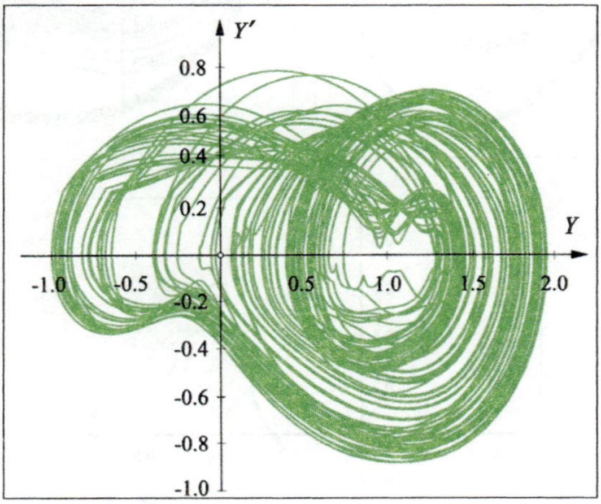

(a)

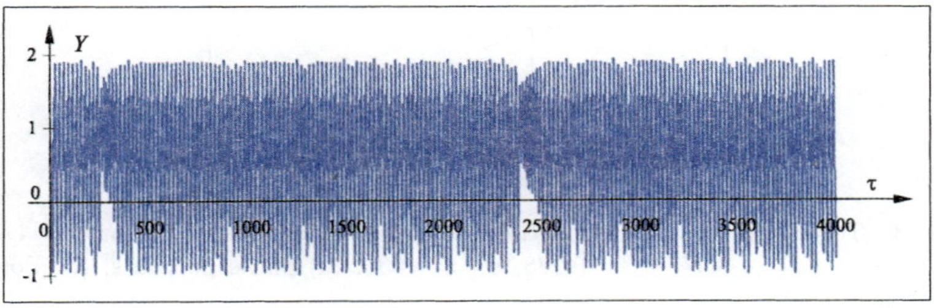

(b)

Fig. 6.22 Chaos after bifurcation of the attractor from Fig. 6.23 through an intermittency phenomenon for $\gamma = 1.65$ and $y_0 = 0.442$, $y_0' = -0.621$, $\psi_0 = 0.179$, $\psi_0' = -0.077$; (a) phase portrait y'/y; (b) time history $y(\tau)$.

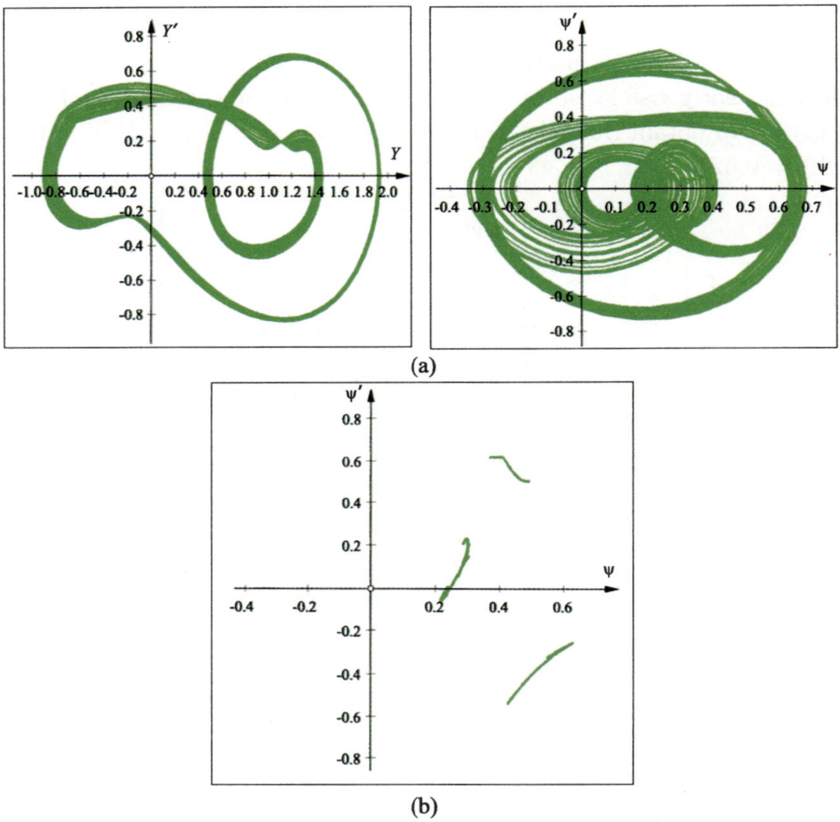

Fig. 6.23 Transition to chaos via period doubling bifurcations of P3 for $\gamma = 1.7$ and $y_0 = -0.895$, $y_0' = -0.07$, $\psi_0 = 0.288$, $\psi_0' = 0.035$; (a) phase portraits; (b) Poincaré map $\psi'/\psi(y' = 0)$.

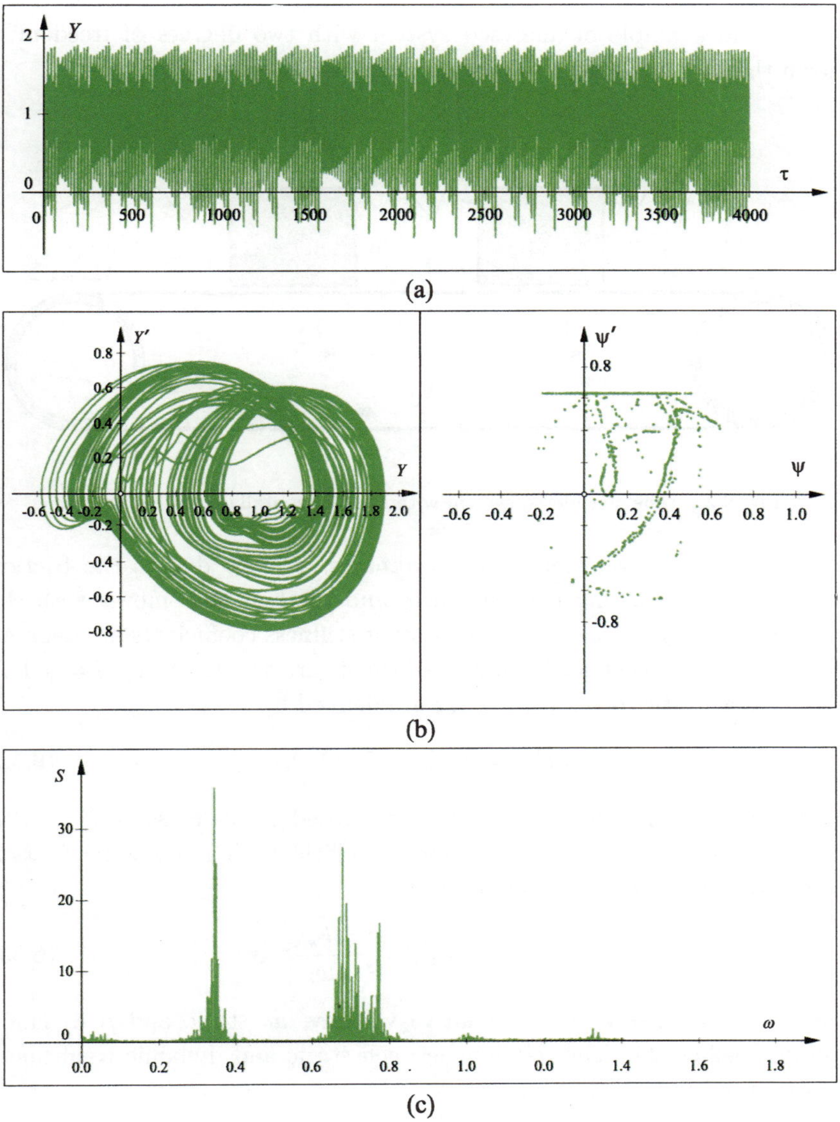

Fig. 6.24 Chaotic motion in the transition region between P3 and P4 periodic orbit for $\gamma = 1.812$ and $y_0 = -0.237$, $y_0' = -0.139$, $\psi_0 = -0.114$, $\psi_0' = -0.042$: (a) time history $y(\tau)$; (b) phase portrait and Poincaré map for $(y' = 0)$; (c) power spectrum of $y(\tau)$.

6.3 Two Horizontally Situated Masses

The second example of analysed system with two degrees of freedom is shown the Fig. 6.25 (see also [Awrejcewicz and Olejnik (2003)]).

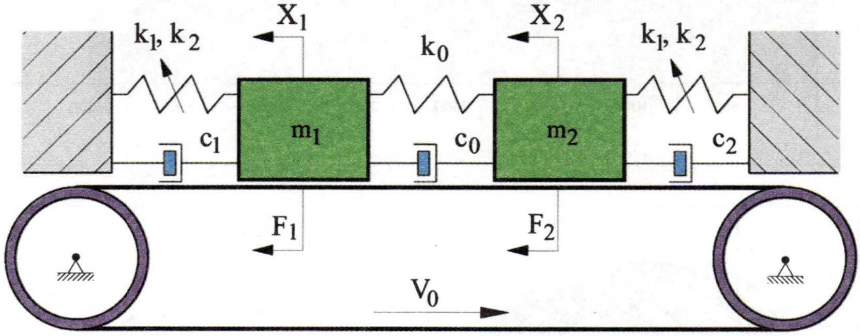

Fig. 6.25 The considered system with two horizontally situated masses.

Two masses move along the coordinates X_1, X_2 due to the friction forces F_1, F_2 occurring between them and the belt that moves with the constant velocity. As usually, the constant stiffness coefficients are denoted by k_i, $i = 0, 1, 2$, whereas damping coefficients are denoted by c_j, $j = 0, 1, 2$. The friction static forces F_i, $i = 1, 2$ are defined by

$$F_{s,i} = \mu_0 F_{N,i} \quad i = 1, 2, \tag{6.52}$$

where $F_{N,i} = m_i g$ is the pressing force generated by the mass m_i $(i = 1, 2)$, and μ_0 is the value of the static force coefficient. The dynamic friction forces are governed by the equations

$$F_i = -\mu F_{N,i} \text{sign} v_{w,i} = -\frac{\mu F_{s,i}}{\mu_0} \text{sign} v_{w,i}, \tag{6.53}$$

where $v_{w,i} = \dot{x}_i - V_0$ is the relative velocity, μ_0 static and μ dynamic friction coefficients. The relation between static and dynamic coefficients is introduced in the following way

$$\mu = \frac{\mu_0}{1 + \delta |v_{w,i}|}. \tag{6.54}$$

In the above, coefficient δ characterizes a way of dynamic coefficient decrease which accompanies an increase of the relative velocity. The static forces occur when the relative velocity is equal to zero. Therefore, one gets

$$\begin{cases} |F_i| \leq F_{s,i} & v_{w,i} = 0, \\ F_i = -\text{sign} v_{w,i} \dfrac{F_{s,i}}{1+\delta|v_{w,i}|} & v_{w,i} \neq 0, \end{cases} \tag{6.55}$$

where $F_{s,i}$ $(i = 1, 2)$ are defined by Eq. (6.52). The following approximation to the sgn function is applied

$$\text{sgn}(v_{w,i}) = \frac{2}{\pi} \arctan(\varepsilon v_{w,i}), \tag{6.56}$$

where $\varepsilon > 0$. The estimation of the Eq. (6.56) is shown in the Fig. 6.26. It is seen that for $\varepsilon = 10^3$ arctan almost exactly describes the properties of

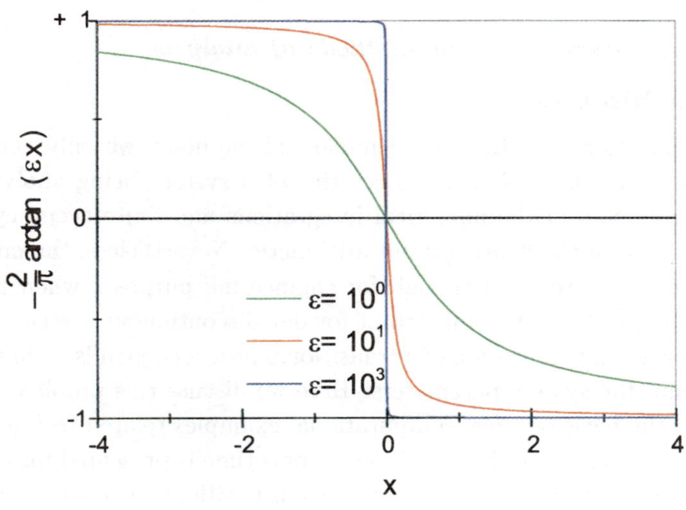

Fig. 6.26 Influence of ε for the approximation given in (6.56).

the function sign. Taking into account the Eq. (6.56) we get

$$F_i = -\frac{2}{\pi} \frac{F_{s,i} \arctan(\varepsilon v_{w,i})}{(1+\delta|v_{w,i}|)}, \quad i = 1, 2, \tag{6.57}$$

which is further used.

The following nondimensional equations govern dynamics of this two body dynamical system:

$$\xi_1\ddot{y}_1 + \alpha_0(\dot{y}_1 - \dot{y}_2) + \alpha_1\dot{y}_1 + (1+\beta_1)y_1 - \beta_2 y_1^3 - y_2 = -\frac{2\arctan(\varepsilon\overline{v}_{w,1})}{\pi\,(1+\gamma|\overline{v}_{x,1}|)},$$

$$\xi_2\ddot{y}_2 + \alpha_0(\dot{y}_1 - \dot{y}_2) + \alpha_2\dot{y}_2 + (1+\beta_1)y_2 - \beta_2 y_2^3 - y_1 = -\frac{2\beta\arctan(\varepsilon\overline{v}_{w,2})}{\pi\,(1+\gamma|\overline{v}_{x,2}|)}. \tag{6.58}$$

The relations between physical values of parameters, coordinates in time, and their nondimensional adequates are as follows, $\xi_i = m_i\omega_1^2/k_0$, $\alpha_j = c_j\omega_1/k_0$, $\beta_1 = k_1/k_0$, $\gamma = F_{s,1}\delta/\sqrt{k_0 m_1}$, $\overline{V_0} = \sqrt{k_0 m_1}V_0/F_{s,1}$, $\overline{v}_{w,i} = \dot{y}_1 - \overline{V_0}$, $\beta = F_{s,2}/F_{s,1}$, $\beta_2 = k_2 F_{s,1}^2/k_0^3$, $i = 1, 2$, $j = 0, 1, 2$. An introduction of dimensionless quantities has led to reduction of the parameters from 13 to 11.

6.3.1 *An overview of the methods of analysis*

(i) Time Histories

One has to realize that using numerical methods, we only get an approximation of the real true trajectories of a system being analysed due to the finite step of the numerical integrations and finite accuracy of the numbers used in the floating-point arithmetic. Nevertheless, the numerical approximations are good enough for engineering purposes when they are properly applied which is important for our discontinuous system.

It is clear that a duration of a transitional process depends on initial conditions and the system parameters. Here we discuss this problem in more detail on the basis of a few computational examples related to Eqs. (6.58). In the Fig. 6.27, one of the coordinate versus time is presented for $t = 0$. In this case, the masses are in the equilibrium positions, whereas their initial velocities are equal to the belt velocity. In practise, the system from the beginning starts to move on an attractor. When the initial conditions are changed, a transitional process occurs of a duration equal to $t = 4$ (see the Fig. 6.28). Our numerical analysis shows that in some cases the transitional state can be even 10 times larger. In both previously discussed cases, we have dealt with the periodic attractor. A situation changes dramatically if a strange chaotic attractor appears. For this case it is rather difficult to define the beginning of observation, and therefore first hundreds "periods" of oscillations are neglected. The corresponding example is given in the Fig. 6.29. It is seen from both Figs. 6.28–6.29 that a decreasing part of time histories is more steep that an increasing part. This is a typical

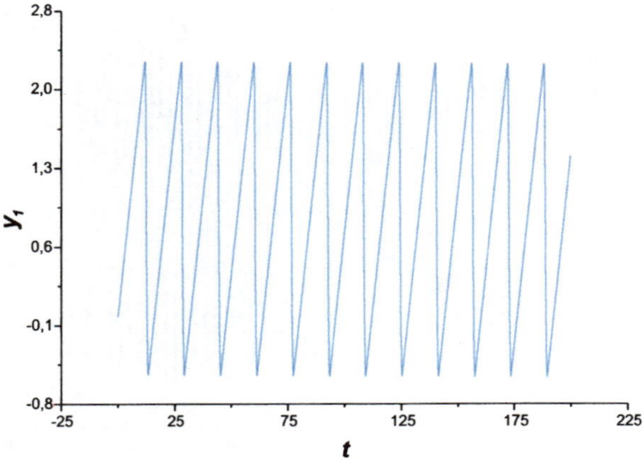

Fig. 6.27 Time history of displacement y_1 with very short transitional process for the parameters: $\gamma = 3,\ \beta = 1, V_0 = 0.2,\ \alpha_0 = \alpha_1 = \alpha_2 = 0,\ \beta_1 = \beta_2 = \xi_1 = \xi_2 = 0.1$ and the initial conditions: $y_1 = y_2 = 0,\ \dot{y}_1 = \dot{y}_2 = 0.2$.

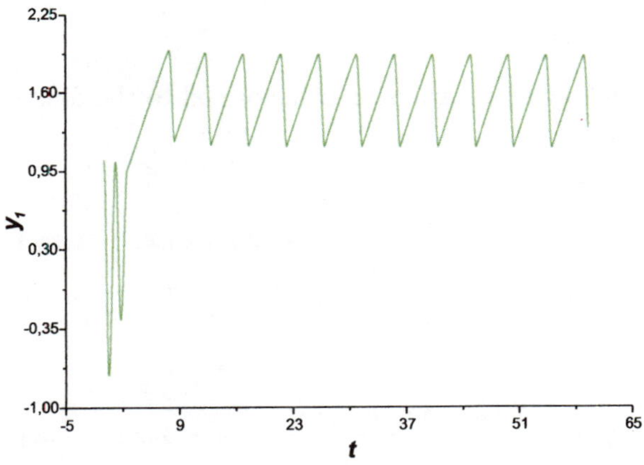

Fig. 6.28 Time history of displacement y_1 (with a clearly visible transitional process) for the initial conditions: $y_1 = y_2 = 0,\ \dot{y}_1 = \dot{y}_2 = 0.2$ (other parameters are the same as in Fig. 6.27).

behaviour of our stick-slip system with friction. The first mentioned part corresponds to a slip, whereas the second one (slower) to a stick between a mass and the tape.

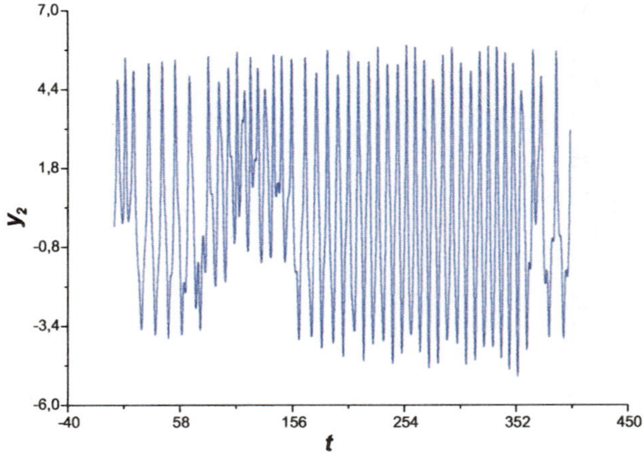

Fig. 6.29 Chaotic time history of $y_2(t)$ for the parameters: $\gamma = 0.03$, $\beta = 1$, $V_0 = 5$, $\alpha_0 = 0.01$, $\alpha_1 = 0$, $\alpha_2 = 0.03$, $\beta_1 = 1$, $\beta_2 = 0.1$, $\xi_1 = 1.12$, $\xi_2 = 1$ and the initial conditions: $y_1 = 0.4$, $y_2 = -0.12$, $\dot{y}_1 = -0.56$, $\dot{y}_2 = 0.12$.

(ii) Phase Spaces

The analysed set of equations is transformed to the following one:

$$\begin{cases} \dfrac{d\tilde{x}_1}{d\tau} = \tilde{x}_2, \\ \dfrac{d\tilde{x}_2}{d\tau} = \dfrac{1}{\xi_1}\Big(-\dfrac{2}{\pi}\dfrac{\arctan(\varepsilon\tilde{v}_{w,1})}{(1+\gamma|\tilde{v}_{w,1}|)} - \alpha_0\tilde{x}_2 + \alpha_0\tilde{x}_4 - \alpha_1\tilde{x}_2 + \\ \qquad - (1+\beta_1)\tilde{x}_1 + \beta_2\tilde{x}_1^3 + \tilde{x}_3)\Big), \\ \dfrac{d\tilde{x}_3}{d\tau} = \tilde{x}_4, \\ \dfrac{d\tilde{x}_4}{d\tau} = \dfrac{1}{\xi_2}\Big(-\dfrac{2\beta}{\pi}\dfrac{\arctan(\varepsilon\tilde{v}_{w,2})}{(1+\gamma|\tilde{v}_{w,2}|)} - \alpha_0\tilde{x}_2 + \alpha_0\tilde{x}_4 - \alpha_2\tilde{x}_4 + \\ \qquad - (1+\beta_1)\tilde{x}_3 + \beta_2\tilde{x}_3^3 + \tilde{x}_1)\Big), \end{cases} \tag{6.59}$$

where: $\tilde{v}_{w,i} = \tilde{x}_{2i} - \overline{V_0}$, $i = 1, 2$, $\tilde{x}_1 = y_1$, $\tilde{x}_2 = \dot{y}_1$, $\tilde{x}_3 = y_2$, $\tilde{x}_4 = \dot{y}_2$. For further considerations we take $\tilde{x}_j \equiv x_j$, $j = 1\ldots 4$, where now x_i correspond to the displacement of m_i, whereas $\dot{x}_i$ are the corresponding velocities $(i = 1, 2)$.

A typical projection of a trajectory associated with our system with friction is shown in the Fig. 6.30. Two parts are easy distinguishable: The stick part (where $\dot{x}_1 = V_0 = 0.2$, which is represented by the horizontal line)

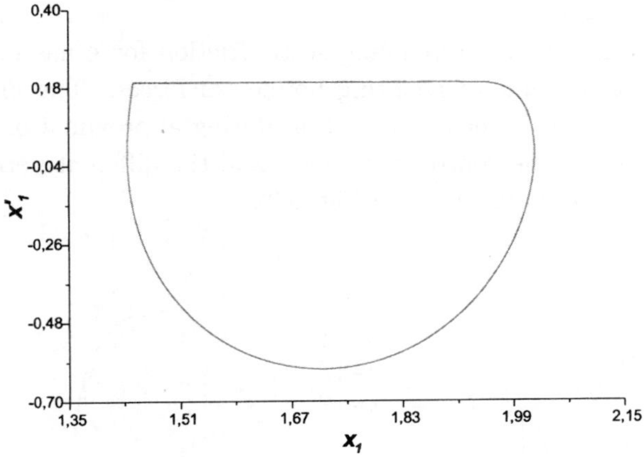

Fig. 6.30 Phase plane $\dot{x}_1(x_1)$ of the periodic motion for the parameters: $\gamma = 3$, $\beta = 1, V_0 = 0.2$, $\alpha_0 = \alpha_1 = \alpha_2 = 0$, $\beta_1 = \beta_2 = 0.1$, $\xi_1 = \xi_2 = 0.5$ and the initial conditions: $x_1 = x_2 = 0$, $\dot{x}_1 = \dot{x}_2 = 0.2$.

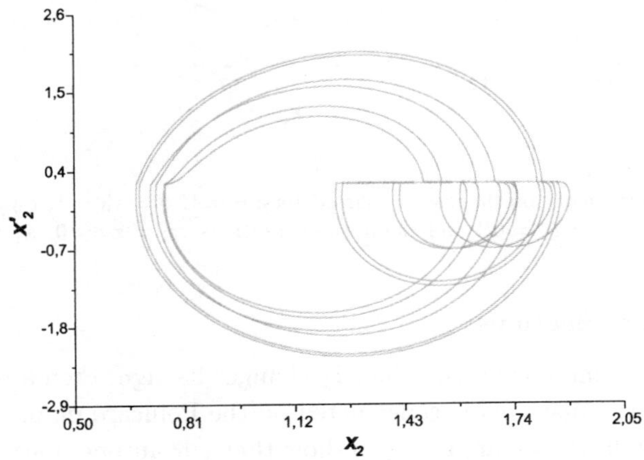

Fig. 6.31 Phase plane $\dot{x}_2(x_2)$ during stick-slip chaos for the parameters: $\gamma = 3$, $\beta = 0.5$, $V_0 = 0.2$, $\alpha_0 = 0.1$, $\alpha_1 = \alpha_2 = 0$, $\beta_1 = \beta_2 = \xi_1 = \xi_2 = 0.1$ and the initial conditions: $x_1 = 0$, $x_2 = 0.12$, $\dot{x}_1 = \dot{x}_2 = 0$.

and the slip part. A more complicated motion is presented in Fig. 6.31. A time occurrence and a duration of a stick are unpredictable. The phases appear with different velocities, which are represented by small and large arcs in the phase plane. For small x_2 values $(0.7 - 0.9)$ a stick does not occur when $\dot{x}_2 = V_0$.

It means that the corresponding static friction force has been smaller then the absolute value of resulting horizontal forces. The phase planes can also be used for an error estimation during approximation of sign by arctan. ε serves as the control parameter and the different periodic orbits for different ε values are shown in Fig. 6.32.

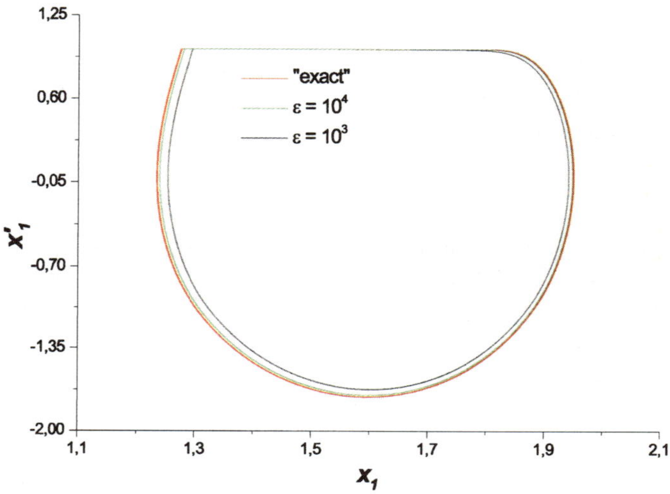

Fig. 6.32 Influence of ε on the results (parameters: $\gamma = 1$, $\beta = V_0 = 1$, $\alpha_0 = \alpha_1 = \alpha_2 = 0.02$, $\beta_1 = \beta_2 = \xi_1 = \xi_2 = 0.1$ and the initial conditions: $x_1 = x_2 = 0$, $\dot{x}_1 = \dot{x}_2 = 1$).

(iii) Poincaré Sections

For our autonomous case, when $\dot{x}_1$ changes its sign, then a point corresponding to the mass m_2 is constructed on the Poincaré map.

In what follows, we are going to show that our autonomous system can exhibit stick-slip chaotic (Fig. 6.33), stick-slip periodic (Fig. 6.34) as well as stick-slip quasi-periodic dynamics (Fig. 6.35). Because of the introduced symmetry, only a behaviour of one mass has been presented.

(iv) Bifurcation Diagrams

The bifurcation diagrams have been constructed in two ways. For instance, by changing a parameter in the interval (0.1, 0.5) with the step 0.001, we get 400 Poincar maps. Then, one of the phase axes is taken and all results are presented versus the parameter. Another way is that

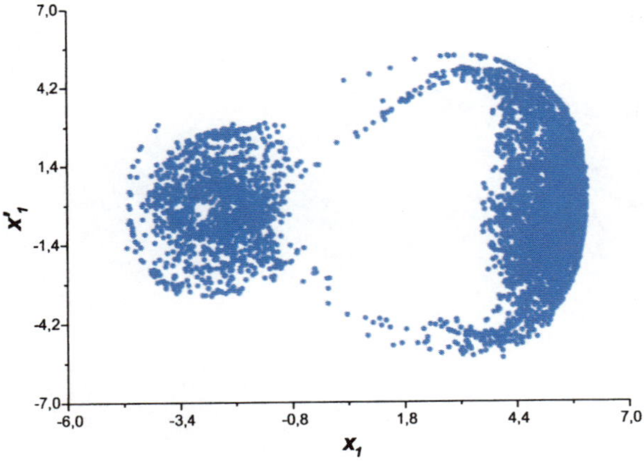

Fig. 6.33 Poincaré map $\dot{x}_1(x_1)$ for the parameters: $\gamma = 0.03$, $\beta = 1$, $V_0 = 5$, $\alpha_0 = 0.01$, $\alpha_1 = 0$, $\alpha_2 = 0.03$, $\beta_1 = 1$, $\beta_2 = 0.1$, $\xi_1 = 1.12$, $\xi_2 = 1$ and the initial conditions: $x_1 = 0.4$, $x_2 = -0.12$, $\dot{x}_1 = -0.56$, $\dot{x}_2 = 0.12$ (chaotic orbit).

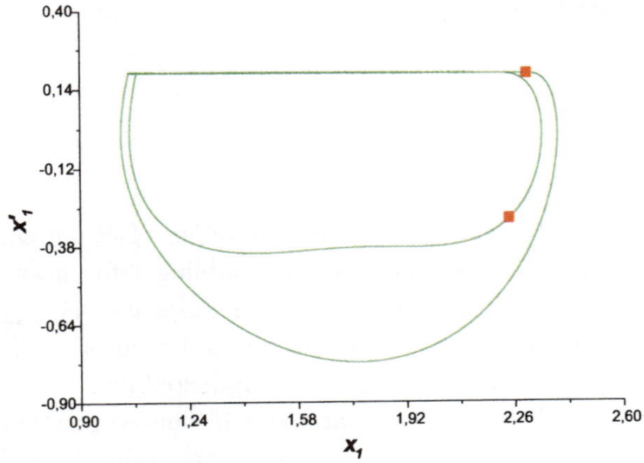

Fig. 6.34 Phase portrait and the corresponding Poincaré map (red squares) $\dot{x}_1(x_1)$ for the parameters: $\gamma = 3$, $\beta = 1$, $V_0 = 0.2$, $\alpha_0 = \alpha_1 = \alpha_2 = \beta_1 = \beta_2 = 0.1$, $\xi_1 = \xi_2 = 1$ and the initial conditions: $x_1 = x_2 = 0 = \dot{x}_1 = \dot{x}_2 = 0$ (periodic orbit).

for increasing the parameter values we change the initial conditions, and contrary to the first case, we leave an attractor (in the previous case we were a whole time on an attractor). An example of bifurcation diagram is shown in Fig. 6.36.

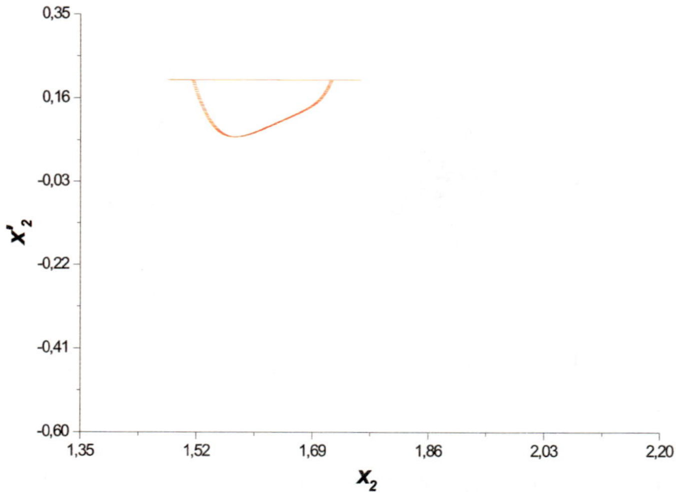

Fig. 6.35 Poincaré map $\dot{x}_2(x_2)$ for the parameters: $\gamma = 3$, $\beta = 1$, $V_0 = 0.2$, $\alpha_0 = \alpha_1 = \alpha_2 = 0$, $\beta_1 = \beta_2 = 0.1$, $\xi_1 = \xi_2 = 1$ and the initial conditions: $x_1 = -0.1$, $x_2 = 0.2$, $\dot{x}_1 = \dot{x}_2 = 0$ (quasi-periodic orbit).

Beginning from the smallest considered values of ξ_2 we observe different multiple periodic motion and period doubling bifurcations occur. In the interval $\xi_2 \in (0.5,\ 0.6)$ period-6 window appears. For $\xi_2 = 0.6$ and $x_2 \approx 1.0$ the saddle-node bifurcation occurs, and then other local classical bifurcations appear. For $\xi_2 \approx 0.9$ the period doubling bifurcation occur (with a decrease of the bifurcation parameter). This corresponds to a route to chaos and to a route from chaos to regular (periodic) behaviour.

It should be emphasized that for $\xi_2 \approx 1.22$ a crisis between period-6 orbit and chaos occurs, and its enlargement is shown in the Fig. 6.36 b. It is clearly seen how the successive period doubling (accompanying a decrease of ξ_2) leads to a periodic motion, which exists for $\xi_2 \in (1.15,\ 1.22)$.

(v) Lyapunov Exponents

The Lyapunov exponents have been estimated from the Eq. (6.59) using a standard calculating procedure.

In order to verify the developed code the well known Lorenz set of

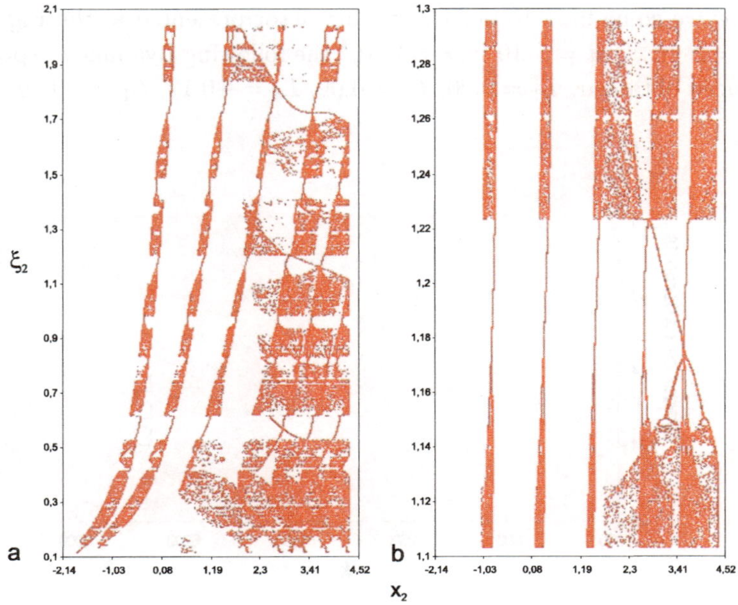

Fig. 6.36 Bifurcation diagram for the parameter ξ_2: a) $\xi_2 \in (0.1,\ 2.1)$, b) $\xi_2 \in (1.1,\ 1.3)$. The following parameters are fixed: $\gamma = 2.03$, $\beta = 21$, $V_0 = 0.2$, $\alpha_0 = 0.0045$, $\alpha_1 = 0.01$, $\alpha_2 = 0.19$, $\beta_1 = 0.12$, $\beta_2 = 0.2$. The initial conditions: $x_1 = -0.1$, $x_2 = 0.2$, $\dot{x}_1 = 0$, $\dot{x}_2 = 0$.

equations (see Eq. (6.60)) has been tested

$$\begin{cases} \dot{x}_1 = -\sigma x_1 + \sigma x_2, \\ \dot{x}_2 = r x_1 - x_2 - x_1 x_3, \\ \dot{x}_3 = x_1 x_2 - b x_3. \end{cases} \tag{6.60}$$

In the reference [Wolf *et. al.* (1985)] for $\sigma = 16$, $r = 40$, $b = 4$ the following Lyapunov exponents have been computed: $L_1 = 1.37$, $L_2 = 0.00$, $L_3 = -22.37$. Using our method we have $L_1 = 1.36$, $L_2 = 0.00$, $L_3 = -22.37$ for $h = 2 \cdot 10^{-3}$, $dt = 4 \cdot 10^{-2}$, $k = 7 \cdot 10^3$. The computation process is illustrated in the Fig. 6.37.

In order to judge about a strangeness of a chaotic attractor we introduce the following Lyapunov dimension

$$d_L = i + (\lambda_1 + \lambda_2 + \lambda_3)/|\lambda_4|, \tag{6.61}$$

where i is the index corresponding to a smallest nonnegative Lyapunov exponent. For the Lorenz system we have obtained $d_L = 2.06$.

Now we come back to our chaotic attractor presented in the Fig. 6.33. For $h = 2 \cdot 10^{-3}$, $dt = 5 \cdot 10^{-2}$, $k = 2 \cdot 10^{4}$ the following Lyapunov exponents have been obtained: $L_1 = 0.20$, $L_2 = 0.09$, $L_3 = -0.10$, $L_4 = -0.29$.

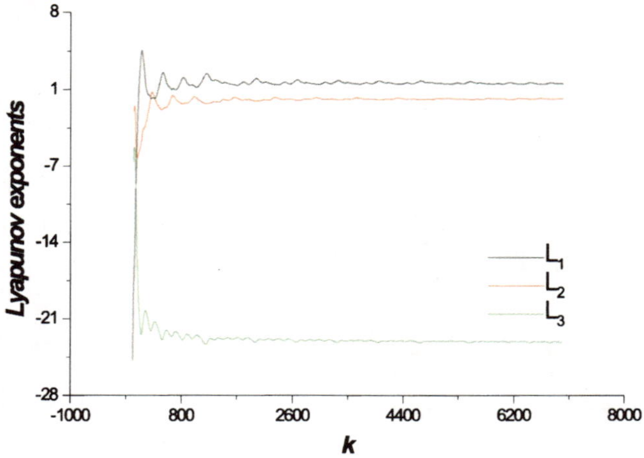

Fig. 6.37 A convergence of the Lyapunov exponents for the Lorenz equations.

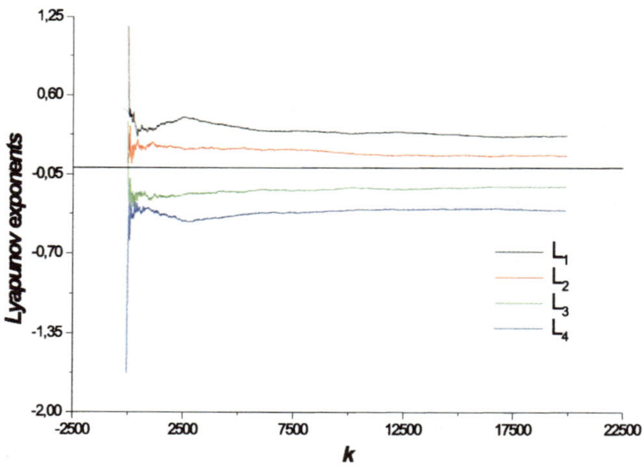

Fig. 6.38 A convergence of the Lyapunov exponents for the chaotic strange attractor illustrated in Fig. 6.33.

Therefore, the hyper-chaotic strange stick-slip attractor has been detected for our system with friction ($d_L = 2.65$). Computation process in this case is illustrated in Fig. 6.38.

In Fig. 6.38, convergence is achieved after 1000 nondimensional time units. For the Lorenz system, the similar like convergence have been achieved after 280 (in both cases an accuracy was limited to two numbers after a dot). In spite of the approximation by the continuous function, the friction still possesses an essential influence on time needed to estimate the Lyapunov exponents.

6.3.2 *Numerical analysis and results*

(i) Algorithm

The standard Runge-Kutta method with variable integration step has been used. Numerical experiments during investigation of our system have shown that very small Δt must be taken in order to make computational time more economical, and that Δt it must be linked with the increase step of the variables Δy.

(ii) Results

The computational results are presented in Figs. 6.39–6.49. Chaotic motions presented in Figs. 6.39, 6.40 are similar to that exhibited by a sinusoidal excited pendulum. In Figs. 6.41, 6.44 the phase planes together with the Poincaré maps (black squares) are presented. It is interesting to note (Fig. 6.41) that a special dynamics can be realized when one of the masses exhibits stick-slip periodic process, whereas the second only slip one. In Fig. 6.44, an example of very complicated stick-slip periodic behaviour is reported. Figs. 6.42, 6.43 illustrate a stick-slip chaos, which has been archived via period doubling bifurcations.

This example contains even more information, when one considers moments of sticks for the mass m_2 (see enlargement given in the Fig. 6.50). It is seen how complicated dynamics can occur during stick-slip processes. In Fig. 6.46, the time histories of masses and chaotic motions are reported. For this case we have hyper-chaotic dynamics because two of the Lyapunov exponents are positive and the Lyapunov dimension $d_L = 2.35$ testifies the chaotic attractor as the strange one. In addition, both masses move in different manner: The mass m_1 jumps between two potential-well very quickly, whereas the second mass m_2 exhibits a similar like jumps, which occur slowly. The quasi-periodic motion is presented in Figs. 6.48, 6.49.

The mass m_1 moves on a torus, but the Poincaré map associated with the second mass possesses two symmetric arc form attractors.

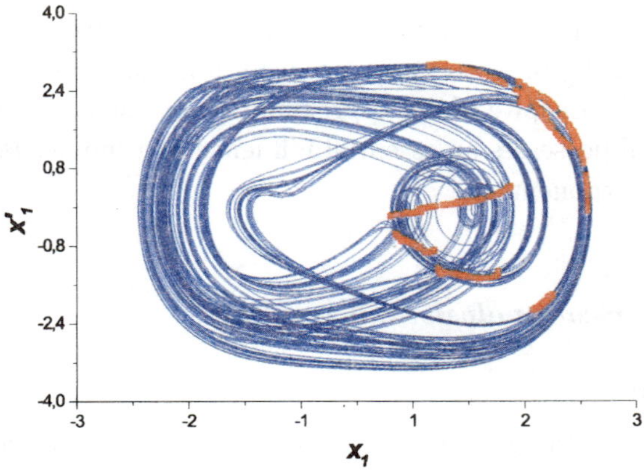

Fig. 6.39 Phase portrait and the corresponding Poincaré map (red points) $\dot{x}_1(x_1)$ for the parameters: $\gamma = 0.03$, $\beta = 1.7$, $V_0 = 2.5$, $\alpha_0 = 0$, $\alpha_1 = 0.002$, $\alpha_2 = 0.09$, $\beta_1 = 0.1$, $\beta_2 = 1.11$, $\xi_1 = 1$, $\xi_2 = 2.4$ and the initial conditions: $x_1 = -0.1$, $x_2 = 0.2$, $\dot{x}_1 = \dot{x}_2 = 0$ (chaotic orbit, Lyapunov exponents: 0.23, -0.01, -0.15, -0.28).

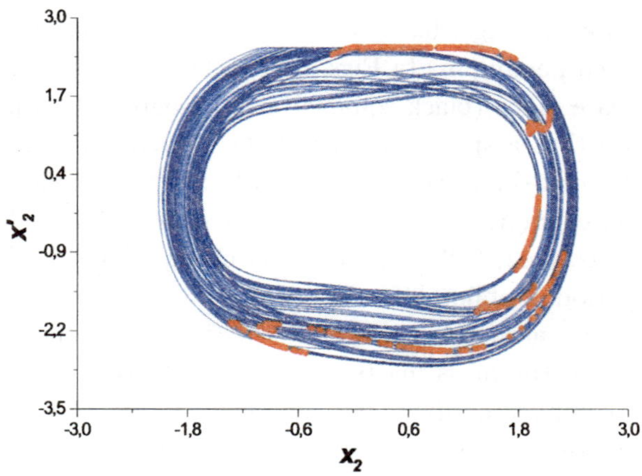

Fig. 6.40 Phase portrait and the corresponding Poincaré map (red points) $\dot{x}_2(x_2)$. Parameters are the same as in Fig. 6.39 (chaotic orbit).

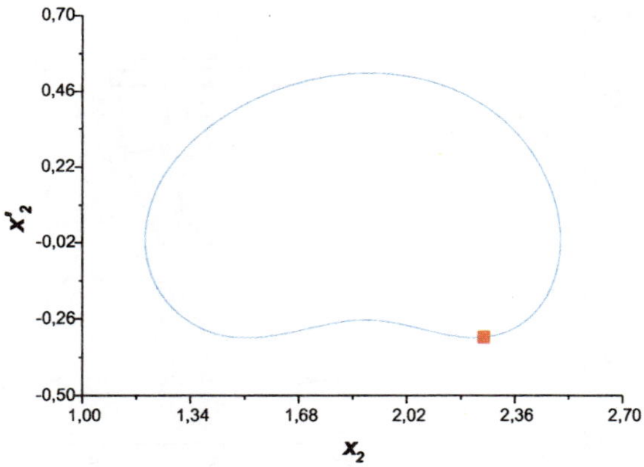

Fig. 6.41 Phase portrait and the corresponding Poincaré map (red square) $\dot{x}_2(x_2)$ for the parameters: $\gamma = 3$, $\beta = 0.04$, $V_0 = 0.2$, $\alpha_0 = 0.1$, $\alpha_1 = 0.03$, $\alpha_2 = 0$, $\beta_1 = 0.31$, $\beta_2 = 0.1$, $\xi_1 = 0.12$, $\xi_2 = 2$ and the initial conditions: $x_1 = x_2 = 0$, $\dot{x}_1 = 0.7$, $\dot{x}_2 = 0$ (periodic orbit, Lyapunov exponents: -0.01, -0.04, -0.19, -0.19).

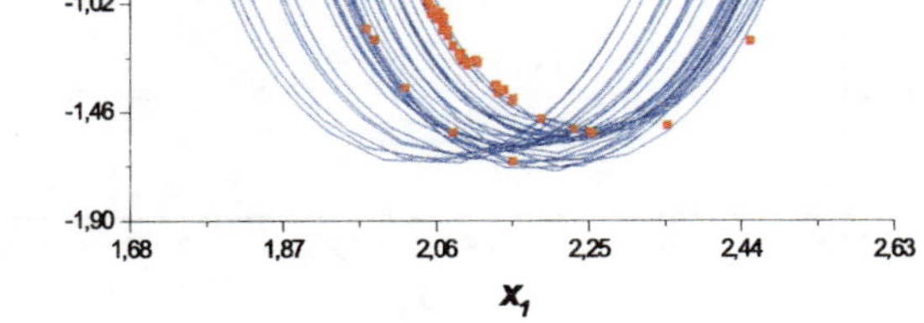

Fig. 6.42 Phase portrait and the corresponding Poincaré map (red squares) $\dot{x}_1(x_1)$ for the parameters: $\gamma = 2.09$, $\beta = 0.984$, $V_0 = 0.043$, $\alpha_0 = 0.01$, $\alpha_1 = 0$, $\alpha_2 = 0.015$, $\beta_1 = 0.31$, $\beta_2 = 0.108$, $\xi_1 = 0.07$, $\xi_2 = 2.23$ and the initial conditions: $x_1 = x_2 = 0$, $\dot{x}_1 = 5$, $\dot{x}_2 = -5$ (chaotic orbit, Lyapunov exponents: 0.48, -0.08, -1.42, -6.31).

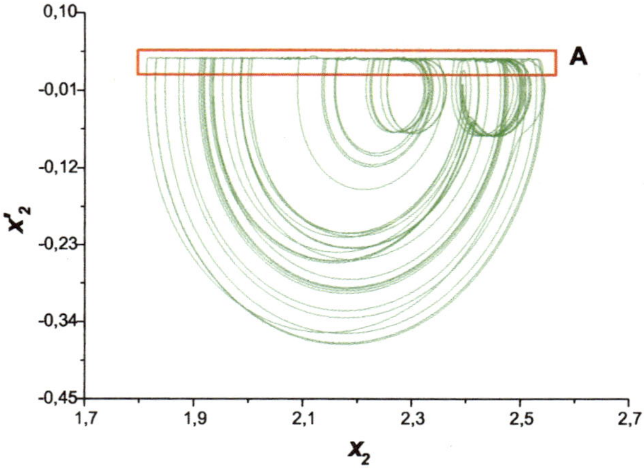

Fig. 6.43 Phase portrait $\dot{x}_2(x_2)$. Parameters are the same as in Fig. 6.42 (chaotic orbit).

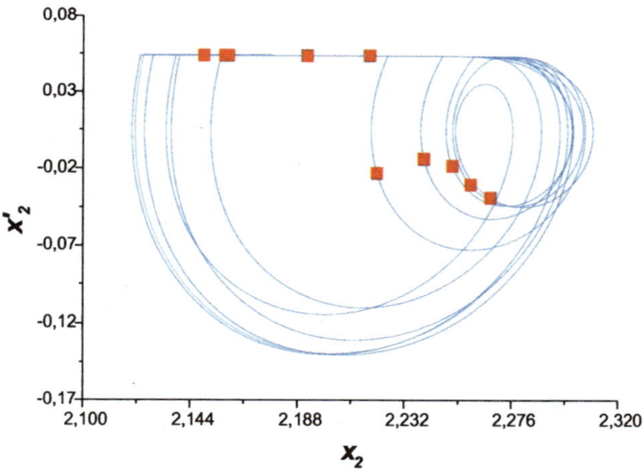

Fig. 6.44 Phase portrait and the corresponding Poincaré map (red squares) $\dot{x}_2(x_2)$ for the parameters: $\gamma = \beta = 1$, $V_0 = 0.05$, $\alpha_0 = \alpha_1 = \alpha_2 = 0.03$, $\beta_1 = \beta_2 = 0.1$, $\xi_1 = 0.12$, $\xi_2 = 1$ and the initial conditions: $x_1 = 0.4$, $x_2 = -0.12$, $\dot{x}_1 = -0.56$, $\dot{x}_2 = 0.12$ (periodic orbit, Lyapunov exponents: -0.16, -0.17, -0.31, -0.33).

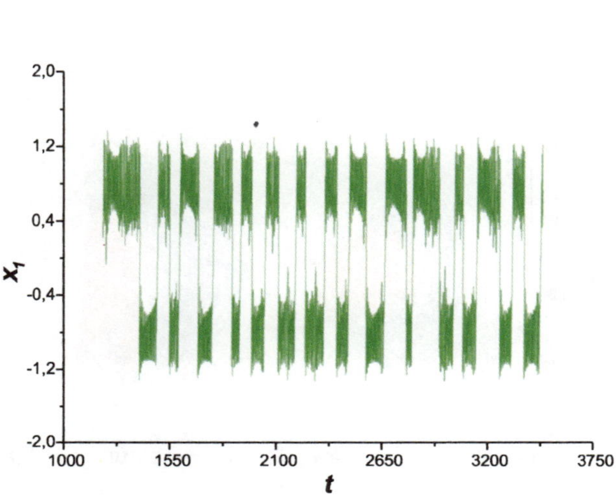

Fig. 6.45 Time history $x_1(t)$ for the parameters: $\gamma = 6$, $\beta = 1$, $V_0 = 3$, $\alpha_0 = 0.03$, $\alpha_1 = 0.05$ $\alpha_2 = 0$, $\beta_1 = 2$, $\beta_2 = 2.34$, $\xi_1 = 0.1$, $\xi_2 = 0.2$ and the initial conditions: $x_1 = 1$, $x_2 = -2.32$, $\dot{x}_1 = -0.78$, $\dot{x}_2 = 3$.

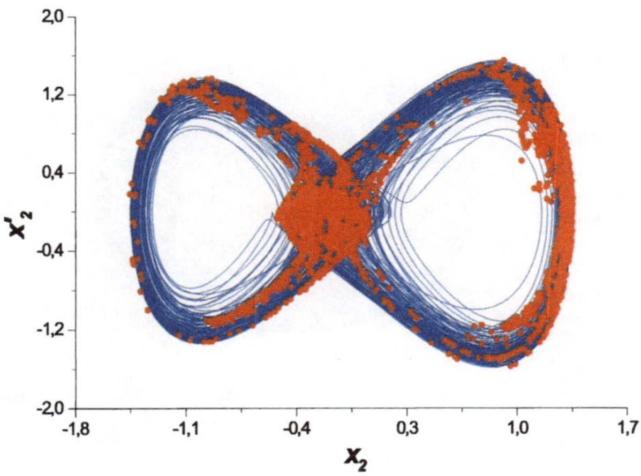

Fig. 6.46 Phase portrait and the corresponding Poincaré map (red points) $\dot{x}_2(x_2)$. Parameters and initial conditions are the same as in Fig. 6.45 (chaotic orbit, Lyapunov exponents: 0.36, 0.21, -0.42, -0.43).

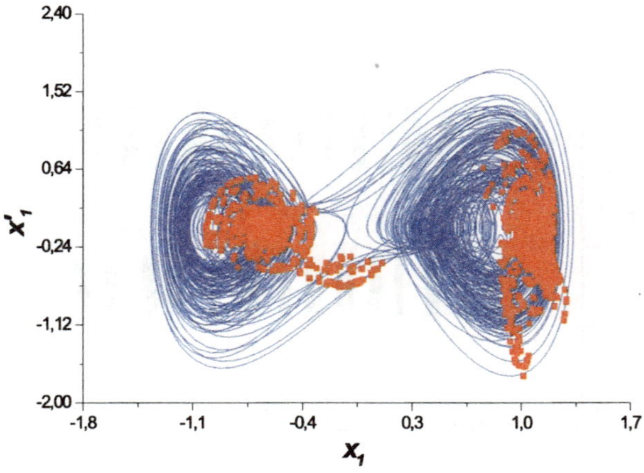

Fig. 6.47 Phase portrait and the corresponding Poincaré map (red points) $\dot{x}_1(x_1)$. Parameters and initial conditions are the same as in Fig. 6.45 (chaotic orbit).

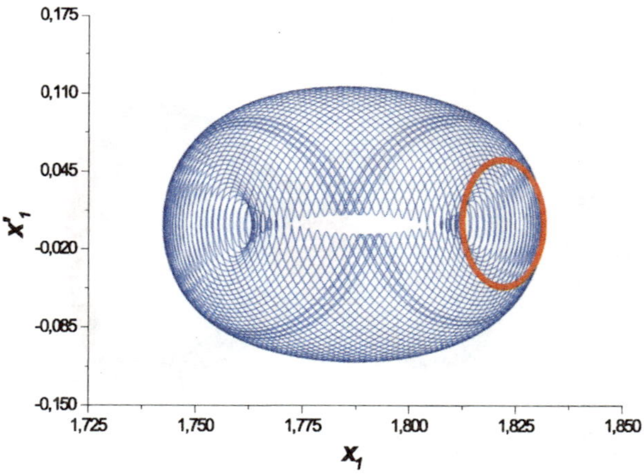

Fig. 6.48 Phase portrait and the corresponding Poincaré map (red oval) $\dot{x}_1(x_1)$ for the parameters: $\gamma = 0.03\beta = 0.77$, $V_0 = 0.2$, $\alpha_0 = 0$, $\alpha_1 = 0.01$, $\alpha_2 = 0$, $\beta_1 = 0.12\beta_2 = 0.2$, $\xi_1 = 0.1$, $\xi_2 = 0.7$ and the initial conditions: $x_1 = 0$, $x_2 = 0.2$, $\dot{x}_1 =$, $\dot{x}_2 = 0$ (quasi-periodic orbit, Lyapunov exponents: 0.00, -0.03, -0.20, -0.20).

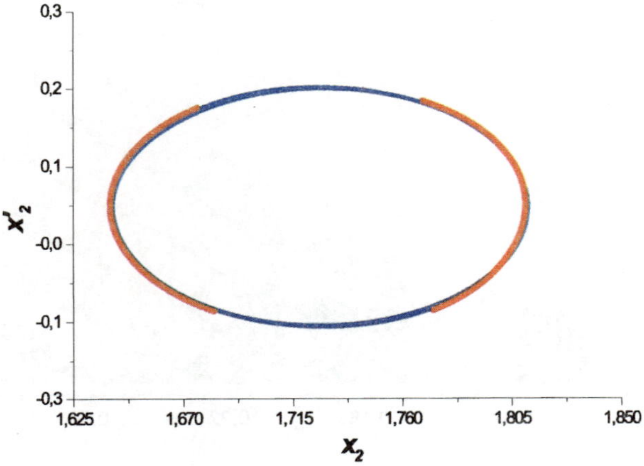

Fig. 6.49 Phase portrait and the corresponding Poincaré map (red lines) $\dot{x}_2(x_2)$. Parameters are the same as in Fig. 6.48 (quasi-periodic orbit).

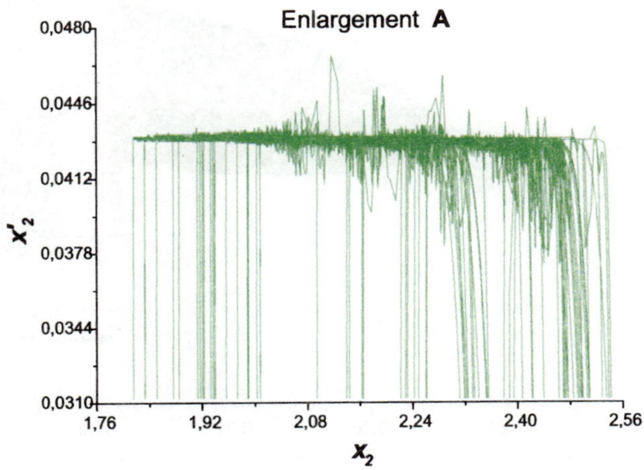

Fig. 6.50 Oscillations of the mass m_2.

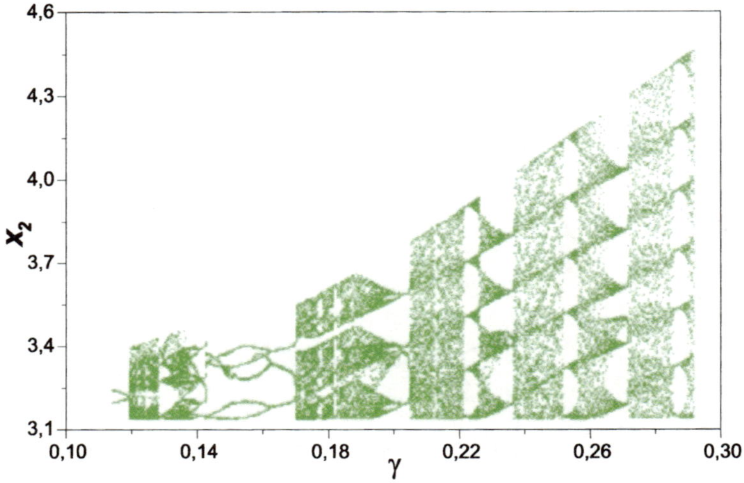

Fig. 6.51 Bifurcation diagram for the parameter $\gamma \in (0.2,\ 0.4)$. The following parameters are fixed: $\beta = 1$, $V_0 = 3$, $\alpha_0 = 0.003$, $\alpha_1 = 0.05$, $\alpha_2 = 0$, $\beta_1 = 0.2$, $\beta_2 = 2.34$, $\xi_1 = 0.45$, $\xi_2 = 0.76$ and the initial conditions: $x_1 = 1$, $x_2 = -2.32$, $\dot{x}_1 = -0.78$, $\dot{x}_2 = 3$.

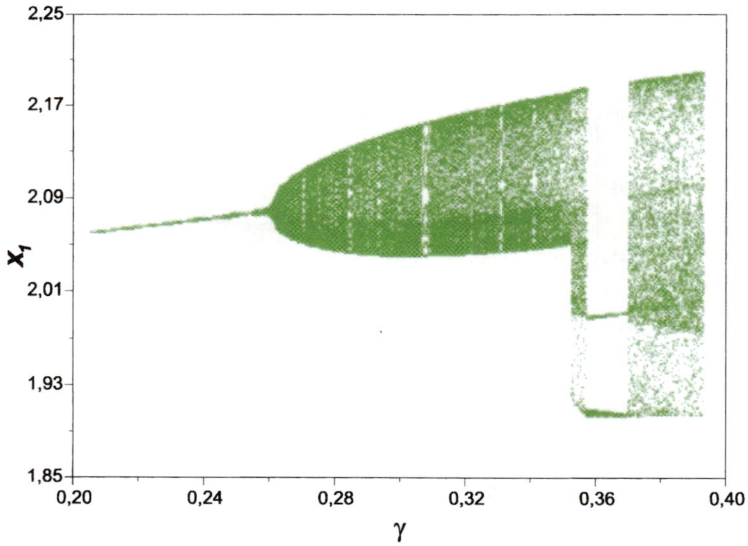

Fig. 6.52 Bifurcation diagram for the parameter $\gamma \in (0.1,\ 0.3)$. The following parameters are fixed: $\beta = 21$, $V_0 = 0.2$, $\alpha_0 = 0.0045$, $\alpha_1 = 0.01$, $\alpha_2 = 0.19$, $\beta_1 = 0.12$, $\beta_2 = 0.2$, $\xi_1 = 0.1$, $\xi_2 = 0.7$ and the initial conditions: $x_1 = -0.1$, $x_2 = 0.2$, $\dot{x}_1 = \dot{x}_2 = 0$.

Finally, we give two examples of bifurcation diagrams. Among others, it is shown that the analysed system with friction exhibits simple bubbles (Fig. 6.51) for $\gamma \approx 0.14$. Another bifurcation diagram (which is rather typical for simple dynamical system), is presented in Fig. 6.52. Both of them show many jumps between periodic and chaotic attractors, and vice versa.

6.3.3 *Concluding remarks*

The classical two-degree-of freedom self-excited system with friction has been analysed using numerical methods. The main purpose of this section has been focussed on analysis of stick-slip regular (periodic and quasi-periodic) and chaotic dynamics. Another goal was to apply a smoothing procedure to model discontinuous friction using the arctan function. It possesses two main advantages.

(1) Slightly modified Runge-Kutta method with variable integration step can be used (during stick phase a larger integration step is recommended).
(2) The smoothness introduced by the arctan function allowed to use classical tools of nonlinear dynamics, and to keep all behaviour related to discontinuous effects.

The sign function is not defined when the relative velocity is equal to zero, whereas the arc tan is a unique function yielding both stick and slip phenomena. During the analysis, all standard technique has been applied, i.e. time histories, phase planes, Poincaré maps, the Lyapunov exponents and the Lyapunov dimensions.

Very rich nonlinear nonsmooth dynamics has been detected. The period doubling route to stick-slip chaos, and from stick-slip chaos to regular motion (Fig. 6.51), very complicated stick-slip periodic orbits (Fig. 6.44), stick-slip hyper-chaos (Figs. 6.33, 6.45–6.47), various quasi-periodic attractors (Figs. 6.35, 6.48, 6.49), as well as different periodic motions exhibited by the masses, i.e. stick-slip and smooth periodic orbits (Figs. 6.27, 6.30, 6.34, 6.41) have been discussed and illustrated.

In addition, the numerical simulations indicate how complicated nonlinear dynamics occurs during the stick-slip processes (Fig. 6.50).

Chapter 7

Piecewise Linear Approximations

7.1 Introduction

The mechanical nonlinearities taken into account for the study of the dynamic behavior of a system have various well-known origins:

- nonlinearities of a geometrical origin introduced for example by a formulation in great displacements or the expression of the curvature;
- nonlinearities resulting from the constitutive laws (mathematically smooth or not), for example due to nonlinear elasticity;
- nonlinearities at the interface, introduced by joints and resulting for example in phenomena of friction or impact. In addition, nonlinearities resulting due to boundary conditions.

These nonlinearities are often introduced into the model via the experimental data. Fitting of these data in order to study a dynamic behavior leads a "designer" to apply mathematically regular interpolations (polynomial in practice) that enable convenient analytical studies or, at the same time, numerical investigations. It is the case in dynamics of the structures in civil engineering to represent the action of the ground on a foundation starting from an "in situ" experiment by a spring with a return strength expressed as a polynomial function of displacement. In addition, simple systems described by piecewise linear models were abundantly studied in the literature. The system of Chua [Madan (1993)], [Chua *et. al.* (1986a)], [Komuro *et. al.* (1991)] constitutes for example a paradigm for the study of chaos, and the studies of global dynamic behavior.

Here we are interested in a model of an arch subject to vibration [Lamarque and Malasoma (1992)], [Szemplińska-Stupnicka (1969)], [Fung and Kaplan (1952)]: The exact model is described by the equation of the vibrations of the first mode, and the approximate models are built artificially by fitting the nonlinear return force intervening in the exact model with a piecewise linear function.

7.2 Exact and Approximated Models

7.2.1 *Exact model*

Here we introduce what we call the "exact model":

$$\ddot{x} + \delta\dot{x} + (a + f\cos(\omega t))x + \gamma x^3 = f\cos(\omega t), \tag{7.1}$$

([Lamarque and Malasoma (1992)] with $\delta = 0.2$, $a = -0.5$, $f = 0.9$ and ω denoting the control parameter). This one degree-of-freedom (DOF) nonlinear differential system describes the behavior of the first mode of vibration of a shallow arch [Lamarque and Malasoma (1992)], [Szemplińska-Stupnicka (1969)], [Fung and Kaplan (1952)].

7.2.2 *Approximated models*

We consider piecewise linear approximations $g_n(x)$ defined for $f(x) = \dfrac{x^3 - x}{2}$ as follows. Let us set $M > 1$, integer n greater than 1: $x_i = M(\dfrac{i-1}{n} - 1)$, $i = 1, \ldots, 2n+1$.

If $x \in [x_i, x_{i+1}], i = 1, \ldots, 2n$, then:

$$\begin{cases} \text{if } -1, 0, 1 \notin [x_i, x_{i+1}], \text{then} \quad g_n(x) = \dfrac{f(x_{i+1}) - f(x_i)}{x_{i+1} - x_i}(x - x_i), \\ \text{if } y = -1, 0, 1 \in [x_i, x_{i+1}], \text{then} \quad \begin{cases} x_i \le x \le y, \text{then} \\ g_n(x) = \dfrac{-f(x_i)}{y - x_i}(x - x_i), \\ y \le x \le x_{i+1}, \text{then} \\ g_n(x) = \dfrac{f(x_{i+1})}{x_{i+1} - y}(x - y), \end{cases} \end{cases}$$

$$\text{if } x \le -M, \text{ then} \quad g_n(x) = \frac{f(x_2) - f(x_1)}{x_2 - x_1}(x - x_1),$$

$$\text{if } x \ge M, \text{ then} \quad g_n(x) = \frac{f(x_{2n+1}) - f(x_{2n})}{x_{2n+1} - x_{2n}}(x - x_{2n}).$$

Thus we preserve the continuity of the function g_n while preserving the equilibrium positions of the function f. Possibly, the symmetry of f can be broken.

7.3 Approximation and Global Dynamic Behavior

In this part, we consider the following general problem:

$$\begin{cases} \dfrac{dX_n}{dt} = F_n(X_n, t),\, t > 0,\, X_n \in \mathbb{R}^m, \\ \\ X_n(0) = X_0, \end{cases}$$

with F_n being the Lipschitz–continuous function locally on $\mathbb{R}^m \times \mathbb{R}$.

We suppose that F_n converges uniformly on every compact of $\mathbb{R}^m \times \mathbb{R}$, which is in general a reasonable assumption. It is clear that uniform compact convergence of F_n to F leads to the following results:

X_n converges uniformly towards X being the solution of:

$$\begin{cases} \dfrac{dX}{dt} = F(X, t),\, t > 0,\, X \in \mathbb{R}^m, \\ \\ X(0) = X_0, \end{cases}$$

on any interval of limited time. Thus it is clear that without a total uniform convergence, one cannot mathematically capture the global dynamic behavior of the system. On the other hand, if the global dynamic behavior of a particular system that one wishes to study is primarily localized on compact space of the phases, one can hope that an approximation of F_n with n reasonably large is capable of describing global dynamics. An assumption stronger than the uniform compact convergence of F_n towards F is not possible if one wishes to approach the polynomials of degree > 1 by piecewise linear functions and to have functions defined by a segment number rather large but finite.

In the case considered in the section 7.2, it is thus clear that the best result of convergence that one can obtain is uniform compact convergence of g_n to f. Then only numerical investigations inform us about behaviors of oscillators governed by exact models or by various approximate models.

7.4 Numerical Results

7.4.1 *Numerical method*

For the numerical study of the exact and approximate systems we considered:

- the well-known Runge-Kutta numerical scheme of the 4th order (carried out on each segment for the approximate models, for in the theory, this numerical scheme is available only for a regularization of the piecewise linear functions by functions of the classes C^4 at least).
- the following values of the parameters: $a = -0.5$, $\delta = 0.2$, $c = -a$, $\gamma = 0.9$, variable ω.

7.4.2 *Periodic solutions*

We studied Poincaré maps numerically:

$$\begin{cases} P : \mathbb{R}^2 \mapsto \mathbb{R}^2, \\ (x_0, \dot{x}_0) \longrightarrow (x(T), \dot{x}(T)), \end{cases} \tag{7.2}$$

with $T = \dfrac{2\pi}{\omega}$, $(x(T), \dot{x}(T))$ being the solution of:

- Either the exact model with initial conditions $(x_0, \dot{x}_0)$,
- Or any approximated model with initial conditions $(x_0, \dot{x}_0)$.

For a step of integration $\leq \dfrac{T}{400}$ and an area of initial conditions $(x, \dot{x}) \in [-5, 5] \times [-10, 10]$ separated into 400×400 points we obtained fixed points of the Poincaré map corresponding to periodic solutions. Table 7.1 presents the coordinates of the fixed points obtained with a tolerance of 10^{-5} for various values of ω and $n = 15$ (30 segments) then $n = 50$ (100 segments) with $M = 5$. We note a good agreement between the exact and approximate models. Table 7.2 illustrates the study for $\omega = 2.15$ and presents the number of periodic solutions T found according to the number of segments. Table 7.3 indicates the same results for $\omega = 2.06$. Because of the numerical tolerance, some slightly different fixed points of the Poincaré map can be counted as 2 fixed points, whereas they clearly will represent the same limit cycle or period-T-limit cycle appears in fact like $2T$ periodic. This explains for example why for $\omega = 2.15$ and $n = 8$ (16 pieces) one finds more than 3 period-$2T$-limit-cycles.

Table 7.1 Limit cycles for different values of forcing frequency ω for the exact model and approximations with 30 and 100 linear pieces. Coordinates of fixed point of the Poincaré map are provided.

ω	Cycle's order	Exact model	30 pieces	100 pieces
2.15	cycle 1	(-0.93261, 0.07321)	(-0.8579, 0.0693)	(-0.9274, 0.0748)
2.15	cycle 2	(0.51083, 0.01009)	(0.5365, 0.0093)	(0.5137, 0.0100)
2.15		(1.26335, -0.16047)	(1.2106, -0.1408)	(1.2589, -0.1585)
2.15	cycle 2	(-0.75978, 2.82986)	(-0.7640, 2.8132)	(-0.7601, 2.828)
2.15		(-0.20178, -1.78705)	(-0.2015, -1.7866)	(-0.2017, -1.7872)
2.00	cycle 2	(-.75749, 2.72642)	(-.7616, 2.7068)	(-.7580, 2.7247)
2.00		(-.38806, -1.47462)	(-.3868, -1.4747)	(-.3879, -1.4748)

Table 7.2 $\omega = 2.15, T = \dfrac{2\pi}{\omega}$. Periodic solutions are looked for in the 150-periods-T time series. Fixed points of numerically approximate Poincaré map are found with an error of 10^{-5}.

Number of pieces	4	6	8	10
Number of limit cycles of period T	0	1	1	1
Number of limit cycles of period 2T	1	1	0	1
Number of pieces	12	14	16	18
Number of limit cycles of period T	1	1	1	2
Number of limit cycles of period 2T	1	1	≥ 3	1
Number of pieces	20	30	100	exact
Number of limit cycles of period T	≥ 3	2	2	2
Number of limit cycles of period 2T	1	2	2	2

Table 7.3 $\omega = 2.06, T = \dfrac{2\pi}{\omega}$. Periodic solutions are looked for in the 150-periods-T time series. Fixed points of numerically approximate Poincaré map are found with an error of 10^{-5}.

Number of pieces	4	6	8	10
Number of limit cycles of period T	0	1	1	1
Number of limit cycles of period 2T	1	1	1	0
Number of pieces	12	14	16	18
Number of limit cycles of period T	1	1	1	1
Number of limit cycles of period 2T	≥ 3	1	1	3
Number of pieces	20	30	100	exact
Number of limit cycles of period T	1	1	2	2
Number of limit cycles of period 2T	2	2	1	1

7.4.3 *Basins of attraction*

In order to test the capacity of the approximate models to describe accurately the exact global dynamics, we propose the calculation of basins of attraction. For each presented example, the zone of initial conditions studied is $[-5, 5] \times [-10, 10] \ni (x, \dot{x})$. This zone was cut into 400×400 cells. The caption common to all the presented figures is the following one: black pixels correspond to the initial conditions leading to divergent trajectories; blue pixels correspond to the initial conditions which lead to a limit cycle of period T; red pixels correspond to the initial conditions which lead to a limit cycle of period $2T$; white pixels correspond to the initial conditions which lead towards an attractor corresponding to none of the preceding cases (it means that there could be limit cycles with period other than T or $2T$, chaos or quasi-periodic steady-states).

For the value $\omega = 2.15$, Figures 7.1 and 7.2 depict the basins of attraction for the exact model and the approximate model, respectively with $n = 15$. One observes good qualitative and quantitative agreement between the two figures: The good results of the Table 8.2 concerning the determination of the limit cycles are confirmed by a good agreement between the global dynamics.

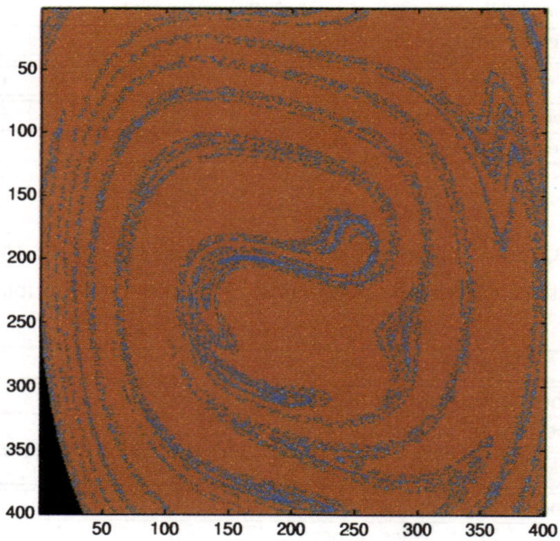

Fig. 7.1 Basin of attraction - exact model $\omega = 2.15$.

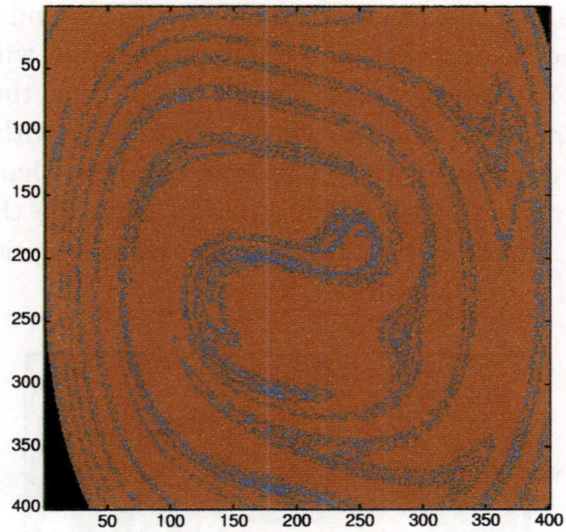

Fig. 7.2 Basin of attraction - piecewise linear model $2n = 30$, $\omega = 2.15$.

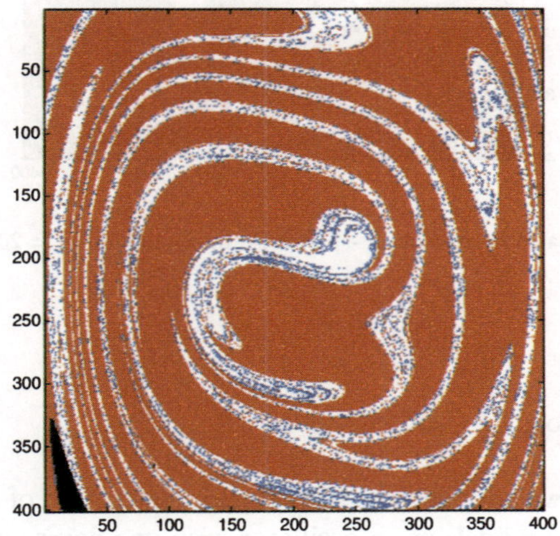

Fig. 7.3 Basin of attraction - exact model $\omega = 2.06$.

For the value $\omega = 2.06$, Figures 7.3 and 7.4 correspond to the basins of attraction for the exact model, and the approximation with $2n = 20$ and $2n = 50$ pieces, respectively. This time it is seen that the cycle of order 1 of the exact model is found to agree qualitatively with $n = 10$. The other basins are described better. This example shows that a "reasonable" approximation ($n = 10$) makes it possible to mark the thin basin of the cycle of order 1. In this case, the quantitative description of the basin would be carried out only with $n > 50$.

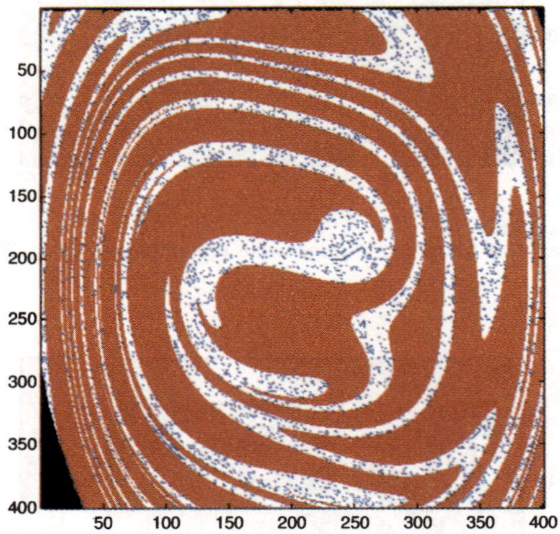

Fig. 7.4 Basin of attraction - piecewise linear model 2n = 20, $\omega = 2.06$.

7.5 Conclusion

In this chapter we consider the numerical investigation of an exact model and approximate models of an oscillator with one degree of freedom. The example shows that a piecewise linear approximation with a reasonable number of pieces (20 to 30) makes it possible to describe qualitatively, or even quantitatively the global dynamics. We underline nevertheless the difficulty, both theoretical and practical, of the choice of the number of pieces for the qualitative and especially quantitative analysis of global dynamics. In fact, from the experimental data, one can be faced with the following difficult choice:

- either one smoothens (for example by a polynomial of degree 3) the data with the risk to calculate "false" coefficients and thus erroneously locate in the diagram of bifurcation corresponding to the evaluation of dynamics related to these parameters,
- or one adopts a piecewise linear approximation based on presumed reliable experimental data. One may assume that good fitting is obtained when the behavior simulated by means of these two types of models is at least qualitatively close to each other.

Chapter 8

Chua's Circuit with Discontinuities

8.1 Introduction

Many works have been devoted to the study of the global behaviour of smooth nonlinear oscillators (Duffing oscillator [Ueda (1979)], shallow arches [Lamarque and Malasoma (1992)], [Malasoma *et. al.* (1994)], [Szemplińska-Stupnicka (1969)], Lorenz's system [Lorenz (1963)], [Sparrow (1982)] or smooth maps (logistic map [Coullet and Tresser (1984)], [Feigenbaum (1978)], etc.): Both periodic, quasi-periodic or chaotic behaviour have been investigated, bifurcations, transitions, universal behaviour have been studied. At the same time a number of papers deal with "simpler, integrable" systems which seem to be "paradigms": unimodal maps [Collet and Eckmann (1980)], [Li and Yorke (1975)], Lozi's attractor [Lozi (1978)], Chua's double scroll circuit [Madan (1993)], etc. Indeed the latter are concerned with piecewise linear dynamics exhibiting chaotic behaviour via Poincaré maps [Chua *et. al.* (1986a)], [Komuro *et. al.* (1991)], [Lozi (1978)], that have been build analytically.

But few studies deal with the bifurcations and the global behaviour of unsmooth systems, i.e. systems with mathematical difficulties such as discontinuities or/and multivalued differential equations. Such models are interesting from the point of view of applications: impacts, friction, shocks and constitutive laws provide unsmooth models of that type [Awrejcewicz and Delfs (1990a)], [Awrejcewicz and Delfs (1990b)], [Capecchi and Vestroni (1995)], [Deimling (1992)], [Dowell and Schwartz (1983a)], [Dowell and Schwartz (1983b)], [Ferri and Bindemann (1995)], [Mahla and Badan Palhares (1993)], [Monteiro Marques (1994)],[Moreau (1988)], [Paoli (1993)], [Paoli *et. al.* (1992)], [Pfeiffer (1988a)], [Popp and Stelter (1990)], [Shaw and Shaw (1989)], [Whiston (1987)].

Mathematical results are sometimes available for studying existence and uniqueness of such nonlinear dynamical systems ([Brezis (1973)], [Deimling (1992)], [Monteiro Marques (1994)], [Moreau (1988)], [Schatzman (1978)],) but generally such results are not followed by correct numerical investigations if the exact integration of the nonlinear system is not possible (except [Paoli (1993)], [Paoli and Schatzman (1993c)], for example).

Our intention here has two main targets. First we present examples of electromechanical realization of Chua's circuit and Chua's *unfolding* circuit. In addition a novel mechanism is proposed for realizing all *unfoldings* of Chua's equations in a purely mechanical way using friction properties. The next target is focused on generalization of the Chua double scroll model to a kind of "non smooth" paradigm. In section 8.2 mechanical realizations of Chua's equations are proposed and discussed [Awrejcewicz and Calvisi (2002)]. In section 8.3 we introduce a model with a finite number of discontinuities. In section 8.3.2 we describe the mechanical point of view. In section 8.3.3 we present the mathematical frame work, study existence and uniqueness of solutions in a general case, and apply the previous results to particular cases of discontinuities (at zero and both -1 and 1). In section 8.3.4 we show how to analytically build the solutions for the two previous cases. In section 8.3.5 we present numerical results for two particular cases. In the first case two discontinuities at -1 and 1 are considered: Bifurcation diagrams are presented and transition to chaos via a bifurcation cascade is investigated. In the second case, discontinuity is located at 0. Bifurcation diagrams are illustrated by phase portraits, Poincaré sections and global behaviour. Trapping areas for the trivial equilibrium are studied. Chaos is pointed out by using the computation of Lyapunov exponents. Then in the last section we draw conclusions from our investigations and point out some extensions.

8.2 Mechanical Realizations of Chua's Circuit

8.2.1 *Introduction*

Chua's circuit (see Fig. 8.1) is one of the simplest physical models which has been widely investigated by mathematical, numerical, and experimental methods. One of the main attractions of Chua's circuit is that it can be easily build with less than a dozen standard circuit components, and has often been referred to as the poor man's chaos generator. A mathematical analysis of the global *unfolding* behaviour of Chua's circuit is given in [Chua

(1993)]. Perhaps one of the most important observation is that by adding a linear resistor in series with the inductor in Chua's circuit, the resulting unfolded Chua's circuit is topologically equivalent to a 21-parameter family of continuous odd-symmetric piecewise linear differential equations in R^3. Any vector field belonging to the "unfolded" topologically conjugate family can be transformed (mapped) via a non singular linear transformation to an unfolded Chua's circuit with only 7 parameters. In addition, it extends the local concept of unfolding to a global one, where all results are valid for the whole space R^3. In other words, any autonomous 3-dimensional system characterized by an odd-symmetric 3-segment continuous piecewise linear function can be mapped to an unfolded Chua's circuit having identical qualitative dynamics.

L. O. Chua [Chua (1993)] stated the following question: since there are several different 3rd-order circuits (which exhibit strange attractors) composed of a continuous odd-symmetric piecewise linear vector field in R^3, does a homeomorphic mapping of such circuits to an unfolded Chua's circuit exists? If such a homeomorphism exists the two circuits are said to be equivalent (or topologically conjugate). The unfolded Chua's circuit is canonical in the sense that the governing equations contain a minimum number of parameters for observing the full generality of dynamical behaviours. We are going to present a mechanical device model of Chua's circuit, as well as an unfolded Chua's circuit.

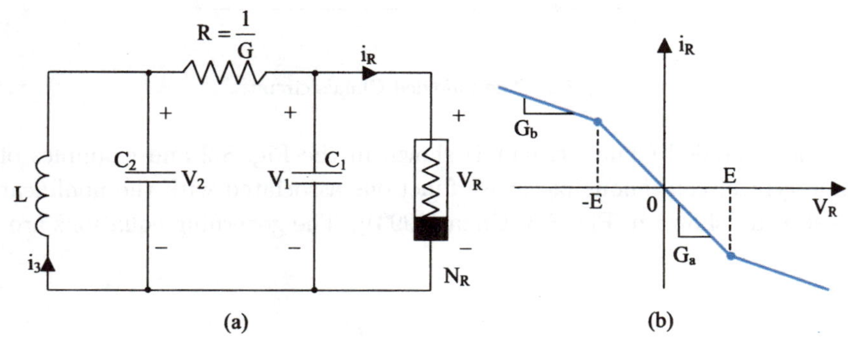

Fig. 8.1 (a) Chua's circuit is made of 4 standard linear circuit components and a nonlinear resistor; (b) $V_R - i_R$ characteristic of the nonlinear resistor, which can be synthesized by 2 standard OP AMPS (operational amplifiers) and 6 linear resistors [Kennedy (1992)].

Chua's circuit is shown in the Fig. 8.1. The governing equations have the form

$$
\begin{aligned}
C_1\frac{dv_{C_1}}{dt} &= \frac{1}{R}(v_{C_2} - v_{C_1}) - g(v_{C_1}),\\
C_2\frac{dv_{C_2}}{dt} &= \frac{1}{R}(v_{C_1} - v_{C_2}) - i_L,\\
L\frac{di_L}{dt} &= -v_{C_2},
\end{aligned}
\tag{8.1}
$$

or in nondimensional form they are described by a two parameter family.

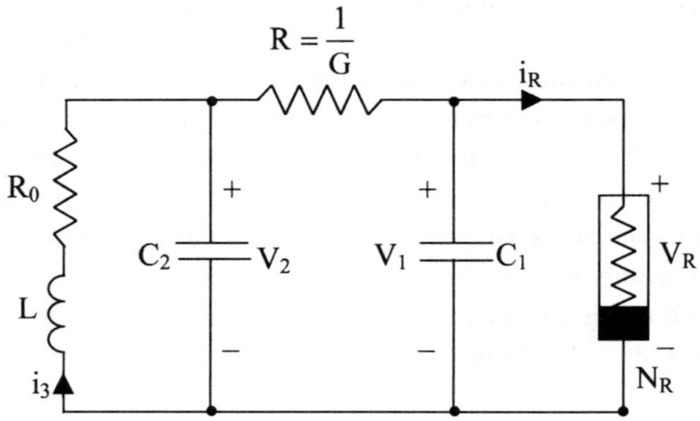

Fig. 8.2 The unfolded Chua's circuit.

The unfolded Chua's circuit is shown in the Fig. 8.2 and examples of some typical continuous piecewise functions associated with the nonlinear resistor are shown in Fig. 8.3 [Chua (1993)]. The governing equations are

$$
\begin{aligned}
\frac{dx}{dt} &= \alpha[y - f(x)],\\
\frac{dy}{dt} &= x - y + z,\\
\frac{dz}{dt} &= -\beta y.
\end{aligned}
\tag{8.2}
$$

$$\frac{dv_1}{dt} = \frac{1}{C_1}[G(v_2 - v_1) - f(v_1)],$$

$$\frac{dv_2}{dt} = \frac{1}{C_2}[G(v_1 - v_2) + i_3], \tag{8.3}$$

$$\frac{di_3}{dt} = -\frac{1}{L}(v_2 + R_0 i_3),$$

where

$$G = \frac{1}{R},$$

and

$$f(v_1) = G_b v_1 + \frac{1}{2}(G_a - G_b)\{|v_1 + E| - |v_1 - E|\}. \tag{8.4}$$

In the nondimensional form, the unfolded Chua's circuit can be reduced to the form

$$\begin{aligned} \dot{x} &= \alpha(y - x - f(x)), \\ \dot{y} &= x - y + z, \\ \dot{z} &= -\beta y + \gamma z, \end{aligned} \tag{8.5}$$

where the relations between the dimensional and nondimensional parameters are given below

$$\begin{aligned} &x = v_1/E,\ y = v_2/E, \quad z = c_3/(EG), \\ &\tau = tG/c_2,\ m_0 = G_a/G,\ \ m_1 = G_b/G, \\ &\alpha = c_2/c_1,\ \beta = c_2/(LG^2),\ \gamma = c_2 R/(LG). \end{aligned} \tag{8.6}$$

At this point, we would like to remark that the piecewise linear characteristics shown in the Fig. 8.3 can appear in a mechanical model in an even more natural way. In particular, some limiting piecewise linear characteristics corresponding to vertical and/or horizontal mechanical positions can be very easily realized.

8.2.2 *Mechanical models of Chua's circuit*

(i) Geometrical Construction of the Piecewise Linear Function

In order to realize Chua's piecewise linear function with mechanical devices, consider the simple geometry shown in the Fig. 8.4.

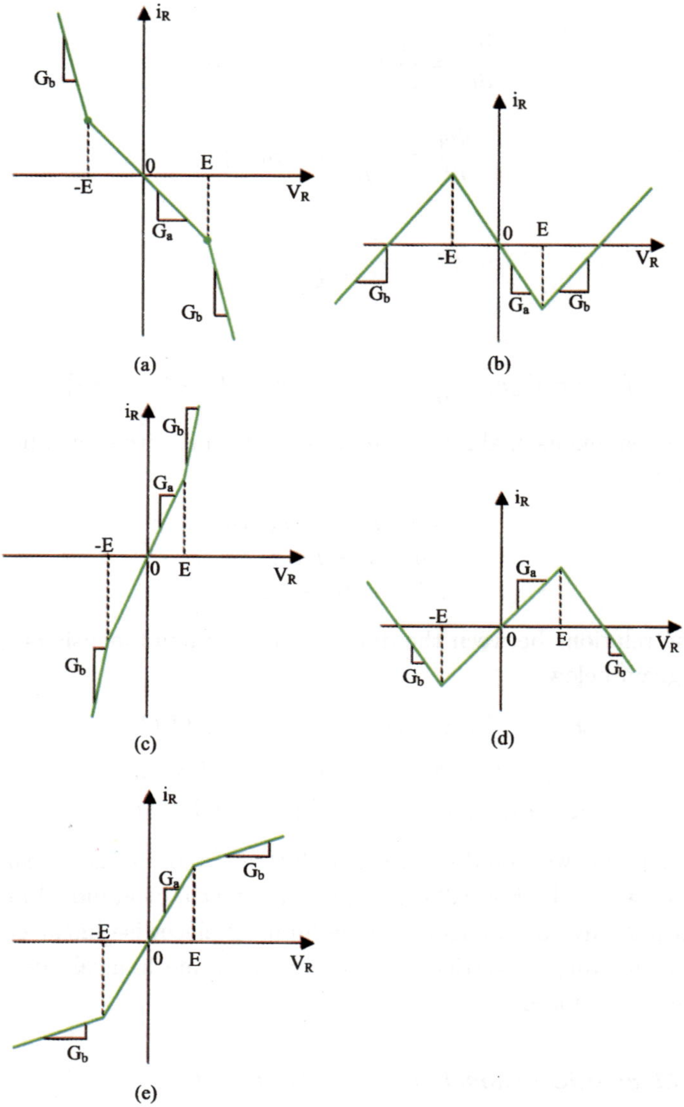

Fig. 8.3 A family of piecewise linear resistor characteristics (a) $G_b < G_a < 0$; (b) $G_a < 0, G_b > 0$; (c) $G_b > G_a > 0$; (d) $G_a > 0, G_b < 0$; (e) $G_a > G_b > 0$.

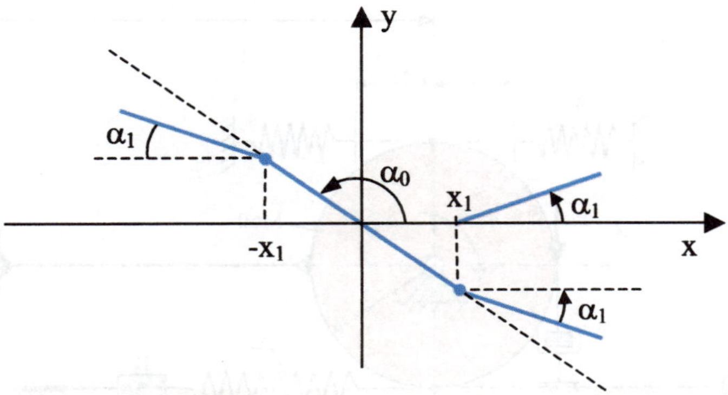

Fig. 8.4 Definition of slope parameters $m_0 = \tan\alpha_0 < 0,\ m_1 = \tan\alpha_1 > 0$.

Assuming $\frac{\pi}{2} \leq \alpha_0 \leq \pi$ and $-x_1 \leq x \leq x_1$, the equation of a straight line through the origin is given by

$$y(x) = m_0 x, \tag{8.7}$$

where $m_0 = \tan\alpha_0$.

The equation of a straight line through $(x_1, 0)$ is given by

$$y(x) = m_1 x + b_1, \tag{8.8}$$

where $m_1 = \tan\alpha_1$. Since $y(x_1) = 0$, it follows that $b_1 = -m_1 x_1$, and hence the equation of the right straight line segment is given by

$$y = m_0 x + m_1(x - x_1) \tag{8.9}$$

for $x > x_1$. In a similar way, the equation of the left straight-line segment is given by

$$y = m_0 x + m_1(x - x_1) \tag{8.10}$$

for $x < -x_1$, where $m_1 > 0$, $m_0 < 0$ and $|m_0| > |m_1|$.

(ii) Mechanical Model of Chua's Circuit

For our first mechanical model of Chua's circuit, we will use three mechanical devices and then couple them via electromechanical devices.

Let us begin with our mechanical realization of a *negative* slope using the device shown in the Fig. 8.5. The mechanism is composed of a rotating disc of radius r whose center is fixed in space and whose inertial moment

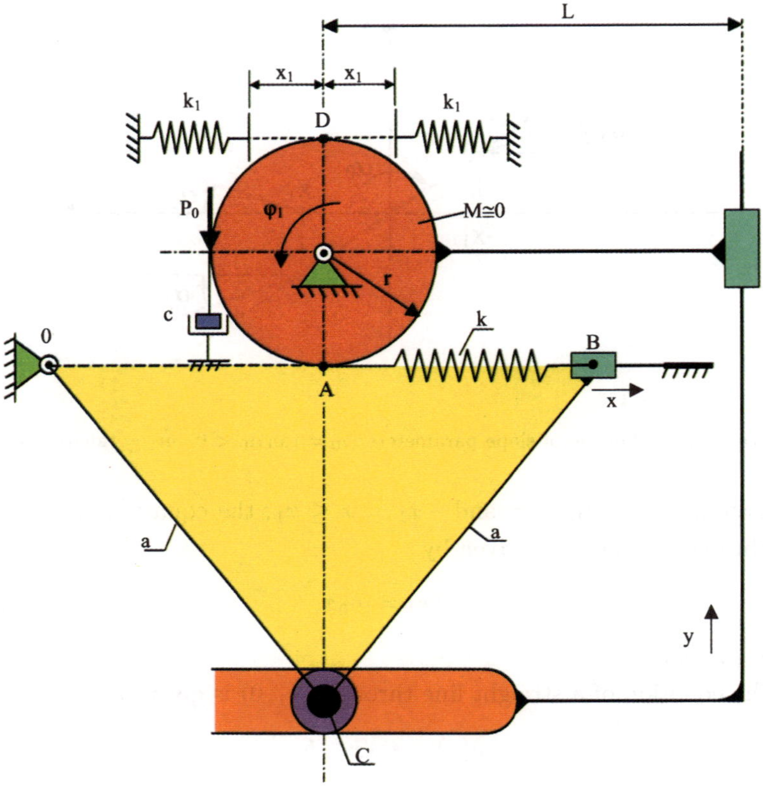

Fig. 8.5 A mechanical device for producing negative stiffness.

is neglected ($M \approx 0$). Its rotation is defined by the angle $\varphi_1(t)$ and is positive in the counterclokwise direction. A dashpot with viscous damping coefficient c as well as a spring with stiffness coefficient k are attached to the disc. It is assumed that all springs and dashpots are linear and mass-less, a classical mechanical assumption. Therefore, the damping force generated by the dashpot is proportional to $-\mathrm{cr}\dot{\varphi}_1$, whereas the conservative force generated by the spring is proportional to $-k(x_A - x_B)$, where x_A and x_B are the displacements of the spring terminals A and B, respectively. At the disc point D a bar perpendicular to the plane of the figure is attached. Observe that the construction is a mechanism, since the degree of freedom of the mechanism $w = 3n - 2p$, where n denotes a number of rigid elements ($n = 5$ in our case), and p is a number of the first class kinematic pairs

($p = 7$ in our case).

Therefore for our mechanism we have $w = 1$, which is satisfied, since its degree of freedom is equal to one (φ_1). When the point D moves within the interval $(-x_1, x_1)$ a force reaction is generated by the dashpot with damping coefficient c, and by the spring with stiffness k. The equations of motion can be easily derived using Newton's law. In order to get a piecewise linear characteristics, two symmetric and linear springs with stiffness k_1 are attached which is activated when $|\varphi_1 r| \geq |x_1|$. When the point D (bar) meets one of the springs we also neglect any possible impact effects and assume that springs are compressed linearly. A rigid rectangular frame is attached to the disc and to point C linked via a rolling bearing which moves along the sliding end of the frame. The bottom pink structure consists of a tube with a ball bearing sliding sidewise. This construction prevents the system against any possible wedging caused by friction. The triangle OCB has two equal and rigid sides OC=CB=a. The rotation of the bearing ring is not included in the derivation of the following equations. Observe that the motion of the system is flat, the end O of the arm OC undergoes only a pure rotation, whereas the point B moves only horizontally, and the points O, A and B lie on a straight horizontal line.

In order to produce a negative stiffness we have constructed the mechanism which for the indicated rotation direction of the disc forces point A and B to move horizontally while satisfying the condition $X_B > X_A$, where X_B and X_A denotes the shifts of points A and B, respectively. According to the marked rotation direction φ_1, as point C moves up, it forces the point B to move to the right. This means that both points A and B move in the same direction.

Thus, a negative stiffness is realized when $X_B > X_A$. Consider now the kinematics related to the triangle OBC in Fig. 8.6. Observe that $CD \perp DE$ and $CE \perp OC$ ($\Delta OCE \sim \Delta CDE$). Defining $y = CD$ and $CK = h$, we obtain

$$\frac{y}{DE} = \frac{OK}{h}, \tag{8.11}$$

and hence

$$DE = \frac{hy}{\sqrt{a^2 - h^2}}, \tag{8.12}$$

where $y = L\varphi_1$ (see the Fig. 8.5). The displacement of the point B is given

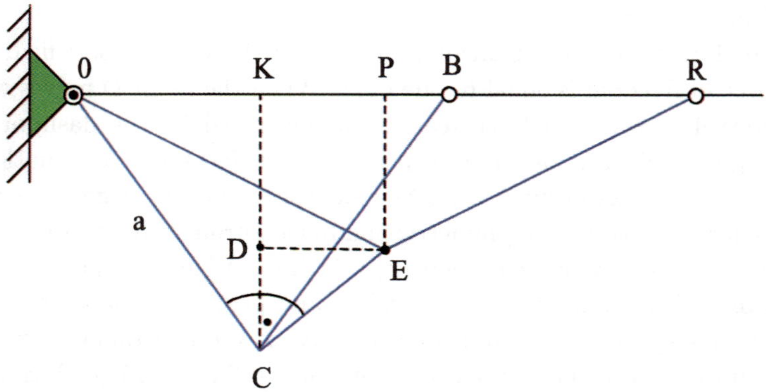

Fig. 8.6 Kinematics of the mechanical device in Fig. 8.6.

by

$$X_B \equiv BR = 2DE. \tag{8.13}$$

Substituting (8.12) into (8.13) we obtain

$$X_B = 2\frac{hL}{\sqrt{a^2 - h^2}}\varphi_1. \tag{8.14}$$

The equation of motion $-x_1 < \varphi_1 r < x_1$ is given by

$$M\ddot{\varphi}_1 + c\dot{\varphi}_1 r^2 - k\left(\frac{2hL}{\sqrt{a^2 - h^2}} - r\right) r\varphi_1 = P_0 r. \tag{8.15}$$

It is easy to design a construction with $2hL(a^2 - h^2)^{-1/2} > r$, thereby realizing a negative stiffness. For example for $r = \frac{h}{2}$ we get $16L^2 > (a^2 - h^2)$, which can be achieved using the device shown in the Fig. 8.5.

The equation of motion of $\varphi_1 r > x_1$ $(\varphi_1 r < -x_1)$ is given by

$$M\ddot{\varphi}_1 + c\dot{\varphi}_1 r^2 - k\left(\frac{2hL}{\sqrt{a^2 - h^2}} - r\right) r\varphi_1 + k_1(\varphi_1 - \varphi_1^*)r^2 = P_0 r, \tag{8.16}$$

where: $x_1 = r\varphi_1^*$. Introducing the notations

$$cr = c_t, \qquad -k\left(\frac{2hL}{\sqrt{a^2 - h^2}} - r\right) = k_{1t}, \qquad k_1 r = k_t, \tag{8.17}$$

in (8.16) and setting $M = 0$, we obtain

$$c_t \dot{\varphi}_1 + k_t(\varphi_1 - \varphi_1^*) + k_{1t}\varphi_1 = P_0, \tag{8.18}$$

where now $k_{1t} < 0$.

Let us examine next the two other devices shown in the Fig. 8.7.

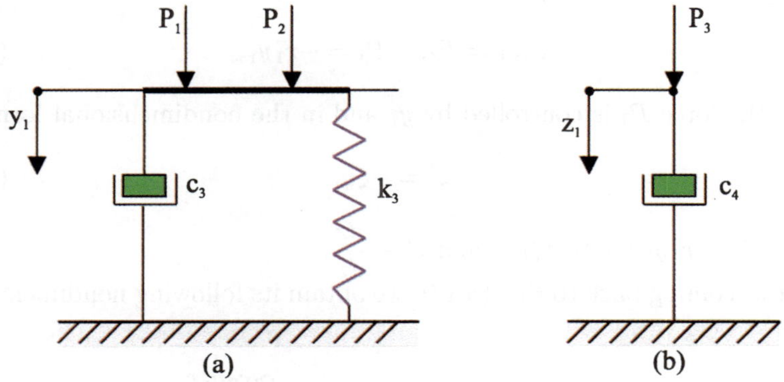

Fig. 8.7 Two mass-less devices with (a) two inputs and (b) one input P_i $(i = 1, 2, 3)$.

The first device (Fig. 8.7a) consists of a dashpot with damping coefficient c_3 and a spring with stiffness coefficient k_3 both being driven from a common force $P_1 + P_2$. It is governed by the equation

$$c_3 \dot{y}_1 + k_3 y_1 = P_1 + P_2, \tag{8.19}$$

where $P_1 = r_3 z_1$, $P_2 = r_4 r \varphi_1$ are electromechanical devices whose forces are controlled by the linear position z_1, defined in the Fig. 8.7b, and by angular position φ_1 defined in the Fig. 8.5, respectively, of other device components of the system. Introducing the nondimensional quantities

$$y_1 = y_1^* y, \quad \varphi_1 = \varphi_1^* \varphi, \quad \tau = \alpha_1 t, \quad \alpha_1 = k_3 c_3^{-1}, \quad z_1 = z_1^* z, \tag{8.20}$$

we get

$$y' + y = \frac{r_3 z_1^*}{y_1^* k_3} z + \frac{r_4 r \varphi_1^*}{y_1^* k_3} \varphi, \tag{8.21}$$

where $y' \equiv \dfrac{dy}{d\tau}$.

Assuming $r_3 z_1^* = y_1^* k_3$, $r_4 r \varphi_1^* = y_1^* k_3$ we obtain the first required equation,

$$y' + y = z + \varphi. \tag{8.22}$$

The system in the Fig. 8.7b is governed by the equation,

$$c_4 \dot{z}_1 = P_3, \quad P_3 = -r_1 y_1, \tag{8.23}$$

where the force P_3 is controlled by y_1 and in the nondimensional form,

$$z' = -\xi y, \tag{8.24}$$

where: $\xi = (r_1 y_1^* c_3)/(c_4 z_1^* k_3)$ and $z' \equiv \dfrac{dz}{d\tau}$.

Now, coming back to Eq. (8.17), we obtain its following nondimensional form

$$\varphi' + \alpha[m_1(\varphi - 1) + m_0\varphi] = \frac{c_3 r_2 y_1^* r}{k_3 c_t \varphi_1^*} y, \tag{8.25}$$

where $m_0 = k_{1t}/k_3$, $\alpha = c_3/c_t$, $m_1 = k_t/k_3$, $P_0 = r_2 y_1^* y$. We require

$$k_3 \varphi_1^* = r_2 y_1^* r, \tag{8.26}$$

which realizes the desired relationship $r_2 = k_3 \varphi_1^* / y_1^* r$.

Denoting

$$h(\varphi) = m_1(\varphi - 1) + m_0\varphi, \tag{8.27}$$

we obtain the following remaining relationship:

$$\varphi' = \alpha[y - h(\varphi)]. \tag{8.28}$$

Observe that Eqs. (8.21), (8.22) and (8.28) describe Chua's circuit.

(iii) Mechanical Model of Unfolded Chua's Circuit

As already stated in the introduction, much richer examples of bifurcation and chaotic dynamics can be observed by an *unfolding* of Chua's circuit. Let us build the corresponding mechanical system. In fact it needs only a slight change. Instead of the dashpot device shown in the Fig. 8.7b we use the spring dashpot combination shown in the Fig. 8.8.

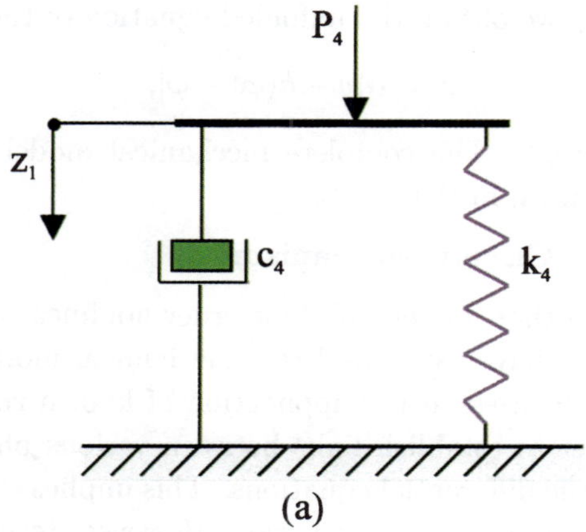

Fig. 8.8 A modified part of Chua's circuit.

The corresponding equation and its dimensionless version are given respectively by

$$c_4\dot{z}_1 + k_4 z_1 = P_4, \tag{8.29}$$

and

$$z' = -\gamma z - \xi y,$$

where, $\gamma = c_3k_4/(c_4k_3)$ and $P_4 = -r_1y_1$. Finally, to obtain the first equation in (8.5), we need to modify the device shown in the Fig. 8.5, which is also easy to realize. If we add a linear stiffness k_2 to Eq. (8.16), we would obtain (for $M = 0$)

$$c\dot{\varphi}_1 r^2 - k\left(\frac{2hL}{\sqrt{a^2-h^2}} - r\right) r\varphi_1 + k_1(\varphi - \varphi_1^*)r^2 + k_2r^2\varphi_1 = P_0 r, \tag{8.30}$$

and,

$$c_t\dot{\varphi}_1 + k_t(\varphi_1 - \varphi^*) + k_{1t}\varphi_1 + k_{2t}\varphi_1 = P_0, \tag{8.31}$$

where, $k_{2t} = k_2 r$ $(k_2 > 0)$. Proceeding in a similar way we get

$$\varphi' + \frac{c_3}{c_t}\left[\frac{k_t}{k_3}(\varphi - 1) + \frac{k_{1t}}{k_3}\varphi + \frac{k_{2t}}{k_3}\varphi\right] = \frac{c_3}{c_t}\frac{rr_2y_1^*}{k_3\varphi_1^*}y. \tag{8.32}$$

Taking $k_{2t} = k_3$ we obtain the unfolded equation of the form

$$\varphi' = \alpha[y - h(\varphi) - \varphi], \tag{8.33}$$

where: $rr_2 y_1^* = k_3 \varphi_1^*$. The complete mechanical model of the unfolded Chua's circuit is shown in Fig. 8.2.2.

(iv) Pure Chua's Circuit Mechanism

It is well known that, in general, first-order nonlinear differential equations do not always have associated real mechanical models. In fact, the classical approach focuses on the application of known rules, laws and/or hypotheses in order to establish links between various physical quantities and then to write the differential equations. This implies that the equations are typically generated in a natural way. However, recent developments from mathematics and computer sciences tend to generate the equations first and then to find various physical realizations from different fields. Our motivation is to transform the well-known Chua's circuit to a possible mechanical device. Earlier, three examples of such mechanical realizations have been proposed.

All of them have used electromechanical feedback devices to transmit signals from one sub-device to another one. The most interesting question, however, remains open. How to build Chua's circuit using only purely mechanical components? We are now ready to address this problem.

Consider the system shown in the Fig. 8.10. Observe that a mechanism is provided to produce negative stiffness (see Fig. 8.5). However our new idea here is related to the application of a friction force, which occurs between the rotor 1 and the ring 2, and linked to a negative stiffness device which possesses its own 1/2 degree of freedom φ_1. The second-order system is represented by a stiff beam, whose oscillations are governed by φ_2. The key idea is that the beam movement changes the normal force transmitted via k_b and c_b, which forces a change in the friction force. Between the disc 2 and the ring there exists a friction force which decreases locally as a function of two mentioned bodies; this is responsible for the occurrence of self-excited oscillations. All other notations used in the Fig. 8.10 are clear (see earlier comments to Fig. 8.5). The mass m of the rigid beam causes an initial deflection of the spring k_b. Let us consider the oscillations around the static equilibrium positions. Observe that in order to realize the piecewise characteristics, the spring k_1 is included into device, what has been omitted in Fig. 8.10.

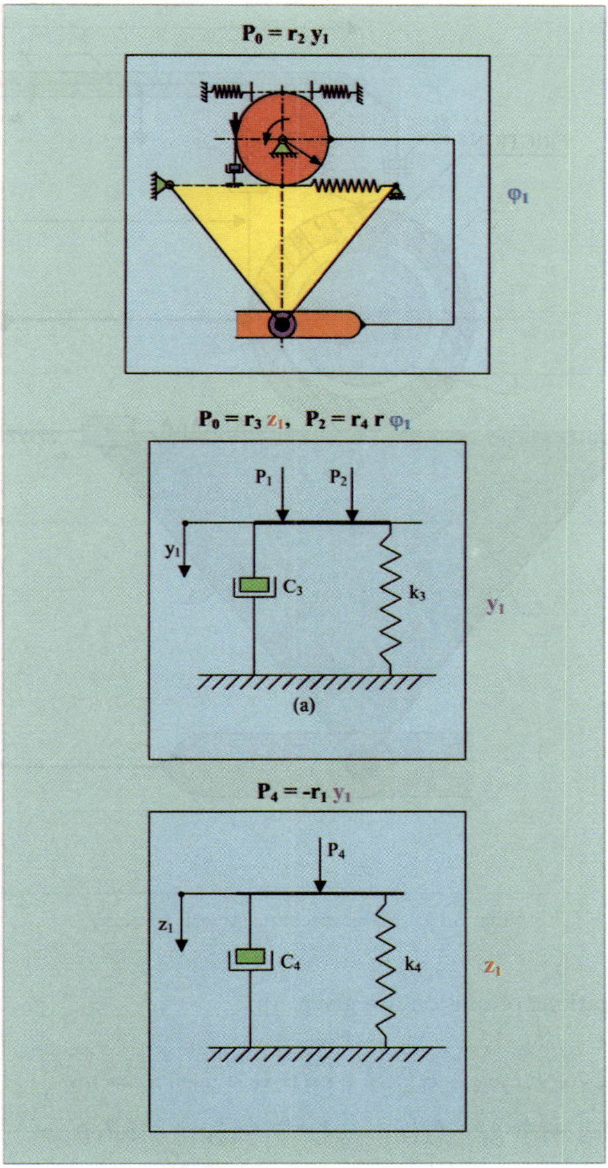

Fig. 8.9 Three components of a mechanical model for unfolding Chua's circuit with their electromechanical couplings.

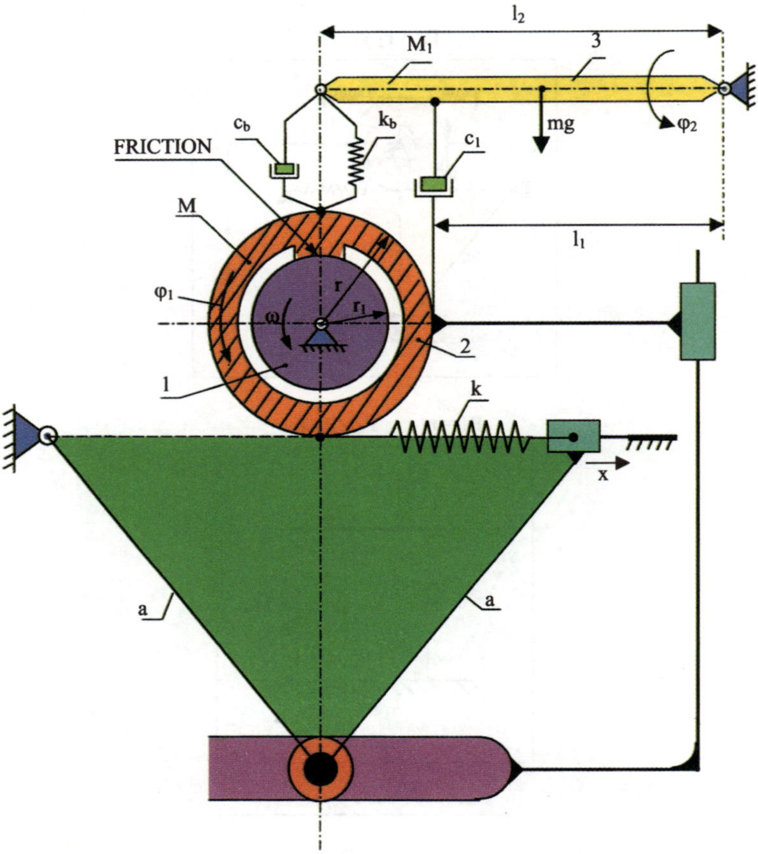

Fig. 8.10 Novel mechanism with friction.

The equations of motion are given by

$$
\begin{aligned}
&M_1\ddot{\varphi}_2 + k_b l_2^2 \varphi_2 + c_b l_2^2 \dot{\varphi}_2 + c_1(\dot{\varphi}_2 l_1 - \dot{\varphi}_1 r_1) l_1 = mg\frac{l_2}{2},\\
&M\ddot{\varphi}_1 + Ar\varphi_1 + k_1(\varphi_1 - \varphi_1^*)r^2 + c_1(\dot{\varphi}_1 r - \dot{\varphi}_2 l_1)r =\\
&= r_1 N(\mu_0 \mathrm{sgn} w - \alpha_1 w + \beta w^3),
\end{aligned}
\tag{8.34}
$$

where:

$$
w = (\omega - \dot{\varphi}_1)r_1, \qquad A = k\left(r - 2\frac{hl}{\sqrt{a^2 - h^2}}\right),
$$

$$N = \frac{mg}{2} + k_b l_2 \varphi_2 + c_b l_2 \dot{\varphi}_2.$$

The parameters μ_0, α and β denote the friction coefficients, and M_1, M denote the moments of inertia. The friction characteristics versus the relative velocity characteristics has been discussed in various papers and books (see, for instance, [Awrejcewicz and Delfs (1990a)], [Awrejcewicz and Holicke (1999)]). However, for our needs, let us assume that the system's regime is associated only with the negative slope portion of the friction versus relative velocity characteristics, and assume simply sgn$w = 1$, and $\beta = 0$. Let us measure the oscillations from the equilibrium positions $-\varphi^* \leq \varphi_{10} \leq \varphi^*$, which means that the spring k_1 is not taken into account. We obtain

$$\begin{aligned} \varphi_{10} &= \frac{r_1 mg(\mu_0 - \alpha\omega r_1)}{rA}, \\ \varphi_{20} &= \frac{mg}{2k_b l_2}. \end{aligned} \tag{8.35}$$

Let us introduce new variables $\varphi_1 = \overline{\varphi}_1 + \varphi_{10}$, $\varphi_2 = \overline{\varphi}_2 + \varphi_{20}$ into the system (8.34).

Assuming $M = 0$ and omitting bars we get

$$\begin{aligned} M_1\ddot{\varphi}_2 + k_b l_2^2 \varphi_2 + (c_b l_2^2 + c_1 l_1^2)\dot{\varphi}_2 &= c_1 r_1 l_1 \dot{\varphi}_1, \\ Ar\varphi_1 + k_1(\varphi_1 - \varphi^*)r^2 + (c_1 r^2 - r_1^2 mg\alpha_1)\dot{\varphi}_1 &= \\ = [r_1(\mu_0 - \alpha_1\omega r_1)c_b l_2 + c_1 l_1 r]\dot{\varphi}_2 &+ r_1(\mu_0 - \alpha_1 \omega r_1)k_b l_2 \varphi_2. \end{aligned} \tag{8.36}$$

Now, introducing the nondimensional time and new variables φ_{1n}, φ_{2n}

$$\tau = \omega^* t, \qquad \omega^* = \frac{c_b l_2^2 + c_1 l_1^2}{M_1}, \qquad \varphi_1 = \varphi_{1n}\varphi^*, \qquad \varphi_2 = \varphi_{2n}\varphi^*, \tag{8.37}$$

we get

$$\begin{aligned} \varphi_{1n}' &= \varphi_{2n}'' + \varphi_{2n}' + \xi\varphi_{2n}, \\ \varphi_{1n}' &= \alpha[\varphi_{2n} - h(\varphi_{1n})], \end{aligned} \tag{8.38}$$

where,

$$
\begin{aligned}
\alpha &= \frac{r_1(\mu_0 - \alpha_1 \omega r_1) k_b l_2}{(c_1 r^2 - r_1^2 m g \alpha_1)\omega^*}, \\
h(\varphi_2) &= m_0 \varphi_1 + m_1(\varphi_1 - 1), \\
m_0 &= -[r_1(\mu_0 - \alpha_1 \omega r_1) k_b l_2]^{-1} A r, \\
m_1 &= -k_1 r^2 [r_1(\mu_0 - \alpha_1 \omega r_1) k_b l_2]^{-1}, \\
r_1(\mu_0 &- \alpha_1 \omega r_1) c_b l_2 + c_1 l_1 r = 0, \\
\xi &= \tfrac{k_b l_2^2}{M_1 \omega^{*2}}, \\
M_1 \omega^* &= c_1 r_1 l_1.
\end{aligned}
\tag{8.39}
$$

Observe that from the two last equations of the system (8.2) we can reduce the variable z. Differentiating the second equation of the system (8.2) in respect to time t and taken into account the third equation of the system (8.2) we get,

$$
\frac{dx}{dt} = \frac{d^2 y}{dt^2} + \frac{dy}{dt} + \beta y, \tag{8.40}
$$

which corresponds to the first equation of the system (8.38).

8.2.3 *Concluding remarks*

The unfolding of Chua's circuit can exhibit various chaotic and bifurcation behaviours which are well-documented in the literature [Madan (1993)]. In this section, the idea of its potential equivalence and homeomorphism to other engineering systems has been discussed and illustrated. Three different examples of Chua's circuit, Chua's unfolded circuit, and a pure Chua's mechanical circuit model have been proposed. First, some "tricks" to produce negative stiffness have been implemented and couplings between three subsystems have been realized using electromechanical devices. The third example (pure Chua's circuit mechanism) includes a proposal to build a Chua's mechanical model using only naturally observed mechanical properties and phenomena. Although friction usually is considered to be a drawback and avoided in industry, here it is exploited to produce some original couplings to synthesize the third-order Chua's equations.

8.3 Generalized Double Scroll Chua's Circuit

8.3.1 *Introduction*

Let us introduce Chua's system and its generalization. Chua's system is a paradigm [Madan (1993)]. It is related to the behaviour of an electrical circuit possessing a diode with a nonlinear response [Chua *et. al.* (1986a)], [Komuro *et. al.* (1991)]. It is a system of 3 differential equations of the first order usually written as

$$\begin{cases} \dot{X} = \alpha(Y - h(X)), \\ \dot{Y} = X - Y + Z, \\ \dot{Z} = -\beta Y, \end{cases} \tag{8.41}$$

with α and β two positive parameters, and $h : \mathbb{R} \longrightarrow \mathbb{R}$ concentrates the nonlinearities. This system can be given in the form:

$$\dot{y} = \mathbf{L}y - \alpha H(y), \tag{8.42}$$

with

$$y = \begin{pmatrix} X \\ Y \\ Z \end{pmatrix}, \mathbf{L} = \begin{pmatrix} 0 & \alpha & 0 \\ 1 & -1 & 1 \\ 0 & -\beta & 0 \end{pmatrix}, H \begin{pmatrix} X \\ Y \\ Z \end{pmatrix} = \begin{pmatrix} h(X) \\ 0 \\ 0 \end{pmatrix}, \tag{8.43}$$

so that we distinguish the linear part from the nonlinear one. The function h is a piecewise linear function that is expressed as follows,

$$h(X) = \begin{cases} m_1 X + (m_0 - m_1) & \text{if} \quad X \geq 1, \\ m_0 X \quad \text{if} \quad | X | \leq 1, \\ m_1 X - (m_0 - m_1) & \text{if} \quad X \leq -1, \end{cases} \tag{8.44}$$

with $m_0 < 0$ and $m_1 > 0$. Here we consider generalizations of Chua's circuit by introducing discontinuities in the frame of multivalued differential equations. In the previous classical Chua circuit, we only modify function h. The system we are dealing with is written as,

$$\begin{cases} \dot{X} = \alpha(Y - \delta(X)), \\ \dot{Y} = X - Y + Z, \\ \dot{Z} = -\beta Y, \end{cases} \tag{8.45}$$

where $\delta(X) = l(X) + m(X)$. Function l denotes a piecewise linear continuous function and m denotes a function with a finite number N of discontinuities. Indeed, at every point of discontinuity X_j, $j = 1, \ldots, N$ we assume that $m(X_j) = [m(X_j^-); m(X_j^+)]$ with

$$\lim_{\substack{X \longrightarrow X_j \\ X < X_j}} m(X) = m(X_j^-),$$

and

$$\lim_{\substack{X \longrightarrow X_j \\ X > X_j}} m(X) = m(X_j^+), \quad \Big(\text{and} \quad m(X_j^+) > m(X_j^-)\Big).$$

Thus we obtain a new differential equation (or rather differential inclusion) of the form:

$$\dot{y} - \mathbf{L}y + \alpha\Delta(y) \ni 0, \tag{8.46}$$

with:

$$\Delta : \begin{cases} \mathbb{R}^3 \longrightarrow \mathbb{R}^3 \\ \begin{pmatrix} X \\ Y \\ Z \end{pmatrix} \mapsto \begin{pmatrix} \delta(X) \\ 0 \\ 0 \end{pmatrix}, \end{cases} \tag{8.47}$$

and $\Delta = L + M$ where:

$$L : \begin{cases} \mathbb{R}^3 \longrightarrow \mathbb{R}^3 \\ \begin{pmatrix} X \\ Y \\ Z \end{pmatrix} \mapsto \begin{pmatrix} l(X) \\ 0 \\ 0 \end{pmatrix}, \end{cases} \qquad M : \begin{cases} \mathbb{R}^3 \longrightarrow \mathbb{R}^3 \\ \begin{pmatrix} X \\ Y \\ Z \end{pmatrix} \mapsto \begin{pmatrix} m(X) \\ 0 \\ 0 \end{pmatrix}. \end{cases} \tag{8.48}$$

For the numerical study, we deal with two particular cases: First, a symmetric discontinuity both in -1 and 1 added to the classical Chua system; and second, a single discontinuity in 0 similarly added to the Chua system. The first case corresponds to

$$l = h, \quad m(X) = \begin{cases} -\varepsilon & \text{if} \quad X < -1, \\ 0 & \text{if} \quad -1 < X < 1, \\ \varepsilon & \text{if} \quad 1 < X. \end{cases} \tag{8.49}$$

The second case corresponds to

$$l(X) = \begin{cases} m_1 X + m_0 - m_1 - 2\varepsilon & \text{if} \quad X \geq 1, \\ (m_0 - \varepsilon)X - \varepsilon & \text{if} \quad -1 \leq X \leq 1, \\ m_1 X - m_0 + m_1 & \text{if} \quad X \leq -1, \end{cases} \tag{8.50}$$

and the discontinuous function m is defined by

$$m(X) = \begin{cases} 0 & \text{if} \quad X < 0, \\ 2\varepsilon & \text{if} \quad X > 0. \end{cases} \tag{8.51}$$

It can be noticed that in the case of discontinuity at 0, the slopes of the initial non perturbed Chua circuit are modified. But in the case of discontinuity at -1 and 1 these initial slopes remain unchanged.

8.3.2 *Mechanical point of view*

Let us show that the Chua circuit corresponds to a special mechanical system with a particular nonlinear constitutive law. Let us consider the following one degree of freedom (1 DOF) mechanical system:

$$m\ddot{w} + g\dot{w} + kw = -\tau(w, \dot{w}). \tag{8.52}$$

Note that $[m] = kg$, $[g] = Ns$, $[\tau] = Nm^{-1}$ and that w is a nondimensional deformation. Assume that

$$\tau(w, \dot{w}) = \tau_0 \Psi(S),$$

where $[\tau_0] = Nm^{-1}$ and $\Psi(S)$ is nondimensional expression of the form:

$$\Psi(S) = \begin{cases} (1+F)S + E - F - \Omega & \text{for} \quad S \geq 1, \\ (1+E)S - \Omega & \text{for} \quad 0 < S \leq 1, \\ (1+E)S + \Omega & \text{for} \quad -1 \leq S < 0, \\ (1+F)S - E + F + \Omega & \text{for} \quad S \geq 1, \end{cases} \tag{8.53}$$

where E, F and Ω are nondimensional. $S(w, \dot{w})$ is a nondimensional form of the tension:

$$S = \frac{\tau_1}{\tau_0} + \frac{\tau_2}{\tau_0} + \frac{\tau_3}{\tau_0},$$

where

$$\tau_1 = a_1' w, \quad \tau_2 = a_2' \dot{w}, \quad \tau_3 = b_1' \int_0^t w(e)de.$$

Thus we have

$$S(w, \dot{w}) = a_1 w + a_2 \dot{w} + b_1 \int_0^t w(e)de,$$

where

$$a_1 = \frac{a_1'}{\tau_0}, \quad a_2 = \frac{a_2'}{\tau_0}, \quad b_1 = \frac{b_1'}{\tau_0},$$

and $[a_1]$ is nondimensional, $[a_2] = s$, $[b_1] = s^{-1}$. After introducing the nondimensional time $q = \sqrt{\frac{k}{m}} t$, we get

$$S(w, \frac{dw}{dq}) = a_1 w + a_2 \sqrt{\frac{k}{m}} \frac{dw}{dq} + \nu_2 \int_0^q w(e)de. \tag{8.54}$$

Setting

$$a_1 = 1, \quad a_2, \sqrt{\frac{k}{m}} = 1, \quad \nu_2 = \frac{b_1}{\sqrt{\frac{k}{m}}}, \quad E = a, \quad F = b, \quad \Omega = c,$$

we obtain function ψ of the Chua's equations:

$$\psi(r) = \begin{cases} (1+b)r + a - b - c, & r \geq 1, \\ (1+a)r - c, & 0 < r \leq 1, \\ (1+a)r + c, & -1 < r \leq 0, \\ (1+b)r - a + b + c, & r \leq -1. \end{cases}$$

Our "rheological" nondimensional equation

$$m\left(\sqrt{\frac{k}{m}}\right)^2 \frac{d^2w}{dq^2} + g\sqrt{\frac{k}{m}}\frac{dw}{dq} + kw = -\tau_0 \Psi(S), \tag{8.55}$$

corresponds to the Chua's equation:

$$\ddot{z} + \dot{z} + (\nu_2 - \nu_1)z = -\nu_1 \psi(\tilde{z}), \tag{8.56}$$

with $\dot{\tilde{z}} = z$. Comparing (8.55) and (8.56) we get

$$\frac{m\sqrt{\frac{k}{m}}}{g} = 1, \quad \frac{k}{\sqrt{\frac{k}{m}}g} = \nu_1 - \nu_2, \quad \frac{\tau_0}{\sqrt{\frac{k}{m}}g} = \nu_1,$$

which leads to the equation

$$k = \tau_0 - gb_1.$$

This mechanical system is described in the Fig. 8.11 and corresponds to the piecewise linear characteristic $\Psi(s)$ presented in the Fig. 8.12.

We have introduced several parameters in order to describe our mechanical model. In the next sections, we deal with the Chua circuit. This Chua circuit is defined by several parameters: α, β, m_0, m_1, ε. Indeed the occurrence of the following relations allows us to transform our mechanical model into the Chua circuit:

$$\alpha = \nu_1, \; \beta = \nu_2, \; \varepsilon = -c, \; m_1 = 1 + b, \; m_0 = 1 + a - c.$$

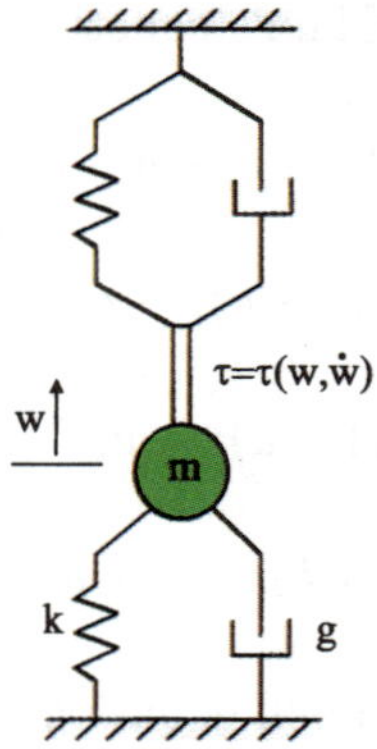

Fig. 8.11 Rheological model corresponding to Chua's circuit.

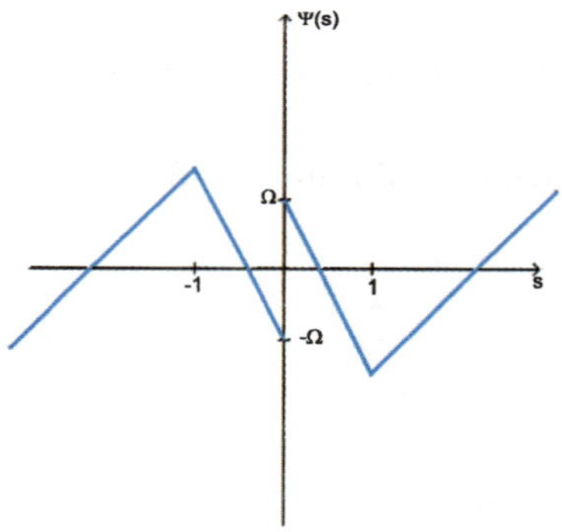

Fig. 8.12 Piecewise linear characteristics $\Psi(s)$.

8.3.3 *Existence and uniqueness of solutions*

8.3.3.1 *Usual Chua's system*

Let $T \in \mathbb{R}^{+\star}$ and $\Omega = [0, T]$. Let us consider the following ordinary differential equation:

$$\begin{cases} \dot{y}(t) = \mathbf{L}y(t) - \alpha H(y(t)) = f(t, y(t)), \quad \forall t \in [0, T], \\ y(0) = y_0 \in \mathbb{R}^3, \end{cases} \tag{8.57}$$

Here $f \in C^0(\Omega \times \mathbb{R}^3)$ because $\mathbf{L}$ is linear and H is continuous. Moreover, h is piecewise linear and then Lipschitz-continuous with the Lipschitz constant

$$K = \max(m_1 - 2m_0, 2m_1 - m_0).$$

So f is Lipschitz-continuous with the Lipschitz constant

$$L = \sqrt{\max(3 + 2\alpha^2 K^2,\ 3 + 2\alpha^2 + \beta^2)}.$$

The system of Eqs. (8.57) verifies assumptions of the Cauchy-Lipschitz theorem that provides existence and uniqueness of the solution of this system for every T.

8.3.3.2 *Existence and uniqueness for the generalized Chua circuit*

The usual Cauchy-Lipschitz theorem does not work in the generalized case. In addition, it is not possible to use the theorem of Caratheodory (see [Crouzeix and Mignot (1984)]) which possesses assumptions weaker than the assumptions of the Cauchy-Lipschitz theorem. We apply the results of the theory of maximal monotone operators to the generalized Chua system. Let us consider the following problem:

$$\begin{cases} \dot{y} = \mathbf{L}y - \alpha M(y) - \alpha L(y), \\ y(0) = y_0, \end{cases} \tag{8.58}$$

where: $\mathbf{L}$ is defined in the section 8.3.1, M denotes a maximal monotone operator on $\mathbb{R}^3$ and L is a Lipschitz-continuous map of $\mathbb{R}^3$. The following claim provides existence and uniqueness of the solution for a system even more general than the system (8.58).

Theorem 8.1 *Let A be a maximal monotone operator on $\mathbb{R}^3$ and F be a Lipschitz-continuous function on $\mathbb{R}^3$. Then the system (8.59) has a unique strong solution on $[0, T]$ for every $T > 0$ where:*

$$\begin{cases} \dot{y} + Ay + F(y) \ni 0, \\ y(0) = y_0. \end{cases} \tag{8.59}$$

Moreover, we have:

$$\forall t_0 \in [0,T], \frac{dy^+}{dt}(t_0) = -Proj_{(A+F)y(t_0)}0. \tag{8.60}$$

The proof is obtained by using the Theorem 2.7 [Brezis (1973)]. Let us define the Lipschitz-continuous operator G on $\mathbb{R}^3$ as follows: $G(y) = F(y) - F(0)$ and let us set $f \equiv -F(0)$. Now the system (8.59) is equivalent to the following system:

$$\begin{cases} \dot{y} + Ay + G(y) \ni f, \\ \\ \qquad y(0) = y_0. \end{cases} \tag{8.61}$$

Assumptions of the Theorem 2.4 are verified ($f \in L^1([0,T], \mathbb{R}^3)$ and $\mathbb{R}^3$ is a Hilbert space of finite dimension!). Moreover, every $t_0 \in [0,T]$ is a Lebesgue point here because f is constant. So we apply the Theorem 2.5 [Brezis (1973)] to the weak solutions we have just obtained.

Let p be the Lipschitz constant of F. G has the same Lipschitz constant. By assumption $G(0) = 0$, hence we have:

$$\forall x \in \mathbb{R}^3 \quad \| G(x) \| \leq p \| x \| . \tag{8.62}$$

Let us set $B = G + pI$. We have:

$$\forall x \in \mathbb{R}^3, \quad \| Bx \| \leq \| G(x) \| + p \| x \|, \tag{8.63}$$

and then

$$\forall x \in \mathbb{R}^3 \quad \| Bx \| \leq 2p \| x \| . \tag{8.64}$$

Let B is a monotone operator: for $x_1,\ x_2 \in \mathbb{R}^3$ we see that

$$(Bx_1 - Bx_2, x_1 - x_2) = (G(x_1) - G(x_2), x_1 - x_2) + p \| x_1 - x_2 \|^2 . \tag{8.65}$$

Because

$$|(G(x_1) - G(x_2), x_1 - x_2)| \leq \| G(x_1) - G(x_2) \| \| x_1 - x_2 \| \leq p \| x_1 - x_2 \|^2, \tag{8.66}$$

we get

$$(G(x_1) - G(x_2), x_1 - x_2) + p \| x_1 - x_2 \|^2 \geq 0, \tag{8.67}$$

and B is monotone. We apply Theorem 2.1 [Brezis (1973)], here with $D(A) = D(B) = H = \mathbb{R}^3$, $k = 0$ and the continuous function θ is defined by

$$\theta : \begin{cases} \mathbb{R}^+ \longrightarrow \mathbb{R}^+, \\ \\ x \mapsto 2px. \end{cases} \tag{8.68}$$

We obtain $A + B$ maximal monotone so $A + G + pI = A_1$ is maximal monotone. Now we apply the Theorem 2.4 [Brezis (1973)] to A_1 and $\omega = p$. There is a unique weak solution of

$$\begin{cases} \dot{y} + A_1 y - py \ni 0, \\ \quad y(0) = y_0, \end{cases} \tag{8.69}$$

which is indeed a strong solution.

We have:

$$\frac{dy^+}{dt}(t_0) = f(t_0^+) + py(t_0^+) - \mathrm{Proj}_{A_1 y(t_0)}(f(t_0^+) + py(t_0^+)). \tag{8.70}$$

Because of the equalities $A_1 = A + G + pI = A + F - F(0) + pI$ and $f \equiv F(0)$, we express the previous relationship in the form:

$$\begin{aligned} \frac{dy^+}{dt}(t_0) &= -F(0) + py(t_0) - \mathrm{Proj}_{(A+F)y(t_0) - F(0) + py(t_0)}(-F(0) + py(t_0)) = \\ &= -\mathrm{Proj}_{(A+F)y(t_0)} 0. \end{aligned} \tag{8.71}$$

Now we can affirm that the problem $\begin{cases} \dot{y} + Ay + F(y) \ni 0 \\ \\ \quad y(0) = y_0 \end{cases}$ has a unique strong solution on $[0, T], \forall T > 0$. This solution verifies:

$$\forall t_0 \in [0, T], \ \frac{dy^+}{dt}(t_0) = -\mathrm{Proj}_{(A+F)y(t_0)} 0. \tag{8.72}$$

We have just proved existence and uniqueness of the solution of the system (8.58).

One can note that indeed we have proved existence and uniqueness of the solution of any problem of the kind of system (8.58) submitted to a Lipschitz-continuous perturbation.

It is clear that because the obtained solution is continuous, (8.58) provides Y and Z in $C^1(\mathbb{R}^+)$ and X only C^1-piecewise because defects of continuity of $\dot{X}$ occur at time t when $X(t)$ is a point of discontinuity of the function m.

8.3.3.3 *Applications to two particular cases*

(i) Chua's System with Discontinuities at -1 and 1

We deal with the system:

$$\begin{cases} \dot{X} = \alpha(Y - \delta(X)), \\ \dot{Y} = X - Y + Z, \\ \dot{Z} = -\beta Y, \end{cases} \tag{8.73}$$

with $\delta(X) = h(X) + m(X)$. m can be written in the following form:

$$m(X) = \begin{cases} \{-\varepsilon\} & \text{if} \quad X < -1, \\ [-\varepsilon, 0] & \text{if} \quad X = -1, \\ \{0\} & \text{if} \quad -1 < X < 1, \\ [0, \varepsilon] & \text{if} \quad X = 1, \\ \{\varepsilon\} & \text{if} \quad X > 1, \end{cases} \tag{8.74}$$

so that it is a multivalued operator. We assume that $0 \leq \varepsilon \leq -m_0$. Thus we deal with the system:

$$\dot{y} - (\mathbf{L} - \alpha H)y + \alpha M y \ni 0, \tag{8.75}$$

where operator $-(\mathbf{L} - \alpha H)$ is Lipschitz-continuous on $\mathbb{R}^3$. It is clear that M is monotone if

$$\forall X_1,\ X_2 \in \mathbb{R}, \quad (m(X_1) - m(X_2),\ X_1 - X_2) \geq 0.$$

We can assume for instance that $X_1 < X_2$. Then M is monotone if

$$\forall S_1 \in m(X_1),\ \forall S_2 \in m(X_2), \quad S_1 - S_2 \leq 0.$$

We can distinguish between five cases: $X_1 \in [-\infty, -1]$, $X_1 = -1$, $X_1 \in [-1, 1]$, $X_1 = 1$, $X_1 \in [1, +\infty]$. Let us discuss the first one for example (the four other cases are being similar). If $X_1 \in [-\infty, -1]$, we have $m(X_1) = \{-\varepsilon\}$. And

$$\forall X \in \mathbb{R}, \quad m(X) \in [-\varepsilon, \varepsilon] \Longrightarrow \forall S_2 \in m(X_2),\ S_2 \geq -\varepsilon.$$

Then it is clear that in this case $m(X_1) - m(X_2) \leq 0$.

Moreover it is easy to see that M is maximal. We have $\mathrm{Im}\,(I + M) = \mathbb{R}^3 = H$. The Theorem 3.2 of Brézis [Brezis (1973)] provides Δ maximal monotone. Because $\alpha > 0$, αM is maximal monotone too.

Thus the system (8.75) verifies all the assumptions of the Theorem 8.1. It possesses a unique strong solution on $[0, T]$ for every $T > 0$.

It is convenient to apply the second part of this theorem to obtain the value of $\dot{y}$ at the crossing of $X = 1$. For t_0 such that $X(t_0) = 1$ we get:

$$\frac{dy^+}{dt}(t_0) = -\mathrm{Proj}_{(\alpha M - \mathcal{L} + \alpha H)y(t_0)} 0 = \begin{pmatrix} S \\ U \\ V \end{pmatrix} \in (-\alpha M + \mathbf{L} - \alpha H)y(t_0). \tag{8.76}$$

Let us set

$$y(t_0) = \begin{pmatrix} 1 \\ Y \\ Z \end{pmatrix}. \tag{8.77}$$

We have

$$\mathbf{L}y(t_0) = \begin{pmatrix} \alpha Y \\ 1 - Y + Z \\ -\beta Y \end{pmatrix}, \quad \alpha M y(t_0) = \begin{pmatrix} \alpha m(1) \\ 0 \\ 0 \end{pmatrix}, \tag{8.78}$$

$$H y(t_0) = \begin{pmatrix} h(1) \\ 0 \\ 0 \end{pmatrix} = \begin{pmatrix} m_0 \\ 0 \\ 0 \end{pmatrix}, \tag{8.79}$$

and then

$$\begin{pmatrix} S - \alpha Y + \alpha m_0 \\ U - 1 + Y - Z \\ V + \beta Y \end{pmatrix} \in -\alpha \begin{pmatrix} [0, \varepsilon] \\ \{0\} \\ \{0\} \end{pmatrix}, \tag{8.80}$$

that yields

$$\begin{pmatrix} \alpha(Y - m_0 - \varepsilon) \leq S \leq \alpha(Y - m_0) \\ U = 1 - Y + Z \\ V = -\beta Y \end{pmatrix}. \tag{8.81}$$

Because of the definition of the projection, $(S, U, V)^t$ has the smallest norm. Here this means that x^2 has to be minimal. We distinguish between three cases:

- if $Y < m_0$, $\min\limits_{\alpha(Y-m_0-\varepsilon)\leq S\leq\alpha(Y-m_0)} S^2$ corresponds to $S = \alpha(Y - m_0)$. So we obtain:

$$\frac{dy^+}{dt}(t_0) = \begin{pmatrix} \alpha(Y(t_0) - m_0) \\ 1 - Y(t_0) + Z(t_0) \\ -\beta Y(t_0) \end{pmatrix}, \tag{8.82}$$

- if $Y > m_0 + \varepsilon$, $\min\limits_{\alpha(Y-m_0-\varepsilon)\leq S\leq\alpha(Y-m_0)} S^2$ corresponds to $S = \alpha(Y - m_0 - \varepsilon)$. So we obtain:

$$\frac{dy^+}{dt}(t_0) = \begin{pmatrix} \alpha(Y(t_0) - m_0 - \varepsilon) \\ 1 - Y(t_0) + Z(t_0) \\ -\beta Y(t_0) \end{pmatrix}, \tag{8.83}$$

- if $m_0 \leq Y \leq m_0 + \varepsilon$, $\min\limits_{\alpha(Y-m_0-\varepsilon)\leq S\leq\alpha(Y-m_0)} S^2$ corresponds to $S = 0$.

So we obtain:

$$\frac{dy^+}{dt}(t_0) = \begin{pmatrix} 0 \\ \\ 1 - Y(t_0) + Z(t_0) \\ \\ -\beta Y(t_0) \end{pmatrix}. \tag{8.84}$$

It can be noticed that the values $\dot{Y}(t_0^+)$ and $\dot{Z}(t_0^+)$ provided by this procedure can be directly derived from the system (8.73). It is necessary to assume that $\alpha > 0$ in order to apply this method.

Note that, one can think of other methods to prove existence and uniqueness results. For example, it would be possible to smooth the discontinuity by using piecewise linear continuous functions or C^∞ functions. But it seems difficult to have prior estimations that allow to control the limits in the smooth systems of differential equations.

Another method is to take into account the particular form of the system. It deals with joining pieces of solutions of linear system of differential equations. The difficulty is to write the crossing of discontinuity plane in order to obtain piecewise C^1 solutions. In our case this method is available. But it is clear that if we slightly modify the Lipschitz-continuous part (small smooth perturbations so that explicit calculation of solutions is not possible), it becomes necessary to use the theory of maximal monotone operators.

(ii) Chua's System with Discontinuity at 0

In this case the starting system of equations is:

$$\begin{cases} \dot{X} = \alpha(Y - \delta(X)), \\ \dot{Y} = X - Y + Z, \\ \dot{Z} = -\beta Y, \end{cases} \tag{8.85}$$

where $\delta(x) = l(x) + m(x)$ (see the section 8.3.2). The multivalued function m is defined according to

$$m(x) = \begin{cases} \{0\} & \text{if} \quad x < 0, \\ \\ [0, 2\varepsilon] & \text{if} \quad x = 0, \\ \\ \{2\varepsilon\} & \text{if} \quad x > 0. \end{cases} \tag{8.86}$$

We write again this system in the form

$$\dot{y} - (\mathcal{L} - \alpha L)y + \alpha M y \ni 0. \tag{8.87}$$

Because $\mathbf{L}$ is linear and then is Lipschitz-continuous, and L is Lipschitz-continuous (l is piecewise linear), the operator $-(\mathbf{L} - \alpha L)$ is Lipschitz-continuous on $\mathbb{R}^3$.

It is easy to prove that M is monotone: This is similar to the previous case. As we did in the previous subsection for the first example, we deduce that αM is a maximal monotone operator.

It is clear that the assumptions of the Theorem 8.1 are verified by the system (8.85), and hence the system possesses a unique strong solution on $[0, T]$ for every $T > 0$.

Moreover, let us pay attention to the second part of the result of the Theorem 8.1 and calculate $\dot{y}$ at $X = 0$ crossing. For t_0 such that $X(t_0) = 0$ we have:

$$\frac{dy^+}{dt}(t_0) = -\mathrm{Proj}_{(\alpha M - \mathbf{L} + \alpha L)y(t_0)} 0 = \begin{pmatrix} S \\ U \\ V \end{pmatrix} \in (\mathbf{L} - \alpha M - \alpha L)y(t_0). \tag{8.88}$$

Let us set:

$$y(t_0) = \begin{pmatrix} 0 \\ Y \\ Z \end{pmatrix}, \tag{8.89}$$

We have

$$\mathbf{L}y(t_0) = \begin{pmatrix} \alpha Y \\ \\ -Y + Z \\ \\ -\beta Y \end{pmatrix}, \quad \alpha M y(t_0) = \begin{pmatrix} m(0) \\ \\ 0 \\ \\ 0 \end{pmatrix}, \tag{8.90}$$

$$L y(t_0) = \begin{pmatrix} l(0) \\ \\ 0 \\ \\ 0 \end{pmatrix} = \begin{pmatrix} -\varepsilon \\ \\ 0 \\ \\ 0 \end{pmatrix}, \tag{8.91}$$

Hence we obtain

$$\begin{pmatrix} S-\alpha Y-\alpha\varepsilon \\ U+Y-Z \\ V+\beta Y \end{pmatrix} \in -\alpha \begin{pmatrix} [0,2\varepsilon] \\ \{0\} \\ \{0\} \end{pmatrix}, \tag{8.92}$$

that yields

$$\begin{cases} \alpha(Y-\varepsilon) \le S \le \alpha(Y+\varepsilon), \\ U=-Y+Z, \\ V=-\beta Y. \end{cases} \tag{8.93}$$

Because of the definition of the projection, the solution $(S,U,V)^t$ has the smallest norm i.e. again S^2 has to be the smallest.

We can divide the calculation into three cases:

- if $Y < -\varepsilon$, $\min\limits_{\alpha(Y-\varepsilon)\le S\le\alpha(Y+\varepsilon)} S^2$ corresponds to $S=\alpha(Y+\varepsilon)$. Hence we get

$$\frac{dy^+}{dt}(t_0) = \begin{pmatrix} \alpha(Y(t_0)+\varepsilon) \\ -Y(t_0)+Z(t_0) \\ -\beta Y(t_0) \end{pmatrix}, \tag{8.94}$$

- if $Y > \varepsilon$, $\min\limits_{\alpha(Y-\varepsilon)\le S\le\alpha(Y+\varepsilon)} S^2$ corresponds to $S=\alpha(Y-\varepsilon)$. Hence we get

$$\frac{dy^+}{dt}(t_0) = \begin{pmatrix} \alpha(Y(t_0)-\varepsilon) \\ -Y(t_0)+Z(t_0) \\ -\beta Y(t_0) \end{pmatrix}, \tag{8.95}$$

- if $-\varepsilon \le Y \le \varepsilon$, $\min\limits_{\alpha(Y-\varepsilon)\le S\le\alpha(Y+\varepsilon)} S^2$ corresponds to $S=0$. Hence we get

$$\frac{dy^+}{dt}(t_0) = \begin{pmatrix} 0 \\ -Y(t_0) + Z(t_0) \\ -\beta Y(t_0) \end{pmatrix}. \tag{8.96}$$

8.3.4 *Analytical calculation of the solution*

We build analytical solutions for the two particular cases that we have considered in the previous section from the point of view of existence and uniqueness.

8.3.4.1 *Discontinuity crossing*

(i) Case 1: Discontinuity at -1 and 1

Because of the symmetry of the system of the Eqs. (8.73) it is sufficient to study the case $X = 1$. Let us set $D_1 = [1, +\infty] \times \mathbb{R}^2$ and $D_0 = [-1, 1] \times \mathbb{R}^2$.

We determine the necessary conditions of crossing into the different areas D_0, D_1 and $X = 1$. Let us assume that for a given t_0 we have $X(t_0) = 1$. One can already notice that if $Y < m_0$ or $Y > m_0 + \varepsilon$, then we can conclude from (8.82) and (8.83):

- *The trajectory runs into D_1:*

In this case one has $X(t_0^+) > 1$ and X is locally increasing after t_0 that yields $\dot{X}(t_0 + \eta) > 0$ for η small enough. We get $\dot{X}(t_0 + \eta) = \alpha(Y(t_0 + \eta) - \delta(X(t_0 + \eta))) > 0$. Because of continuity of X and Y we write

$$Y(t_0 + \eta) \underset{\eta \searrow 0}{\longrightarrow} Y(t_0), \quad X(t_0 + \eta) \underset{\eta \searrow 0}{\searrow} 1,$$

and then

$$\delta(X(t_0 + \eta)) \underset{\eta \searrow 0}{\longrightarrow} m_0 + \varepsilon.$$

So we get $Y(t_0) \geq m_0 + \varepsilon$.

- *The trajectory runs into D_0:*

Here we have $X(t_0^+) < 1$ and X is locally decreasing after t_0 that yields $\dot{X}(t_0 + \eta) < 0$ for η small enough. We get $\dot{X}(t_0 + \eta) = \alpha(Y(t_0 + \eta) - \delta(X(t_0 + \eta))) < 0$. Because of continuity of X and Y we write

$$Y(t_0 + \eta) \underset{\eta \searrow 0}{\longrightarrow} Y(t_0), \quad X(t_0 + \eta) \underset{\eta \searrow 0}{\nearrow} 1,$$

and then

$$\delta(X(t_0 + \eta)) \underset{\eta \searrow 0}{\longrightarrow} m_0.$$

So we get $Y(t_0) \leq m_0$.

- *Trajectory runs into* $X = 1$*:*

X is constant and of C^1 on an open set of non-zero measure. Consequently we have $\dot{X} = 0$ on this open set, hence $Y(t) \in \delta(1) = [m_0, m_0 + \varepsilon]$. So the trajectory stays in $X = 1$ as long as we have $Y(t) \in [m_0, m_0 + \varepsilon]$.

Two cases have to be taken into account: $Y(t_0) = m_0$ and $Y(t_0) = m_0 + \varepsilon$.

- If $Y(t_0) = m_0$, the trajectory can either stay at $X = 1$ or enter into D_0. In order that it stays at $X = 1$ we need $\dot{X}(t_0 + \eta) = 0$ for $\eta > 0$ small enough, hence from (8.73) $Y(t_0 + \eta) \in \delta(X(t_0 + \eta)) = \delta(1) = [m_0, m_0 + \varepsilon]$. The latter leads to $Y(t_0 + \eta) \geq Y(t_0)$ and then $\dot{Y}(t_0) \geq 0$. This inequality provides a condition for Z coordinate: $Z(t_0) \geq m_0 - 1$.

Let us look for a necessary condition of passing into D_0. We have $\dot{X}(t_0^+) = 0$. Let us calculate $\ddot{X}(t_0^+)$. One has $\ddot{X}(t_0^+) = \alpha(\dot{Y}(t_0) - m_0 \dot{X}(t_0^+))$. We conclude that $\ddot{X}(t_0^+) = \alpha(1 - m_0 + Z(t_0))$. It is necessary to have $\dot{X}(t_0 + \eta) < 0 = \dot{X}(t_0^+)$ so $\ddot{X}(t_0^+) \leq 0$. Then we have $Z(t_0) \leq m_0 - 1$.

We have to consider the last situation: $Z(t_0) = m_0 - 1$.

Let us assume that $Z(t_0) = m_0 - 1$, and that the trajectory goes into D_0. We can write

$$\begin{cases} \dot{X}(t_0^+) = 0 \quad \text{and} \quad \ddot{X}(t_0^+) = 0, \\ \dot{Y}(t_0) = 0, \\ \dot{Z}(t_0) = -\beta m_0 > 0. \end{cases} \tag{8.97}$$

Let us calculate $\dddot{X}(t_0^+)$. For $\eta > 0$ small enough we have:

$$\dddot{X}(t_0+\eta) = \alpha(\ddot{Y}(t_0+\eta) - \ddot{X}(t_0+\eta)m_0),$$

and then

$$\dddot{X}(t_0+\eta) = \alpha(\dot{X}(t_0+\eta)-\dot{Y}(t_0+\eta)+\dot{Z}(t_0+\eta)-\alpha m_0(\dot{Y}(t_0+\eta)-m_0\dot{X}(t_0+\eta))).$$

That yields

$$\dddot{X}(t_0^+) = \alpha\big[(1+\alpha m_0^2)\dot{X}(t_0^+) - (1+\alpha m_0)\dot{Y}(t_0) + \dot{Z}(t_0)\big],$$

that is $\dddot{X}(t_0^+) = -\alpha\beta m_0 > 0$. This situation cannot occur because $X(t_0 + \eta) < 0$ has to be verified for every η small enough. Thus it is clear that the trajectory cannot go into D_0, which means that it stays in $X = 1$.

- If $Y(t_0) = m_0 + \varepsilon$, the trajectory can either stay in $X = 1$ or reach D_1.

Let us look for a necessary condition for it to stay in $X = 1$: some calculations lead to $Z(t_0) \leq m_0 + \varepsilon - 1$.

Let us look for a necessary condition for it to run into D_1: we find $Z(t_0) \geq m_0 + \varepsilon - 1$.

Then we have to study the last case $Z(t_0) = m_0 + \varepsilon - 1$.

If $Z(t_0) = m_0 + \varepsilon - 1$ one has

$$\begin{cases} \dot{X}(t_0^+) = 0, \\ \dot{Y}(t_0^+) = 0, \\ \dot{Z}(t_0) = -\beta(m_0+\varepsilon) > 0, \end{cases}$$

and

$$\ddot{Y}(t_0) = \dot{X}(t_0^+) - \dot{Y}(t_0) + \dot{Z}(t_0) = -\beta(m_0+\varepsilon) > 0.$$

Hence $\dot{Y}$ is strictly increasing after t_0, and then $Y(t_0+\eta) > Y(t_0) = m_0+\varepsilon$ for η small enough and the trajectory goes into D_1.

We can sum up the different behaviours of the trajectory at the crossing of $X = 1$:

- The trajectory goes into D_1 if $Y(t_0) > m_0 + \varepsilon$ or ($Y(t_0) = m_0 + \varepsilon$ and $Z(t_0) \geq m_0 + \varepsilon - 1$),

- The trajectory goes into D_0 if $Y(t_0) < m_0$ or $(Y(t_0) < m_0$ and $Z(t_0) < m_0 - 1)$,
- The trajectory stays in $X = 1$ if $m_0 < Y(t_0) < m_0+\varepsilon$ or $(Y(t_0) = m_0+\varepsilon$ and $Z(t_0) < m_0 + \varepsilon - 1)$ or $(Y(t_0) = m_0$ and $Z(t_0) \geq m_0 - 1)$.

Let us notice that we have only established necessary conditions: The same calculations provide the sufficient conditions.

(ii) Case 2: Discontinuity at 0

In this second example it is sufficient to determine the crossing at $X = 0$. Let us introduce $D_0^- = [1, 0] \times \mathbb{R}^2$ and $D_0^+ = [0, 1] \times \mathbb{R}^2$. Starting from a given t_0 so that $X(t_0) = 0$, it is possible to study three cases (the trajectory goes into D_0^- or D_0^+, or it stays in $X = 0$). Calculations are similar to the previous example. The results can be summed up as follows:

- The trajectory runs into D_0^+ if $Y(t_0) > \varepsilon$ or $(Y(t_0) = \varepsilon$ and $Z(t_0) > \varepsilon)$,
- The trajectory runs into D_0^- if $Y(t_0) < -\varepsilon$ or $(Y(t_0) = -\varepsilon$ and $Z(t_0) < -\varepsilon)$,
- The trajectory stays in $X = 0$ if $-\varepsilon < Y(t_0) < \varepsilon$ or $(Y(t_0) = \varepsilon$ and $Z(t_0) \leq \varepsilon)$ or $(Y(t_0) = -\varepsilon$ and $Z(t_0) \geq -\varepsilon)$.

8.3.4.2 *Analytical calculation of the solution*

(i) Discontinuity at -1 and 1

We define u_1, C_1, C_0, $\mathbf{L}_1$ and T_1 in the following way:

$$u_1 = \begin{pmatrix} -\alpha(m_0 - m_1 + \varepsilon) \\ 0 \\ 0 \end{pmatrix},$$

$$C_1 = \begin{pmatrix} -\alpha m_1 & \alpha & 0 \\ 1 & -1 & 1 \\ 0 & -\beta & 0 \end{pmatrix},$$

$$C_0 = \begin{pmatrix} -\alpha m_0 & \alpha & 0 \\ 1 & -1 & 1 \\ 0 & -\beta & 0 \end{pmatrix},$$

Let us assume that the eigenvalues of C_0 are $\sigma_1 \pm i\omega_1$ (complex conjugate) and γ_1 (real). We can find $\mathbf{L}_1$

$$\mathbf{L}_1 = \begin{pmatrix} \sigma_1 & -\omega_1 & 0 \\ \omega_1 & \sigma_1 & 0 \\ 0 & 0 & \gamma_1 \end{pmatrix},$$

and T_1 that are defined by

$$\mathbf{L}_1 = T_1^{-1} C_1 T_1,$$

- The first column of T_1 is the real part of the eigenvector associated with $\sigma_1 + i\omega_1$,
- Its second column is the imaginary part of the same eigenvector,
- Its third column corresponds to the eigenvector associated with γ_1,
- We decide that the first line of T_1 is $[\frac{m_0}{m_1}\ 0\ \frac{m_0}{m_1}]$.

We assume that the numerical values of the parameters provide one real eigenvalue and two complex conjugate ones (see in [Kuznetsov *et. al.* (1996)]) (this is verified by every parameter that will be used for numerical examples).

We have to solve the following problem:

$$\begin{cases} \dot{y} = C_1 y + u_1 \text{ on } & D_1 = [1, +\infty] \times \mathbb{R}^2, \\ \dot{y} = C_0 y \quad\quad\ \text{ on } & D_0 = [-1, 1] \times \mathbb{R}^2, \\ \dot{y} = C_1 y - u_1 \text{ on } & D_{-1} = [-\infty, -1] \times \mathbb{R}^2. \end{cases} \tag{8.98}$$

Each equation is a linear one: it is easy to find the solution whatever the initial condition. The key point is to find successive times when the coordinate $X(t)$ crosses a plane of discontinuity $X = -1$ or $X = 1$. This is done by solving scalar nonlinear equations depending on t and by using the results of the section 8.3.1.

(ii) Discontinuity at 0

We introduce C_0', u_1', $\mathbf{L}_0$ and T_0 as follows.

$$u_1' = \begin{pmatrix} -\alpha(m_0 - m_1) \\ 0 \\ 0 \end{pmatrix},$$

$$u_0 = \begin{pmatrix} -\alpha\varepsilon \\ 0 \\ 0 \end{pmatrix},$$

$$C_0' = \begin{pmatrix} -\alpha(m_0 - \varepsilon) & \alpha & 0 \\ 1 & -1 & 1 \\ 0 & -\beta & 0 \end{pmatrix}.$$

Let us assume that the eigenvalues of C_0' are $\sigma_0 \pm i\omega_0$ (complex conjugate) and γ_0 (real). We can find $\mathbf{L}_0$

$$\mathbf{L}_0 = \begin{pmatrix} \sigma_0 & -\omega_0 & 0 \\ \omega_0 & \sigma_0 & 0 \\ 0 & 0 & \gamma_0 \end{pmatrix},$$

and T_0 that are defined by

$$\mathbf{L}_0 = T_0^{-1} C_0 T_0,$$

- The first column of T_0 is the real part of the eigenvector associated with $\sigma_0 + i\omega_0$,
- Its second column is the imaginary part of the same eigenvector,

- Its third column corresponds to the eigenvector associated with γ_0,
- We decide that the first line of T_0 is $[\frac{m_0}{m_0-\varepsilon} \; 0 \; \frac{m_0}{m_0-\varepsilon}]$.

We assume that the numerical values of the parameters provide one real eigenvalue and two complex conjugate ones [Kuznetsov *et. al.* (1996)] (this is again verified by every parameter that will be used for numerical examples).

We have to solve the following problem:

$$\begin{cases} \dot{y} = C_1 y + u_1' \text{ on } & D_1 = [1, +\infty] \times \mathbb{R}^2, \\ \dot{y} = C_0' y + u_0 \text{ on } & D_0^+ = [0, 1] \times \mathbb{R}^2, \\ \dot{y} = C_0' y - u_0 \text{ on } & D_0^- = [-1, 0] \times \mathbb{R}^2, \\ \dot{y} = C_1 y - u_1' \text{ on } & D_{-1} = [-\infty, -1] \times \mathbb{R}^2. \end{cases} \tag{8.99}$$

Again each linear equation can be solved. Scalar nonlinear equations depending on t have to be solved to determine the crossing of $X = 0$. Results of the section 8.3.1 are used to calculate the solution after the meeting of discontinuity.

8.3.5 *Numerical results*

8.3.5.1 *First example: Discontinuities at -1 and 1*

(i) Bifurcation Diagram

In order to test the occurrence of chaos, we study bifurcation diagrams. In the Fig. 8.13, such a bifurcation diagram is presented for the following parameter values:

$$m_0 = -\frac{1}{7}, \quad m_1 = \frac{2}{7}, \quad \alpha = 3.612, \quad \beta = 4.4,$$

and ε varies from 0 to $\frac{1}{7}$. In this diagram we do not plot transient i.e. we plot $\| y(t) \| = \sqrt{X(t)^2 + Y(t)^2 + Z(t)^2}$ such as $X(t) = 1$ for $3000 \leq t \leq 4000$. Obviously chaos seems to occur and after some chaotic and periodic windows, chaos disappears via a cascade.

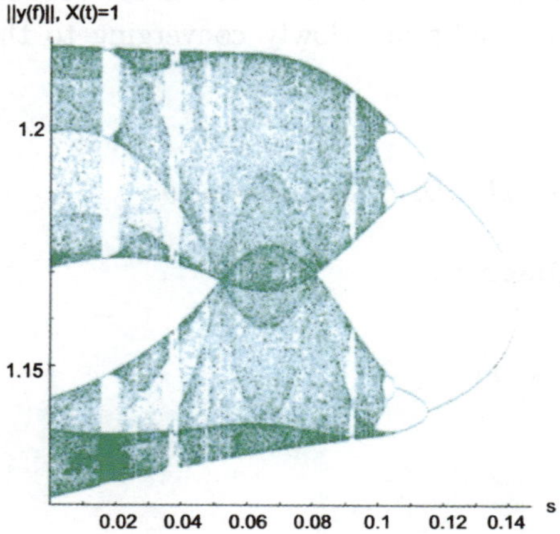

Fig. 8.13 Bifurcation diagram for $\alpha = 3.612$ and $\beta = 4.4$ for the system with discontinuities at -1 and 1.

Table 8.1 Analysis of the bifurcation cascade.

Period	**Parameter** ε_k	**Ratio** δ_{k-2}
1	$\varepsilon_0 = 0.1425252608 \pm 10^{-10}$	
2	$\varepsilon_1 = 0.115436145850 \pm 10^{-12}$	
4	$\varepsilon_2 = 0.105493669499 \pm 10^{-12}$	$\delta_0 = 2.725$
8	$\varepsilon_3 = 0.102879706220 \pm 10^{-12}$	$\delta_1 = 3.804$
16	$\varepsilon_4 = 0.102286178168 \pm 10^{-12}$	$\delta_2 = 4.404$
32	$\varepsilon_5 = 0.102155640398 \pm 10^{-12}$	$\delta_3 = 4.547$
64	$\varepsilon_6 = 0.102126753995 \pm 10^{-12}$	$\delta_4 = 4.519$
128	$\varepsilon_7 = 0.102120342667 \pm 10^{-12}$	$\delta_5 = 4.506$
256	$\varepsilon_8 = 0.10211894039545 \pm 10^{-14}$	$\delta_6 = 4.572$
512	$\varepsilon_9 = 0.10211858239154 \pm 10^{-14}$	$\delta_5 = 3.917$
1024	$\varepsilon_{10} = 0.1021185049383287 \pm 10^{-16}$	$\delta_5 = 4.622$

(ii) Analysis of the Cascade

Here we test numerically the first terms of the series introduced by Feigenbaum [Feigenbaum (1978)] $\delta_{k-1} = \dfrac{\varepsilon_k - \varepsilon_{k-1}}{\varepsilon_{k+1} - \varepsilon_k}$ with ε_{k-1} the value of ε such as a bifurcation from a 2^{k-1} cycle to a 2^k one. The results are presented in the Table 8.1. Periodic solutions are searched for $t \geq 50000$. It seems that these results are slowly converging to the universal value $4.66920\ldots$.

8.3.5.2 *Second example: Discontinuities at 0*

(i) Bifurcation Diagram

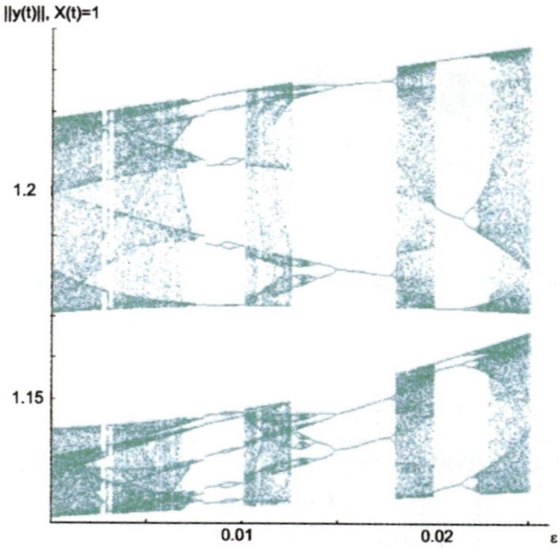

Fig. 8.14 Bifurcation diagram for $\alpha = 3.612$ and $\beta = 4.4$ for the system with discontinuity at 0.

This section is devoted to the bifurcation diagrams obtained for different values of the pair (α, β). In Fig. 8.14, such a bifurcation diagram is presented for $\alpha = 3.612$, $\beta = 4.4$ with ε varying from 0 to 0.025. Again $\| y(t) \|$ is plotted versus ε for $3000 \leq t \leq 4000$ when $X(t) = 1$. In this diagram several attractors seem to coexist: in the first two periodic windows, bifurcation branches show jumps. Just before ε reaches the value 0.025 the point $(0, 0, 0)$ captures the solutions. Indeed the trajectory issued from the initial conditions $(X_0, Y_0, Z_0) = (1.4, -0.3, -1.)$ is trapped by $(0, 0, 0)$ for $t \geq 100000$ if $\varepsilon \geq 0.024882$. Such a coexistence is clearer if we choose $\alpha = 15.6$, $\beta = 28.58$. In Figs. 8.15 and 8.16 where ε varies from 0. to 0.055, we can see two parts of the same diagram: Chaos and periodic solutions coexist for $\varepsilon = 0.0005$ and $\varepsilon = 0.009$ for example. Phase portraits are showed in the next subsection in order to clarify this point.

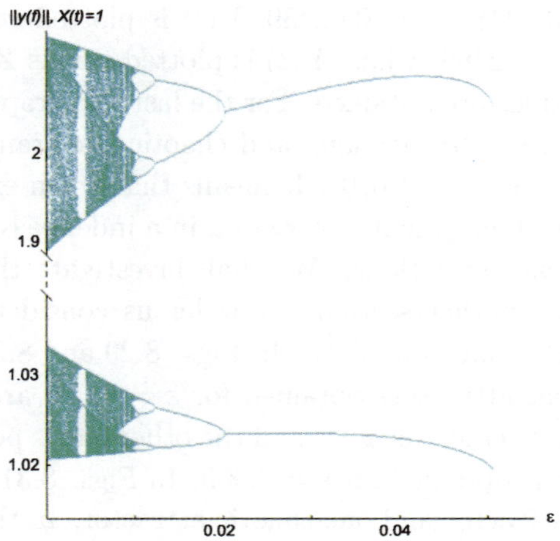

Fig. 8.15 First bifurcation diagram for $\alpha = 15.6$ and $\beta = 28.58$ for the system with discontinuity at 0.

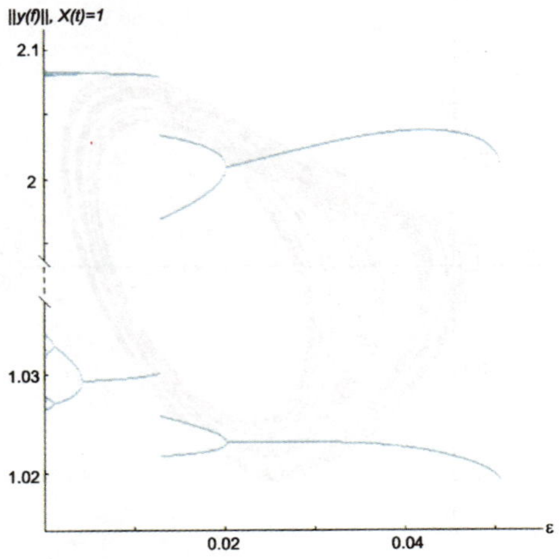

Fig. 8.16 Second bifurcation diagram for $\alpha = 15.6$ and $\beta = 28.58$ for the system with discontinuity at 0.

(ii) Phase Portraits

Let us consider the first case ($\alpha = 3.612$, $\beta = 4.4$): Several phase portraits are presented in the Figs. 8.17 to 8.28 for $\varepsilon = 0$, $\varepsilon = 0.0027$, $\varepsilon = 0.005$, $\varepsilon = 0.008$, $\varepsilon = 0.0103$, $\varepsilon = 0.016$, $\varepsilon = 0.018$, $\varepsilon = 0.0205$,

$\varepsilon = 0.0227$, $\varepsilon = 0.02488$, $\varepsilon = 0.02489$, $Y(t)$ is plotted versus $X(t)$ except for Fig. 8.28 ($\varepsilon = 0.02489$) where $Y(t)$ is plotted versus $Z(t)$. Both chaotic and periodic solutions are obtained. For the last two graphics of this figure, we can see that after a rather long and chaotic-like transient ($t = 41969$) the solution is trapped to $(0,0,0)$. It means that when exerting a friction-like action on the Chua's oscillator we obtain a indexpassive controlpassive control of nonlinear oscillations. We shall investigate the area of control in phase space in another section. Now let us consider the second case associated with $\alpha = 15.6, \beta = 28.58$. In Figs. 8.29 and 8.30 phase portraits of the two different attractors obtained for $\varepsilon = 0.005$ are plotted: $Y(t)$ is plotted versus $X(t)$. One is chaotic and the other one is periodic. The same kind of behaviour is obtained for $\varepsilon = 0.009$: In Figs. 8.31 and 8.32 one can see one periodic attractor and one chaotic attractor. In these two cases the periodic and the chaotic attractors are quite "close" to each other in phase space.

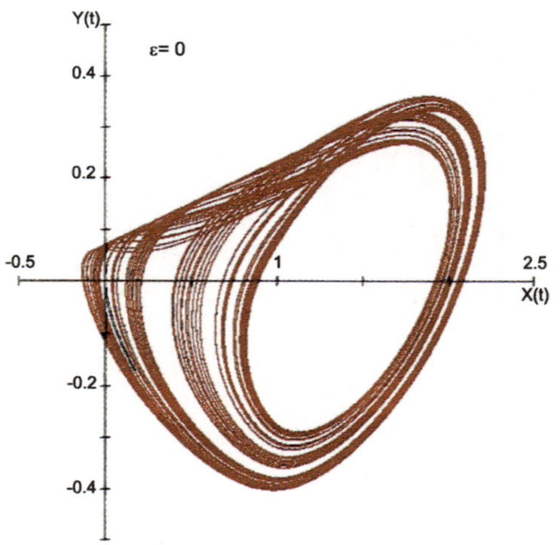

Fig. 8.17 (X, Y)-phase portrait for $\alpha = 3.612$, $\beta = 4.4$ and $\varepsilon = 0$.

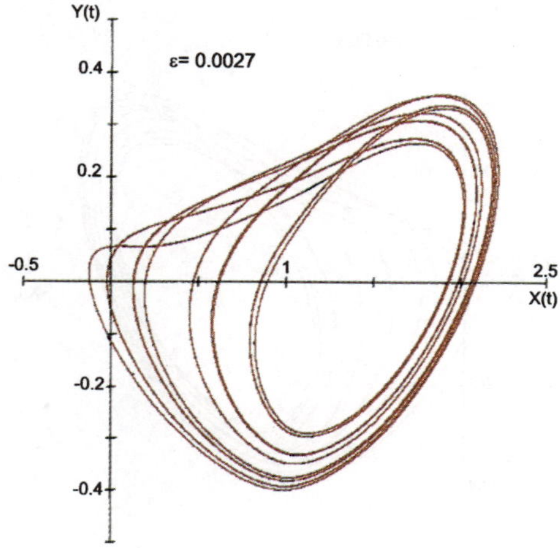

Fig. 8.18 (X, Y)-phase portrait for $\alpha = 3.612$, $\beta = 4.4$ and $\varepsilon = 0.0027$.

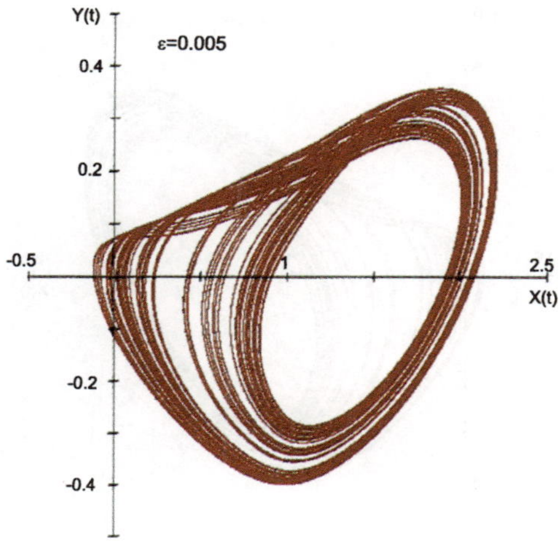

Fig. 8.19 (X, Y)-phase portrait for $\alpha = 3.612$, $\beta = 4.4$ and $\varepsilon = 0.005$.

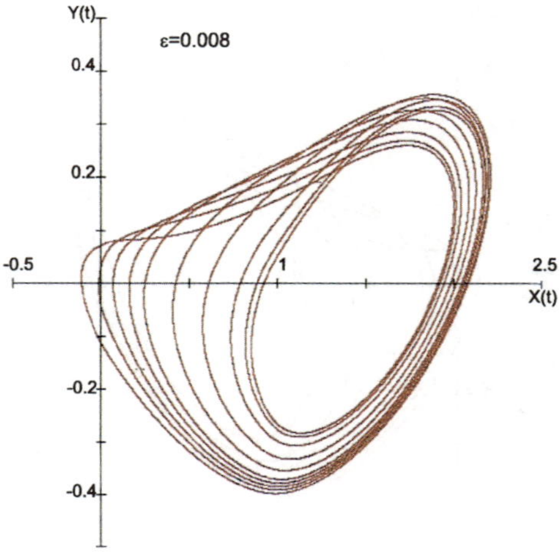

Fig. 8.20 (X, Y)-phase portrait for $\alpha = 3.612$, $\beta = 4.4$ and $\varepsilon = 0.008$.

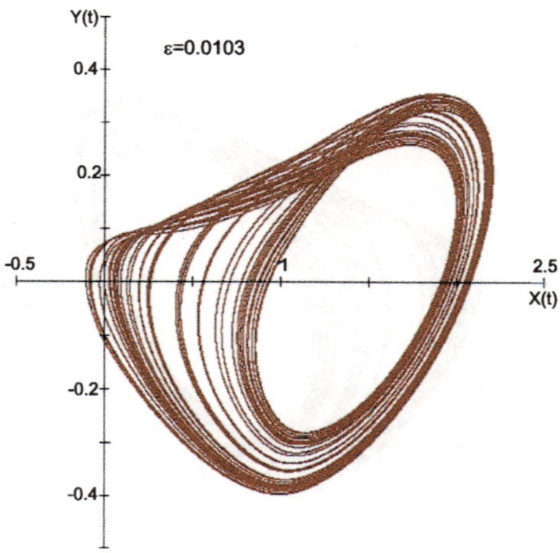

Fig. 8.21 (X, Y)-phase portrait for $\alpha = 3.612$, $\beta = 4.4$ and $\varepsilon = 0.0103$.

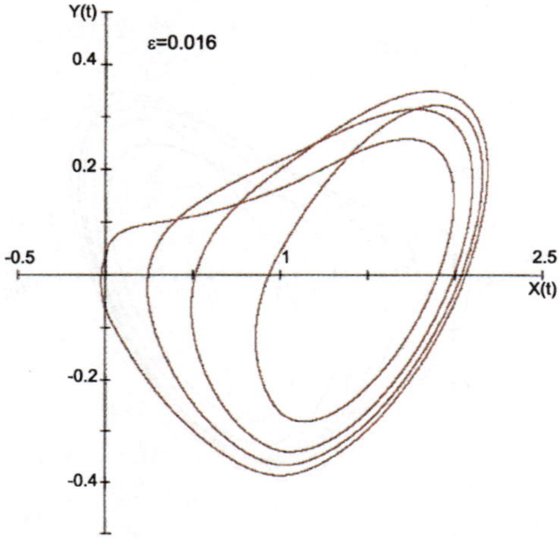

Fig. 8.22 (X, Y)-phase portrait for $\alpha = 3.612$, $\beta = 4.4$ and $\varepsilon = 0.016$.

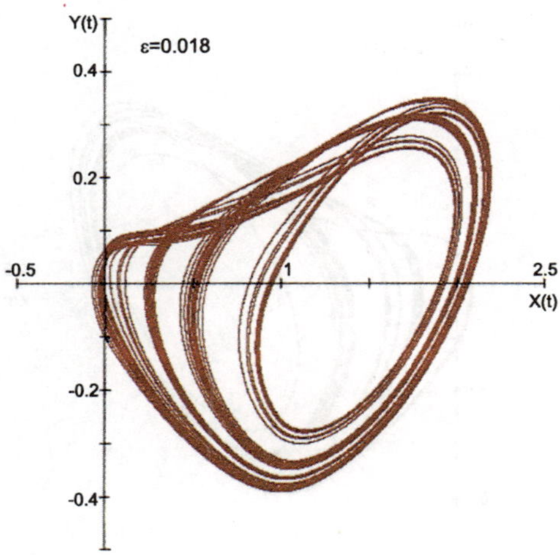

Fig. 8.23 (X, Y)-phase portrait for $\alpha = 3.612$, $\beta = 4.4$ and $\varepsilon = 0.018$.

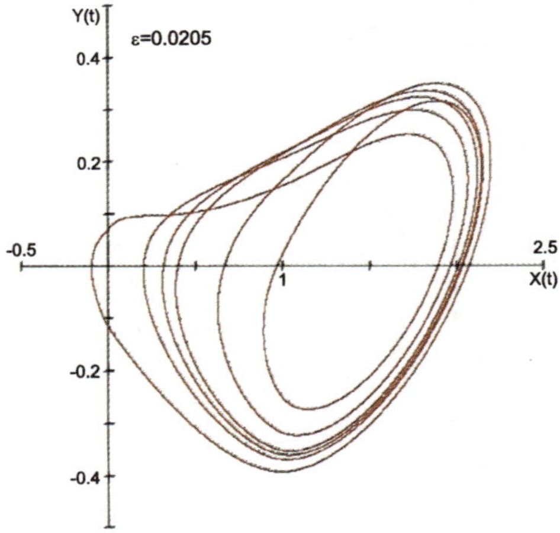

Fig. 8.24 (X, Y)-phase portrait for $\alpha = 3.612$, $\beta = 4.4$ and $\varepsilon = 0.0205$.

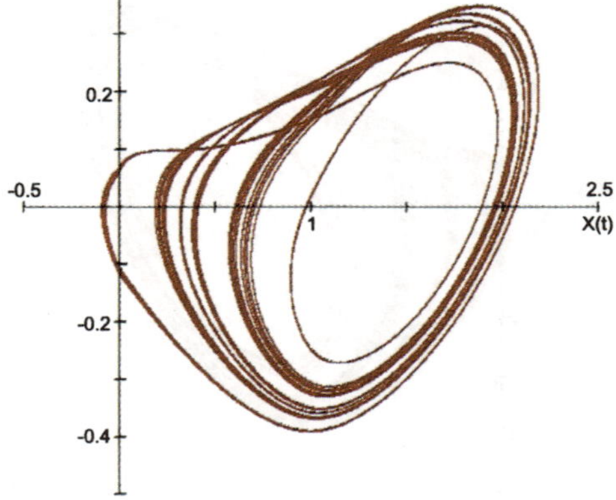

Fig. 8.25 (X, Y)-phase portrait for $\alpha = 3.612$, $\beta = 4.4$ and $\varepsilon = 0.0227$.

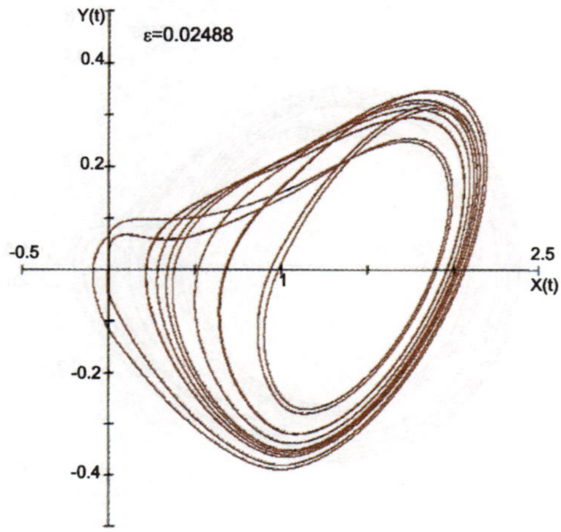

Fig. 8.26 (X, Y)-phase portrait for $\alpha = 3.612$, $\beta = 4.4$ and $\varepsilon = 0.02488$.

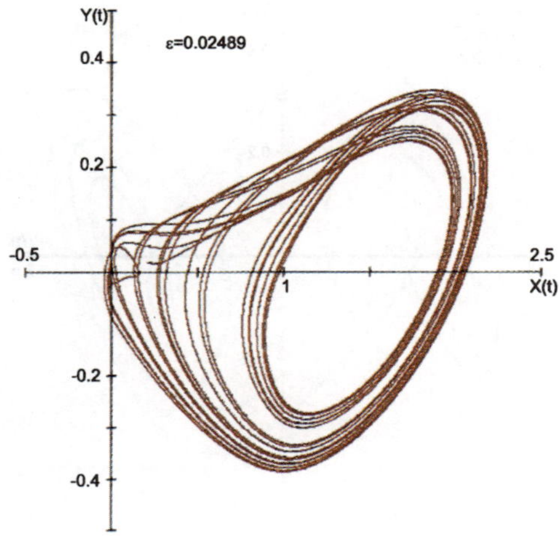

Fig. 8.27 (X, Y)-phase portrait for $\alpha = 3.612$, $\beta = 4.4$ and $\varepsilon = 0.02489$.

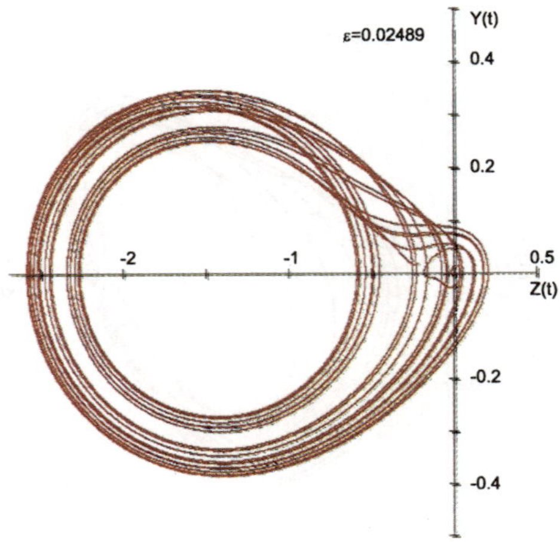

Fig. 8.28 (Y, Z)-phase portrait for $\alpha = 3.612$, $\beta = 4.4$ and $\varepsilon = 0.02489$.

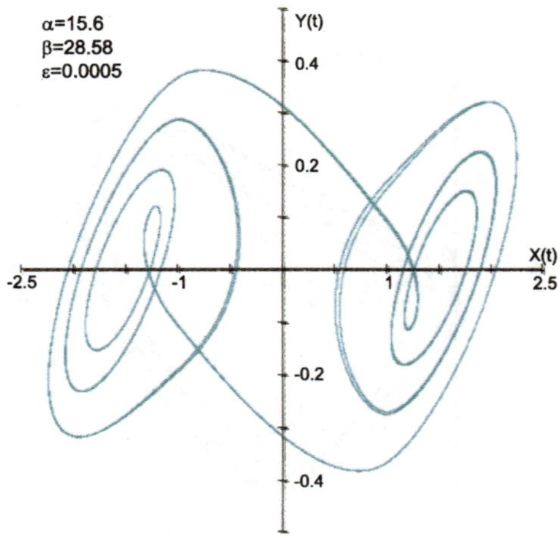

Fig. 8.29 (X, Y)-phase portrait for $\alpha = 15.6$, $\beta = 28.58$ and $\varepsilon = 0.0005$ with initial conditions $X_0 = 1, Y_0 = 0.096, Z_0 = -1.815$.

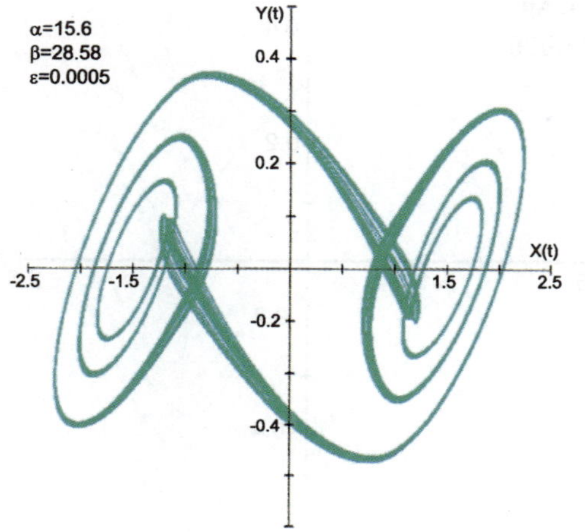

Fig. 8.30 (X,Y)-phase portrait for $\alpha = 15.6$, $\beta = 28.58$ and $\varepsilon = 0.0005$ with initial conditions $X_0 = 1.4, Y_0 = -0.3, Z_0 = -1$ for the system with discontinuity at 0.

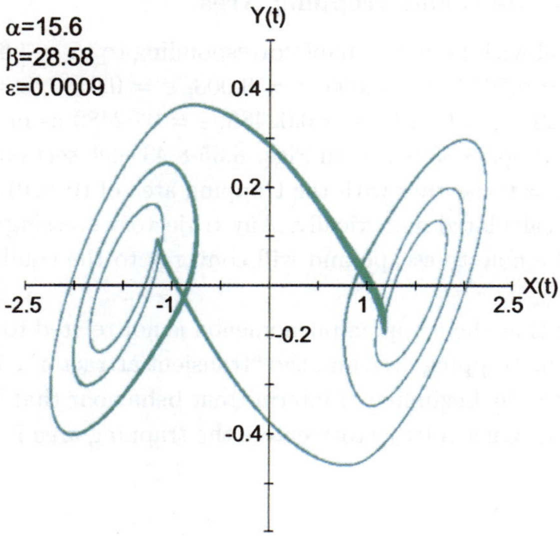

Fig. 8.31 (X,Y)-phase portrait for $\alpha = 15.6$, $\beta = 28.58$ and $\varepsilon = 0.0009$, with initial conditions $X_0 = 1, Y_0 = 0, Z_0 = -1.74$.

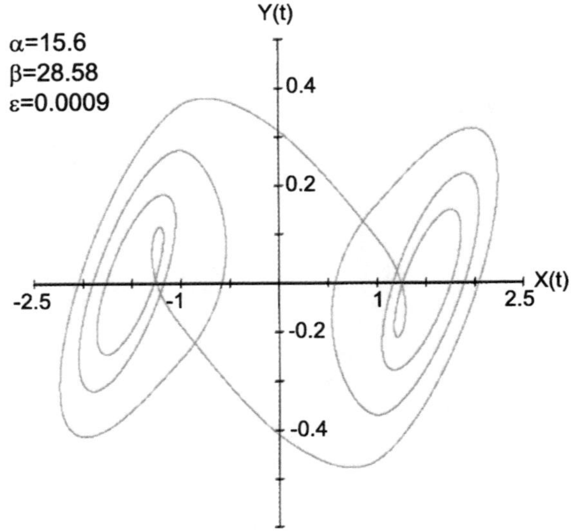

Fig. 8.32 (X, Y)-phase portrait for $\alpha = 15.6$, $\beta = 28.58$ and $\varepsilon = 0.0009$, with initial conditions $X_0 = 1.4, Y_0 = -0.3, Z_0 = -1$ for the system with discontinuity at 0.

(iii) Poincaré Maps and Trapping Area

Here we deal with Poincaré maps corresponding to ($\alpha = 3.612$, $\beta = 4.4$) and $\varepsilon = 0$, $\varepsilon = 0.0027$, $\varepsilon = 0.005$, $\varepsilon = 0.008$, $\varepsilon = 0.0103$, $\varepsilon = 0.016$, $\varepsilon = 0.018$, $\varepsilon = 0.0205$, $\varepsilon = 0.0227$, $\varepsilon = 0.02488$, $\varepsilon = 0.02489$ as in the previous subsection (see Figs. 8.33-8.43). In Figs. 8.35-8.43 such sections are plotted in the plane $X = 0$ together with the trapping area of $(0, 0, 0)$. This yellow area has been calculated numerically. Any trajectory crossing through this area will not be able to escape and will converge to the equilibrium point $(0, 0, 0)$.

We can see that the trapping phenomenon is not related to the intersection between the trapping area and the "transient attractor". It seems that friction provides the beginning of intermittent behaviour that is suppressed by the trapping: Once a trajectory enters the trapping area it can never go out anymore.

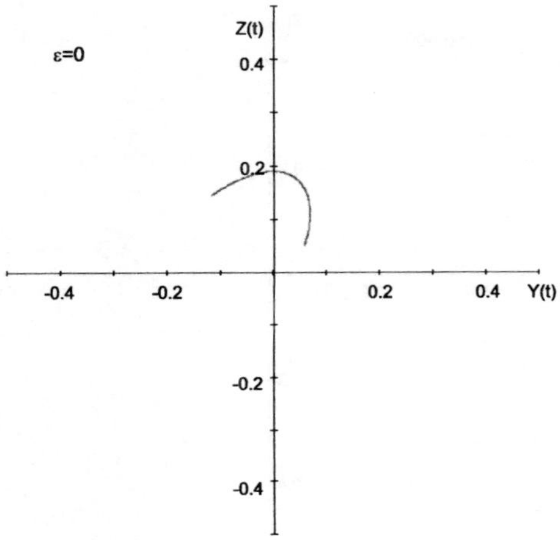

Fig. 8.33 Poincaré section in the plane $X = 0$ with trapping area for $\alpha = 3.612$, $\beta = 4.4$ and $\varepsilon = 0$ for the system with discontinuity at 0.

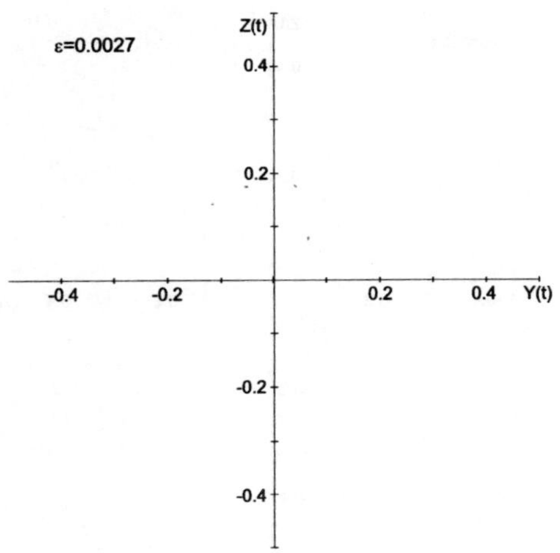

Fig. 8.34 Poincaré section in the plane $X = 0$ with trapping area for $\alpha = 3.612$, $\beta = 4.4$ and $\varepsilon = 0.0027$ for the system with discontinuity at 0.

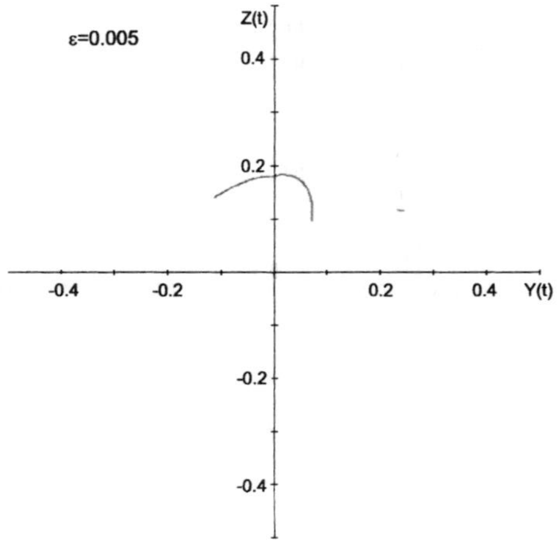

Fig. 8.35 Poincaré section in the plane $X = 0$ with trapping area for $\alpha = 3.612$, $\beta = 4.4$ and $\varepsilon = 0.005$ for the system with discontinuity at 0.

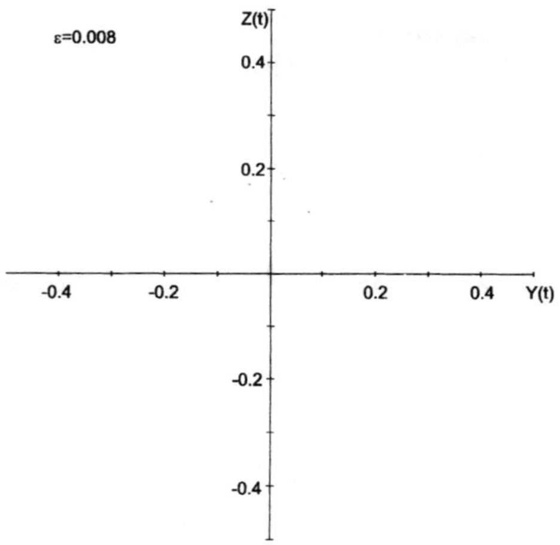

Fig. 8.36 Poincaré section in the plane $X = 0$ with trapping area for $\alpha = 3.612$, $\beta = 4.4$ and $\varepsilon = 0.008$ for the system with discontinuity at 0.

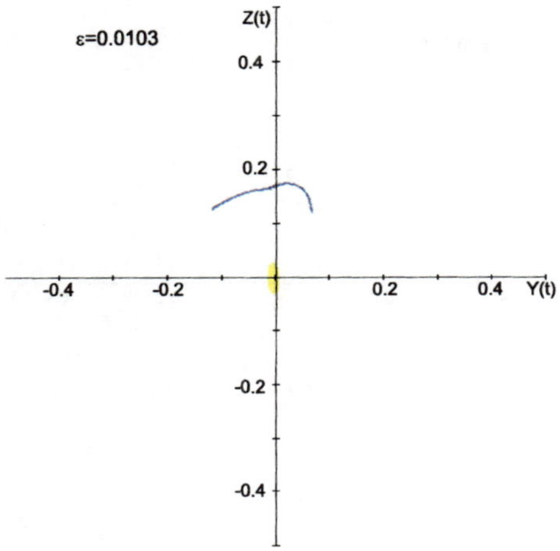

Fig. 8.37 Poincaré section in the plane $X = 0$ with trapping area for $\alpha = 3.612$, $\beta = 4.4$ and $\varepsilon = 0.0103$ for the system with discontinuity at 0.

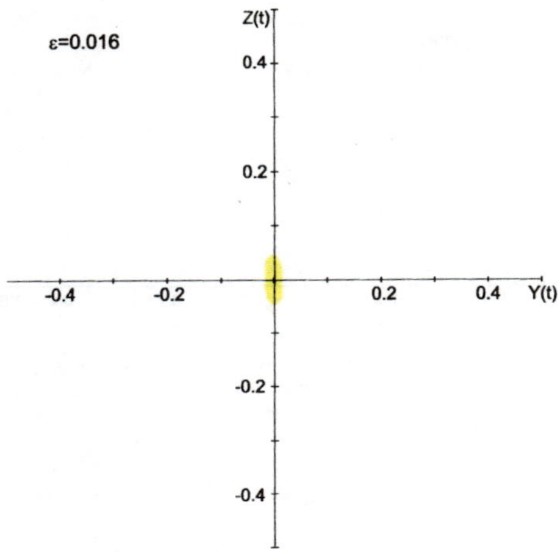

Fig. 8.38 Poincaré section in the plane $X = 0$ with trapping area for $\alpha = 3.612$, $\beta = 4.4$ and $\varepsilon = 0.016$ for the system with discontinuity at 0.

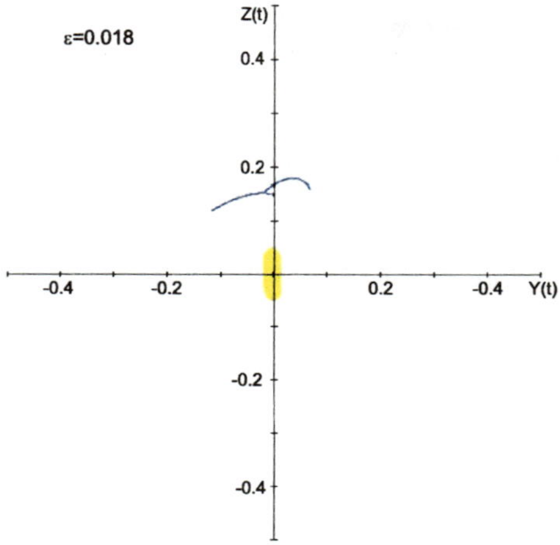

Fig. 8.39 Poincaré section in the plane $X = 0$ with trapping area for $\alpha = 3.612$, $\beta = 4.4$ and $\varepsilon = 0.018$ for the system with discontinuity at 0.

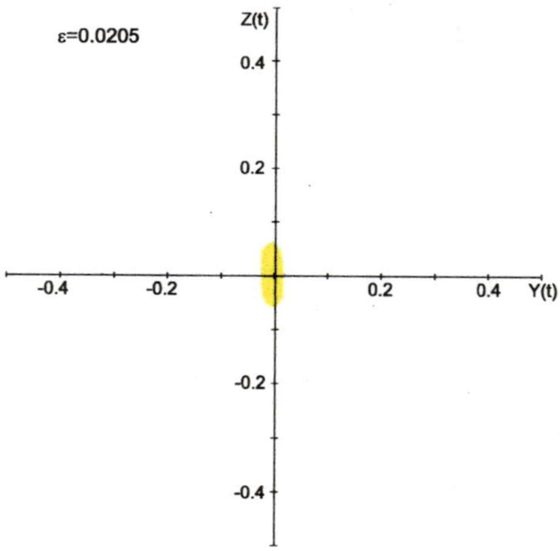

Fig. 8.40 Poincaré section in the plane $X = 0$ with trapping area for $\alpha = 3.612$, $\beta = 4.4$ and $\varepsilon = 0.0205$ for the system with discontinuity at 0.

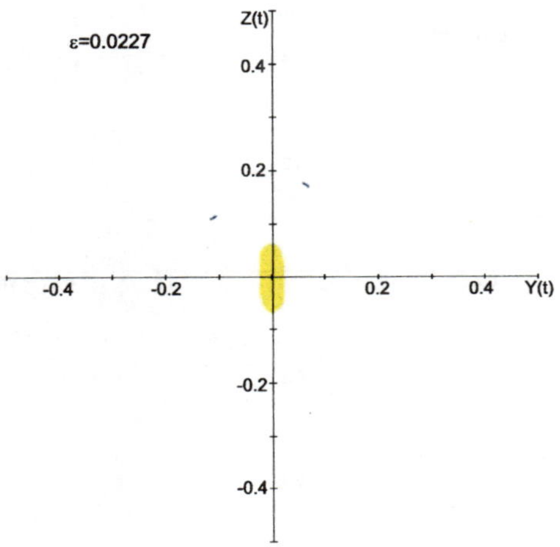

Fig. 8.41 Poincaré section in the plane $X = 0$ with trapping area for $\alpha = 3.612$, $\beta = 4.4$ and $\varepsilon = 0.0227$ for the system with discontinuity at 0.

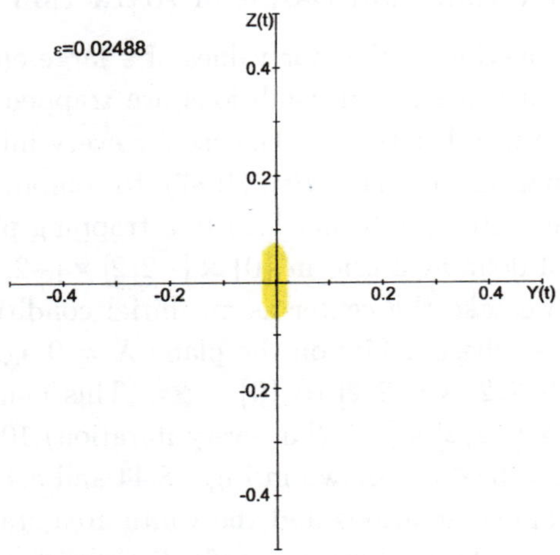

Fig. 8.42 Poincaré section in the plane $X = 0$ with trapping area for $\alpha = 3.612$, $\beta = 4.4$ and $\varepsilon = 0.02488$ for the system with discontinuity at 0.

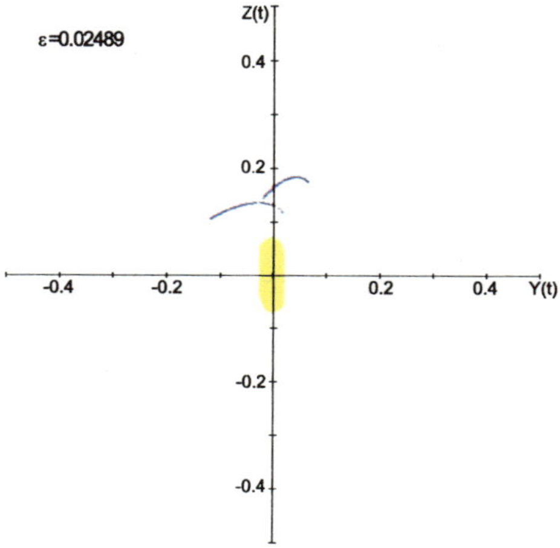

Fig. 8.43 Poincaré section in the plane $X = 0$ with trapping area for $\alpha = 3.612$, $\beta = 4.4$ and $\varepsilon = 0.02489$ for the system with discontinuity at 0.

(iv) Trapping Solutions and Basins of Attraction

We have seen previously that for values of ε large enough, the trajectories starting from some initial conditions are trapped to $(0., 0., 0.)$. In order to determine whether that is the case for every initial condition, we have used a cell-mapping method [Hsu (1987)] to compute the set of initial conditions included into $X = 0$ that lead to a trapping phenomenon. This method consists of defining a grid in $\{0\} \times [-2, 2] \times [-2, 2]$. Then for each cell on this grid we take the center as an initial condition, calculate the flow and determine where it hits on the plane $X = 0$ again. That gives a map from $\{0\} \times [-2, 2] \times [-2, 2]$ to $\{0\} \times \mathbb{R}^2$. This map is iterated (and restrained to $\{0\} \times [-2, 2] \times [-2, 2]$ at every iteration) 1000 times. Results for $\varepsilon = 0.02$ and $\varepsilon = 0.04$ are shown in Figs. 8.44 and 8.45. The black area corresponds to trapped solutions and the white area stands for divergent solutions. The two colours (blue and red) distinguish the solutions that start from $X = 0$ and then go to the domain D_0^+ from solutions that start from $X = 0$ and then go to the domain D_0^-.

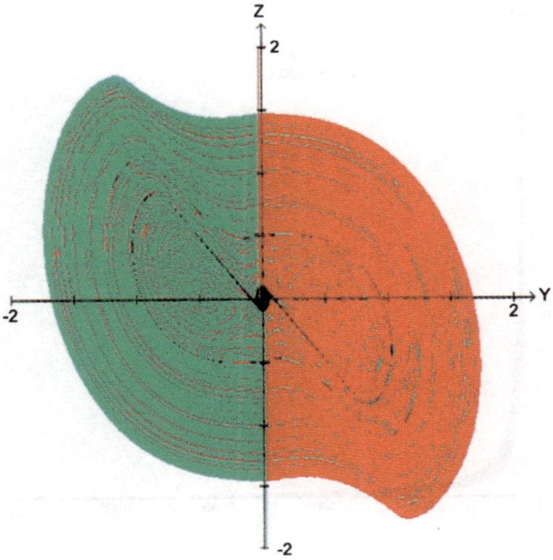

Fig. 8.44 Trapping area in the plane $X = 0$ for $\alpha = 3.612$, $\beta = 4.4$ and $\varepsilon = 0.02$.

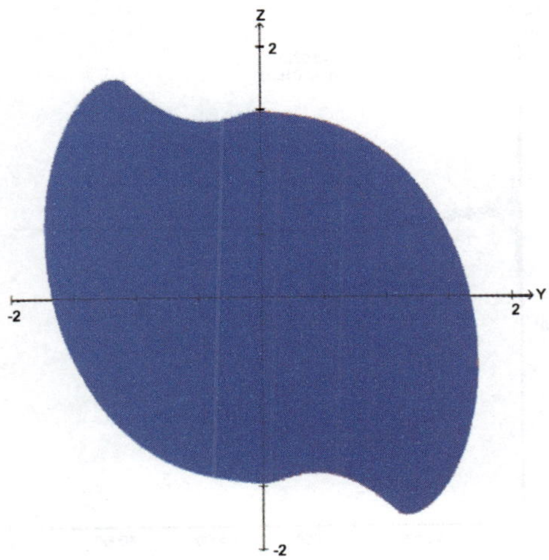

Fig. 8.45 Trapping area in the plane $X = 0$ for $\alpha = 3.612$, $\beta = 4.4$ and $\varepsilon = 0.04$.

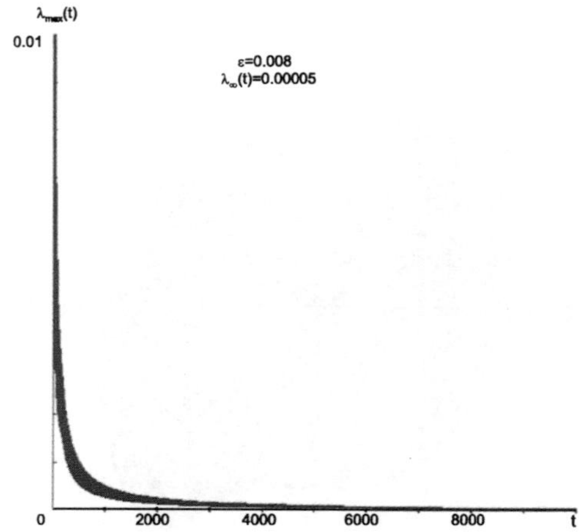

Fig. 8.46 Evolution of the largest Lyapunov exponent versus time for $\alpha = 3.612$, $\beta = 4.4$ and $\varepsilon = 0.008$ for the system with discontinuity at 0.

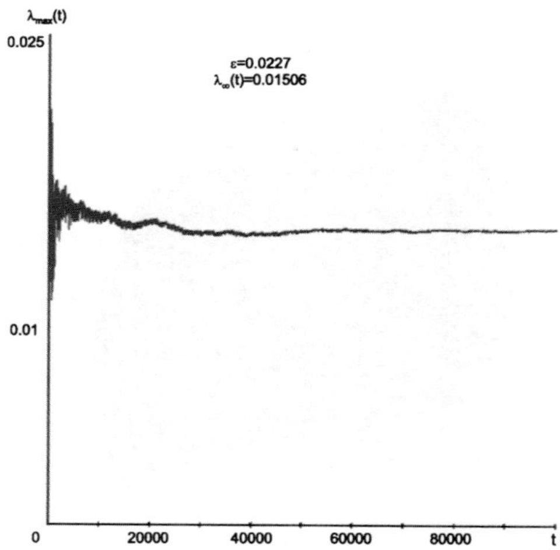

Fig. 8.47 Evolution of the largest Lyapunov exponent versus time for $\alpha = 3.612$, $\beta = 4.4$ and $\varepsilon = 0.0227$ for the system with discontinuity at 0.

It can be seen that before ($\varepsilon = 0.02$) the critical parameter $\varepsilon \simeq 0.02488$, the trapping area is small and concentrated in the neighbourhood of $(0., 0., 0.)$. After ($\varepsilon = 0.04$) the critical parameter, the whole nondivergent area is caught by $(0., 0., 0.)$ (for the same time interval): It means that every trajectory starting from any initial condition in $\mathbb{R}^3$ will either be trapped to $(0., 0., 0.)$ or diverge to infinity.

(v) Lyapunov Exponents

Methods

In the special case of piecewise linear differential systems, four methods at least could be used to compute Lyapunov exponents.

The first one consists in estimating analytically the value $\frac{D\varphi}{DX_0}$ for each area where the system is linear, the flow φ can be expressed analytically as a function of the initial condition.

With the second method, the flow is still calculated analytically but the differentiation is this time a numerical one. The third method uses numerical schemes to build the flow, but then it requires an analytical differentiation of the estimated flow. The last method is fully numerical: Both the trajectory and the differential $\frac{D\varphi}{DX_0}$ are estimated by numerical devices.

The method that we have chosen here is the second one. Indeed, the first method is very time expensive, and the last two were not used because they do not benefit from the piecewise linearity of the system, which allows analytical calculation of the trajectories.

The method used consists in starting with two very close initial conditions and building analytically their trajectories. At regular intervals, the trajectories are re-scaled and a computation of $\frac{D\varphi}{DX_0}$ is performed. This method provides an estimation of the largest Lyapunov exponent of the dynamical system, which is enough to check the presence of chaos.

Results

The largest Lyapunov exponent $\lambda_{\max}$ was computed for several parameter values. Table 8.2 shows the results corresponding to the Figs. 8.17 to 8.26: We can see a good correlation between phase portrait and Lyapunov exponents, except for the case $\varepsilon = 0.02488$ where phase portrait and Poincaré section suggest that there is a periodic attractor, whereas $\lambda_{\max}$ seems to be clearly positive (see the Fig. 8.48).

Table 8.2 Largest Lyapunov exponent for $\alpha = 3.612$ and $\beta = 4.4$ after $T = 100000$.

ε	λ_{max}
0	0.04777
0.0027	0.00027
0.005	0.03800
0.008	0.00005
0.0103	0.03970
0.016	0.00002
0.018	0.02792
0.0205	0.00008
0.0227	0.01507
0.02488	0.03226

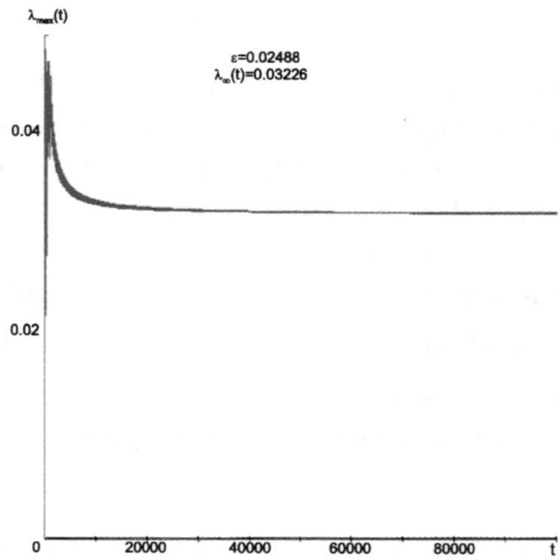

Fig. 8.48 Evolution of the largest Lyapunov exponent versus time for $\alpha = 3.612$, $\beta = 4.4$ and $\varepsilon = 0.02488$ for the system with discontinuity at 0.

In Figs. 8.46 to 8.48 we have plotted the value of $\lambda_{\max}$ as a function of time to check the stabilization of calculations. It appears that at $t = 100000$, $\lambda_{\max}$ has only small variations so that we can give its final value with a 10^{-3} uncertainty at most. Finally, in Fig. 8.49 the value of $\lambda_{\max}$ for $t = 40000$ was plotted as a function of ε, with the same parameter values as in Fig. 8.14. Both figures show good correlation, every periodic window corresponding to values $\lambda_{\max} \leq 0.001$ and every chaotic one to values more clearly positive.

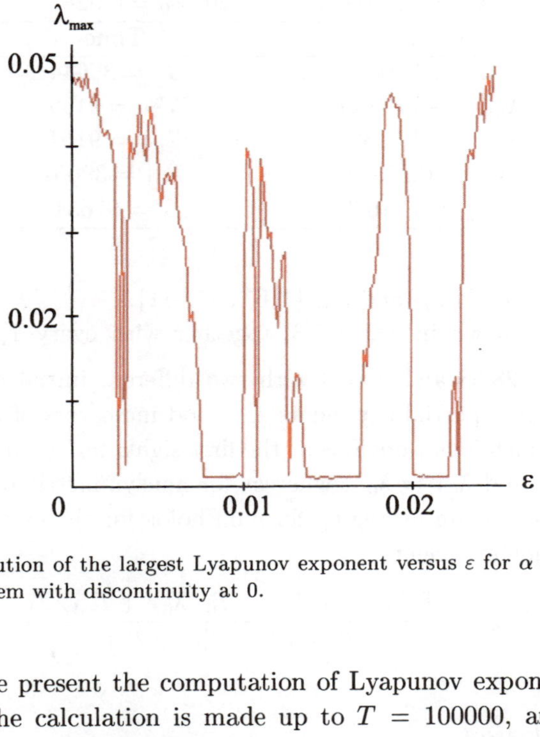

Fig. 8.49 Evolution of the largest Lyapunov exponent versus ε for $\alpha = 3.612$, $\beta = 4.4$ and for the system with discontinuity at 0.

Finally, we present the computation of Lyapunov exponents split into four parts: the calculation is made up to $T = 100000$, and every time the trajectory enters a domain (D_1, D_{0^+}, D_{0^-} or D_{-1}) we keep the time spent in that domain and the contribution to the calculation of the largest Lyapunov exponent. Thus we define λ_1, λ_{0^+}, λ_{0^-} or λ_{-1} as follows:

$$\lambda_i = \frac{1}{\sum_n t_n^i} \ln \left(\prod_n \frac{\Delta y_n^i}{\Delta y_{n-1}^i} \right), \tag{8.100}$$

where t_n is the nth time spent in the domain D_i, $\dfrac{\Delta y_n^i}{\Delta y_{n-1}^i}$ is the correspond-

Table 8.3 Partial Lyapunov exponents for $\alpha = 15.6$, $\beta = 28.58$ and $\varepsilon = 0$.

$X_0 = -0.002$, $Y_0 = 0.014$, $Z_0 = 0.010$	
Exponent	Time
$\lambda_1 = -0.09435$	$T_1 = 40181$
$\lambda_{0+} = -0.04497$	$T_{0+} = 9591$
$\lambda_{0-} = -0.21074$	$T_{0-} = 10179$
$\lambda_{-1} = 0.22839$	$T_{-1} = 40053$
$\lambda_{\max} = 0.02780$	$T = 100004$
$X_0 = -0.008$, $Y_0 = 0.020$, $Z_0 = 0.028$	
Exponent	Time
$\lambda_1 = 0.38000$	$T_1 = 39644$
$\lambda_{0+} = -1.01647$	$T_{0+} = 9169$
$\lambda_{0-} = -1.00361$	$T_{0-} = 9161$
$\lambda_{-1} = 0.37334$	$T_{-1} = 39657$
$\lambda_{\max} = 0.11632$	$T = 97631$

ing contribution to λ_{max}, and $i \subset \{1, 0^+, 0^-, -1\}$.

Results are shown in Table 8.3, together with every $T_i = \sum_n t_n^i$, for $\alpha = 15.6$, $\beta = 28.58$ and $\varepsilon = 0$ with two different initial conditions. It appears that these partial exponents are good indicators of the symmetry of attractors which is not obvious at the first sight: for symmetric ones, we have $\lambda_1 = \lambda_{-1}$ and $\lambda_{0+} = \lambda_{0-}$, whereas for nonsymmetric ones, all λ_i are different. Moreover, an averaging formula holds for the calculation of the largest Lyapunov exponent:

$$\lambda_{max} = \frac{T_1\lambda_1 + T_{0+}\lambda_{0+} + T_{0-}\lambda_{0-} + T_{-1}\lambda_{-1}}{T_1 + T_{0+} + T_{0-} + T_{-1}}. \tag{8.101}$$

8.3.6 *Conclusion*

In this section, we have investigated a mechanical model which can be modelled as a Chua circuit with discontinuities. Existence and uniqueness have been obtained in the frame of monotone maximal operators. Then analytical procedures have been developed in order to track the discontinuity times. Thus numerical results are provided including the detection of discontinuity with a very high accuracy. Discontinuities seem to provide symmetry breaking. Chaos has been observed. It has been characterized by positive Lyapunov exponent.

Discontinuities may correspond to a kind of friction from the mechanical point of view. We have shown that it is possible to control Chua's circuit: If the discontinuity is strong enough, every initial condition starting from a neighbourhood of $(0, 0, 0)$ is trapped to this equilibrium point. Analytical estimates have been obtained in [Janin (2001)] for trapping areas, and control parameters ε.

Now some extension can be easily pointed out. From the mechanical point of view, one could consider external excitation. Hence it would be interesting to deal with Eq. (8.55) modified by an external field of the form $\gamma \cos(\Omega t)$ and to investigate control. Another point of view has not been taken into account here: Our numerical procedures are based upon analytical calculation and they are not pure numerical schemes. Thus it is possible to locate "exactly" discontinuity times and to calculate the values of the coordinates at these times. It would be interesting to investigate numerical schemes that can be provided by the mathematical frame and to compare the results with those of the previous sections. Indeed, analytical calculations are not always possible, for instance when smooth polynomial nonlinearities perturb the equations of the Chua system. Then numerical accurate detection of domain transitions would be necessary.

Chapter 9

Mechanical System with Impacts and Modal Approaches

9.1 Introduction

Analysis of the response of structures is convenient if a linear model can fully describe the structure. In this frame, it is useful to introduce in a finite or infinite dimension (Hilbertian case) the notion of eigenmodes of the structure. They are either normal modes (defined by adding conservative conditions to the model) or complex modes (taking into account viscous damping for example) [Caughey (1965); Meirovitch (1967)]. The linear theory of differential systems provides the response of the structure to an external elementary sinusoidal solicitation under an interesting form: The full response is simply the superposition of the responses of each mode to the solicitation. Such a formula is well known; this is the superposition formula which is the basis of modal synthesis [Jézéquel (1985); Meirovitch (1967)]. The notion of modal synthesis can be extended to the case of substructures by using linear operator theory [Bourquin (1991); Jézéquel (1985)].

In the nonlinear case, the notion of nonlinear modes had been considered first. In the case of mathematically smooth nonlinearities and for a finite number of degrees of freedom with particular polynomial nonlinearities, Rosenberg first introduced natural modes [Rosenberg and Atkinson (1959)] and then nonlinear normal modes [Pak and Rosenberg (1968)], [Rosenberg (1961); Rosenberg (1962); Rosenberg (1964)], and investigated their stability. Until now, many methods have been used to introduce modes (natural, nonlinear, nonlinear normal, minimal normal, nonlinear similar normal, etc.) in the case of nonlinear structures: Methods derived from the works of Rosenberg [Anand (1972); Cooke and Struble (1966); Pak (1989); Stuart (1971)], stroboscopic method [Christopher (1979)], methods based on averaging and modal truncation [Szemplińska-Stupnicka (1969); Szemplińska-Stupnicka (1983); Szemplińska-Stupnicka (1990)], direct or geometrical methods for conservative systems [Mikhlin (1996); Rand (1974); Vakakis (1992)] or in the Hamiltonian frame [Hyams and Month (1984); Johnson and Rand (1979); Month and Rand (1977); Weinstein (1973); Zevin (1984)], Padé approximation [Mikhlin (1995); Vakakis *et. al.* (1996)], multi-spectral, Volterra-series and HFRFs [Chen *et. al.* (1997); Gifford and Tomlinson (1989)], integral transforms [Bargmann (1961)], Lie series [Claude (1986)], methods based on normal forms in the hamiltonian [Arnold (1976a); Gustavson (1966); Month and Rand (1980)] or general frame [Hsu (1977); Hsu (1983a); Hsu (1983b); Poincaré (1987); Smith (1986b); Smith (1986a)], methods using center manifold theory [Boivin *et. al.* (1995); Holmes (1981); Holmes and Rand (1980); Pierre and Shaw (1991); Shaw and Shaw (1989); Shaw and Pierre (1992); Shaw and Pierre (1994)] or amplitude equations [Coullet and Spiegel (1985)]. Jézéquel and Lamarque extended the modal superposition starting from nonlinear modes built via normal form, for systems with a few degrees of freedom and smooth nonlinearities [Jézéquel and Lamarque (1991); Jézéquel and Lamarque (1992); Lamarque (1992)], even in the case of complex modes [Lamarque and Jézéquel (1992)]: this method is valid for small enough nonlinear oscillations. The obtained nonlinear modes and generalized masses depend on the amplitudes of the normal coordinates. They are built in order to agree at best with a resonance equation. The modes do not always verify the reciprocity condition as in the linear case.

The question of nonlinear modes is obviously related to the search of periodic solutions to nonlinear dynamical systems and study of their stability [Month and Rand (1977); Rand *et. al.* (1992)]. So it is related to a huge literature dealing with numerous analytical methods, perturbation methods, methods of bifurcation analysis for instance [Arnold (1976a); Birkhoff (1927); Bogaevski and Povzner (1991); Brjuno (1989); Carr (1981); Giacaglia (1972); Hale and Chow (1982); Iooss and Joseph (1980); Nayfeh (1993); Nayfeh and Mook (1979); Urabe (1967); Verhulst (1996)].

In the field of stochastic behaviour, the question of nonlinear modes has been already examined [Sri Namachchivaya and Lin (1991)] and tools are available [Boxler (1991)]. In the case of nonsmooth nonlinearities, only particular cases have been investigated: normal modes for piecewise linear systems [Chen and Shaw (1996); Zuo and Curnier (1992)]. Sometimes, in order to deal with localized or weak nonlinear nonsmooth phenomena using linear methods, one can introduce a modified dissipation or stiffness matrix.

From the point of view of dynamics, vibro-impact systems have been thoroughly studied in the literature: global behaviours and periodic solutions have been investigated in the single degree of freedom case (both analytically and numerically in [Peterka and Vacik (1992); Shaw and Holmes (1983)], [Shaw and Rand (1989); Whiston (1987)] or by means of a change of variables in [Ivanov (1994)]) and in the two degree of freedom case (double impact oscillator in [Peterka (1999a); Peterka and Szöllös (1996)] or impact damper in [Chatterjee and Mallik (1996); Sung and Yu (1992)]). Singularities in the dynamics of such systems have been pointed out in [Molenaar *et. al.* (1999); Nordmark (1991); Whiston (1992)], and some authors have examined the effect of dry friction on mechanical systems with impacts in [Cipera *et. al.* (1996); Peterka (1996); Peterka (1999b)], [Cone and Zadoks (1995); Pfeiffer (1988a)]. General results about vibro-impact systems can also be found in [Babitsky (1998a); Brogliato (1996)]. Moreover, a modal approach has been introduced in [Mikhlin (1995)] to deal with direct and inverse problems in discrete systems with impacts, based on the theory of nonlinear normal modes [Vakakis *et. al.* (1996)].

Nevertheless, to our knowledge no attempt of building a modal superposition similar to the linear case exists in the case of hard nonsmooth nonlinearities such as friction or impact. The main aim of this chapter is to carry out some answers to the question: is it possible to build a modal superposition and a modal synthesis in the case of structures exhibiting a nonlinearity of impact type? We consider the problem in the case of simple systems with one and two degrees of freedom.

In section 9.2, we consider a single degree of freedom system. Using a piecewise exact integration (section 9.2.1), we study the periodic responses under sinusoidal excitation (section 9.2.2). Then we test the building of a modal superposition in section 9.2.3 by introducing successively a generalized eigenfrequency from free vibrations, a generalized mode and a generalized mass associated with forced oscillations.

In section 9.3, we consider two degrees of freedom systems. We deal first with the case of a weak coupling and impact of a mass against a rigid stop (section 9.3.1). Then we examine a case of strong coupling, again with impacts against a external rigid stop (section 9.3.2). We apply the results obtained to the case of direct impacts between two rigid solids (section 9.3.3). Finally we conclude in section 9.4 on the relevance of the obtained modal superposition formula.

9.2 Single Degree of Freedom System

The studied system consists of a single degree of freedom damped harmonic oscillator with a unilateral constraint, for which an impact law is defined (see Fig. 4.1). The impact process is considered to be instantaneous and the behaviour of the system at impact time is described using the coefficient of restitution $e \in [0, 1]$, characterizing the energy loss during impact. The equations governing the dynamics of the system are then:

$$\begin{cases} \ddot{x} + a\dot{x} + \omega_1^2 x = f\cos(\omega t), \\ x(t) \leq x_{max}, \\ \dot{x}(t^+) = -e\dot{x}(t^-) \quad \text{if} \quad x(t) = x_{max}. \end{cases} \tag{9.1}$$

If the velocity immediately before the impact at $\bar{t}$ is zero, several cases can occur: If at $\bar{t}$ the acceleration is negative, then the system is still described by (9.1) after the impact and the trajectory is tangent to the stop at $\bar{t}$. If on the other hand the acceleration is positive, then the system remains in contact with the stop on a nonzero time interval. It can be shown that sticking never occurs if $f < \omega_1^2 x_{max}$. In the following we will assume that the system's parameters satisfy this condition.

9.2.1 *Analytical solution*

The system with sinusoidal forcing can be written:

$$\begin{cases} \ddot{x} + a\dot{x} + \omega_1^2 x = f\cos(\omega t), \\ \dot{x}(t^+) = -e\dot{x}(t^-) \quad \text{if} \quad x(t) = x_{max}, \\ x(0) = x_0, \quad \dot{x}(0) = \dot{x_0}. \end{cases} \tag{9.2}$$

Since this system is linear between two consecutive impacts, it is possible to determine a piecewise analytical form of the solution on $\mathbb{R}^+$. We set $\tilde{\omega_1} = \sqrt{\omega_1^2 - \frac{a^2}{4}}$ and $\eta = \frac{a}{2\tilde{\omega_1}}$. $\forall k \in \mathbb{N}^*$, the solution on $[t_{k-1}, t_k]$ can be written in the form:

$$\begin{cases} x(t) = \mathrm{e}^{-at/2}[A_k \cos(\tilde{\omega_1} t), +B_k \sin(\tilde{\omega_1} t)] + f_1 \cos(\omega t) + f_2 \sin(\omega t), \\ x(t_k^+) = x(t_k^-) = x_{max}, \\ \dot{x}(t_k^+) = -e\dot{x}(t_k^-), \end{cases} \tag{9.3}$$

where $t_0 = 0$ and

$$\begin{cases} f_1 = f\dfrac{\omega_1^2 - \omega^2}{(\omega_1^2 - \omega^2)^2 + a^2\omega^2}, \\ f_2 = f\dfrac{a\omega}{(\omega_1^2 - \omega^2)^2 + a^2\omega^2}. \end{cases}$$

From Eqs. (9.3) we get the following recursive relation giving the value of the constants A_k and B_k:

$$\begin{cases} \begin{pmatrix} A_1 \\ B_1 \end{pmatrix} = \begin{pmatrix} x_0 - f_1 \\ \dfrac{\dot{x_0} - f_2\omega}{\tilde{\omega_1}} + \eta(x_0 - f_1) \end{pmatrix}, \\ \begin{pmatrix} A_{k+1} \\ B_{k+1} \end{pmatrix} = \begin{pmatrix} A_k \\ B_k \end{pmatrix} + (1+e)u(A_k, B_k, t_k) \begin{pmatrix} -\sin(\tilde{\omega_1} t_k) \\ \cos(\tilde{\omega_1} t_k) \end{pmatrix}, \\ u(A_k, B_k, t_k) = [\sin(\tilde{\omega_1} t_k) + \eta\cos(\tilde{\omega_1} t_k)]A_k + \\ \quad +[-\cos(\tilde{\omega_1} t_k) + \eta\sin(\tilde{\omega_1} t_k)]B_k + \\ \quad + \dfrac{\omega}{\tilde{\omega_1}} e^{a t_k/2}[f_1 \sin(\omega t_k) - f_2\cos(\omega t_k)]. \end{cases} \quad (9.4)$$

9.2.2 *Periodic solutions*

Thanks to the analytical form of the solution given by (9.3), (9.4) it is possible to analytically seek periodic solution, similarly to what has been done in [Shaw and Holmes (1983)] or [Peterka and Vacik (1992)] for example. We will call (n, k)-periodic solution of period nT with k impacts per cycle, where $T = \dfrac{2\pi}{\omega}$ is the period of the external excitation.

9.2.2.1 $(\mathbf{n}, \mathbf{0})$-*periodic solutions*

The simplest case consists in looking for nT-periodic solutions which never impact against the stop. Such a case implies $n = 1$ and the initial conditions leading to a $(1,0)$-periodic solution are $\begin{cases} x_0 = f_1, \\ \dot{x}_0 = f_2\omega. \end{cases}$ Let us set:

$$\begin{cases} \omega_+ = \sqrt{\tilde{\omega_1}^2 - \dfrac{a^2}{4} + \sqrt{\dfrac{f^2}{x_{max}^2} - a^2\tilde{\omega_1}^2}}, \\ \\ \omega_- = \sqrt{\tilde{\omega_1}^2 - \dfrac{a^2}{4} - \sqrt{\dfrac{f^2}{x_{max}^2} - a^2\tilde{\omega_1}^2}}. \end{cases} \tag{9.5}$$

As we assumed that $f < \omega_1^2 x_{max}$, it can then be shown that $(n,0)$-periodic solutions exist if and only if $\omega \leq \omega_-$ or $\omega \geq \omega_+$.

The stability of these periodic solutions can be determined using a Poincaré map defined by a constant phase plane $Z = T$ in the coordinates $(X, Y, Z) = (x, \dot{x}, t \bmod T)$. As pointed out in [Nordmark (1991)], such a mapping is not everywhere continuous nor differentiable. Therefore, for each periodic solution corresponding to a fixed point of the Poincaré map where it is continuously differentiable, the stability can be investigated. We obtain that $(1,0)$-periodic solutions are always stable.

9.2.2.2 $(\mathbf{n}, \mathbf{1})$-*periodic solutions*

Similarly to subsection 9.2.2.1, we can seek nT-periodic solutions with one impact per cycle by using (9.3) and (9.4). In this case, the impact time can be analytically determined, and the initial conditions leading to $(n,1)$-periodic solutions are then given by,

$$\begin{cases} x_0 = A_1(t_1) + f_1, \\ \dot{x_0} = \tilde{\omega_1}[B_1(t_1) - \eta A_1(t_1)] + \omega f_2, \end{cases}$$

where $A_1(t_1)$ and $B_1(t_1)$ depend analytically on the system's parameters. An example of $(1,1)$-periodic solution is shown in Fig. 9.1. Existence of $(1,1)$-periodic solutions is investigated in Fig. 9.2 versus frequency ω.

As in 9.2.2.1, we can use the Poincaré map (when it is defined and of class C^1 locally) in order to determine the type of the obtained periodic solutions. The Jacobian matrix can be analytically calculated from the analytical form of the Poincaré map by taking into account the influence of the partial derivatives of the impact time t_1 with respect to the initial conditions.

We can show that such a stability study is valid only if $\omega \neq \omega_+$ and

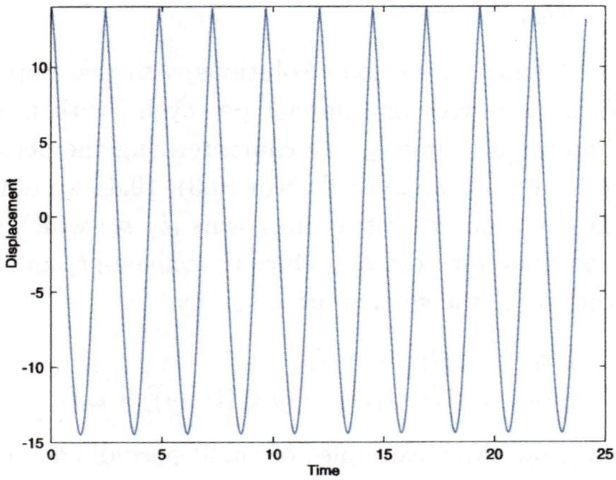

Fig. 9.1 (1,1)-periodic solution to the system $\ddot{x} + a\dot{x} + \omega_1^2 x = f\cos(\omega t)$, $\dot{x}(t^+) = -e\dot{x}(t^-)$ if $x(t) = x_{max}$; $x(0) = x_0$, $\dot{x}(0) = \dot{x_0}$ for $\omega = 2.6$, $x_0 = 13.32968$ and $\dot{x_0} = 19.36619$.

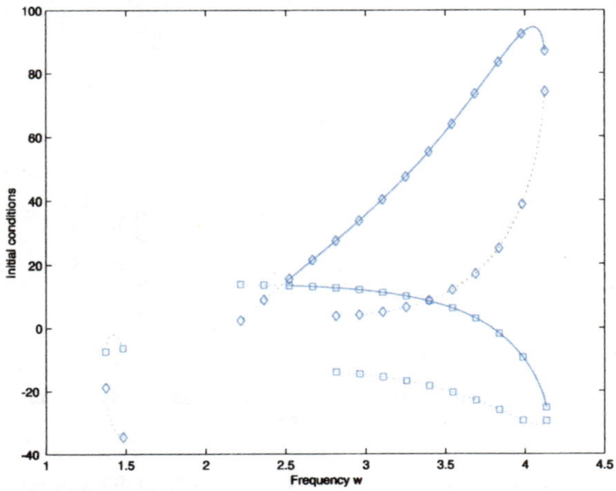

Fig. 9.2 Existence of (1,1) periodic solutions. Stable (square: x_0, diamond: $\dot{x_0}$) and unstable (dotted curve) solutions.

$\omega \neq \omega_-$ for in that case the Poincaré map is not differentiable at the considered fixed point.

9.2.2.3 $(\mathbf{n}, \mathbf{2})$-*periodic solutions*

The method for seeking nT-periodic solutions with two impacts per cycle is identical to the case with one impact per cycle, with a new unknown due to the second impact time t_2. By characterizing the periodicity of the solution and by using the analytical form (9.3), (9.4) we obtain a system of two nonlinear equations with two unknowns (t_1, t_2) which can be solved using Newton's method for example. Once the value of t_1 and t_2 are known, the initial conditions of the system are given by

$$\begin{cases} x_0 = A_1(t_1, t_2) + f_1, \\ \dot{x_0} = \tilde{\omega_1}(B_1(t_1, t_2) - \eta A_1(t_1, t_2)) + \omega f_2. \end{cases}$$

Figs. 9.3 to 9.5 show three examples of $(n, 2)$-periodic solutions that can be obtained analytically.

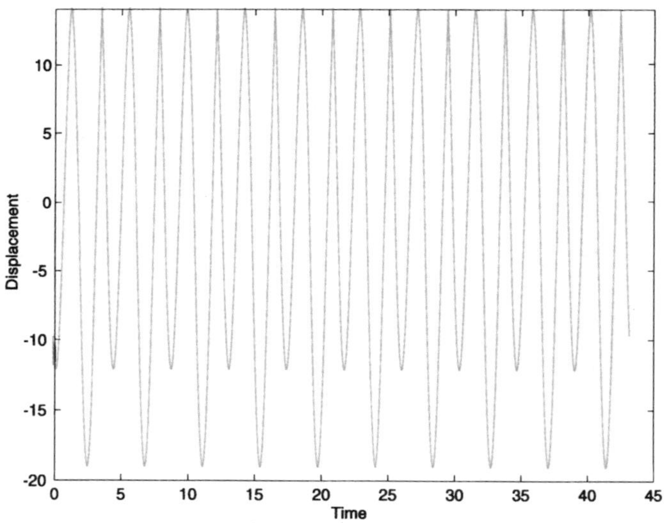

Fig. 9.3 $(1, 2)$-periodic solution for $\omega = 1.45$, $x_0 = -9.89493$ and $\dot{x_0} = -20.71102$.

As in the case of $(n, 1)$-periodic solutions, we can study the stability of the $(n, 2)$-periodic solutions by using the Poincaré map (see in Fig. 9.6 to 9.8): when $\omega \notin \{\omega_+, \omega_-\}$, it is possible to calculate the jacobian matrix of the Poincaré map, analytically and its eigenvalues determine the stability or instability of the periodic solution.

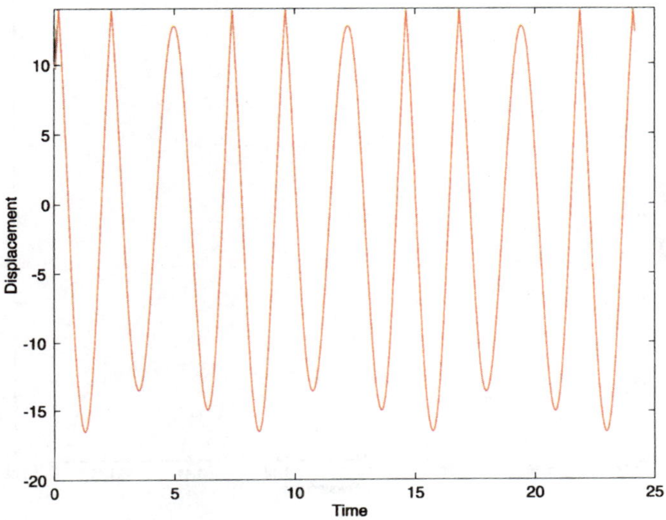

Fig. 9.4 (3, 2)-periodic solution for $\omega = 2.6$, $x_0 = 9.77211$ and $\dot{x_0} = 35.46891$.

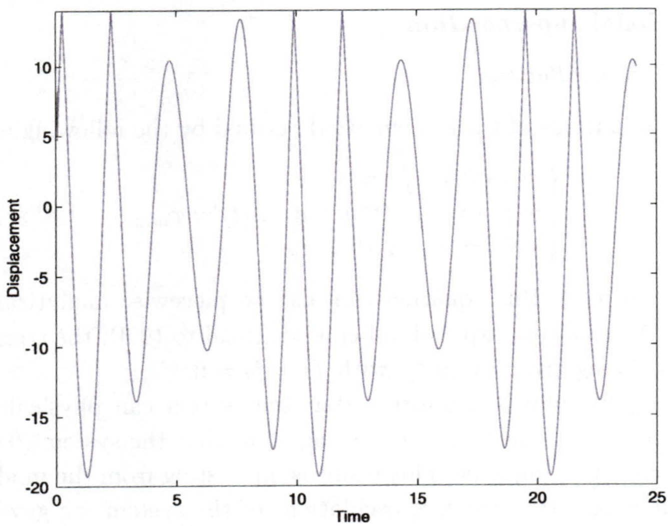

Fig. 9.5 (4, 2)-periodic solution for $\omega = 2.6$, $x_0 = 4.76445$ and $\dot{x_0} = 47.91445$.

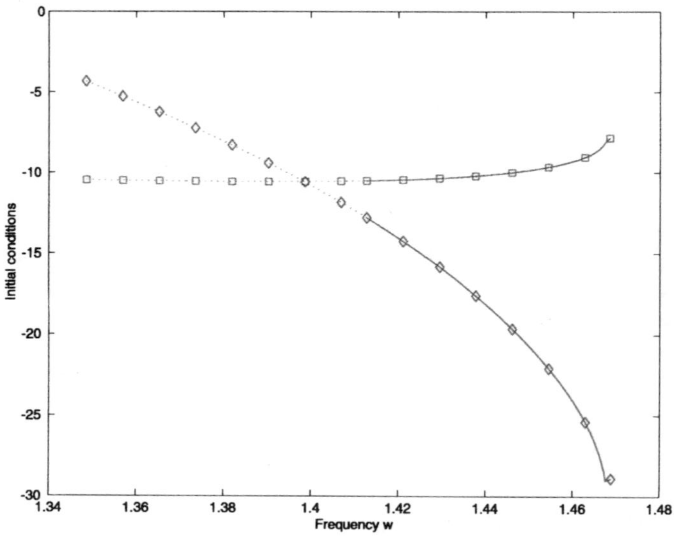

Fig. 9.6 Existence of (1, 2) periodic solutions. Stable (square: x_0, diamond: $\dot{x_0}$) and unstable (dotted curve) solutions.

9.2.3 *Modal superposition*

9.2.3.1 *Free oscillations*

The free oscillations of the system are described by the following equation:

$$\begin{cases} \ddot{x} + a\dot{x} + \omega_1^2 x = 0, \\ \dot{x}(t^+) = -e\dot{x}(t^-) \quad \text{if} \quad x(t) = x_{max}, \\ x(0) = x_0, \quad \dot{x}(0) = \dot{x_0}. \end{cases} \tag{9.6}$$

The solution of this equation can also be piecewise analytically written, with the resulting expression being identical to (9.3), the constants of integration being given by (9.4), with $f_1 = f_2 = 0$.

If the equilibrium is a position that the system can physically reach, namely if $x_{max} > 0$, and if $a \neq 0$, we can show that the system (9.6) has a finite number K of impacts. This result is interesting from the modal point of view: it means that the free oscillations of the system are governed by the frequency ω_1, except for a bounded time span. Hence we will consider later on that the natural frequency of the system is ω_1.

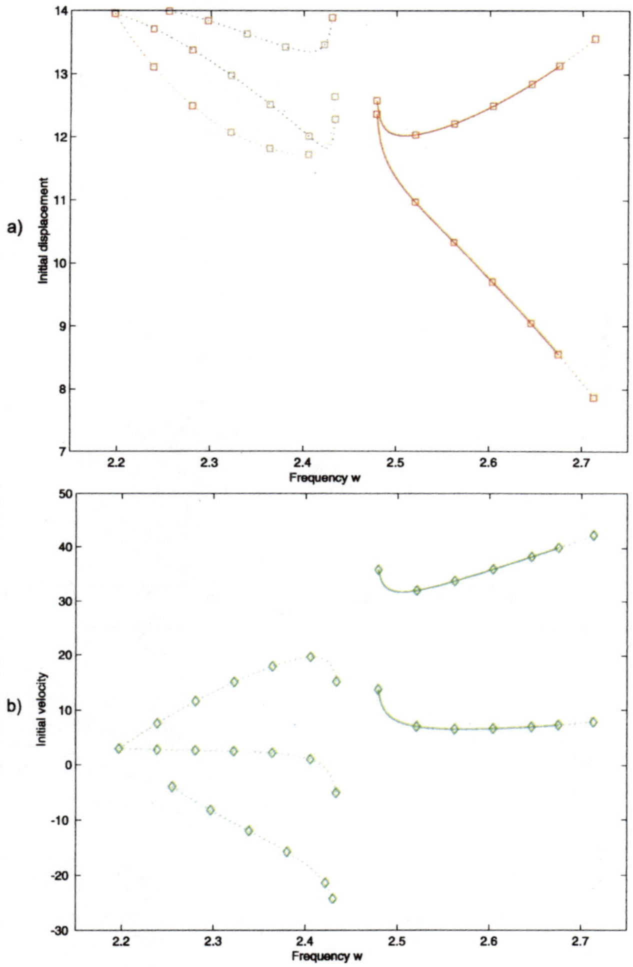

Fig. 9.7 Existence of (3, 2)-periodic solutions: stable (solid curve) and unstable solutions (dotted curve). Initial displacements (a). Initial velocities (b).

Remark: When $a = 0$, the number of impacts is infinite and we have $\lim_{k \to +\infty} t_{k+1} - t_k = \frac{2\pi}{\tilde{\omega_1}}$. If $x_{max} = 0$, there also is an infinite number of impacts and $t_{k+1} - t_k = \frac{\pi}{\tilde{\omega_1}}\ \forall k$. Finally, if $x_{max} < 0$ then $\lim_{k \to +\infty} t_{k+1} - t_k = 0$.

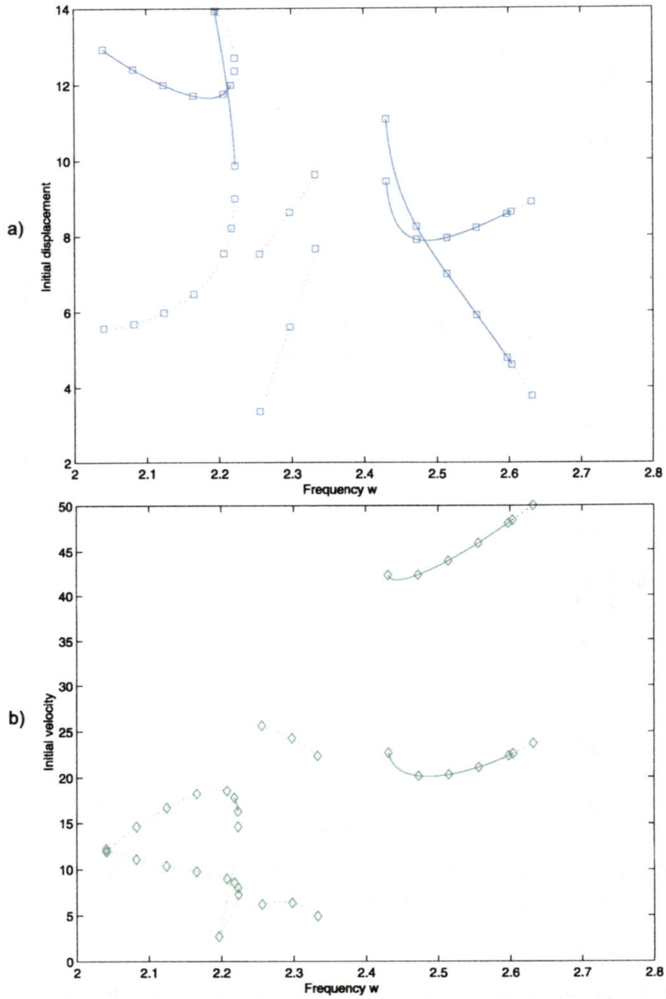

Fig. 9.8 Existence of (4, 2)-periodic solutions: stable (solid curve) and unstable solutions (dotted curve). Initial displacements (a). Initial velocities (b).

9.2.3.2 *Generalized mass and modal superposition*

In this section, we intend to establish a modal superposition formula for the irregular nonlinear model previously introduced. We consider the case $x_{max} > 0$ for which the natural frequency is ω_1, and we try to obtain a modal superposition formula similar to the linear case, connecting the nth harmonic amplitude of the response to the amplitude of the forcing.

Let us consider a (n, k)-periodic solution, i.e., a solution with period nT and k impacts per cycle. Thanks to the nT-periodicity of the response, we can define the Fourier coefficients; for all $j \in \mathbb{Z}$ we have:

$$\begin{aligned} c_j(\omega) &= \frac{1}{nT}\int_0^{nT} x(t)\exp\big(-i\frac{j}{n}\omega t\big)\mathrm{d}t, \\ &= \frac{1}{nT}\Big[\int_0^{t_1} x(t)\exp\big(-i\frac{j}{n}\omega t\big)\mathrm{d}t + \sum_{m=1}^{k-1}\int_{t_m}^{t_{m+1}} x(t)\exp\big(-i\frac{j}{n}\omega t\big)\mathrm{d}t + \\ &\quad + \int_{t_k}^{nT} x(t)\exp\big(-i\frac{j}{n}\omega t\big)\mathrm{d}t\Big]. \end{aligned}$$

We know $x(t)$ piecewise via (9.3). These $k+1$ integrals can then be calculated analytically. Let us set for $k \neq 0$

$$\begin{aligned} H^j_{n,k}(\omega) = \frac{2\tilde{\omega}_1}{nT}\{&B_1 + A_{k+1}\mathrm{e}^{-anT/2}\sin(n\tilde{\omega}_1 T) - B_{k+1}\mathrm{e}^{-anT/2}\cos(n\tilde{\omega}_1 T) + \\ &+(\eta + i\gamma_j)[A_1 - A_{k+1}\mathrm{e}^{-anT/2}\cos(n\tilde{\omega}_1 T) - B_{k+1}\mathrm{e}^{-anT/2}\sin(n\tilde{\omega}_1 T)] + \\ &+(1+e)\sum_{m=1}^{k}\mathrm{e}^{\frac{-at_m}{2}}\exp\big(-i\frac{j}{n}\omega t_m\big)u(A_m, B_m, t_m)\}, \end{aligned} \tag{9.7}$$

where: $\gamma_j = \dfrac{j\omega}{n\tilde{\omega}_1}$, and if $k = 0$, $H^j_{n,0}(\omega) = 0$. We obtain then

$$c_j(\omega) = \frac{f_1 - if_2}{2}\delta^n_j + \frac{f_1 + if_2}{2}\delta^{-n}_j + \frac{H^j_{n,k}(\omega)}{2}. \tag{9.8}$$

Let us set:

$$\Delta I_1(\omega) = \omega_1^2 - \omega^2 + ai\omega. \tag{9.9}$$

The nth Fourier coefficient is then given by

$$c_n(\omega) = \frac{f_1 - if_2}{2} + \frac{H^n_{n,k}(\omega)}{2\Delta I_1(\omega)} = \frac{f + H^n_{n,k}(\omega)}{2\Delta I_1(\omega)}. \tag{9.10}$$

Let us define the modal mass $m_{n,k}$ by:

$$m_{n,k} = \frac{1}{1 + \dfrac{H^n_{n,k}(\omega)}{f}}. \tag{9.11}$$

Because the free damped steady-state oscillations are linear, the generalized mode is here represented by the scalar 1. The nth harmonic amplitude can then be written in the form:

$$\mathbf{A_n}(\omega) = \sqrt{\mathbf{2}(|\ \mathbf{c_n}\ |^2 + |\ \mathbf{c_{-n}}\ |^2)} = \mathbf{2}\ |\ \mathbf{c_n}\ | = \left| \frac{\mathbf{f}}{\mathbf{m_{n,k}\Delta I_1}(\omega)} \right|. \tag{9.12}$$

We have just established a modal superposition formula connecting the nth harmonic amplitude of the forced response to the amplitude of the forcing via the free response. This formula is similar to the formula that we obtain in the linear case, but the mass (which should be equal to 1) is replaced by a modal mass (9.11).

Due to this definition, the mass is a complex: To give it a more physical meaning, it is necessary to consider its module. We have thus modelled the unitary mass system with impact, subjected to a sinusoidal forcing of frequency ω, like a linear system without impact of mass $|\ m_{n,k}\ |$, subjected to the same forcing. This modelling holds in term of spectral amplitude for a (n, k)-periodic response: the spectral amplitude is the same one for both systems.

Remark:

- In the case of a nT-periodic response without any impacts, we saw that only the case $n = 1$ was possible. Thus, $m_{1,0} = 1$ which is coherent since in this case the classical modal superposition formula applies and gives
  $\mathbf{A_1}(\omega) = \left| \dfrac{\mathbf{f}}{\mathbf{\Delta I_1}(\omega)} \right|.$
- We do not have any more a unique modal superposition formula as in the linear case, but an infinity *a priori*: It depends on the period and number of impacts per cycle of the periodic solution.
- Preceding calculations give us access to the whole Fourier series of a (n, k)-periodic response.
- We can get an analytical expression of the module of the modal mass if and only if $k = 1$, i.e., when the periodic solution has only one impact

by period. Setting

$$X_1 = \frac{2(1+e)\omega}{nfT}[f_1 \sin(\omega t_1) - f_2 \cos(\omega t_1)] \times$$
$$\times \frac{-1 + 2\mathrm{e}^{anT/2}\cos(n\tilde{\omega}_1 T) - \mathrm{e}^{anT}}{e + \mathrm{e}^{anT/2}\{(1-e)\cos(n\tilde{\omega}_1 T) + (1+e)\eta \sin(n\tilde{\omega}_1 T)\} - \mathrm{e}^{anT}},$$

we obtain for the modal mass:

$$m_{n,1} = \frac{1}{1 + X_1 \mathrm{e}^{-i\omega t_1}}. \tag{9.13}$$

Since t_1 is analytically known in the case of a $(n,1)$-periodic response, the modal mass $|m_{n,1}|$ can be expressed analytically as a function of the parameters of the system.

9.2.3.3 *Examples of modal superposition*

We are now able to show some applications of the modal superposition formula previously established, using the analytical search for periodic solutions carried out in subsection 9.2.2. We first consider the following set of parameters $\omega_1 = 2.5$, $a = 0.05$, $x_{max} = 14$, $e = 0.9$ and $f = 20$. Let us notice that for these values of parameters we have $\omega_1^2 x_{max} > f$ and $e > 0$, so that sticking never occurs.

The Fig. 9.9 outlines a very important difference with the linear case. Indeed, the peak of amplitude corresponding to resonance does no longer exist around ω_1. Consequently, for ω close to the natural frequency of the system, the modal mass is the largest because there is no resonance between the external forcing and the system. The maximum amplitude occurs at a frequency higher than the natural frequency, but the maximum reached is much weaker than in the linear case: there is not a true peak of amplitude in the usual sense.

In addition, the modal superposition formula previously established enables us to compute the amplitude of nth harmonic of a solution, whose period is nT. In the linear case (without impact), this is sufficient to know the whole spectral response of the system: only the case $n = 1$ occurs and the Fourier coefficients are all zero except the first one. In our case it is not sufficient any more, since the occurrence of impacts leads to periodic solutions with plenty of harmonics, and the nth harmonic amplitude can be a bad approximation to the spectral amplitude of the response. Fourier coefficients are reported in Figs. 9.12 - 9.16.

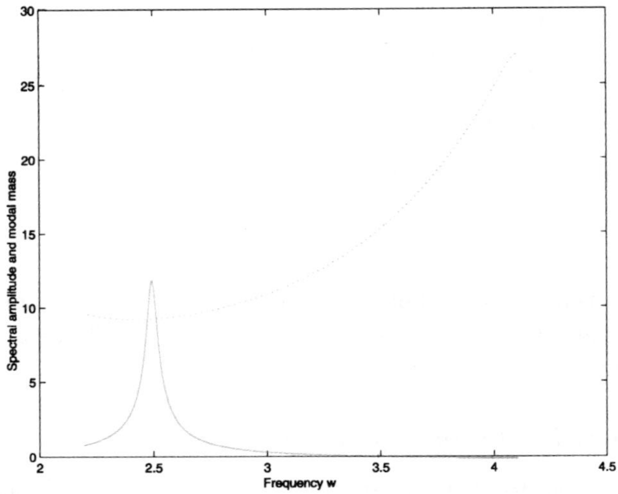

Fig. 9.9 First harmonic amplitude (dotted curve) and modal mass (solid curve) of the (1, 1)-periodic solutions.

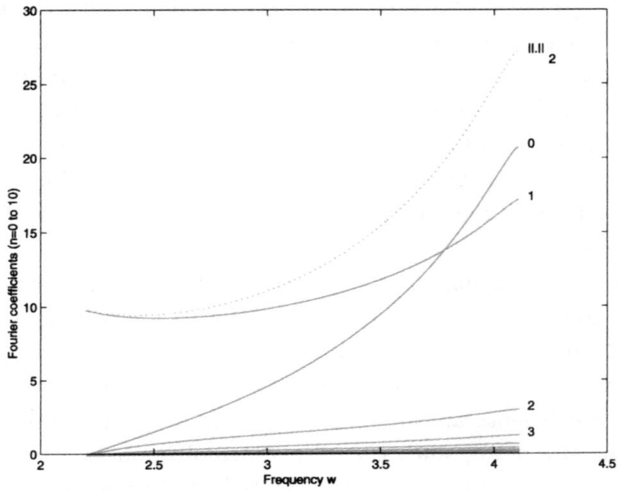

Fig. 9.10 Fourier coefficients of the (1, 1)-periodic responses.

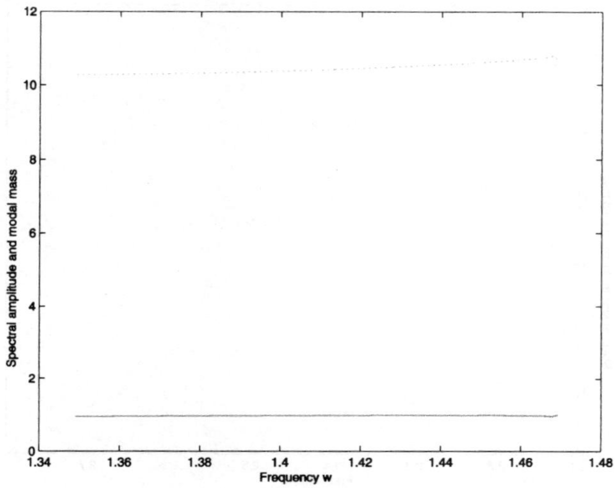

Fig. 9.11 First harmonic amplitude (dotted curve) and modal mass (solid curve) of the (1, 2)-periodic solutions.

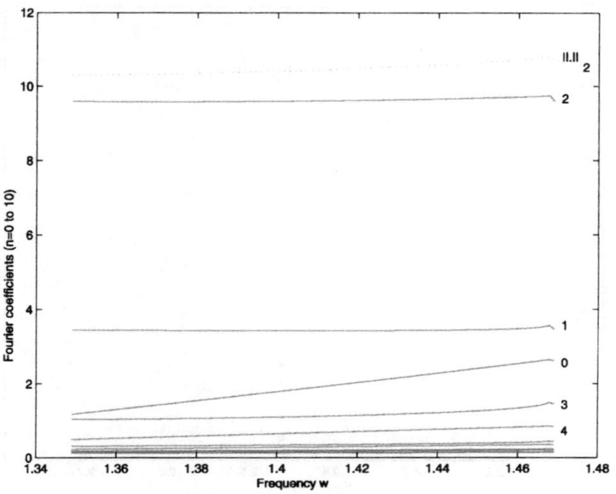

Fig. 9.12 Fourier coefficients of the (1, 2)-periodic responses.

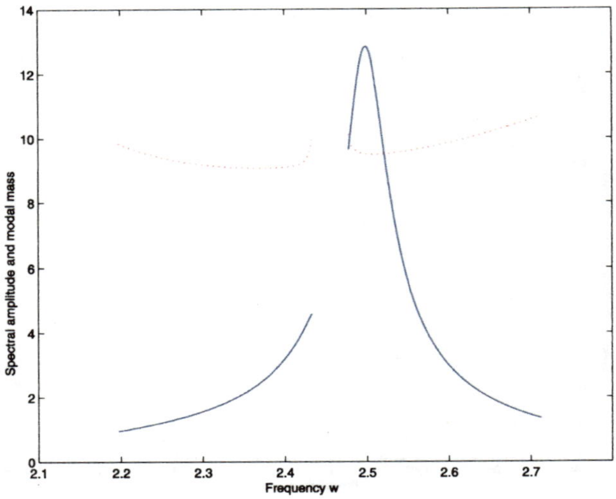

Fig. 9.13 Third harmonic amplitude (dotted curve) and modal mass (solid curve) of the (3, 2)-periodic solutions.

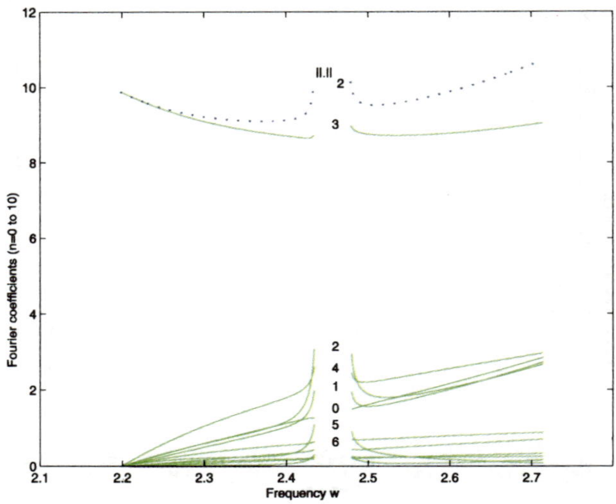

Fig. 9.14 Fourier coefficients of the (3, 2)-periodic response.

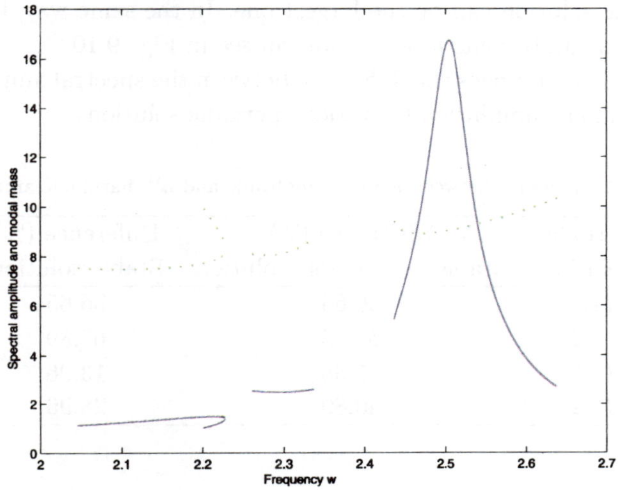

Fig. 9.15 Fourth harmonic amplitude (dotted curve) and modal mass (solid curve) of the (4, 2)-periodic solutions.

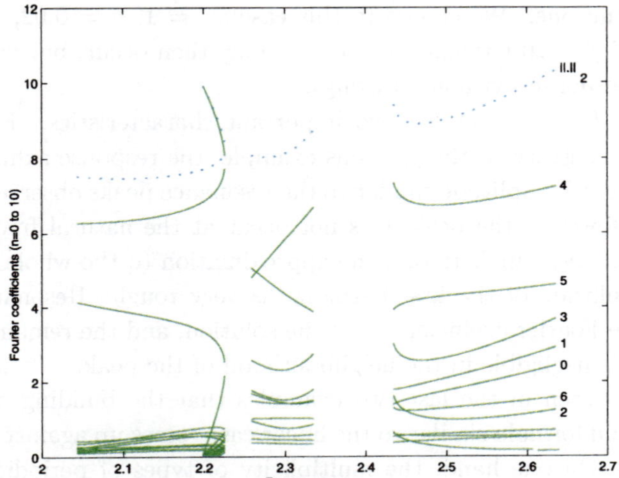

Fig. 9.16 Fourier coefficients of the (4, 2)-periodic responses.

For example, we see in Fig. 9.12 that the second harmonic amplitude for a T-periodic solution can be the largest one. In the same way, the Fourier coefficient c_0 can become large, as we can see in Fig. 9.10.

Table 9.1 summarizes the difference between the spectral amplitude and the nth harmonic amplitude for various periodic solutions.

Table 9.1 Difference between Fourier amplitude and n^{th} harmonic amplitude.

Periodic solution	Difference (%) Stable or unstable solution	Difference (%) Stable solution
(1,1)	36.65	36.65
(1,2)	67.89	67.89
(3,2)	15.56	13.96
(4,2)	40.89	28.96

For this example, a modal superposition formula can be built and is worthwhile as long as ω remains close to the "primary resonance". Nevertheless, this frequency area does not always correspond to the largest amplitudes.

Let us consider an example showing a spectral behaviour different from the preceding one. We choose in this case $\omega_1 = 1$, $a = 0.02$, $x_{max} = 1$, $e = 0.9$ and $f = 20$ (sticking to the stop may then occur, but we will only deal with responses without sticking).

In Fig. 9.17 we can see two important characteristics. First of all, and on the contrary to the previous example, the response exhibits a true peak of spectral amplitude similar to the resonance peaks observed in linear systems. However, the peak does not occur at the natural frequency ω_1, but for $\omega \approx 2\omega_1$. Furthermore, the approximation to the whole amplitude by the amplitude of the first harmonic is very rough: Resonance occurs through the Fourier coefficient c_0 of the solution, and the remainder of the harmonics is negligible in the neighbourhood of the peak.

We have seen in the last two examples that the building of a modal superposition formula similar to the linear case comes up against two major difficulties. On one hand, the multiplicity of types of periodic solutions prevents from giving a unique formula holding in any case: It is necessary to prior know the period of the response and the number of impacts per cycle. On the other hand, taking into account only one harmonic in this superposition may not be sufficient any more. It would be necessary to

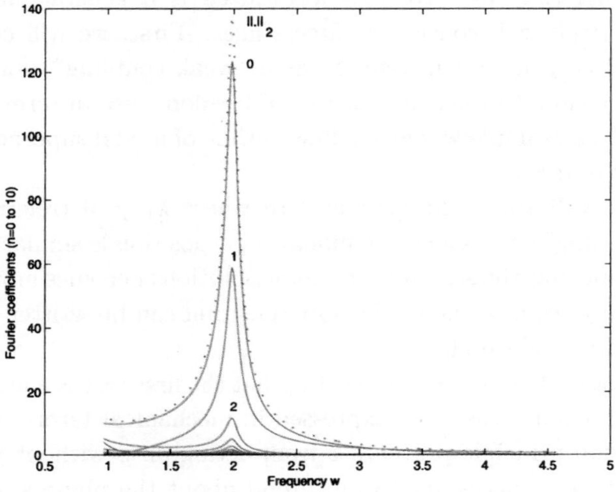

Fig. 9.17 Fourier coefficients of the (1, 1)-periodic responses.

enclose at least the first four harmonics in the formula in order to get a closer approximation in the shown examples.

9.3 Two Degrees of Freedom Systems

In this section we deal with a system with two degrees of freedom, one of them being constrained by a stop:

$$\begin{cases} \ddot{x} + a\dot{x} + \omega_1^2 x + k_1 y = f_1 \cos(\omega t), \\ x \in \mathbf{C^0}(\mathbb{R}^+, [-\infty, \mathbf{x_{max}}]), \\ x(t) = x_{max} \Rightarrow \dot{x}(t^+) = -e\dot{x}(t^-), \\ \ddot{y} + a\dot{y} + \omega_2^2 y + k_2 x = f_2 \cos(\omega t), \\ y \in \mathbf{C^1}(\mathbb{R}^+, \mathbb{R}). \end{cases}$$

This system can be written in the form:

$$\ddot{X} + A\dot{X} + KX = F\cos(\omega t) \; + \; \text{“impact”}, \tag{9.14}$$

with $X = \begin{pmatrix} x \\ y \end{pmatrix}$, $A = \begin{pmatrix} a & 0 \\ 0 & a \end{pmatrix}$, $K = \begin{pmatrix} \omega_1^2 & k_1 \\ k_2 & \omega_2^2 \end{pmatrix}$, $F = \begin{pmatrix} f_1 \\ f_2 \end{pmatrix}$ and the term "+ impact" represents the constraint induced on x by the occurrence of impacts.

The model that we have just introduced is a general mathematical model. Its study will consist of three stages. First, we will consider the case $k_1 = 0$ (which we will refer to as a "weak coupling") for which the study is very close to the single degree of freedom case in term of periodic solutions. This example will give a first outline of modal superposition with two degrees of freedom.

Then we will study the general case where $k_1 \neq 0$ (referred to as a "strong coupling"), for which the modal superposition is similar to the case $k_1 = 0$, except that the search for periodic solutions becomes more complex. We will finally see how the modal superposition can be written in the case of two colliding rigid bodies.

Let us notice, before everything else, that the first case is a mathematical model which cannot easily be expressed in mechanical terms. Indeed, the action-reaction principle prevents x from acting on y without y acting on x. Therefore, it is necessary to be careful about the physical conclusions that could be stated from this model.

9.3.1 *Weak coupling*

We deal with the system (9.14) in the special case where $k_1 = 0$, which is close to the single degree of freedom case previously studied.

9.3.1.1 *Analytical solution of the system*

Decoupling Equations. System (9.14) can be decoupled in order to obtain explicit solutions for x and y. If we set $v_2 = \dfrac{k_2}{\omega_1^2 - \omega_2^2}$ and $P = \begin{pmatrix} 1 & 0 \\ v_2 & 1 \end{pmatrix}$, and multiply the system (9.14) without impacts by P^{-1}, we get a system in the new coordinates $\begin{pmatrix} x_1 \\ x_2 \end{pmatrix} = P^{-1}X$:

$$\begin{cases} \ddot{x_1} + a\dot{x_1} + \omega_1^2 x_1 = f^1 \cos(\omega t), \\ \ddot{x_2} + a\dot{x_2} + \omega_2^2 x_2 = f^2 \cos(\omega t), \end{cases} \tag{9.15}$$

where:

$$\begin{cases} f^1 = f_1, \\ f^2 = -v_2 f_1 + f_2. \end{cases}$$

Written in that form, the system is easy to solve and we obtain as long as there is no impact:

$$\begin{cases} x_1(t) = \mathrm{e}^{-at/2}[A^1 \cos(\tilde{\omega_1} t) + B^1 \sin(\tilde{\omega_1} t)] + f_1^1 \cos(\omega t) + f_2^1 \sin(\omega t), \\ x_2(t) = \mathrm{e}^{-at/2}[A^2 \cos(\tilde{\omega_2} t) + B^2 \sin(\tilde{\omega_2} t)] + f_1^2 \cos(\omega t) + f_2^2 \sin(\omega t), \end{cases} \tag{9.16}$$

where $\tilde{\omega_1} = \sqrt{\omega_1^2 - \frac{a^2}{4}}$, $\tilde{\omega_2} = \sqrt{\omega_2^2 - \frac{a^2}{4}}$ and:

$$\begin{cases} f_1^1 = f^1 \dfrac{\omega_1^2 - \omega^2}{(\omega_1^2 - \omega^2)^2 + a^2\omega^2}, \\ f_2^1 = f^1 \dfrac{a\omega}{(\omega_1^2 - \omega^2)^2 + a^2\omega^2}, \\ f_1^2 = f^2 \dfrac{\omega_2^2 - \omega^2}{(\omega_2^2 - \omega^2)^2 + a^2\omega^2}, \\ f_2^2 = f^2 \dfrac{a\omega}{(\omega_2^2 - \omega^2)^2 + a^2\omega^2}. \end{cases}$$

Moreover, we will set $\eta_1 = \frac{a}{2\tilde{\omega_1}}$ and $\eta_2 = \frac{a}{2\tilde{\omega_2}}$. The solution is given in the initial coordinate system by:

$$\begin{cases} x = x_1, \\ y = v_2 x_1 + x_2. \end{cases} \tag{9.17}$$

Gluing at Impact Times. An impact occurs when $x(t) = x_{max}$, which is equivalent to $x_1(t) = x_{max}$. Let us assume that there is an impact at t_k. On $[t_{k-1}, t_k]$, the solution (9.16) is given by:

$$\begin{cases} x_1(t) = \mathrm{e}^{-at/2}[A_k^1 \cos(\tilde{\omega_1} t) + B_k^1 \sin(\tilde{\omega_1} t)] + f_1^1 \cos(\omega t) + f_2^1 \sin(\omega t), \\ x_2(t) = \mathrm{e}^{-at/2}[A_k^2 \cos(\tilde{\omega_2} t) + B_k^2 \sin(\tilde{\omega_2} t)] + f_1^2 \cos(\omega t) + f_2^2 \sin(\omega t). \end{cases}$$

From this equation we infer that t_k is solution of the equation:

$$f(t_k) = \mathrm{e}^{\frac{-at_k}{2}}[A_k^1 \cos(\tilde{\omega_1} t_k) + B_k^1 \sin(\tilde{\omega_1} t_k)] + f_1^1 \cos(\omega t_k) + f_2^1 \sin(\omega t_k) - x_{max} = 0.$$

Moreover, by assumptions on x and y, we have at impact time:

$$\begin{cases} x(t_k^+) = x(t_k^-), \\ \dot{x}(t_k^+) = -e\dot{x}(t_k^-), \\ y(t_k^+) = y(t_k^-), \\ \dot{y}(t_k^+) = \dot{y}(t_k^-), \end{cases}$$

which yields for the new variables:

$$\begin{cases} x_1(t_k^+) = x_1(t_k^-), \\ \dot{x_1}(t_k^+) = -e\dot{x_1}(t_k^-), \\ x_2(t_k^+) = x_2(t_k^-), \\ \dot{x_2}(t_k^+) = \dot{x_2}(t_k^-) + (1+e)v_2\dot{x_1}(t_k^-). \end{cases}$$

We thus obtain the following relations that provide the constants of integration:

$$\begin{cases} A_{k+1}^1 = A_k^1 - (1+e)\sin(\tilde{\omega_1}t_k)u(A_k^1, B_k^1, t_k), \\ B_{k+1}^1 = B_k^1 + (1+e)\cos(\tilde{\omega_1}t_k)u(A_k^1, B_k^1, t_k), \\ A_{k+1}^2 = A_k^2 + (1+e)v_2\dfrac{\tilde{\omega_1}}{\tilde{\omega_2}}\sin(\tilde{\omega_2}t_k)u(A_k^1, B_k^1, t_k), \\ B_{k+1}^2 = B_k^2 - (1+e)v_2\dfrac{\tilde{\omega_1}}{\tilde{\omega_2}}\cos(\tilde{\omega_2}t_k)u(A_k^1, B_k^1, t_k), \end{cases} \tag{9.18}$$

where u is given by:

$$\begin{aligned} u(A,B,t) = {} & A[\sin(\tilde{\omega_1}t) + \eta_1\cos(\tilde{\omega_1}t)] + B[-\cos(\tilde{\omega_1}t) + \eta_1\sin(\tilde{\omega_1}t)] \\ & + \mathrm{e}^{at/2}\frac{\omega}{\tilde{\omega_1}}[f_1^1\sin(\omega t) - f_2^1\cos(\omega t)]. \end{aligned}$$

9.3.1.2 *Search for periodic solutions*

The search for periodic solutions (x, y) is equivalent to the search for periodic solutions (x_1, x_2). According to (9.17), we have $x_1 = x$, therefore the search for periodic solutions for x_1 is similar to the one carried out in the case of a single degree of freedom system. We can thus determine x_1^0 and $\dot{x}_1^0$ so that x_1 is (n, k)-periodic where $k \in \{0, 1, 2\}$. As regards x_2, the method is identical, only the recursion that gives A_2^2 and B_2^2 is different. We obtain 2×2 linear systems in $(A_1^2(t_1), B_1^2(t_1))$. When the determinant is nonzero, we can get x_2^0 and $\dot{x}_2^0$ so that x_2 is (n, k)-periodic where $k \in \{0, 1, 2\}$.

The initial conditions for the original coordinates system (x, y) are given by (9.17):

$$\begin{cases} x_0 = x_1^0, \\ y_0 = v_2x_1^0 + x_2^0, \\ \dot{x_0} = \dot{x}_1^0, \\ \dot{y_0} = v_2\dot{x}_1^0 + \dot{x}_2^0. \end{cases}$$

9.3.1.3 *Modal superposition*

Free Oscillations of the System. We saw in section 9.2.3.1 that, in the case of a single degree of freedom system without external forcing, the number of impacts is finite and the steady-state response is periodic with frequency ω_1. Hence, in the case of weak coupling, the free response exhibits also a finite number of impacts, and the steady-state response for x_1 is periodic with frequency ω_1. As for x_2, since there are no more impacts once the steady-state response is reached, the equation of its movement is given by (9.15):

$$\ddot{x_2} + a\dot{x_2} + \omega_2^2 x_2 = 0.$$

This is the equation of a classical damped oscillator without external forcing, whose steady-state response has frequency ω_2. In order to establish a modal superposition formula, we will then start from the natural frequencies ω_1 and ω_2. Moreover, in that case again, the generalized modes correspond to linear modes.

Generalized Masses and Modal Superposition. Let us consider a (n, k)-periodic solution. We can calculate the Fourier coefficients of the functions x_1 and x_2. Those are quite similar to the coefficients that we found for the single degree of freedom system. Let us set

$$\begin{cases} H^j_{n,k}(\omega) = \dfrac{2\tilde{\omega_1}}{nT}\dfrac{1}{\omega_1^2 - \frac{j^2\omega^2}{n^2} + ia\frac{j\omega}{n}} \times \\ \qquad \{B_1^1 + A_{k+1}^1 \mathrm{e}^{-\frac{anT}{2}} \sin(n\tilde{\omega_1}T) - B_{k+1}^1 \mathrm{e}^{-\frac{anT}{2}} \cos(n\tilde{\omega_1}T) + \\ \qquad + (\eta_1 + i\gamma_j^1)[A_1^1 - A_{k+1}^1 \mathrm{e}^{-\frac{anT}{2}} \cos(n\tilde{\omega_1}T) - B_{k+1}^1 \mathrm{e}^{-\frac{anT}{2}} \sin(n\tilde{\omega_1}T)] + \\ \qquad + (1+e)\displaystyle\sum_{m=1}^{k} \mathrm{e}^{-\frac{at_m}{2}} \exp\big(-i\frac{j}{n}\omega t_m\big) u(A_m^1, B_m^1, t_m)\}, \\ \\ G^j_{n,k}(\omega) = \dfrac{2\tilde{\omega_2}}{nT}\dfrac{1}{\omega_2^2 - \frac{j^2\omega^2}{n^2} + ia\frac{j\omega}{n}} \times \\ \qquad \{B_1^2 + A_{k+1}^2 \mathrm{e}^{-\frac{anT}{2}} \sin(n\tilde{\omega_2}T) - B_{k+1}^2 \mathrm{e}^{-\frac{anT}{2}} \cos(n\tilde{\omega_2}T) + \\ \qquad + (\eta_2 + i\gamma_j^2)[A_1^2 - A_{k+1}^2 \mathrm{e}^{-\frac{anT}{2}} \cos(n\tilde{\omega_2}T) - B_{k+1}^2 \mathrm{e}^{-\frac{anT}{2}} \sin(n\tilde{\omega_2}T)] + \\ \qquad - (1+e)v_2\dfrac{\tilde{\omega_1}}{\tilde{\omega_2}}\displaystyle\sum_{m=1}^{k} \mathrm{e}^{-\frac{at_m}{2}} \exp\big(-i\frac{j}{n}\omega t_m\big) u(A_m^1, B_m^1, t_m)\}, \end{cases}$$

with $\gamma_j^1 = \dfrac{j\omega}{n\tilde{\omega_1}}$ and $\gamma_j^2 = \dfrac{j\omega}{n\tilde{\omega_2}}$. The Fourier coefficients are then given by:

$$\begin{cases} c_j^1(\omega) = \dfrac{f_1^1 - if_2^1}{2}\delta_j^n + \dfrac{f_1^1 + if_2^1}{2}\delta_j^{-n} + \dfrac{H_{n,k}^j(\omega)}{2}, \\ c_j^2(\omega) = \dfrac{f_1^2 - if_2^2}{2}\delta_j^n + \dfrac{f_1^2 + if_2^2}{2}\delta_j^{-n} + \dfrac{G_{n,k}^j(\omega)}{2}. \end{cases}$$

We infer the Fourier coefficient corresponding to the nth harmonic using the coordinate transformation (9.17):

$$\begin{cases} c_n^x(\omega) = c_n^1(\omega) = \dfrac{f^1}{2m_1^{n,k}\Delta I_1}, \\ c_n^y(\omega) = v_2 c_n^1(\omega) + c_n^2(\omega) = v_2\dfrac{f^1}{2m_1^{n,k}\Delta I_1} + \dfrac{f^2}{2m_2^{n,k}\Delta I_2}, \end{cases} \tag{9.19}$$

where we define $\Delta I_1 = \omega_1^2 - \omega^2 + ai\omega$, $\Delta I_2 = \omega_2^2 - \omega^2 + ai\omega$ and

$$\begin{cases} m_1^{n,k} = \dfrac{1}{1 + \dfrac{H_{n,k}^n(\omega)}{f^1}}, \\ m_2^{n,k} = \dfrac{1}{1 + \dfrac{G_{n,k}^n(\omega)}{f^2}}. \end{cases} \tag{9.20}$$

The nth Fourier coefficients are then in the original basis:

$$\begin{cases} c_n^x(\omega) = \dfrac{f_1}{2m_1^{n,k}\Delta I_1}, \\ c_n^y(\omega) = v_2\dfrac{f_1}{2m_1^{n,k}\Delta I_1} + \dfrac{-v_2 f_1 + f_2}{2m_2^{n,k}\Delta I_2}. \end{cases}$$

The contribution of the nth harmonic to the Fourier spectrum is given (as we saw in (9.12)) by

$$\begin{cases} \mathbf{A_n^X}(\omega) = |\mathbf{X_n}(\omega)| = \left|\dfrac{\mathbf{f_1}}{\mathbf{m_1^{n,k}\Delta I_1}}\right|, \\ \mathbf{A_n^Y}(\omega) = |\mathbf{Y_n}(\omega)| = \left|\mathbf{v_2}\dfrac{\mathbf{f_1}}{\mathbf{m_1^{n,k}\Delta I_1}} + \dfrac{\mathbf{-v_2 f_1 + f_2}}{\mathbf{m_2^{n,k}\Delta I_2}}\right|, \end{cases} \tag{9.21}$$

which can also be written as:

$$\begin{pmatrix} X_n(\omega) \\ Y_n(\omega) \end{pmatrix} = \frac{\begin{pmatrix} 1 & 0 \\ v_2 & 0 \end{pmatrix}}{m_1^{n,k}\Delta I_1} \begin{pmatrix} f_1 \\ f_2 \end{pmatrix} + \frac{\begin{pmatrix} 0 & 0 \\ -v_2 & 1 \end{pmatrix}}{m_2^{n,k}\Delta I_2} \begin{pmatrix} f_1 \\ f_2 \end{pmatrix}. \tag{9.22}$$

Let us set $A_1 = \begin{pmatrix} 1 & 0 \\ v_2 & 0 \end{pmatrix}$ and $A_2 = \begin{pmatrix} 0 & 0 \\ -v_2 & 1 \end{pmatrix}$. Our aim is to express A_1 in the form $T_1^t T_1^{'}$, and A_2 in the form $T_2^t T_2^{'}$ where T_1, $T_1^{'}$, T_2 and $T_2^{'}$ denote the generalized modes of the system.

Let us set $T_1 = (a_1, b_1)$ and $T_1^{'} = (a_1^{'}, b_1^{'})$: then $T_1^t T_1^{'} = \begin{pmatrix} a_1 a_1^{'} & a_1 b_1^{'} \\ b_1 a_1^{'} & b_1 b_1^{'} \end{pmatrix}$.

We obtain three equations with four unknowns:

$$\begin{cases} b_1^{'} = 0, \\ a_1 a_1^{'} = 1, \\ b_1 a_1^{'} = v_2. \end{cases}$$

It is thus possible to arbitrarily set one of the unknowns: for example, as in the case of linear modes, let us choose $a_1^{'} = 1$. We obtain then:

$$\begin{cases} a_1 = 1, \\ b_1 = v_2. \end{cases}$$

Hence

$$\begin{cases} T_1 = (1, v_2), \\ T_1^{'} = (1, 0). \end{cases} \tag{9.23}$$

The same kind of calculations and assumptions for A_2 lead to

$$\begin{cases} T_2 = (0, 1), \\ T_2^{'} = (-v_2, 1). \end{cases} \tag{9.24}$$

Using (9.23) and (9.24), the relation (9.22) can then be written in the form:

$$\begin{pmatrix} X_n(\omega) \\ Y_n(\omega) \end{pmatrix} = \frac{T_1^t T_1^{'}}{m_1^{n,k}\Delta I_1(\omega)} \begin{pmatrix} f_1 \\ f_2 \end{pmatrix} + \frac{T_2^t T_2^{'}}{m_2^{n,k}\Delta I_2(\omega)} \begin{pmatrix} f_1 \\ f_2 \end{pmatrix}. \tag{9.25}$$

We have just established a modal superposition formula holding for a two degrees of freedom system with weak coupling. This formula links the nth harmonic amplitude to the free response and the forcing by means of a modal mass. The vectors T_1, $T_1^{'}$, T_2 and $T_2^{'}$ represent the modes (left and right respectively). We can notice that T_1 and T_2 are merely the eigenvectors associated with ω_1^2 and ω_2^2.

Finally, as in the case of the single degree of freedom system, we do not have a unique modal superposition formula holding in all cases, but according to the period of the response and the number of impacts per cycle, it is necessary to choose the right formula.

Remark: The modal superposition formula seems to exhibit a reciprocity breaking compared to the linear case: T_1 and T_1' are different whatever the choice we make when setting one of the unknowns. Indeed, the second component of T_1' is always zero whereas that of T_1 is always nonzero. This is in fact due to the choice of the model, for which $k_1 = 0$ and $k_2 \neq 0$.

Examples of Modal Superposition. For the first example, we choose the following parameters: $\omega_1 = 2.5$, $\omega_2 = 3.8$, $a = 0.05$, $e = 0.9$, $x_{max} = 14$, $f_1 = 20$, $f_2 = 18$ and $k_2 = 1$. Results are given in Fig. 9.18.

The second example is derived from the single degree of freedom case and the chosen parameters are the following: $\omega_1 = 1$, $\omega_2 = 3.8$, $a = 0.02$, $e = 0.9$, $x_{max} = 1$, $f_1 = 20$, $f_2 = 18$ and $k_2 = 1$. The spectral amplitude for x is the same one as that obtained in the single degree of freedom case. The continuous degree of freedom y exhibits a resonance for $\omega \simeq \omega_2$ similar to the linear case, but several additional secondary resonances occur for $\omega \simeq \dfrac{\omega_2}{2}$ and $\omega \simeq \dfrac{\omega_2}{3}$ (see Fig. 9.19). Furthermore, for the frequency associated with a spectral amplitude peak for x, we find a small peak for y, which can become large if k_2 is sufficiently large. Lastly, if ω_1 is close to $\dfrac{\omega_2}{2}$ and k_2 is rather large, then the main peak of resonance can occur in the neighborhood of $\dfrac{\omega_2}{2}$.

9.3.2 *Strong coupling*

In this section, we study the system (9.14) with $k_1 \neq 0$.

9.3.2.1 *Analytical solution of the system*

Decoupling Equations. We are going to decouple the system (9.14) in order to be able to write the solutions x and y explicitly. The matrix K has the following characteristic polynomial:

$$P_K(\lambda) = \lambda^2 - (\omega_1^2 + \omega_2^2)\lambda + \omega_1^2\omega_2^2 - k_1 k_2,$$

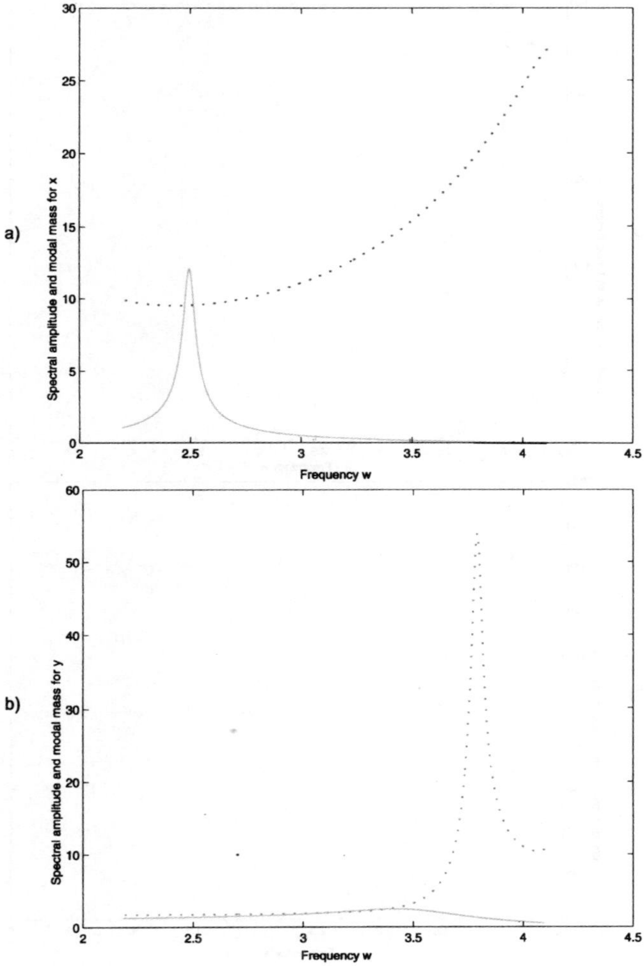

Fig. 9.18 First harmonic amplitude (dotted curve) and modal mass (solid curve) of the $(1,1)$-periodic responses. First degree of freedom x (a). Second degree of freedom y (b).

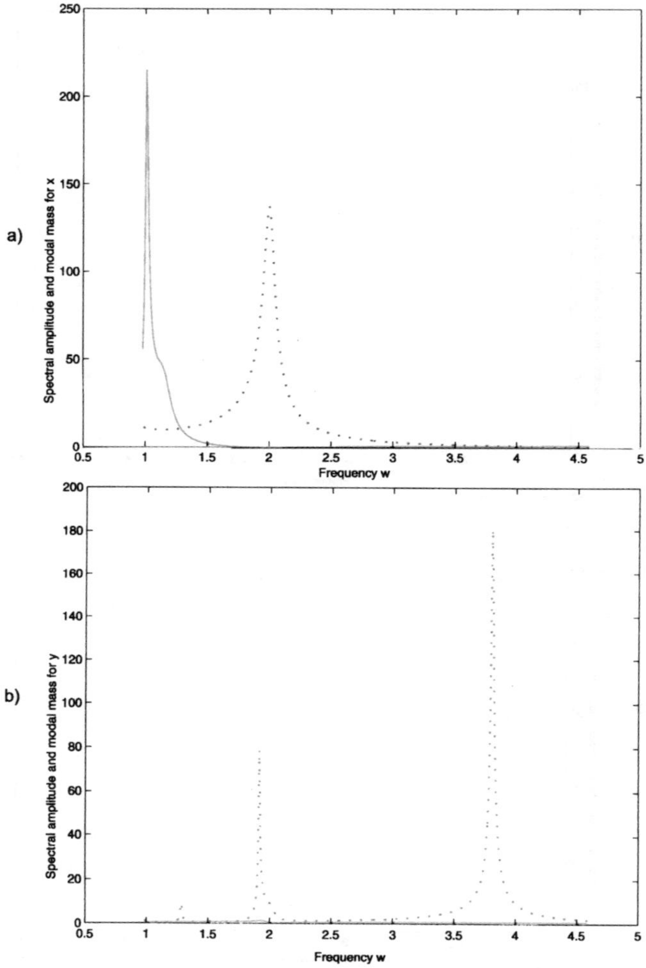

Fig. 9.19 First harmonic amplitude (dotted curve) and modal mass (solid curve) of the (1, 1)-periodic responses. First degree of freedom x (a). Second degree of freedom y (b).

with discriminant $\Delta = (\omega_1^2 - \omega_2^2)^2 + 4k_1k_2$.

The system parameters will be chosen so that $\Delta > 0$. Then the matrix K admits two distinct real eigenvalues:

$$\begin{cases} \lambda_1 = \dfrac{\omega_1^2 + \omega_2^2 - \sqrt{\Delta}}{2}, \\ \\ \lambda_2 = \dfrac{\omega_1^2 + \omega_2^2 + \sqrt{\Delta}}{2}. \end{cases}$$

Let us set, assuming $\omega_1 \neq \omega_2$, $v_1 = \dfrac{k_1}{\omega_1^2 - \omega_2^2}$ and $v_2 = \dfrac{k_2}{\omega_1^2 - \omega_2^2}$. Two eigenvectors associated with λ_1 and λ_2 are respectively $\begin{pmatrix} 1 \\ v_2 \end{pmatrix}$ and $\begin{pmatrix} -v_1 \\ 1 \end{pmatrix}$, which define the matrix $P = \begin{pmatrix} 1 & -v_1 \\ v_2 & 1 \end{pmatrix}$. The assumption $\Delta > 0$ implies that its determinant is nonzero, and P is invertible with $P^{-1} = \dfrac{1}{1 + v_1v_2}\begin{pmatrix} 1 & v_1 \\ -v_2 & 1 \end{pmatrix}$.

Then we set $X = PY$ and we multiply the system (9.14) by P^{-1} in order to obtain

$$\ddot{Y} + a\dot{Y} + P^{-1}KPY = P^{-1}F\cos(\omega t) \ + \ \text{“impact”}.$$

Let us set

$$\begin{cases} f^1 = \dfrac{f_1 + v_1 f_2}{1 + v_1 v_2}, \\ f^2 = \dfrac{f_2 - v_2 f_1}{1 + v_1 v_2}. \end{cases}$$

The new system obtained using the new basis is given by

$$\begin{cases} \ddot{x_1} + a\dot{x_1} + \lambda_1 x_1 = f^1\cos(\omega t), \\ \ddot{x_2} + a\dot{x_2} + \lambda_2 x_2 = f^2\cos(\omega t). \end{cases} \tag{9.26}$$

as long as there is no impact. We can thus easily solve these equations:

$$\begin{cases} x_1(t) = \mathrm{e}^{-at/2}[A^1\cos(\tilde{\omega_1}t) + B^1\sin(\tilde{\omega_1}t)] + f_1^1\cos(\omega t) + f_2^1\sin(\omega t), \\ x_2(t) = \mathrm{e}^{-at/2}[A^2\cos(\tilde{\omega_2}t) + B^2\sin(\tilde{\omega_2}t)] + f_1^2\cos(\omega t) + f_2^2\sin(\omega t), \end{cases} \tag{9.27}$$

with $\tilde{\omega_1} = \sqrt{\lambda_1 - \frac{a^2}{4}}$, $\tilde{\omega_2} = \sqrt{\lambda_2 - \frac{a^2}{4}}$ and

$$\begin{cases} f_1^1 = f^1 \dfrac{\lambda_1 - \omega^2}{(\lambda_1 - \omega^2)^2 + a^2\omega^2}, \\ f_2^1 = f^1 \dfrac{a\omega}{(\lambda_1 - \omega^2)^2 + a^2\lambda^2}, \\ f_1^2 = f^2 \dfrac{\lambda_2 - \omega^2}{(\lambda_2 - \omega^2)^2 + a^2\omega^2}, \\ f_2^2 = f^2 \dfrac{a\omega}{(\lambda_2 - \omega^2)^2 + a^2\lambda^2}. \end{cases}$$

The solution in the original basis is finally given by

$$\begin{cases} x = x_1 - v_1 x_2, \\ y = v_2 x_1 + x_2. \end{cases} \tag{9.28}$$

Gluing Solutions at Impact Times. An impact occurs when $x(t) = x_{max}$, namely when $x_1(t) - v_1 x_2(t) = x_{max}$. Let us assume that there is an impact at t_k. For $t \in [t_{k-1}, t_k]$, the decoupled solution is given by

$$\begin{cases} x_1(t) = \mathrm{e}^{-at/2}[A_k^1 \cos(\tilde{\omega_1} t) + B_k^1 \sin(\tilde{\omega_1} t)] + f_1^1 \cos(\omega t) + f_2^1 \sin(\omega t), \\ x_2(t) = \mathrm{e}^{-at/2}[A_k^2 \cos(\tilde{\omega_2} t) + B_k^2 \sin(\tilde{\omega_2} t)] + f_1^2 \cos(\omega t) + f_2^2 \sin(\omega t). \end{cases}$$

From these equations we infer the equation verified by t_k:

$$\begin{aligned} f(t_k) = \mathrm{e}^{-at_k/2} \left\{ A_k^1 \cos(\tilde{\omega_1} t_k) + B_k^1 \sin(\tilde{\omega_1} t_k) - v_1[A_k^2 \cos(\tilde{\omega_2} t_k) + B_k^2 \sin(\tilde{\omega_2} t_k)] \right\} \\ + \left(f_1^1 - v_1 f_1^2\right) \cos(\omega t_k) + \left(f_2^1 - v_1 f_2^2\right) \sin(\omega t_k) - x_{max} = 0. \end{aligned}$$

Yet by assumptions on x and y, we have at the impact time:

$$\begin{cases} x(t_k^+) = x(t_k^-), \\ \dot{x}(t_k^+) = -e\dot{x}(t_k^-), \\ y(t_k^+) = y(t_k^-), \\ \dot{y}(t_k^+) = \dot{y}(t_k^-), \end{cases}$$

which yields, according to (9.28):

$$\begin{cases} x_1(t_k^+) = x_1(t_k^-), \\ \dot{x}_1(t_k^+) = \left(1 - \dfrac{1+e}{1+v_1v_2}\right)\dot{x}_1(t_k^-) + \dfrac{(1+e)v_1}{1+v_1v_2}\dot{x}_2(t_k^-), \\ x_2(t_k^+) = x_2(t_k^-), \\ \dot{x}_2(t_k^+) = \dfrac{(1+e)v_2}{1+v_1v_2}\dot{x}_1(t_k^-) + \left(\dfrac{1+e}{1+v_1v_2} - e\right)\dot{x}_2(t_k^-). \end{cases}$$

Let us set for $i \in \{1,2\}$:

$$u_i(k) = A_k^i[\sin(\tilde{\omega_i}t_k) + \eta_i\cos(\tilde{\omega_i}t_k)] + B_k^i[-\cos(\tilde{\omega_i}t_k) + \eta_i\sin(\tilde{\omega_i}t_k)] + \\ + \frac{\omega}{\tilde{\omega_i}}e^{at_k/2}[f_1^i\sin(\omega t_k) - f_2^i\cos(\omega t_k)].$$

We obtain:

$$\begin{cases} A_{k+1}^1 = A_k^1 - \dfrac{1+e}{1+v_1v_2}\sin(\tilde{\omega_1}t_k)\left[u_1(k) - v_1\dfrac{\tilde{\omega_2}}{\tilde{\omega_1}}u_2(k)\right], \\ B_{k+1}^1 = B_k^1 + \dfrac{1+e}{1+v_1v_2}\cos(\tilde{\omega_1}t_k)\left[u_1(k) - v_1\dfrac{\tilde{\omega_2}}{\tilde{\omega_1}}u_2(k)\right], \\ A_{k+1}^2 = A_k^2 + \dfrac{(1+e)v_2}{1+v_1v_2}\sin(\tilde{\omega_2}t_k)\left[\dfrac{\tilde{\omega_1}}{\tilde{\omega_2}}u_1(k) - v_1u_2(k)\right], \\ B_{k+1}^2 = B_k^2 - \dfrac{(1+e)v_2}{1+v_1v_2}\cos(\tilde{\omega_2}t_k)\left[\dfrac{\tilde{\omega_1}}{\tilde{\omega_2}}u_1(k) - v_1u_2(k)\right]. \end{cases} \tag{9.29}$$

9.3.2.2 *Search for periodic solutions*

The search for periodic solutions (x, y) is equivalent to the search for periodic solutions (x_1, x_2). We will only look for $(n, 0)$ and $(n, 1)$-periodic solutions. The following calculations are similar to what can be found in [Peterka and Szöllös (1996)], with the addition of damping.

(n, 0)-Periodic Solutions. As for the single degree of freedom system, it is easy to prove that the system admits a $(n, 0)$-periodic solution if and only if x_1 and x_2 are given by:

$$\begin{cases} x_1(t) = f_1^1\cos(\omega t) + f_2^1\sin(\omega t), \\ x_2(t) = f_1^2\cos(\omega t) + f_2^2\sin(\omega t). \end{cases}$$

According to (9.28) we have:

$$x(t) = (f_1^1 - v_1f_1^2)\cos(\omega t) + (f_2^1 - v_1f_2^2)\sin(\omega t).$$

It is thus possible to get results similar to those obtained for a single degree of freedom system in 9.2.1. This gives a condition so that a $(n,0)$-periodic solution exists: it requires $(f_1^1 - v_1 f_1^2)^2 + (f_2^1 - v_1 f_2^2)^2 \leq x_{max}^2$. From this inequality, we can obtain a fourth degree polynomial in ω^2, allowing us to determine the values of ω for which the system admits a $(n,0)$-periodic response. The remark made in 9.2.1 still holds: The existence of $(n,0)$-periodic solutions requires $x_{max} > 0$.

(n, 1)-Periodic Solutions. nT-periodicity requires:

$$\begin{cases} x_1(nT) = x_1^0, \\ \dot{x}_1(nT) = \dot{x}_1^0, \\ x_2(nT) = x_2^0, \\ \dot{x}_2(nT) = \dot{x}_2^0, \end{cases}$$

or using linear combination, for a solution with one impact per cycle:

$$\begin{cases} A_2^1 \cos(n\tilde{\omega}_1 T) + B_2^1 \sin(n\tilde{\omega}_1 T) - \mathrm{e}^{anT/2} A_1^1 = 0, \\ A_2^2 \cos(n\tilde{\omega}_2 T) + B_2^2 \sin(n\tilde{\omega}_2 T) - \mathrm{e}^{anT/2} A_1^2 = 0, \\ A_2^1 \sin(n\tilde{\omega}_1 T) - B_2^1 \cos(n\tilde{\omega}_1 T) + \mathrm{e}^{anT/2} B_1^1 = 0, \\ A_2^2 \sin(n\tilde{\omega}_2 T) - B_2^2 \cos(n\tilde{\omega}_2 T) + \mathrm{e}^{anT/2} B_1^2 = 0. \end{cases} \tag{9.30}$$

Using (9.29), we obtain a 4×4 linear system in $(A_1^1, B_1^1, A_1^2, B_1^2)$. The resolution of this system leads to $A_1^1(t_1)$, $B_1^1(t_1)$, $A_1^2(t_1)$ and $B_1^2(t_1)$.

It remains to determine t_1 via the equation:

$$\begin{aligned} f(t_1) = \mathrm{e}^{-at_1/2}\{&A_1^1(t_1)\cos(\tilde{\omega}_1 t_1) + B_1^1(t_1)\sin(\tilde{\omega}_1 t_1) + \\ &- v_1[A_1^2(t_1)\cos(\tilde{\omega}_2 t_1) + B_1^2(t_1)\sin(\tilde{\omega}_2 t_1)]\} + \\ &+ (f_1^1 - v_1 f_1^2)\cos(\omega t_1) + (f_2^1 - v_1 f_2^2)\sin(\omega t_1) - x_{max} = 0. \end{aligned}$$

Initial conditions of the system leading to a $(n,1)$-periodic solution are then given by

$$\begin{cases} x_0 = A_1^1(t_1) + f_1^1 - v_1(A_1^2(t_1) + f_1^2), \\ y_0 = v_2(A_1^1(t_1) + f_1^1) + A_1^2(t_1) + f_1^2, \\ \dot{x}_0 = \tilde{\omega}_1[B_1^1(t_1) - \eta_1 A_1^1(t_1)] + f_2^1\omega - v_1\tilde{\omega}_2[B_1^2(t_1) - \eta_2 A_1^2(t_1)] - v_1 f_2^2\omega, \\ \dot{y}_0 = v_2\tilde{\omega}_1[B_1^1(t_1) - \eta_1 A_1^1(t_1)] + v_2 f_2^1\omega + \tilde{\omega}_2[B_1^2(t_1) - \eta_2 A_1^2(t_1)] + f_2^2\omega. \end{cases}$$

9.3.2.3 *Modal superposition*

Free Oscillation of the System. We can show that the system (9.14) without external forcing has a finite number of impacts. Thus the steady-state of the system (9.26) with $f^1 = f^2 = 0$ is periodic with frequency λ_1 for x_1 and λ_2 for x_2. These two frequencies will be used as natural frequencies for the forced system.

Generalized Masses and Modal Superposition. Let us consider a (n, k)-periodic solution and calculate the Fourier coefficients of x_1 and x_2. The calculation of $c_j^1(\omega)$ and $c_j^2(\omega)$ is similar to the one carried out in the weak coupling case: only the recursive relation (9.29) changes. We set in this case:

$$\begin{cases} H_{n,k}^j(\omega) = \dfrac{2\tilde{\omega}_1}{nT} \dfrac{1}{\omega_1^2 - \frac{j^2\omega^2}{n^2} + ia\frac{j\omega}{n}} \times \\ \quad \times \left\{B_1^1 + A_{k+1}^1 e^{-\frac{anT}{2}} \sin(n\tilde{\omega}_1 T) - B_{k+1}^1 e^{-\frac{anT}{2}} \cos(n\tilde{\omega}_1 T) + \right. \\ \quad + (\eta_1 + i\gamma_j^1)[A_1^1 - A_{k+1}^1 e^{-\frac{anT}{2}} \cos(n\tilde{\omega}_1 T) - B_{k+1}^1 e^{-\frac{anT}{2}} \sin(n\tilde{\omega}_1 T)] + \\ \quad \left. + \dfrac{1+e}{1+v_1 v_2} \displaystyle\sum_{m=1}^{k} e^{-\frac{at_m}{2}} \exp\left(-i\frac{j}{n}\omega t_m\right) \left[u_1(m) - v_1 \frac{\tilde{\omega}_2}{\tilde{\omega}_1} u_2(m)\right] \right\}, \\[2ex] G_{n,k}^j(\omega) = \dfrac{2\tilde{\omega}_2}{nT} \dfrac{1}{\omega_2^2 - \frac{j^2\omega^2}{n^2} + ia\frac{j\omega}{n}} \times \\ \quad \times \left\{B_1^2 + A_{k+1}^2 e^{-\frac{anT}{2}} \sin(n\tilde{\omega}_2 T) - B_{k+1}^2 e^{-\frac{anT}{2}} \cos(n\tilde{\omega}_2 T) + \right. \\ \quad + (\eta_2 + i\gamma_j^2)[A_1^2 - A_{k+1}^2 e^{-\frac{anT}{2}} \cos(n\tilde{\omega}_2 T) - B_{k+1}^2 e^{-\frac{anT}{2}} \sin(n\tilde{\omega}_2 T)] + \\ \quad \left. - \dfrac{(1+e)v_2}{1+v_1 v_2} \displaystyle\sum_{m=1}^{k} e^{-\frac{at_m}{2}} \exp\left(-i\frac{j}{n}\omega t_m\right) \left[\frac{\tilde{\omega}_1}{\tilde{\omega}_2} u_1(m) - v_1 u_2(m)\right] \right\}, \end{cases}$$

with $\gamma_j^1 = \dfrac{j\omega}{n\tilde{\omega}_1}$ and $\gamma_j^2 = \dfrac{j\omega}{n\tilde{\omega}_2}$. We have then:

$$\begin{cases} c_j^1(\omega) = \dfrac{f_1^1 - if_2^1}{2}\delta_j^n + \dfrac{f_1^1 + if_2^1}{2}\delta_j^{-n} + \dfrac{H_{n,k}^j(\omega)}{2}, \\[2ex] c_j^2(\omega) = \dfrac{f_1^2 - if_2^2}{2}\delta_j^n + \dfrac{f_1^2 + if_2^2}{2}\delta_j^{-n} + \dfrac{G_{n,k}^j(\omega)}{2}. \end{cases}$$

We deduce the nth Fourier coefficients of the system by the basis change (9.28):

$$\begin{cases} c_n^x(\omega) = c_n^1(\omega) - v_1 c_n^2(\omega) = \dfrac{f^1}{2m_1^{n,k}\Delta I_1} - v_1 \dfrac{f^2}{2m_2^{n,k}\Delta I_2}, \\ c_n^y(\omega) = v_2 c_n^1(\omega) + c_n^2(\omega) = v_2 \dfrac{f^1}{2m_1^{n,k}\Delta I_1} + \dfrac{f^2}{2m_2^{n,k}\Delta I_2}. \end{cases} \tag{9.31}$$

where we defined $\Delta I_1 = \lambda_1 - \omega^2 + ai\omega$, $\Delta I_2 = \lambda_2 - \omega^2 + ai\omega$ and:

$$\begin{cases} m_1^{n,k} = \dfrac{1}{1 + \frac{H_{n,k}^n(\omega)}{f^1}}, \\ m_2^{n,k} = \dfrac{1}{1 + \frac{G_{n,k}^n(\omega)}{f^2}}. \end{cases} \tag{9.32}$$

We have in the original basis:

$$\begin{cases} c_n^x(\omega) = \dfrac{1}{1+v_1v_2}\left[\dfrac{f_1 + v_1 f_2}{2m_1^{n,k}\Delta I_1} - v_1 \dfrac{f_2 - v_2 f_1}{2m_2^{n,k}\Delta I_2}\right], \\ \\ c_n^y(\omega) = \dfrac{1}{1+v_1v_2}\left[v_2 \dfrac{f_1 + v_1 f_2}{2m_1^{n,k}\Delta I_1} + \dfrac{f_2 - v_2 f_1}{2m_2^{n,k}\Delta I_2}\right]. \end{cases}$$

The contribution of the nth harmonic is (as we did for the weak coupling in (9.21)) given by:

$$\begin{pmatrix} X_n(\omega) \\ Y_n(\omega) \end{pmatrix} = \frac{1}{1+v_1v_2}\left\{ \frac{\begin{pmatrix} 1 & v_1 \\ v_2 & v_1v_2 \end{pmatrix}}{m_1^{n,k}\Delta I_1}\begin{pmatrix} f_1 \\ f_2 \end{pmatrix} + \frac{\begin{pmatrix} v_1v_2 & -v_1 \\ -v_2 & 1 \end{pmatrix}}{m_2^{n,k}\Delta I_2}\begin{pmatrix} f_1 \\ f_2 \end{pmatrix}\right\}. \tag{9.33}$$

Let us set $A_1 = \dfrac{1}{1+v_1v_2}\begin{pmatrix} 1 & v_1 \\ v_2 & v_1v_2 \end{pmatrix}$ and $A_2 = \dfrac{1}{1+v_1v_2}\begin{pmatrix} v_1v_2 & -v_1 \\ -v_2 & 1 \end{pmatrix}$, and let us try to express A_1 in the form $T_1^t T_1^{'}$ and A_2 in the form $T_2^t T_2^{'}$.

Let us set $T_1 = (a_1, b_1)$ and $T_1^{'} = (a_1^{'}, b_1^{'})$: then $T_1^t T_1^{'} = \begin{pmatrix} a_1a_1^{'} & a_1b_1^{'} \\ b_1a_1^{'} & b_1b_1^{'} \end{pmatrix}$.

We have four equations with four unknowns:

$$\begin{cases} a_1 a_1' = \dfrac{1}{1+v_1 v_2}, \\ a_1 b_1' = \dfrac{v_1}{1+v_1 v_2}, \\ b_1 b_1' = \dfrac{v_1 v_2}{1+v_1 v_2}, \\ b_1 a_1' = \dfrac{v_2}{1+v_1 v_2}. \end{cases}$$

As in the case of weak coupling, it is possible to fix one of the unknowns: for example, in order to respect a reciprocity condition as long as possible, let us set $a_1 = \dfrac{1}{\sqrt{1+v_1 v_2}}$. We have $1 + v_1 v_2 > 0$, for the parameters of the system verify $(\omega_1^2 - \omega_2^2)^2 + 4k_1 k_2 > 0$. Then we obtain

$$\begin{cases} a_1' = \dfrac{1}{\sqrt{1+v_1 v_2}}, \\ b_1' = \dfrac{v_1}{\sqrt{1+v_1 v_2}}, \\ b_1 = \dfrac{v_2}{\sqrt{1+v_1 v_2}}, \end{cases}$$

and:

$$\begin{cases} T_1 = \dfrac{1}{\sqrt{1+v_1 v_2}}(1, v_2), \\ T_1' = \dfrac{1}{\sqrt{1+v_1 v_2}}(1, v_1). \end{cases} \tag{9.34}$$

In the same way, we obtain for A_2:

$$\begin{cases} T_2 = \dfrac{1}{\sqrt{1+v_1 v_2}}(-v_1, 1), \\ T_2' = \dfrac{1}{\sqrt{1+v_1 v_2}}(-v_2, 1). \end{cases} \tag{9.35}$$

Using (9.34) and (9.35), the relation (9.33) then becomes:

$$\begin{pmatrix} X_n(\omega) \\ Y_n(\omega) \end{pmatrix} = \frac{T_1^t T_1'}{m_1^{n,k} \Delta I_1} \begin{pmatrix} f_1 \\ f_2 \end{pmatrix} + \frac{T_2^t T_2'}{m_2^{n,k} \Delta I_2} \begin{pmatrix} f_1 \\ f_2 \end{pmatrix}. \tag{9.36}$$

We have just established a modal superposition formula for the two degrees of freedom system with strong coupling.

<u>Remark:</u>

- T_1 and T_2 thus defined are in fact eigenvectors associated with λ_1 and λ_2.

- T_1' and T_2 are orthogonal, just as T_2' and T_1.
- We do not have reciprocity when $k_1 \neq k_2$. But when the matrix K is symmetrical, we have the reciprocity property since then $T_1 = T_1'$ and $T_2 = T_2'$.

9.3.4 *Two colliding rigid bodies*

In this section, we study a mechanical system consisting of two oscillating rigid bodies which can collide during their movement (see Fig. 9.20).

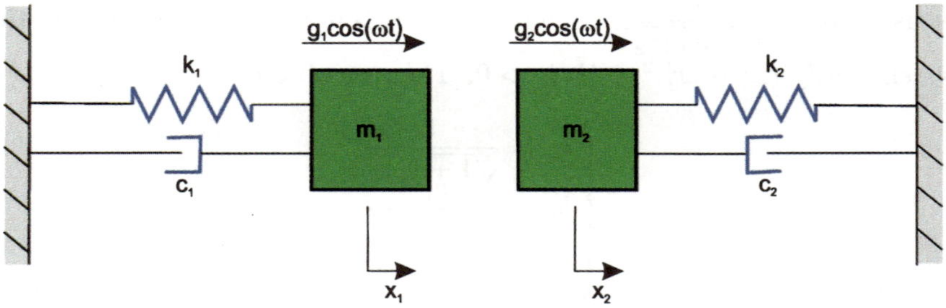

Fig. 9.20 Mechanical model of two colliding oscillating rigid bodies.

We will assume that the equilibrium positions x_1^0 and x_2^0 of two solids are such that $x_2^0 - x_1^0 = x_{max} > 0$ (system without preload). The equations of the system in relative displacements are then:

$$\begin{cases} m_1\ddot{x_1} + c_1\dot{x_1} + \kappa_1 x_1 = g_1\cos(\omega t), \\ m_2\ddot{x_2} + c_2\dot{x_2} + \kappa_2 x_2 = g_2\cos(\omega t), \\ x_2 - x_1 \geq x_{max}. \end{cases}$$

Let us set $a_1 = \dfrac{c_1}{m_1}$, $\lambda_1 = \dfrac{\kappa_1}{m_1}$, $f^1 = \dfrac{g_1}{m_1}$ and similarly $a_2 = \dfrac{c_2}{m_2}$, $\lambda_2 = \dfrac{\kappa_2}{m_2}$, $f^2 = \dfrac{g_2}{m_2}$. We assume hereafter that $a_1 = a_2 = a$. The system becomes:

$$\begin{cases} \ddot{x_1} + a\dot{x_1} + \lambda_1 x_1 = f^1\cos(\omega t), \\ \ddot{x_2} + a\dot{x_2} + \lambda_2 x_2 = f^2\cos(\omega t), \\ x_2 - x_1 \geq x_{max}. \end{cases} \tag{9.37}$$

When $x_2(t) - x_1(t) = x_{max}$, an impact occurs at t and the restitution law leads the relative velocity between the two solids to be multiplied by a factor $-e$ at the impact time:

$$\dot{x_2}t^+ - \dot{x_1}t^+ = -e(\dot{x_2}t^- - \dot{x_1}t^-).$$

Moreover, the conservation of the momentum provides the second equation to be able to determine post-impact velocities:

$$m_2\dot{x_2}^{t^+} + m_1\dot{x_1}^{t^+} = m_2\dot{x_2}^{t^-} + m_1\dot{x_1}^{t^-}.$$

We can apply a simple change of coordinates in order to express the system (9.14) in the form previously studied. Let us set:

$$\begin{cases} x = x_1 - x_2, \\ y = \dfrac{m_1}{m_2}x_1 + x_2. \end{cases} \tag{9.38}$$

For this new system of variables, the equations become:

$$\begin{cases} \ddot{x} + a\dot{x} + \omega_1^2 x + k_1 y = f_1 \cos(\omega t) \ + \text{"impact"}, \\ \ddot{y} + a\dot{y} + \omega_2^2 y + k_2 x = f_2 \cos(\omega t), \\ x_2 - x_1 \geq 0, \end{cases} \tag{9.39}$$

where we set $\omega_1^2 = \dfrac{m_1\lambda_2 + m_2\lambda_1}{m_1 + m_2}$, $\omega_2^2 = \dfrac{m_1\lambda_1 + m_2\lambda_2}{m_1 + m_2}$, $k_1 = \dfrac{m_2}{m_1 + m_2}(\lambda_1 - \lambda_2)$, $k_2 = \dfrac{m_1}{m_1 + m_2}(\lambda_1 - \lambda_2)$, $f_1 = f^1 - f^2$ and $f_2 = \dfrac{m_1}{m_2}f^1 + f^2$.

Thus we deal with a system of the type (9.14); the results of the preceding study are then applicable. We can first of all notice that k_1 and k_2 are proportional, so that we are always in the case of the strong coupling studied in section 9.3. As regards the solutions of the system between two impacts, there is no work to do in order to decouple the system, since our initial system is already written in decoupled form. Thus the eigenvalues of the system are always real, and are exactly λ_1 and λ_2.

We have in this case $v_1 = 1$ and $v_2 = \dfrac{m_1}{m_2}$. According to (9.34) and (9.35), the modes of the system are given by

$$\begin{aligned} T_1 &= \alpha_{1,2}\left(1, \frac{m_1}{m_2}\right), \\ T_1' &= \alpha_{1,2}(1, 1), \\ T_2 &= \alpha_{1,2}(-1, 1), \\ T_2' &= \alpha_{1,2}\left(-\frac{m_1}{m_2}, 1\right), \end{aligned}$$

where we set $\alpha_{1,2} = \dfrac{1}{\sqrt{1+\dfrac{m_1}{m_2}}}$.

Using these modes, we deduce the modal superposition formula as in (9.36), the expression of the modal masses being given by (9.32). It is interesting to notice that in general there is no reciprocity, except if $m_1 = m_2$.

Finally, we can deal with the periodic responses of the system. First of all, according to subsection 9.3.2.2, there are $(n, 0)$-periodic solutions because we assumed $x_{max} > 0$. In the same way, calculations of 9.3.2.2 allow to search for $(n, 1)$-periodic solutions. We obtain diagrams of existence of periodic solutions versus the frequency of the external excitation. From these periodic solutions, we can then test the modal superposition formula, plotting the difference between the spectral amplitude of the system's response and the nth harmonic amplitude given by (9.36).

For the first example presented, we choose $m_1 = 1$, $m_2 = 0.7$, $\lambda_1 = 5$, $\lambda_1 = 13$, $a = 0.05$, $e = 0.9$, $f^1 = 20$, $f^2 = 18$ and $x_{max} = 14$. Absolute displacements of the $(1, 1)$-periodic solution is given in Fig. 9.21. The $(1, 0)$ and $(1, 1)$-periodic solutions are studied versus frequency ω in Figs. 9.22 and 9.23, respectively.

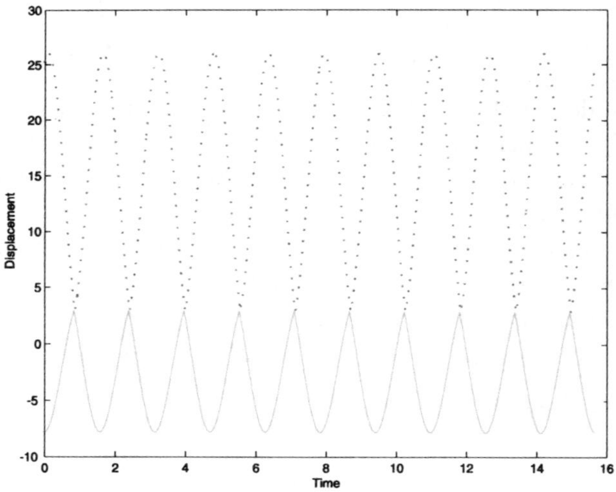

Fig. 9.21 $(1, 1)$-periodic solution for $\omega = 4$, $x_0^1 = -7.90482$, $\dot{x}_0^1 = -2.07935$, $x_0^2 = 11.20441$, and $\dot{x}_0^2 = 15.14127$: absolute displacement.

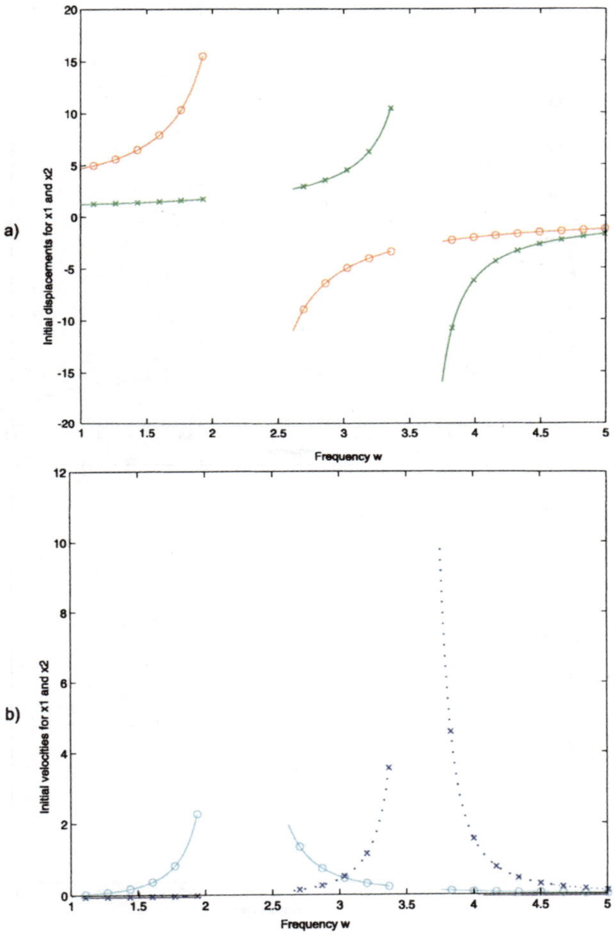

Fig. 9.22 Existence of (1, 0)-periodic solutions. Initial displacements for x_1 (circle) and x_2 (x-mark) (a). Initial velocities for x_1 (circle) and x_2 (x-mark) (b).

As for the single degree of freedom system, we notice in Fig. 9.24 that the usual resonance of linear systems does not occur any more. Indeed, when the frequency of the external excitation is equal to the natural frequency of the system, the modal mass corresponding to the excited mode goes through a local maximum, so that the associated spectral amplitude does not present a particular peak.

Nevertheless, spectral amplitude peaks appear for values of ω far away

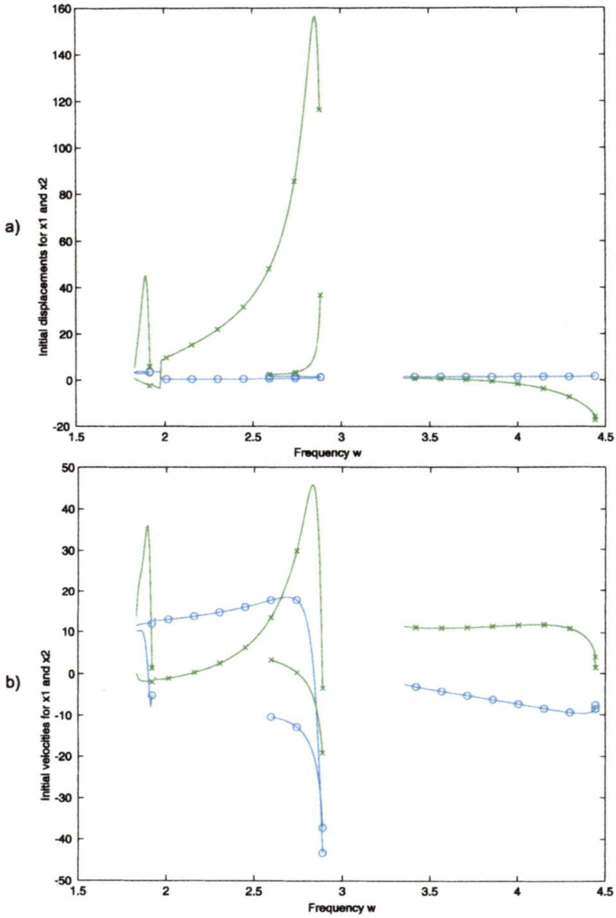

Fig. 9.23 Existence of $(1, 1)$-periodic solutions. Initial displacements for x_1 (circle) and x_2 (x-mark) (a). Initial velocities for x_1 (circle) and x_2 (x-mark) (b).

from the natural frequencies.

The main peak is located in the neighbourhood of ω_2 (which is one of the natural frequencies of the system after the change of variables), but it is difficult to show a resonance in ω_2 for the peak is appreciably shifted from this frequency. From the modal point of view, the occurrence of such peaks does not correspond to a classical resonance (where the generalized natural frequency of the system is close to the frequency of external excitation): Resonance can be interpreted like the locus of frequency where the

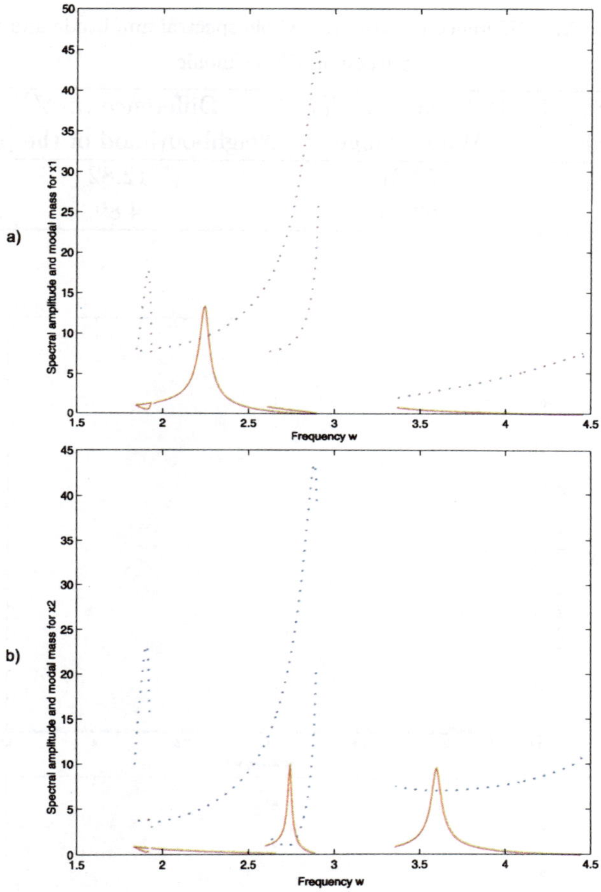

Fig. 9.24 First harmonic amplitude (dotted curve) and modal mass (solid curve) of the $(1, 1)$-periodic solutions. First degree of freedom x_1 (a). Second degree of freedom x_2 (b).

generalized modal mass is minimum with respect to ω.

As regards the closeness of the approximation to the spectral amplitude by the nth harmonic amplitude, we note, as in the single-degree-of-freedom case, that the first harmonic is not always the leading term in the amplitude of a $(1, 1)$-periodic response (see Fig. 9.25). Thus, the constant coefficient c_0 overrides all the others for some values of ω. Yet if we consider ω close to the peaks in the spectral response, the first harmonic gives a close approximation to the total amplitude (see Table 9.2).

Table 9.2: Difference between the whole spectral amplitude and the amplitude of nth harmonic.

Degree of freedom	Difference (in %) Whole range	Difference (in %) Neighbourhood of the peak
x_1	77.31	12.82
x_2	82.75	4.89

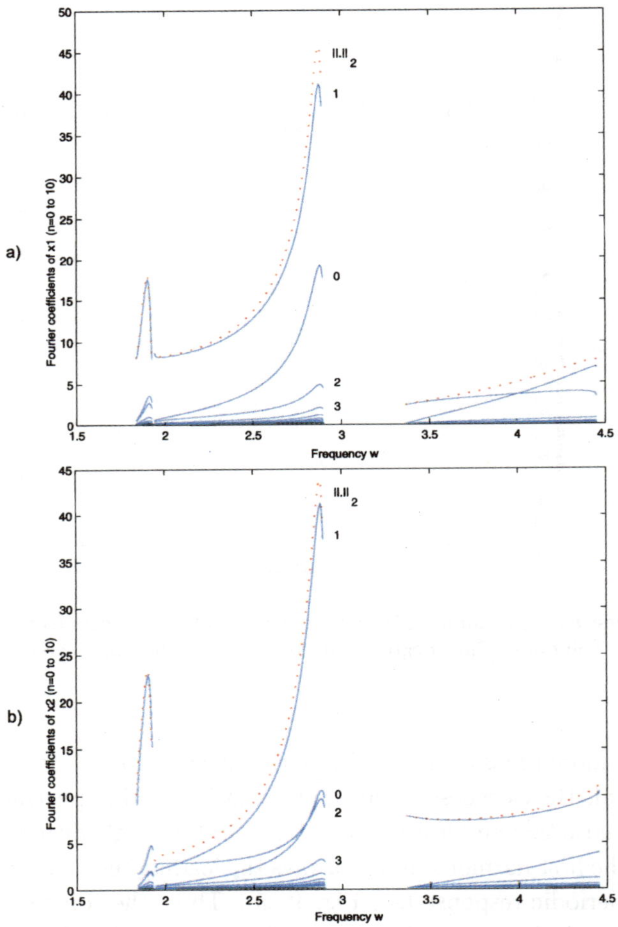

Fig. 9.25 Fourier coefficients of the $(1,1)$-periodic responses. First degree of freedom x_1 (a). Second degree of freedom x_2 (b).

For the last example, we choose a set of parameters close to the second example in the single degree of freedom system: $m_1 = 1$, $m_2 = 0.7$, $\lambda_1 = 1$, $\lambda_2 = 13$, $a = 0.02$, $e = 0.9$, $f^1 = 20$, $f^2 = 18$ et $x_{max} = 1$. Looking at Figs. 9.26 and 9.27 we first note that amplitude peaks appear at the same frequencies for x_1 and x_2, which is due to energy transmission between the two bodies through impacts. Moreover, the first harmonic amplitude can once again be far lower than the spectral amplitude: most of amplitude peaks in x_1 come from $\mathbf{A}_0$, and for x_2 the second harmonic is often overriding.

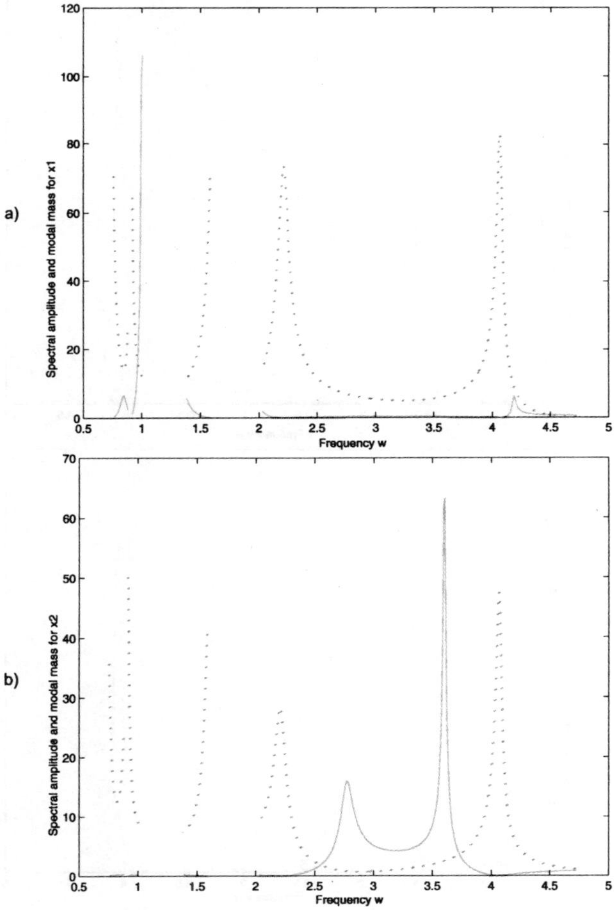

Fig. 9.26 First harmonic amplitude (dotted curve) and modal mass (solid curve) of the (1,1)-periodic solutions. First degree of freedom x_1 (a). Second degree of freedom x_2 (b).

We have just studied the $(1,1)$-periodic responses of the system. Nevertheless, we must keep in mind that many other types of periodic solutions are possible, as we saw for the single-degree-of-freedom system, depending on the time period and the number of impacts per cycle. Theoretically, nothing can prevent us from writing a modal superposition formula for any (n,k)-periodic response, but in practice such a response is hard to find by analytical means when $k > 1$.

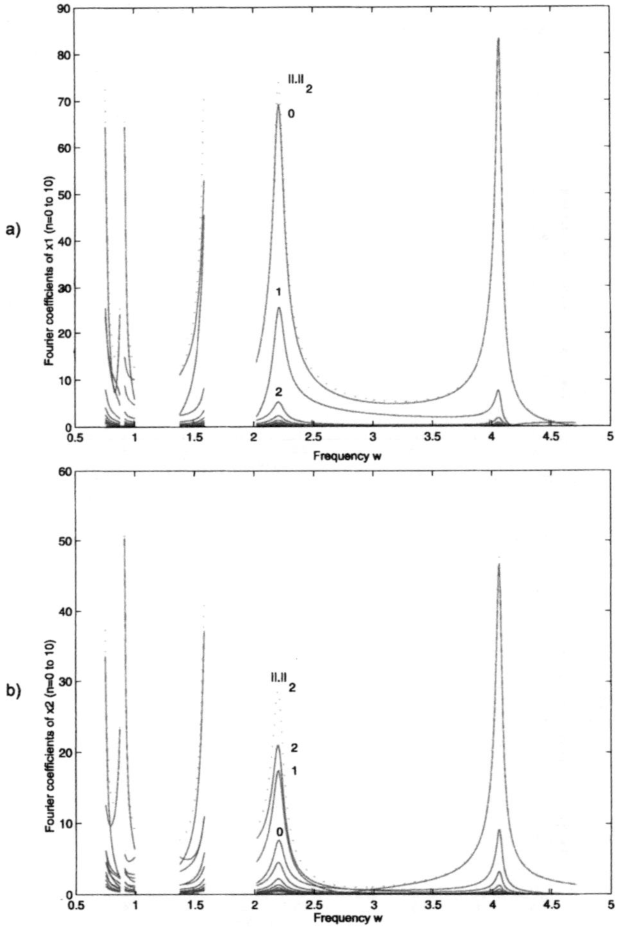

Fig. 9.27 Fourier coefficients of the $(1,1)$-periodic responses. First degree of freedom x_1 (a). Second degree of freedom x_2 (b).

9.4 Conclusion

We have investigated the feasibility of building a modal superposition formula for systems with irregular nonlinearities of impact type, imitating the procedure used in the smooth nonlinear case [Jézéquel and Lamarque (1992); Lamarque (1992)]. The formula has been built for simple single and two degrees of freedom systems with unilateral constraint and restitution law. The generalized modes and frequencies obtained turn out to be identical to the linear case, the nonlinearity of the system being concentrated into the modal masses.

The considered examples show that the formula is valid in the case of a primary resonance for which the spectral amplitude is given by the Fourier coefficient corresponding to the periodicity of the forced obtained solution. Nevertheless, these examples have illustrated above all the obstructions to such a building. Firstly, the multiplicity of periodic solutions, with different periodicity or number of impacts per cycle, compels to build several potential formulas, and it is not possible to know in advance which one has to be used. Furthermore, main amplitude peaks may appear away from any a priori clearly identifiable resonance, for which some unusual harmonics may override and consequently cause the modal superposition formula to fail.

Therefore, it is not possible in a general case to build a modal superposition formula using only the usual sequence definition of generalized frequencies, definition of generalized modes and then definition of generalized modal masses: the nonlinearities of impact type produce an obstruction to the building of a general formula following the usual procedure.

Chapter 10

One DOF Mechanical System with Friction

10.1 Introduction

Many nonlinear mechanical systems have been extensively studied since thirty years: many of these systems possess a finite number of degrees of freedom and almost all of them correspond to a smooth mathematical model ([Ueda (1979)], [Lorenz (1963)], [Sparrow (1982)], [Lamarque and Malasoma (1992)]) except in the case of systems with plasticity, impacts or friction forces. Smooth nonlinearities are generally due to large displacements (geometrical nonlinearities) and nonlinear elasticity (constitutive laws of materials).

Impacts and friction have been introduced because of their many engineering applications: Study of joints [Paoli *et. al.* (1992)], impacts in gearboxes [Pfeiffer (1988a)], models of beams [Dowell and Schwartz (1983a)], [Dowell and Schwartz (1983b)], [Whiteman and Ferri (1996)], rotor-casing [Li and Yorke (1975)], etc.

Many numerical works (for example, let us cite the works of [Whiston (1987)] (numerical study of a single degree of freedom vibro-impact-system), [Shaw and Shaw (1989)], [Shaw and Rand (1989)], [Awrejcewicz and Delfs (1990a)], [Awrejcewicz and Delfs (1990b)], [Foale and Bishop (1994)], [Popp and Stelter (1990)], [Cone and Zadoks (1995)]) are available in the case of one or two degrees of freedom systems.

Let us notice that in each case the problem of existence and uniqueness is solved by a direct quadrature of equations so that the mathematical difficulty disappears. Exact equations have to be solved in order to obtain impact-time and/or stick or slip time.

In general cases it is not always possible to obtain a piecewise-first-integral corresponding to the model. So two problems are considered:

- Find existence and uniqueness results, at least local results and (better) global results,
- Find a correct numerical procedure associated with differential inclusion.

Theoretical background leads to such results ([Deimling (1992)], [Moreau (1988)], [Brezis (1973)], [Schatzman (1978)]) in the frame of maximal monotone operators or graphs, as it has been recalled in the chapter 1.5. Paoli [Paoli (1993)], [Paoli and Schatzman (1993a)] studied a one degree of freedom system with impacts and loss of energy or without loss of energy. Paoli and Schatzman [Paoli and Schatzman (1993b)], [Paoli and Schatzman (1994)] provided a numerical scheme adapted to this dynamical impact problem.

One dimensional friction has been studied by Monteiro Marques [Monteiro Marques (1994)]. Results of existence and uniqueness are provided for a Coulomb-like damping and numerical schemes are explained.

Recently non uniqueness results have been pointed out for varying time friction coefficients ([Schatzman *et. al.* (1999)]).

Chaotic behavior occurs in many examples. One of them is especially interesting: The double scroll or Chua oscillator [Chua *et. al.* (1986a)] is a simple piecewise linear oscillator extensively used as a paradigm for chaos. One can notice that it is defined by C^1-piecewise functions and it is not globally "smooth".

Nonlinear oscillations of forced pendulum have already been extensively investigated [Doerner *et. al.* (1994)]. But here we intend to study oscillations of a pendulum submitted to friction, viscous damping and external sinusoidal excitation: this is a non smooth system that can not be studied by joining "exact piecewise solutions". So we study the behavior of such a system via a well-defined numerical approximation.

This chapter is organized as follows. In section 10.2 we describe the model of the pendulum with friction and we deal with mathematical problem: a numerical scheme is presented and its performances are tested numerically. In section 10.3 we give numerical examples of results due to friction for free and forced behavior of the pendulum. Some chaotic oscillations are analyzed. In section 10.4 we present a limit Melnikov analysis based on the classical method for the pendulum without friction. Finally, in section 10.5 we state conclusion of this work.

10.2 Modelling the Pendulum with Friction

Let us consider a forced pendulum with a viscous damping and Coulomb friction. This pendulum corresponds to the model:

$$\ddot{x} + a\dot{x} + \lambda \sin(x) + \alpha\sigma(\dot{x}) - f(t) \ni 0,$$

where : $\cdot = \frac{d}{dt}$, $\lambda \in \mathbb{R}$, $a \in \mathbb{R}, \alpha \in \mathbb{R}^+$ and $\sigma(\cdot)$ denotes the graph of the function sign:

$$\sigma(u) = -1 \quad \text{if} \quad u < 0, \; \sigma(u) = +1 \quad \text{if} \quad u > 0, \; \sigma(u) = [-1, 1] \quad \text{if} \quad u = 0.$$

This model has to be understood as a differential inclusion. $\alpha\sigma(\dot{x})$ is the expression of a Coulomb friction applied to the pendulum. Hereafter, we choose a particular expression of the external forcing $f(t)$:

$$f(t) = f \sin(\omega t), \; f, \; \omega \in \mathbb{R}.$$

10.2.1 *Existence and uniqueness*

The previous model is written in the form of a first order differential inclusion:

$$\dot{Y} + \begin{pmatrix} y_2 \\ ay_2 + \lambda \sin(y_1) - f(t) \end{pmatrix} + \begin{pmatrix} 0 \\ \alpha\sigma(y_2) \end{pmatrix} \ni \begin{pmatrix} 0 \\ 0 \end{pmatrix}$$
$$\Longleftrightarrow \dot{Y} + F(Y,t) + H(Y) \ni 0, \qquad (10.1)$$

with $Y = \begin{pmatrix} y_1 \\ y_2 \end{pmatrix} = \begin{pmatrix} x \\ \dot{x} \end{pmatrix}$ and initial condition $Y(t_0) = Y_0 = \begin{pmatrix} x_0 \\ \dot{x}_0 \end{pmatrix}$. It is enough to deal with $x_0 \in [-\pi, \pi] \ni \dot{x}_0$.

F is clearly a Lipschitz-continuous function: for usual Euclidean scalar product $(\cdot, \cdot)$ and its associated norm $|| \cdot ||$ of $\mathbb{R}^2$ we have:

$$\forall t \in \mathbb{R}, \; \forall (Y, Z) \in \mathbb{R}^2 \times \mathbb{R}^2, \; Y = \begin{pmatrix} y_1 \\ y_2 \end{pmatrix}, \; Z = \begin{pmatrix} z_1 \\ z_2 \end{pmatrix},$$

$$|| F(Y,t) - F(Z,t) || \leq (1 + | a | + | \lambda |) \, || Y - Z || \, .$$

H is a monotone operator: for the same scalar product of $\mathbb{R}^2$ we have:

$$\forall Y = \begin{pmatrix} y_1 \\ y_2 \end{pmatrix} \in \mathbb{R}^2, \quad \forall Z = \begin{pmatrix} z_1 \\ z_2 \end{pmatrix} \in \mathbb{R}^2,$$

$$\forall U = \begin{pmatrix} u_1 \\ u_2 \end{pmatrix} \in \mathbb{R}^2, \ \forall V = \begin{pmatrix} v_1 \\ v_2 \end{pmatrix} \in \mathbb{R}^2,$$

$$U \in H(Y), \ V \in H(Z) \Longrightarrow (V - U, Z - Y) = \alpha(v_2 - u_2, z_2 - y_2) \geq 0,$$

because σ is monotone.

H is maximal because $(I_{\mathbb{R}^2} + \mu H)$ is invertible for every real $\mu \geq 0$ with

$$(I_{\mathbb{R}^2} + \mu H)^{-1} \begin{pmatrix} x_1 \\ x_2 \end{pmatrix} = \begin{pmatrix} x_1 \\ (I_{\mathbb{R}} + \mu\alpha\sigma)^{-1}(x_2) \end{pmatrix},$$

and

$$(I_{\mathbb{R}} + \mu\alpha\sigma)^{-1}(x) = \begin{cases} x - \mu\alpha & \text{if} \quad x \geq \mu\alpha, \\ x + \mu\alpha & \text{if} \quad x \leq -\mu\alpha, \\ 0 & \text{if} \quad \mid x \mid \geq \mu\alpha. \end{cases}$$

Results by Brézis ([Brezis (1973)] page 107), Monteiro-Marques [Monteiro Marques (1994)] or of chapter 1.5 address existence and uniqueness of the solution of differential inclusion problem (10.1).

The proof is based on piecewise linear approximation σ_μ of σ:

$$\sigma_\mu(z) = \begin{cases} \text{sgn}(z) & \text{if} \quad \mid z \mid \geq \mu, \\ \dfrac{z}{\mu} & \text{if} \quad \mid z \mid \leq \mu. \end{cases}$$

In this case we deal with the family of problems:

$$\ddot{\hat{x}} + a\dot{\hat{x}} + \lambda \sin(\hat{x}) - f \sin(\omega t) + \alpha\sigma_\mu(\dot{\hat{x}}) = 0, \tag{10.2}$$

and each of this problem possesses exactly one smooth solution $(x_\mu, \dot{x}_\mu)$ with strong convergence of x_μ to x and $\dot{x}_\mu$ to $\dot{x}$ in $C^0([t_0, t_0 + T_0], \mathbb{R})$ when

$\mu \longrightarrow 0$: x and $\dot{x}$ correspond to the solution of the exact problem and $T_0 > 0$.

10.2.2 *Numerical scheme*

According to previous chapter it is possible to build at least two different correct numerical schemes. Here we use an implicit-like-Euler numerical scheme [Schatzman *et. al.* (1999)] defined by:

$$\begin{cases} \dfrac{x_{n+1} - x_n}{\Delta t} = y_n, \\ \dfrac{y_{n+1} - y_n}{\Delta t} + a y_n + \lambda \sin(x_n) - f(t_n) + \alpha\sigma(y_{n+1}) \ni 0. \end{cases}$$

with constant time step Δt. Clearly this corresponds to Euler-implicit method with inversion of the maximal monotone graph $(I_{\mathbb{R}} + \Delta t \alpha \sigma)$.

Monteiro-Marques proved in such a case that the approximated solution obtained by the numerical scheme converges strongly in $C^0(t_0, t_0 + T_0)$ to the exact solution of the problem.

The detailed numerical scheme is written as

$$\begin{cases} \Delta t, x_0, y_0 \quad \text{given}, \\ \text{for} \quad n \geq 0,\ x_{n+1} = x_n + \Delta t y_n, \\ \text{for} \quad n \geq 0, \\ \text{if} \quad \Delta t[f(n\Delta t) - \lambda \sin(x_n) - a y_n] + y_n > \alpha \Delta t \\ y_{n+1} = y_n + \Delta t(f(n\Delta t) - \lambda \sin(x_n) - \alpha - a y_n), \\ \text{if} \quad \mid \Delta t[f(n\Delta t) - \lambda \sin(x_n) - a y_n] + y_n \mid \leq \mid \alpha \Delta t \mid \\ y_{n+1} = 0, \\ \text{if} \quad \Delta t[f(n\Delta t) - \lambda \sin(x_n) - a y_n] + y_n < -\alpha \Delta t \\ y_{n+1} = y_n + \Delta t(f(n\Delta t) - \lambda \sin(x_n) + \alpha - a y_n). \end{cases}$$

In order to test the behavior of the numerical scheme we can first estimate numerically its order at least if oscillations of the pendulum do not exhibit chaos (even in the chaotic case, the discrete trajectory converges to the exact continuous one when time step decreases to 0).

Now let us notice an important point. On the one hand it is impossible to obtain analytically piecewise exact solutions (as in the case of a linear oscillator with friction for example), so numerical schemes are needed. On the other hand, it is not useful to refine the time step only around phase changes because during a phase, order of the correct numerical scheme is 1 (order of Euler's method).

10.2.3 *Numerical estimation of order*

In practice, the numerical scheme has a good behavior. We obtain roughly order 1, for transient behavior followed by periodic behavior as we can see in Fig. 10.1: we present the graph $(-\log(h), || x_h - x_{h/2}||_\infty)$ for different constant steps h. There x_h denotes the numerical solution obtained for the constant time step h and $x_{h/2}$ denotes the numerical solution obtained for the constant time step $h/2$. Norm $|| x_h - x_{h/2}||_\infty$ is computed for a finite given time interval. Indeed, by analogy with the smooth case, we investigate close to $h = 0$ the relation:

$$|| x_h - x_{h/2}||_\infty \leq|| x_h - x||_\infty + || x_{h/2} - x||_\infty \leq K_1 h^p + K_1 (h/2)^p = K_2 h^p, \tag{10.3}$$

because in a smooth case, for a scheme of order p, there is a constant K so that

$$|| x_h - x||_\infty \leq K h^p. \tag{10.4}$$

In the case of a chaotic behavior, it is clear that sensitivity to initial condition does not allow us to test any estimation of order for a given finite "long" time. But for short time interval, order is again roughly 1 (see in Fig. 10.2).

10.3 Numerical Results

Now we present numerical results corresponding to different choice of parameters [Doerner *et. al.* (1994)] related to free or forced vibrations of the pendulum with friction.

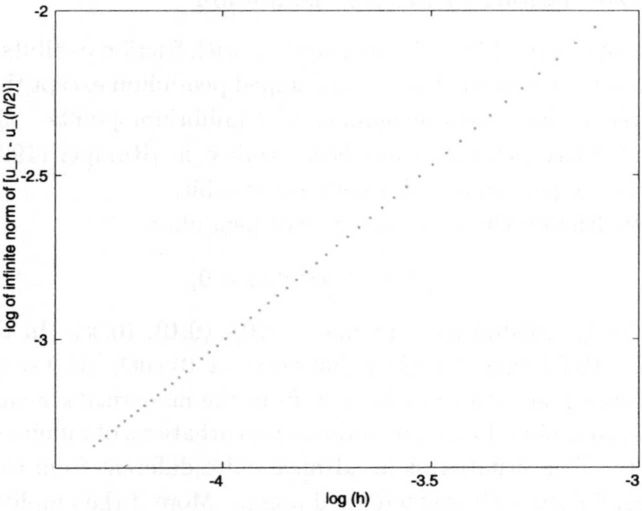

Fig. 10.1 Values of 50 time steps from 0.00094352 to 0.04740810. Mean square slope = order estimate = 1.059: computation of 100 periods $2\pi/\omega$ for each value of time step for $\alpha = 0.5, \lambda = 0.87, a = 0.052, \omega = 0.666, f = 0.586$.

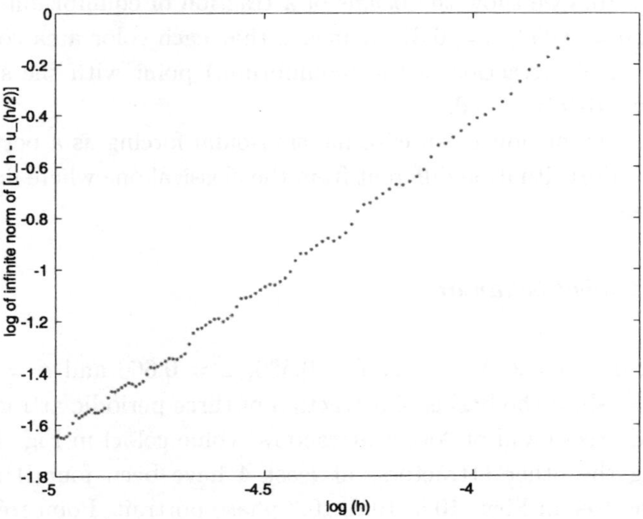

Fig. 10.2 Values of 100 time steps from 0.02364464 to 0.00023586. Mean square slope = order estimate = 1.223: computation of 1 period $2\pi/\omega$ for each time step for $\alpha = 0.144$, $\lambda = 0.87$, $a = 0.052$, $\omega = 0.666$, $f = 0.586$.

10.3.1 *Oscillations of the free pendulum*

In the case of free oscillations the pendulum with friction exhibits obviously the same kind of behavior than a free damped pendulum except that friction term is able to increase the number of equilibrium points. Qualitative behavior of a free pendulum has been studied in [Rumpel (1996)] in the point of view of persistence of a homoclinic orbit.

Without friction the unperturbed free pendulum

$$\ddot{x} + a\dot{x} + \lambda \sin(x) = 0, \tag{10.5}$$

exhibits clearly isolated fixed points $(-\pi, 0)$, $(0, 0)$, $(0, \pi)$. In the case of pendulum with friction, it is clear that every $(x, 0)$ such as $\mid \lambda \sin(x) \mid \leq \alpha$ is an equilibrium point (this can be seen from the numerical scheme too). So an external sinusoidal forcing introduces perturbations of infinite number of fixed points. This situation is in advance quite different from the classical damped pendulum with isolated fixed points. More if the amplitude of the sinusoidal perturbation is small enough, indeed if

$$\mid \lambda \mid\mid \sin(x) \mid + \mid f \mid \leq \alpha,$$

equilibrium can persist.

In Fig. 10.3 we show the basins of attraction of equilibrium points for $\alpha = 0.84$, $a = 0.052$, $\lambda = 0.87$: it means that each color area corresponds to the basin of attraction of the (equilibrium) point with the same color located on axis $y = \dot{x} = 0$.

Now if one introduces an external sinusoidal forcing as a perturbation, we have a priori situation different from the classical one where fixed points are isolated.

10.3.2 *Global behavior*

Let us fix $a = 0.052$, $\lambda = 0.87$, $f = 0.586$, $\omega = 0.666$ and $\alpha = 0.144$. In this case we show the basins of attraction of three periodic attractors (red, yellow and green) and of "other attractors" (blue color) in Fig. 10.4.

Among the other attractors (at least 4 have been found) there is a chaotic one (see in Figs. 10.5, 10.6, 10.7 phase portrait, Poincaré map and velocity versus time) after transient corresponding to 380 000Δt with again $\frac{2\pi}{1000\omega}$: we plot 20 000 steps in Figs. 10.5 and 10.7 and $y(kT)$ versus $x(kT)$ with $800 \geq k \geq 380$ in Fig. 10.6.

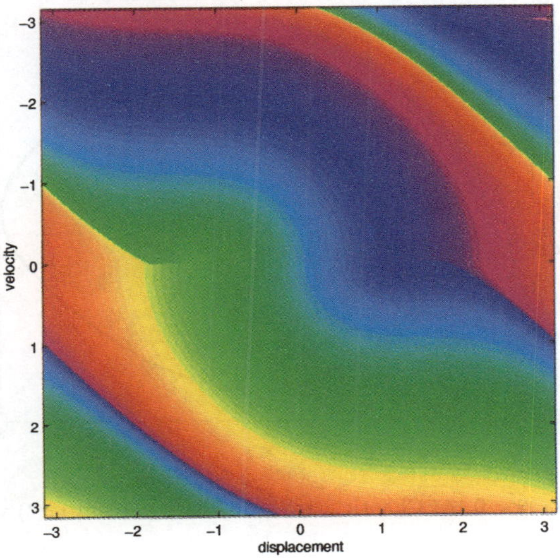

Fig. 10.3 Basin of attraction of equilibrium points for $\alpha = 0.84$, $\lambda = 0.87$, $a = 0.052$, $\omega = 0.666$, $f = 0$. Horizontally $x \in [-\pi, \pi]$, vertically $\dot{x} \in [-\pi, \pi]$.

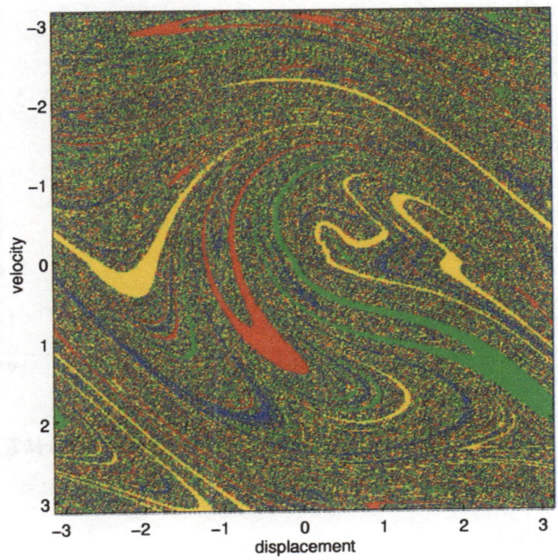

Fig. 10.4 Basins of attraction for $\alpha = 0.144$, $\lambda = 0.87$, $a = 0.052$, $\omega = 0.666$, $f = 0.586$. Horizontal axis $x \in [-\pi, \pi]$, vertical axis $\dot{x} \in [-\pi, \pi]$. Red, yellow, green correspond to $\dfrac{2\pi}{\omega}$-periodic attractors. Blue corresponds to other attractors.

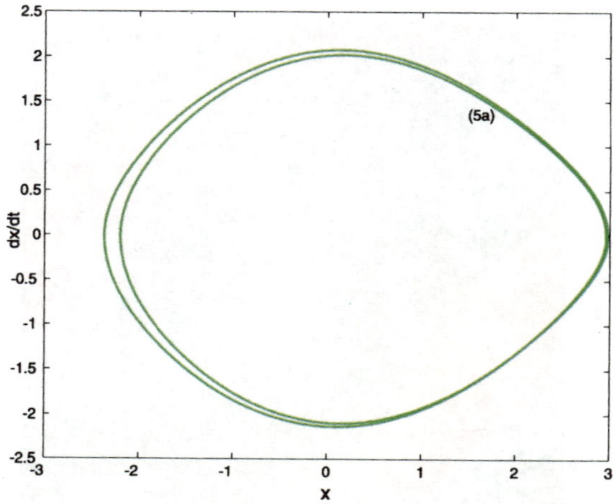

Fig. 10.5 $\alpha = 0.144$ Chaotic attractor. Phase portrait ($\dot{x}(t)$ vs $x(t)$).

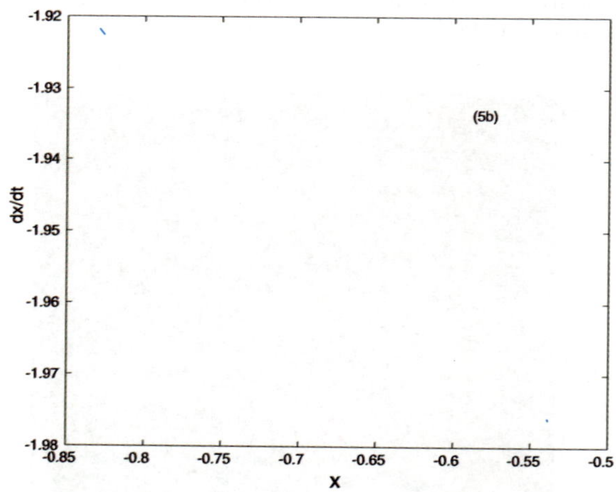

Fig. 10.6 $\alpha = 0.144$ Chaotic attractor. Poincaré section $((x(kT), \dot{x}(kT))$.

One can see that:

- Friction provides identical but shifted solutions: such multiple shifted solutions correspond to the effect of the many equilibria obtained in section 2.4
- Analysis of chaos is improved in the next section.

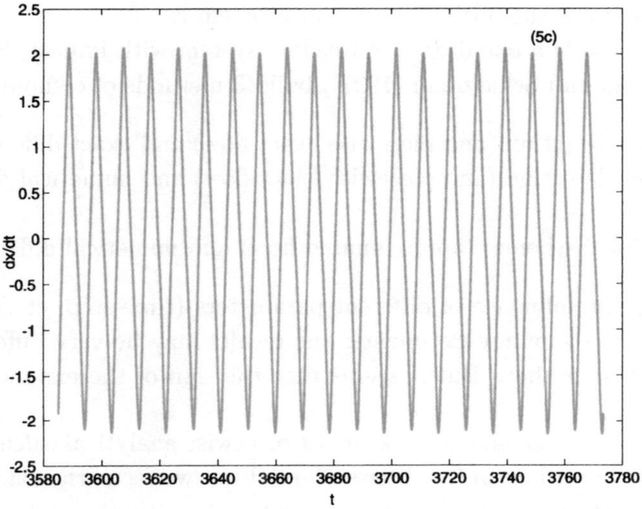

Fig. 10.7 Chaotic attractor. Velocity ($\dot{x}(t)$) versus time for $\alpha = 0.144$.

10.3.3 *Lyapunov exponents*

We deal here with the calculation of the largest Lyapunov exponent. This calculation provides a criterion for the "chaotic" nature of the behavior of the forced pendulum with friction: If the largest Lyapunov exponent is strictly positive it means that the pendulum oscillations are chaotic. Nevertheless, one has to notice the following difficulty: usual definition of the Lyapunov exponents needs a C^1 flow of the differential equation. This is not the case here. But, practically the flow is piecewise C^1. Consequently, we first explain how to compute the largest Lyapunov exponent, then we provide numerical examples of computations.

10.3.3.1 *Computation method of the largest Lyapunov exponent*

Müller [Müller (1995)] gives a method to calculate Lyapunov exponents for systems with discontinuities: calculation is given from the terms of differential equation. He studies a linear oscillator with friction so that piecewise analytical calculations can be made. Here we shall use a similar idea to process the numerical scheme associated with the initial differential problem and not the initial differential problem itself.

In the case of a one degree of freedom system with impact, Schatzman and Nqi [Nqi and Schatzman (1997)] build 3 methods of computation:

- Impact detection (and piecewise exact flow) and exact differentiation,
- Impact detection (and piecewise exact flow) and numerical differentiation,
- Numerical scheme (approximated flow) and numerical differentiation,

and study the influence of different parameters (time-step, etc.) over the values of the Lyapunov exponents: the results may be very different from a method to another. But it seems that the sign of the exponents is not modified.

Here it is not possible to make exact piecewise analytical calculation. So we suggest to work with the discrete flow. First we recall the calculation in a C^1 case and then we explain how to modify this situation in the piecewise C^1 case. Let us assume

$$\begin{cases} x_{n+1} = \tilde{\phi_1}(t_n, x_n, y_n, t_{n+1}) = \phi_1(t_0, X_0, t_{n+1}), \\ \\ y_{n+1} = \tilde{\phi_2}(t_n, x_n, y_n, t_{n+1}) = \phi_2(t_0, X_0, t_{n+1}), \end{cases} \tag{10.6}$$

with $X_0 = (x_0, y_0)$ and $\tilde{\phi_1}$, $\tilde{\phi_2}$, ϕ_1, ϕ_2: $\mathbb{R}^4 \mapsto \mathbb{R}$ 4 functions C^1 according to their second and third variables. Let us write the condensed vector form of the previous definition:

$$X_{n+1} = \tilde{\phi}(t_n, X_n, t_{n+1}) = \phi(t_0, X_0, t_{n+1}), \tag{10.7}$$

with $\tilde{\phi}$ and ϕ: $\mathbb{R} \times \mathbb{R}^2 \times \mathbb{R} \mapsto \mathbb{R}$ two C^1 functions according to their second and third variables. The classical situation of divergence from initial trajectory is presented in Fig. 10.8.

In such a case, one wishes to compute for t_n large enough the largest Lyapunov exponent:

$$\lambda_{max,n} = \frac{1}{t_n - t_0} \log\Big(|| \frac{D\phi}{DX_0}(t_0, X_0, t_n) || \Big). \tag{10.8}$$

Practically we shall use a chain rule: as a matter of fact if one has series $X_0, X_1, X_2, \ldots, X_n$ such that

$$X_{j+1} = \tilde{\phi}(t_j, X_j, t_{j+1}) \underset{\text{(notation)}}{=} \tilde{\phi}(X_j), \tag{10.9}$$

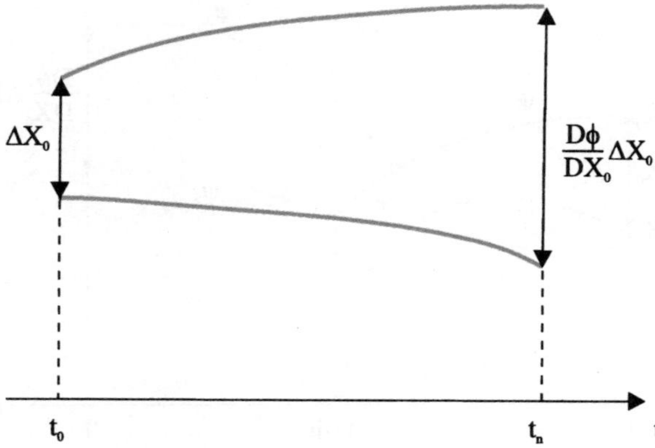

Fig. 10.8 Classical divergence for smooth flows $\phi(X_0)$ and Lyapunov exponent computation.

one can calculate in practice the differential versus X_0 with

$$D\tilde{\phi}^n(X_0) = D\tilde{\phi}(X_{n-1}), \dots, D\tilde{\phi}(X_0). \tag{10.10}$$

So for a given $v_0 \in \mathbb{R}^2$ let us set:

$$\begin{cases} u_j = D\tilde{\phi}(X_{j-1})v_{j-1}, \quad j = 1, \dots, n, \\ \\ \qquad v_j = \dfrac{u_j}{\| u_j \|}. \end{cases} \tag{10.11}$$

We obtain by the chain rule

$$D\tilde{\phi}^n(X_0) = \prod_{j=1}^{n-1} \| u_j \| u_n, \tag{10.12}$$

such as

$$\lambda_{max,n} = \frac{1}{t_n - t_0} \sum_{j=1}^{n} \log(\| u_j \|). \tag{10.13}$$

In our case, it is possible to make piecewise calculations of $D\tilde{\phi}(X_j)$ and to add correction terms due to the phase transitions. Let us consider the general situation presented in Fig. 10.9.

The flow is changing according to the realization of the equation:

$$E(t_0, X_0, t) = 0. \tag{10.14}$$

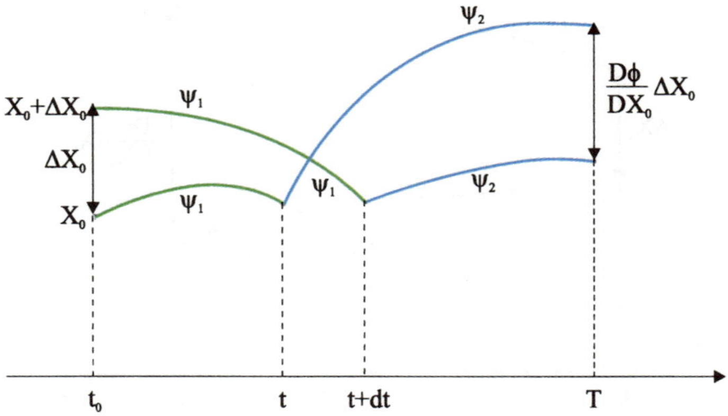

Fig. 10.9 Divergence for nonsmooth flows $\phi(X_0)$ and Lyapunov exponent computation.

Before such an equality occurs let us denote by ψ_1 the C^1 function which defines the flow, and after the occurrence of the same equation, let us denote by ψ_2 the new C^1 function which defines the flow. As long as no phase transition occurs it is sufficient to calculate the usual quantity:

$$D\psi_1 = \frac{D\psi}{DX_0}. \tag{10.15}$$

After the occurrence of a phase transition one has to modify the calculation according to the expression:

$$D\psi_1 = \frac{D\psi}{DX_0} - \frac{1}{\dfrac{\partial E}{\partial t}}\left(\frac{\partial \psi_1}{\partial t} - \frac{\partial \psi_2}{\partial t}\right)\frac{\partial E}{\partial X_0}. \tag{10.16}$$

We apply this result to the calculation of every term $D\tilde{\phi}(X_j)$. In our case we have

$$\tilde{\phi}(t_n, X_n, t_{n+1}) = \begin{pmatrix} x_n + \Delta t y_n \\ \\ \tilde{\phi}_2^{\varepsilon}(t_n, x_n, y_n, t_{n+1}) \end{pmatrix}, \tag{10.17}$$

with $\varepsilon = +, 0$ or $-$. The second component of the flow can be written as:

$$\begin{cases} \tilde{\phi}_2^- = y_n + \Delta t.(g(t_n, x_n, y_n) + \alpha) & \text{if} \quad y_n + \Delta t.(g(t_n, x_n, y_n) + \alpha) < 0, \\ \tilde{\phi}_2^0 = 0 & \text{if} \quad | y_n + \Delta t.g(t_n, x_n, y_n) | \leq \alpha \Delta t, \\ \tilde{\phi}_2^+ = y_n + \Delta t.(g(t_n, x_n, y_n) - \alpha) & \text{if} \quad y_n + \Delta t.(g(t_n, x_n, y_n) - \alpha) > 0. \end{cases} \tag{10.18}$$

with $g(t_n, x_n, y_n) = f \sin(\omega t_n) - a y_n - \lambda \sin(x_n)$. We have two phase transition equations:

$$E_+(t_n, x_n, y_n) = y_n + \Delta t(g(t_n, x_n, y_n) + \alpha), \tag{10.19}$$

$$E_-(t_n, x_n, y_n) = y_n + \Delta t(g(t_n, x_n, y_n) - \alpha). \tag{10.20}$$

For the smooth part $\tilde{\phi}^r$ of $\tilde{\phi}$ (indeed everything except the term due to friction) we have:

$$D\tilde{\phi}^r(X_n) = \begin{pmatrix} 1 & \Delta t \\ -\lambda \Delta t \cos(x_n) & 1 - a\Delta t \end{pmatrix}, \tag{10.21}$$

except when the second component of $\tilde{\phi}$ is $\tilde{\phi}_2^0$. In this case we have:

$$D\tilde{\phi}^r(X_n) = \begin{pmatrix} 1 & 0 \\ 0 & 0 \end{pmatrix}. \tag{10.22}$$

These terms have to be modified by the 2×2 matrices given in the Table 10.3.3.2 which correspond to the correction terms of the phase transition: $\tilde{\phi}_2^- \longrightarrow \tilde{\phi}_2^-$, $\tilde{\phi}_2^- \longrightarrow \tilde{\phi}_2^0$, $\tilde{\phi}_2^- \longrightarrow \tilde{\phi}_2^+$, etc. All the transitions are taken into account because of the discrete time: in order to make computations with a small (but not too small) time step we consider transitions $\tilde{\phi}_2^- \longrightarrow \tilde{\phi}_2^+$ (rare in practice).

10.3.3.2 *Numerical results*

For the set of parameters $a = 0.052$, $\lambda = 0.87$, $f = 0.586$, $\omega = 0.666$ and different values of α we computed the largest Lyapunov exponents λ_{max}. We present these quantities calculated with 200 000 time steps $\Delta t = \dfrac{2\pi}{1000\omega}$ in Table 10.2

Table 10.1 Calculation of (piecewise) differential according to phase transition.

Phase trans	Correction term	Full differential
$\tilde{\phi}_2^- \longrightarrow \tilde{\phi}_2^-$	$\begin{pmatrix} 0 & 0 \\ 0 & 0 \end{pmatrix}$	$\begin{pmatrix} 1 & \Delta t \\ -\lambda \Delta t \cos(x_n) & 1 - a\Delta t \end{pmatrix}$
$\tilde{\phi}_2^- \longrightarrow \tilde{\phi}_2^0$	$\begin{pmatrix} 0 & 0 \\ \lambda \Delta t \cos(x_n) & a\Delta t - 1 \end{pmatrix}$	$\begin{pmatrix} 1 & \Delta t \\ 0 & 0 \end{pmatrix}$
$\tilde{\phi}_2^- \longrightarrow \tilde{\phi}_2^+$	$\begin{pmatrix} 0 & 0 \\ 0 & 0 \end{pmatrix}$	$\begin{pmatrix} 1 & \Delta t \\ -\lambda \Delta t \cos(x_n) & 1 - a\Delta t \end{pmatrix}$
$\tilde{\phi}_2^0 \longrightarrow \tilde{\phi}_2^-$	$\begin{pmatrix} 0 & 0 \\ -\lambda \Delta t \cos(x_n) & 1 - a\Delta t \end{pmatrix}$	$\begin{pmatrix} 1 & \Delta t \\ -\lambda \Delta t \cos(x_n) & 1 - a\Delta t \end{pmatrix}$
$\tilde{\phi}_2^0 \longrightarrow \tilde{\phi}_2^0$	$\begin{pmatrix} 0 & 0 \\ 0 & 0 \end{pmatrix}$	$\begin{pmatrix} 1 & \Delta t \\ 0 & 0 \end{pmatrix}$
$\tilde{\phi}_2^0 \longrightarrow \tilde{\phi}_2^+$	$\begin{pmatrix} 0 & 0 \\ -\lambda \Delta t \cos(x_n) & 1 - a\Delta t \end{pmatrix}$	$\begin{pmatrix} 1 & \Delta t \\ -\lambda \Delta t \cos(x_n) & 1 - a\Delta t \end{pmatrix}$
$\tilde{\phi}_2^+ \longrightarrow \tilde{\phi}_2^-$	$\begin{pmatrix} 0 & 0 \\ 0 & 0 \end{pmatrix}$	$\begin{pmatrix} 1 & \Delta t \\ -\lambda \Delta t \cos(x_n) & 1 - a\Delta t \end{pmatrix}$
$\tilde{\phi}_2^+ \longrightarrow \tilde{\phi}_2^0$	$\begin{pmatrix} 0 & 0 \\ \lambda \Delta t \cos(x_n) & a\Delta t - 1 \end{pmatrix}$	$\begin{pmatrix} 1 & \Delta t \\ 0 & 0 \end{pmatrix}$
$\tilde{\phi}_2^+ \longrightarrow \tilde{\phi}_2^+$	$\begin{pmatrix} 0 & 0 \\ 0 & 0 \end{pmatrix}$	$\begin{pmatrix} 1 & \Delta t \\ -\lambda \Delta t \cos(x_n) & 1 - a\Delta t \end{pmatrix}$

The sign of each Lyapunov exponent is in agreement with the behavior announced by the Poincaré sections and phase portraits of Sections 10.3.1 and 10.3.2.

In Figs. 10.10 and 10.11 evolution of the greatest Lyapunov exponent is plotted versus time. Fig. 10.10 corresponds to $\alpha = 0.0294$ and the Fig.

10.11 to $\alpha = 0.0207$. Time is varying from 0 to $350T$. Horizontal scales are artificial: They correspond to the number of points equally picked up in this interval. We can see, especially in Fig. 10.10 a good stabilization of the λ_{max} value near 0.1.

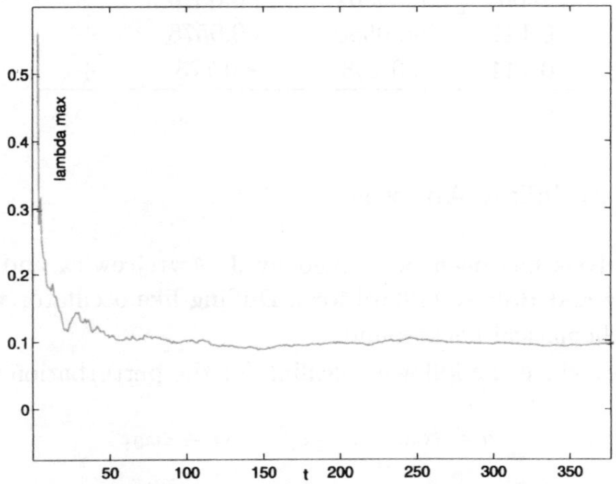

Fig. 10.10 Variation of the largest Lyapunov exponent versus time $(0, 350T)$, $\alpha = 0.0294$.

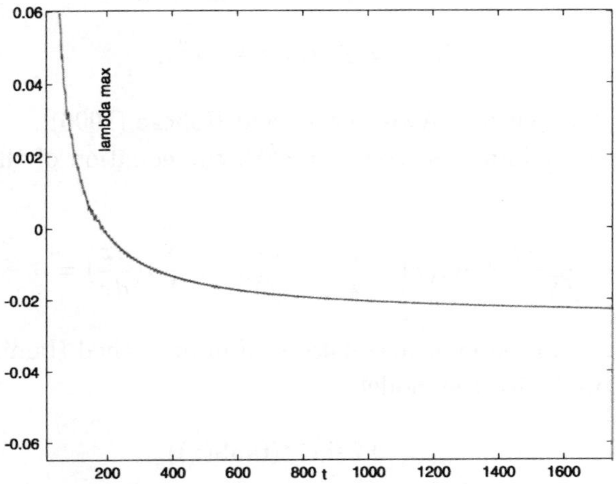

Fig. 10.11 Variation of the largest Lyapunov exponent versus time $(0, 350T)$, $\alpha = 0.0207$.

Table 10.2 Computation of Lyapunov exponents for some α.

α	λ_{max} 200T	λ_{max} 350T	**Sign**
0.0207	−0.00645	−0.0141	−
0.0294	+0.113	+0.105	+
0.06	+0.0867	+0.0845	+
0.141	+0.0536	+0.0578	+
0.144	+0.168	+0.173	+

10.4 The Melnikov Analysis

Such an analysis has been performed by J. Awrejcewicz and M. Holicke [Awrejcewicz and Holicke (1996)] for a Duffing like oscillator with friction without mathematical background.

Let us introduce the following scaling for the perturbation terms:

$$a = \varepsilon a_0, \quad f = \varepsilon f_0, \quad \alpha = \varepsilon \alpha_0. \tag{10.23}$$

If $\varepsilon = 0$, $(\pi, 0)$ is a hyperbolic fixed point which possesses stable and unstable manifolds and generates homoclinic situation.

Let us assume that for ε sufficiently small, such a situation persists when a friction term is working and that stable and unstable manifolds exist.

The Melnikov function $\Delta(t_0)$ ([Wiggins (1988)]) can be obtained with

$$\Delta(t_0) = \Delta^1(t_0)\varepsilon + o(\varepsilon^2), \tag{10.24}$$

and work of Awrejcewicz [Awrejcewicz and Holicke (1996)].

By using a rescaling relation $\tau = \sqrt{\lambda}t$, the equation of the pendulum becomes:

$$\frac{d^2\hat{x}}{d\tau^2} + \frac{a}{\sqrt{\lambda}}\frac{d\hat{x}}{d\tau} + \sin(\hat{x}) - \frac{f}{\lambda}\sin(\frac{\omega}{\sqrt{\lambda}}\tau) + \frac{\alpha}{\lambda}\sigma(\frac{d\hat{x}}{d\tau}) = 0. \tag{10.25}$$

By using the usual parametric equations of unperturbed Hamiltonian $\dot{x}^2 - \cos(x)$ associated with the model

$$x^{\pm} = \pm 2\sin^{-1}(\tanh(\tau)), \tag{10.26}$$

$$\dot{x}^{\pm} = \pm 2\frac{1}{\cosh(\tau)}, \tag{10.27}$$

we obtain

$$\Delta^1(t_0) = -\frac{8a_0}{\sqrt{\lambda}} + \frac{f_0}{\lambda}\sin(\frac{\omega}{\sqrt{\lambda}}t_0)\cosh(\frac{\pi\omega}{2\sqrt{\lambda}}) - \frac{2\alpha_0\pi}{\lambda}. \tag{10.28}$$

Critical curves are given by the relation:

$$-8a\sqrt{\lambda} + f\cosh(\frac{\pi\omega}{2\sqrt{\lambda}}) \leq 2\alpha\pi, \tag{10.29}$$

because a, λ, α, $f \geq 0$ are positive parameters. From a mathematical point of view, the critical relation is obtained as the limit of relations that are obtained for smoothed approximations of the graph σ.

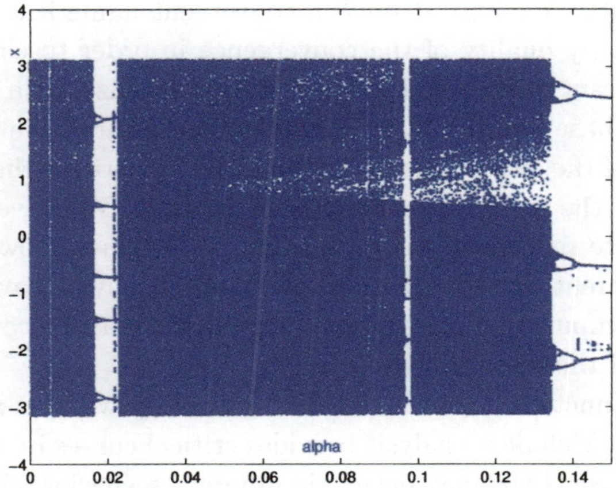

Fig. 10.12 Bifurcation diagram: $\omega = 0.666$, $a = 0.052$, $f = 0.586$, $\lambda = 0.87$. Parameter α is varying from 0 to 0.15 according to 1000 equal steps.

Here with $\lambda = 0.87$, $\omega = 0.666$, $f = 0.586$ and $a = 0.052$ we obtain $\alpha \geq 0.0968$: It means that no chaos can occur because of the studied critical situation if α is large enough. It means that the chaotic window that occurs for $\alpha \geq 0.1$ in the bifurcation diagram of the Fig. 10.12, is either a transient chaos or chaos arising from another critical situation [Lamarque *et. al.* (1998)]: This has not been clarified now. In this diagram α is varying from 0.0 to 0.15 according to 1000 equal steps. We computed for every parameter α 1200000 time steps: the constant time step has been chosen as $\dfrac{2\pi}{1000\omega}$. First we eliminated a transient area corresponding to 800000 steps. Then for every α we retained the 400 points of the Poincaré section associated with the 400 last periods (800001 to 1200000 steps). Initial conditions are $x_0 = 1$ and $y_0 = 1.5$ for every parameter (we did not use continuation plot).

10.5 Conclusion

In this chapter we have analyzed an approximate solution of differential inclusion model of pendulum with friction. We have given existence and uniqueness results. We have defined a convergent numerical scheme. We studied numerically quality of the convergence in order to choose a small convenient constant time step. So we do not localize with a very high accuracy each phase transition since it is not necessary: We have a good approximation of the exact behavior. Using this numerical scheme we studied behaviors of the forced pendulum with friction. We have shown that friction can create special coexisting attractors. We have shown that chaos may occur. Different characterizations of chaotic behavior have been given including the computation of Lyapunov exponents with piecewise discrete differentiation of the discrete flow.

A Melnikov function has been derived by analogy without a full mathematical support. Melnikov analysis provides critical curves in α, ω, f, a, λ space. But because chaos may occur via different scenarios, chaos may exist beyond the critical curve because of transition different from the critical case pointed out for this Melnikov analysis. A proof of validity of this generalized Melnikov analysis in such a non smooth case has to be made based on persistence of homoclinic orbits [Rumpel (1996)].

Chapter 11

Modelling the Dynamical Behaviour of Elasto-Plastic Systems

11.1 Rheological Systems with "Friction"

11.1.1 *Introduction*

In the literature, many studies about the behaviour of nonlinear models can be found within the last twenty years. Brogliato [Brogliato (1996)] and Palmov [Palmov (1998)] give in their respective books numerous nonlinear mechanical models. Among these models, we are interested in those involving friction laws.

Authors especially study mechanical models with a finite number of degrees of freedom involving frictions terms and submitted to dynamical solicitations. Some of these papers provide mathematical results of existence and uniqueness [Jean and Pratt (1985); Laghdir and Monteiro Marques (1995); Monteiro Marques (1994); Matrosov and Finogenko (1995); Matrosov and Finogenko (1996b); Matrosov and Finogenko (1996a); Trinkle *et. al.* (1997)]. Some others investigate physical behaviours without dealing with theoretical results: some works present friction laws issued from experiments [Anderson and Ferri (1990)] and [Ferri and Bindemann (1995)] ; stick-slip phenomena is the main topic investigated in the references [Awrejcewicz and Delfs (1990a); Awrejcewicz and Delfs (1990b); Baumberger *et. al.* (1995); Ionescu and Paumier (1993); Pratt and Williams (1981)] and [Stelter (1992)]. The study of friction may be based upon analytical calculation of solutions [Capecchi and Vestroni (1995)]. Some works describe experiments, identification and modelling of friction [Dowell and Schwartz (1983a); Dowell and Schwartz (1983b)] and [Tomlinson and Chen (1996)]. Numerical experiments have been made by many authors using classical numerical schemes [Stewart (1996); Stewart and Trinkle (1996)] and [Stewart and Trinkle (1997)]. Chaotic behaviour have been exhibited by nonlinear models including friction [Chua *et. al.* (1986c); Chua *et. al.* (1986b); Madan (1993); Popp and Stelter (1990)] and [Shaw (1986)]. Except the paper of Monteiro Marques [Monteiro Marques (1994)], these works do not present convergence results of the approximate solution.

In an older literature, we can read works coping with continuous models involving elastic linear behaviour and friction terms at boundary conditions [Duvaut and Lions (1972)]. In this case, one has existence and uniqueness results for the mechanical problems which is expressed via variational inequalities. These inequalities are discretized by numerical schemes with good mathematical properties; one has results of convergence of the approximate solution described by the numerical schemes to the exact solution ([Głowiński *et. al.* (1976a)] and [Głowiński *et. al.* (1976b)]). Nevertheless these works are related to stationary problems. So they are useless for the study of our rheological problems.

In order to study micro-plasticity, Sidoroff [Fougeres and Sidoroff (1989)] investigates the classical elastoplastic Masing model, with a finite or infinite number of degrees of freedom; but, this model is a stationary one.

In this section, we intend to generalize the study by M. D. P. Monteiro Marques (see [Monteiro Marques (1994)]): He examines the model composed of one linear spring and one St-Venant element connected in parallel to a material point; existence and uniqueness results are provided. This proof is based on the properties of two numerical schemes. This model can be used to process the case of a n-degrees-of-freedom model of elements connected in parallel. But his results can not be used in order to deal with connections in series of spring and St-Venant elements. So we describe a general mathematical frame which permits us to process many rheological models including springs and St-Venant elements connected either in series or in parallel. Moreover this mathematical frame provides well adapted numerical schemes and their properties of convergence to the exact solution of the model. In a mathematical point of view, linear viscosity and linear elasticity are similar so, our developments are valid for both viscoplastic or elastoviscoplastic models.

In this section, we will describe some models involving springs, dashpots and St-Venant elements connected in series or in parallel; we give a mathematical theory of the dynamics of this rheological models. In section 11.1.2, we can see that the motion of these dynamical systems is governed by differential inclusions of the form

$$\dot{X}(t) + MA(X(t)) \ni F(t, X(t)), \tag{11.1}$$

where M is an invertible matrix, X is a function from $[0, T]$ in $\mathbb{R}^N$, A is a maximal monotone graph on $\mathbb{R}^N$ and F a function from $[0, T] \times \mathbb{R}^N$ in $\mathbb{R}^N$. The existence and uniqueness of solutions are consequences of very classical results. In section 11.1.3, we give a numerical scheme and we report the results of numerical simulations for elastoplastic models. We find in particular that under periodic loading, the solutions tend to a limit cycle which in the displacement-force plane is a hysteresis cycle. We will prove that there is a one-to-one correspondence between the geometrical shapes of these cycles and the mechanical characteristics of the elastoplastic models.

11.1.1.1 *The physical models*

In this section, we will study combinations of springs, dry friction elements, dashpots and material points. We study two elementary models. First, the model composed of one spring and one St-Venant element connected in parallel, which has been studied by Monteiro Marques in [Monteiro Marques (1994)]. Then, we present the association of one spring, one St-Venant element and one dashpot, connected in series. We consider elementary associations (spring, St-Venant element and material points) and their combinations.

Two Elementary Models

One Spring and One St–Venant Element Connected in Parallel

A material point of mass m submitted to an external force F is connected in parallel to a spring with stiffness k and a St-Venant element with threshold α; let x be the abscissa of this material point. This system is governed by the equation:

$$m\ddot{x} + kx + \alpha\sigma(\dot{x}) \ni F, \tag{11.2}$$

with initial data:

$$x(0) = x_0, \quad \dot{x}(0) = \dot{x}_0, \tag{11.3}$$

the graph σ is defined by (see Fig. 11.1a):

$$\sigma(x) = \begin{cases} -1 & \text{if } x < 0, \\ 1 & \text{if } x > 0, \\ [-1, 1] & \text{if } x = 0. \end{cases} \tag{11.4}$$

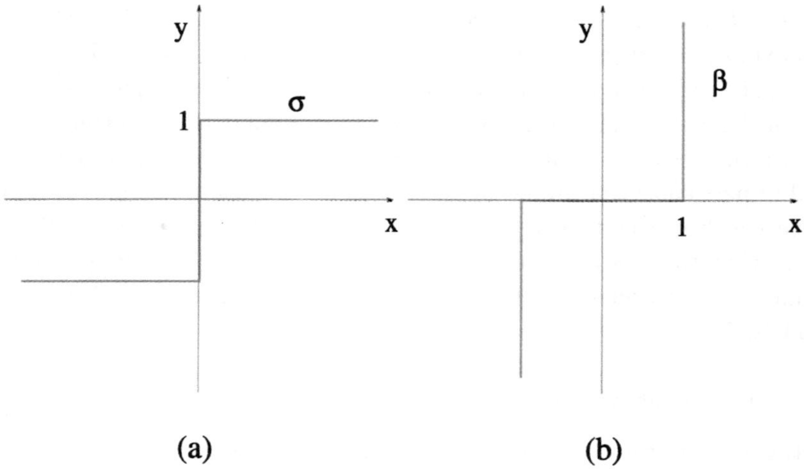

Fig. 11.1 The graph σ (a) and the graph β (b).

A Viscoelastoplastic Model and the Prandtl Rheological Model

We present now a viscoelastoplastic model, described in Fig. 11.2: A material point of mass m is submitted to an external force F and is connected in series to a spring with stiffness k, a St-Venant element with threshold α and a dashpot with viscosity c. Let u be the displacement of the extremity A of the spring, relatively to a reference position, let v be the difference between AB and a reference length and let w be the difference between BC and a reference length. Denote x the abscissa of the material point of mass m, relatively to a reference position, so that

$$x = u + v + w. \tag{11.5}$$

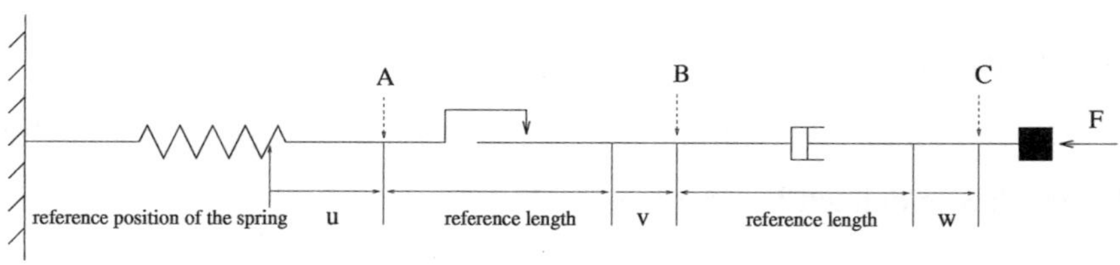

Fig. 11.2 A viscoelastoplastic model.

Denote f the force exerted by A on B. We make the assumption that $k \neq 0$, $c \neq 0$ and $\alpha \neq 0$. We write the constitutive law of the spring under the form:

$$f = -ku, \tag{11.6}$$

the constitutive law of the St-Venant element under the form:

$$f \in -\alpha\sigma(\dot{v}), \tag{11.7}$$

and the constitutive law of the dashpot under the form:

$$f = -c\dot{w}. \tag{11.8}$$

The fundamental theorem of dynamics and (11.6) give:

$$m\ddot{x} = F - ku. \tag{11.9}$$

We rewrite (11.5), (11.6), (11.7), (11.8) and (11.9) to understand better mathematical structure of the problem. Differentiating (11.5) with respect to time, we infer from (11.6), (11.7) and (11.8) the following relation:

$$ku \in \alpha\sigma(\dot{x} - \dot{u} - ku/c). \tag{11.10}$$

We can remark that (11.10) implies:

$$ku \in [-\alpha, \alpha]. \tag{11.11}$$

Let us define the graph β by (see Fig. 11.1b):

$$\beta(x) = \begin{cases} \emptyset & \text{if } x \in [-\infty, -1] \cup [1, +\infty], \\ \{0\} & \text{if } x \in [-1, 1], \\ \mathbb{R}_- & \text{if } x = -1, \\ \mathbb{R}_+ & \text{if } x = 1. \end{cases} \tag{11.12}$$

With this notation, the graph β is the inverse of σ and (11.10) is equivalent to

$$\dot{u} + \beta(ku/\alpha) \ni \dot{x} - ku/c. \tag{11.13}$$

The initial data $x(0)$, $u(0)$ and $\dot{x}(0)$ are given. Therefore, the system describing the mechanical setup of Fig. 11.2 is (11.9) together with (11.13) and the initial data:

$$x(0) = x_0, \quad \dot{x}(0) = \dot{x}_0, \quad u(0) = u_0. \tag{11.14}$$

Conversely, from the knowledge of (u, x) we can recover the functions (v, w) thanks to (11.5) and (11.8). By setting

$$\eta = \alpha/k \quad \text{and} \quad y = \dot{x}, \tag{11.15}$$

we can see that the system (11.9), (11.13) and (11.14) is equivalent to the system:

$$\begin{cases} \dot{x} = y, \\ \dot{y} = (F - ku)/m, \\ \dot{u} + \beta(u/\eta) \ni y - ku/c, \end{cases} \tag{11.16}$$

with the initial data:

$$x(0) = x_0, \quad y(0) = y_0, \quad u(0) = u_0 \in [-\eta, \eta]. \tag{11.17}$$

For the system (11.16) and (11.17), the limit case $c \longrightarrow +\infty$ corresponds to the Prandtl rheological model, governed by the Eq. (11.17) and

$$\begin{cases} \dot{x} = y, \\ \dot{y} = (F - ku)/m, \\ \dot{u} + \beta(u/\eta) \ni y. \end{cases} \tag{11.18}$$

This can be explained on physical grounds: if the viscosity of the dashpot is very large, it works as a solid and the model described in Fig. 11.2 is equivalent to the association of one spring and one St-Venant element (see Fig. 11.3).

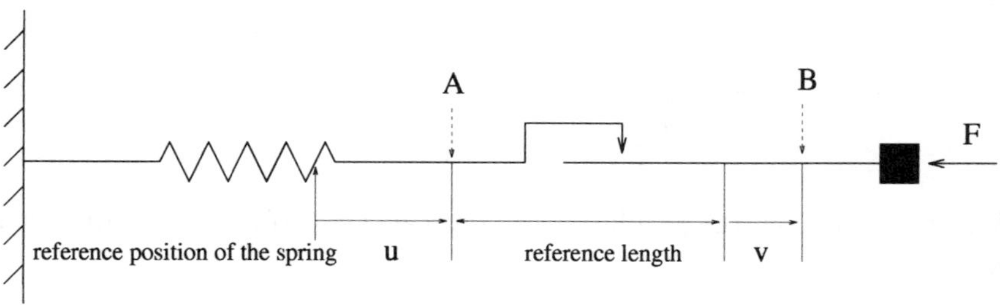

Fig. 11.3 The Prandtl rheological model.

The addition of dashpots does not change the mathematical theory; thus, we study henceforth only elastoplastic models, composed of springs and St-Venant elements.

11.1.1.2 *Other models*

Now let us consider models made out of a number of Hooke and St-Venant elements and material points. In general, the elements that we consider satisfy the following constitutive laws:

$$f_i = -k_i u_i, \tag{11.19}$$

for the i-th Hooke element with $k_i > 0$ and

$$g_i \in -\alpha_i \sigma(\dot{v}_i), \tag{11.20}$$

for the i-th St-Venant element, with $\alpha_i > 0$. Observe that in order to obtain interesting and new models, we cannot associate arbitrarily these components (see section 2.4.2.2), moreover, we show the equivalence between certain classes of models, and this reduces the number of cases to consider. Then, in section 11.1.1, we present all the models that we study in this chapter.

The Studied Models

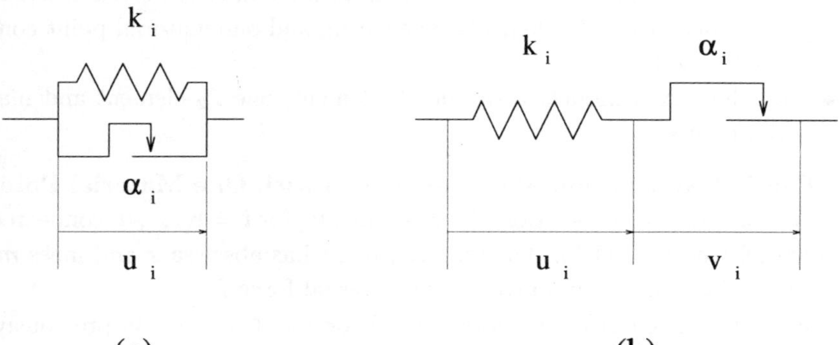

Fig. 11.4 P_i-element (a) and S_i-element (b).

The foregoing study shows that we may restrict ourselves to two kinds of elementary pairs: the P_i-elements (see Fig. 11.4a) are parallel associations of one spring of stiffness $k_i > 0$ and one St-Venant element of threshold $\alpha_i > 0$; the constitutive law of P_i is mathematically described by:

$$f_i = -k_i u_i \text{ and } g_i \in -\alpha_i \sigma(\dot{u}_i). \tag{11.21}$$

The S_i-elements (see Fig. 11.4b) are serial associations of one spring of stiffness $k_i > 0$ and one St-Venant element of threshold $\alpha_i > 0$; the constitutive law of S_i is mathematically described by:

$$f_i = -k_i u_i \in -\alpha_i \sigma(\dot{v}_i). \tag{11.22}$$

We assume that there is no secular term of deformation when the external force f_i is vanishing: according to the model described here, we have chosen a reference position of the material point, and a reference position of the St-Venant element such as f_i is vanishing when u_i is equal to zero.

We present now some models with P_i-elements, S_i-elements and several material points. According to the foregoing study of the examples in sections 2.4.2.2, we consider particular combinations of P_i-elements or S_i-elements and material points. We will give successively:

- The model with n P_i-elements connected in parallel with one material point,
- The model with n S_i-elements connected in parallel with one material point,
- The model with n S_i-elements and n material points connected in series,
- The model with n P_i-elements and n material points connected in series,
- The model with n P_i-elements, one spring and one material point connected in series,
- And the mixed models with one S_1-element, one P_2-element and material points.

Parallel Association of n $\mathbf{P_i}$-elements with One Material Point

We consider the association of n P_i-elements, for $i = 1, \ldots, n$, connected in parallel (see Fig. 11.5). The material point has abscissa x and mass m. This material point is submitted to an external force F.

We have the constitutive laws (11.21) for $i = 1, \ldots, n$. As previously, we have the differential system:

$$m\ddot{x} + \left(\sum_{i=1}^{n} k_i\right) x + \left(\sum_{i=1}^{n} \alpha_i\right) \sigma(\dot{x}) \ni F, \tag{11.23}$$

which is the inclusion (11.2). Thus, we have the equivalence between n P_i-elements connected in parallel and one P_i-element.

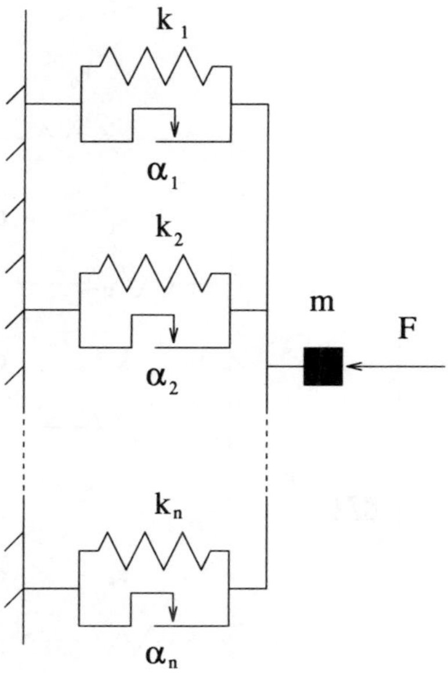

Fig. 11.5 Parallel association of n P_i-elements and one material point.

Parallel Association of n S_i-elements and One Material Point
This system is described in Fig. 11.6a; the material point has abscissa x and mass m.

This mechanical system is described by the differential inclusion:

$$\begin{cases} m\ddot{x} = -\sum_{i=1}^{n} k_i u_i + F, \\ \forall i \in \{1, ..., n\}, \quad \dot{u}_i + \beta(k_i u_i/\alpha_i) \ni \dot{x}. \end{cases} \tag{11.24}$$

Setting

$$y = \dot{x} \text{ and } \forall i \in \{1, ..., n\}, \quad \eta_i = \alpha_i/k_i, \tag{11.25}$$

we write Eq. (11.24) under the form:

$$\begin{cases} \dot{x} = y, \\ \dot{y} = F/m - (1/m)\sum_{i=1}^{n} k_i u_i, \\ \forall i \in \{1, ..., n\}, \quad \dot{u}_i + \beta(u_i/\eta_i) \ni y, \end{cases} \tag{11.26}$$

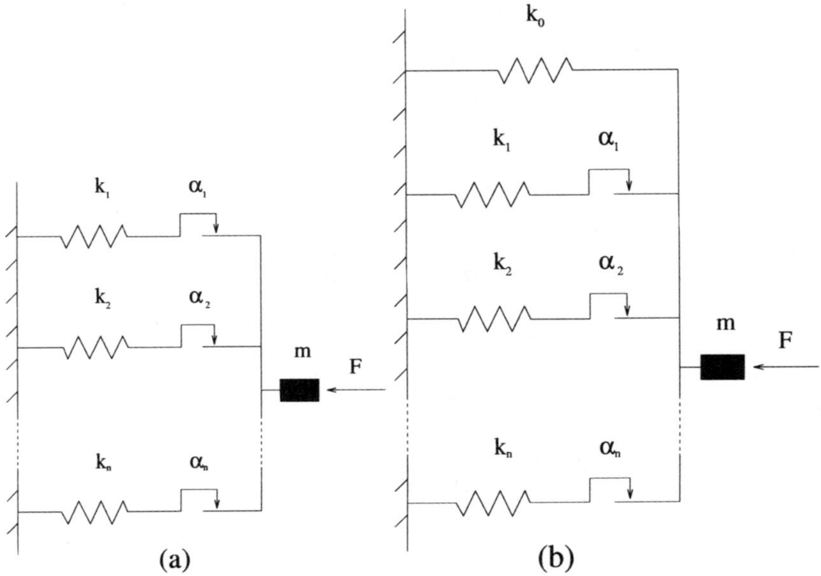

Fig. 11.6 The generalized Prandtl rheological model (a) and the generalized Prandtl rheological model with linear hardening (b).

the initial data are:

$$x(0) = x_0, \quad y(0) = y_0, \quad \forall i \in \{1, ..., n\}, \quad u_i(0) = u_{i,0} \in [-\eta_i, \eta_i]. \tag{11.27}$$

This system is the generalized Prandtl rheological model.

Setting one of the α_i equal to infinity allows to say that one of the S_i-elements is a pure spring, this hardens the model. We obtain the equation:

$$\begin{cases} \dot{x} = y, \\ \dot{y} = F/m - (k_0/m)x - (1/m)\displaystyle\sum_{i=1}^{n} k_i u_i, \\ \forall i \in \{1, \ldots, n\}, \quad \dot{u}_i + \beta(u_i/\eta_i) \ni y, \end{cases} \tag{11.28}$$

The relevant mechanical system is described in Fig. 11.6b (with n S_i-elements and one spring), it is a generalized Prandtl rheological model with linear hardening. This model is also called the discrete Masing model.

Association in Series of n S_i-elements and n Material Points

This system is described in Fig. 11.7; x_i, for $i = 1, \ldots, n$, denotes the abscissa of a material point of mass m_i.

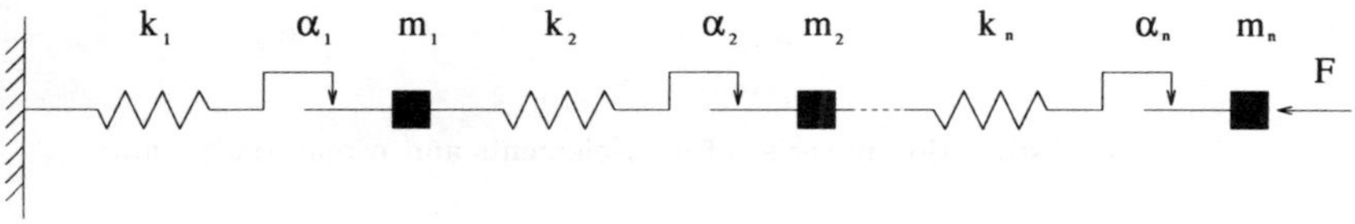

Fig. 11.7 Association in series of n S_i-elements and n material points.

By setting

$$\forall i \in \{1, \ldots, n\}, \quad y_i = \dot{x}_i \text{ and } \eta_i = \alpha_i / k_i, \tag{11.29}$$

we obtain the system

$$\left\{\begin{array}{l} \left\{\begin{array}{r} \dot{x}_1 = y_1, \\ \dot{y}_1 = (-k_1 u_1 + k_2 u_2)/m_1, \\ \dot{u}_1 + \beta(u_1/\eta_1) \ni y_1, \end{array}\right. \\ \forall i \in \{2, \ldots, n-1\}, \left\{\begin{array}{r} \dot{x}_i = y_i, \\ \dot{y}_i = (-k_i u_i + k_{i+1} u_{i+1})/m_i, \\ \dot{u}_i + \beta(u_i/\eta_i) \ni y_i - y_{i-1}, \end{array}\right. \\ \left\{\begin{array}{r} \dot{x}_n = y_n, \\ \dot{y}_n = (-k_n u_n + F)/m_n, \\ \dot{u}_n + \beta(u_n/\eta_n) \ni y_n - y_{n-1}, \end{array}\right. \end{array}\right. \tag{11.30}$$

with the initial data:

$$\forall i \in \{1, ..., n\} \quad x_i(0) = x_{i,0}, \quad y_i(0) = y_{i,0}, \quad u_i(0) = u_{i,0} \in [-\eta_i, \eta_i]. \tag{11.31}$$

Association in Series of m P_i-elements and n Material Points

This system is described in Fig. 11.8; where x_i, $i = 1, \ldots, n$, denotes the abscissa of the material point of mass m_i.

We have the system

$$\left\{\begin{array}{r} \dot{X} = Y, \\ \dot{Y} + D^{-1} A(Y) \ni D^{-1} H - D^{-1} K X, \end{array}\right. \tag{11.32}$$

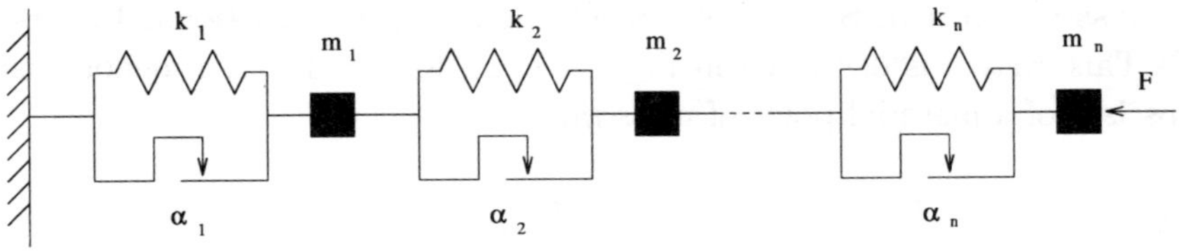

Fig. 11.8 Association in series of n P_i-elements and n material points.

with the initial data:

$$X(0) = X_0 \quad \text{and} \quad Y(0) = Y_0. \tag{11.33}$$

We have set:

$$X = (x_1, \ldots, x_n), \quad Y = (y_1, \ldots, y_n), \quad D = \text{diag}(m_1, \ldots, m_n), \tag{11.34}$$

$$H = \begin{pmatrix} 0 \\ \vdots \\ 0 \\ F \end{pmatrix}, \ K = \begin{pmatrix} k_1 + k_2 & -k_2 & 0 & \cdots & & 0 & 0 \\ -k_2 & k_2 + k_3 & -k_3 & \cdots & & 0 & 0 \\ & \ddots & \ddots & \ddots & & & \\ 0 & \cdots & -k_i & k_i + k_{i+1} & -k_{i+1} & \cdots & \\ & & & \ddots & \ddots & \ddots & \\ 0 & & \cdots & 0 & & -k_n & k_n \end{pmatrix}. \tag{11.35}$$

The maximal monotone operator A is defined on $\mathbb{R}^n$ by, for all $Y = (y_1, ..., y_n) \in \mathbb{R}^p$:

$$A(Y) = \Big\{ (g_i - g_{i+1})_{1 \le i \le n} : \quad g_1 \in \alpha_1 \sigma(y_1), \quad \forall i \in \{2, ..., n\}, \\ g_i \in \alpha_i \sigma(y_i - y_{i-1}), \quad g_{n+1} = 0 \Big\}. \tag{11.36}$$

Association in Series of n $\mathbf{P_i}$-elements, one Spring and One Material Point

The model described in Fig. 11.9 is a discrete Masing model constituted by the combination in series of n P_i elements and one spring. In the static setting, it is equivalent to the parallel combination of n P_i-elements and one spring, described in Fig. 11.6b (see [Fougeres and Sidoroff (1989)]). We shall not consider here the problem of the dynamical equivalence of these two kinds of models.

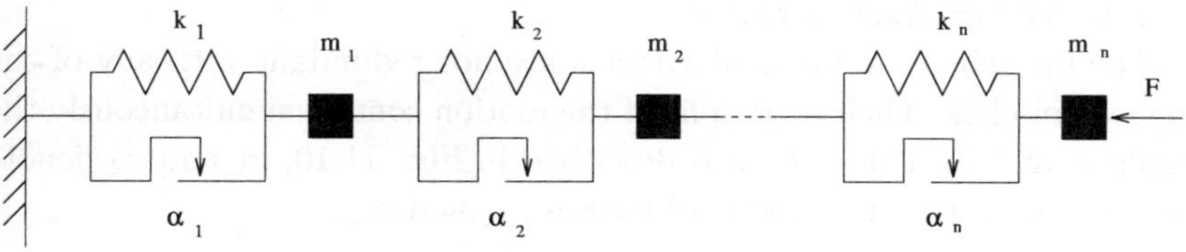

Fig. 11.9 Association in series of n P_i-elements, one spring and one material point.

We assume $k_0 > 0$. We denote by x the abscissa of the material point of mass m. By choosing appropriate auxiliary functions $v_1, \ldots, v_n$, we can prove that this mechanical system is governed by:

$$\begin{cases} \dot{x} = y, \\ \dot{y} = F/m + \gamma \left(\sum_{i=1}^{n} v_i/k_i - x \right), \\ \dot{V} + KB(V) \ni -k_0 y W, \end{cases} \tag{11.37}$$

with the initial data:

$$x(0) = x_0,\ y(0) = y_0,\ \forall i \in \{1, \ldots, n\},\ v_i(0) = v_{i,0} \in [-\alpha_i, \alpha_i]. \tag{11.38}$$

The number γ is defined by

$$\gamma = \left(m \sum_{i=0}^{n} \frac{1}{k_i} \right)^{-1}. \tag{11.39}$$

The vectors V and W are given by

$$V = (v_1, \ldots, v_n) \text{ and } W = (1, \ldots, 1). \tag{11.40}$$

The coefficients of the matrix K are:

$$\forall i, j \in \{1, \ldots, n\}, \quad K_{i,j} = \begin{cases} k_0 + k_i & \text{if } i = j, \\ k_0 & \text{if } i \neq j, \end{cases} \tag{11.41}$$

this matrix is symmetric, positive, definite. The maximal monotone operator B is defined on $\mathbb{R}^n$ by:

$$\forall X = (x_1, \ldots, x_n) \in \mathbb{R}^n, \quad B(X) = \beta(x_1/\alpha_1) \times \beta(x_2/\alpha_2) \times ... \times \beta(x_n/\alpha_n). \tag{11.42}$$

Two Other Associations

The following two kinds of systems are not redundant with any of the previous models. Their equations of the motion contain simultaneously the graphs σ and β. This system is described in Fig. 11.10, x_1 and x_2 denote the abscissa of material points of masses m_1 and m_2.

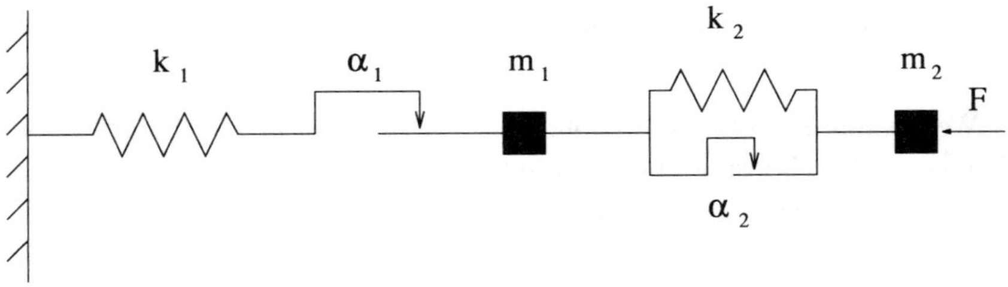

Fig. 11.10 One S_1-element, one P_2-element and two material points connected in series.

Let us set $y_1 = \dot{x}_1$, $y_2 = \dot{x}_2$ and $\eta_1 = \alpha_1/k_1$, then we obtain:

$$\left\{\begin{array}{c} \dot{x}_1 = y_1, \\ \dot{y}_1 - (1/m_1)\, g_2 = (-z_1 + k_2(x_2 - x_1))/m_1, \\ \dot{x}_2 = y_2, \\ \dot{y}_2 + (1/m_2)\, g_2 = (F - k_2(x_2 - x_1))/m_2, \\ g_2 \in \alpha_2 \sigma(y_2 - y_1), \\ \dot{u}_1 + \beta(u_1/\eta_1) \ni y_1, \end{array}\right. \tag{11.43}$$

with the initial data

$$u_1(0) = u_{1,0} \in [-\eta_1, \eta_1] \text{ and for } i = 1, 2 \quad x_i(0) = x_{i,0}, \quad y_i(0) = y_{i,0}. \tag{11.44}$$

When we choose the S_1-element and the P_2-element connected in parallel, then we obtain the system (with $\eta_1 = \alpha_1/k_1$):

$$\left\{\begin{array}{c} \dot{x} = y, \\ \dot{y} + (\alpha_2/m)\sigma(y) \ni (F - z - x)/m, \\ \dot{u} + \beta(u/\eta_1) \ni y, \end{array}\right. \tag{11.45}$$

with the initial data

$$x(0) = x_0, \quad y(0) = y_0 \text{ and } z(0) = z_0 \in [-\eta_1, \eta_1], \tag{11.46}$$

where x denotes the abscissa of the material point of mass m.

11.1.2 *Existence and uniqueness results*

We observe that all the models introduced in section 11.1.1.1, contrary to the one in Fig. 2.1, can be subsummed under one form which is conveniently described in the language of maximal monotone operators. In this section here, we recall some properties of maximal monotone operators. Then, we give existence and uniqueness results for these differential inclusions.

11.1.2.1 *Maximal monotone graphs σ and β*

The reader is referred to [Brezis (1973)] and chapter 2. Let $<,>$ be a scalar product on $\mathbb{R}^p$. If ϕ is a convex proper and lower semi-continuous function from $\mathbb{R}^p$ to $[-\infty,+\infty]$, we can define its subdifferential $\partial\phi$ by:

$$\begin{cases} y \in \partial\phi(x) \iff \forall h \in \mathbb{R}^p, \quad \phi(x+h)-\phi(x) \geq < y, h >, \\ \qquad\qquad D(\partial\phi) = \{x : \partial\phi(x) \neq \emptyset\}, \end{cases} \tag{11.47}$$

moreover, $\partial\phi$ is a maximal monotone graph in $\mathbb{R}^p \times \mathbb{R}^p$. The maximal monotone graphs σ and β of section 11.1.1 are subdifferentials of proper semi-continuous convex functions $|x|$ and $\psi_{[-1,1]}$ defined by:

$$\forall x \in \mathbb{R}, \quad \psi_{[-1,1]}(x) = \begin{cases} 0 & \text{if } x \in [-1,1], \\ +\infty & \text{if } x \notin [-1,1], \end{cases} \tag{11.48}$$

and the choice of the canonical scalar product in $\mathbb{R}$. Therefore:

$$\forall x \in \mathbb{R}, \quad \sigma(x) = \partial|x| \text{ and } \beta(x) = \partial\psi_{[-1,1]}(x). \tag{11.49}$$

We observe that if $\mathbb{R}^p$ is equipped with its canonical scalar product, and with another scalar product

$$< x, y >_M = x^T M^{-1} y, \tag{11.50}$$

where M is symmetric positive definite, then we can relate the subdifferential $\partial\phi$ of ϕ relatively to the canonical scalar product and the subdifferential $\partial_M\phi$ relatively to $<,>_M$ by

$$\partial_M\phi(x) = M\partial\phi(x). \tag{11.51}$$

11.1.2.2 *Mathematical study of differential systems*

We give now the general mathematical formulation of our problem. We assume that T is strictly positive and that G is a function from $[0,T] \times \mathbb{R}^p$

to $\mathbb{R}^p$ which is Lipschitz-continuous with respect to its second argument, i.e. there exists $\omega \geq 0$ such that:

$$\forall t \in [0,T], \quad \forall X_1, X_2 \in \mathbb{R}^p, \quad \|G(t,X_1) - G(t,X_2)\| \leq \omega \|X_1 - X_2\|. \tag{11.52}$$

Moreover, we assume that:

$$\forall Y \in \mathbb{R}^p, \quad G(\cdot, Y) \in L^\infty(0,T,\mathbb{R}^p). \tag{11.53}$$

The matrix M is symmetric positive definite, and ϕ is convex proper and lower semi-continuous on $\mathbb{R}^p$.

Proposition 11.1 *Under the previous assumptions, for all $\xi \in D(\partial\phi)$, there exists a unique function X in $W^{1,1}(0,T,\mathbb{R}^p)$ such that:*

$$\begin{cases} \dot{X}(t) + M\partial\phi(X(t)) \ni G(t, X(t)) \quad \text{a.e. on } [0,T], \\ X(0) = \xi. \end{cases} \tag{11.54}$$

Proof *We observed above that $\partial_M\phi(x) = M\partial\phi(x)$, therefore proposition 3.13 page 107 of [Brezis (1973)] applies.*

Thus, all the systems of section 11.1.1 can be written under the form (11.54) and have a unique solution. For all systems, Table 11.1 provides the corresponding integer p, function ϕ and matrix M. It is easy to prove that ϕ are convex proper and lower semi-continuous functions on $\mathbb{R}^p$.

With this table, we can observe that there are three classes of mechanical systems:

- In the first class, the function ϕ is a linear combinations of $|x_i|$,
- In the second class, the function ϕ is a sum of functions $\psi_{[-\eta_i,\eta_i]}$,
- In the third class, ϕ involves the both functions $|x|$ and $\psi_{[-\eta_i,\eta_i]}$.

11.1.3 *Numerical simulations*

In this section, we present some numerical simulations for the Prandtl model, for the Prandtl model with linear hardening and for the generalized Prandtl model with linear hardening in order to show that they can describe materials with elastoplastic constitutive laws. In this section, we give several simulations of viscoelastoplastic model. All the obtained curves are presented successively. We have also simulated numerically other models, governed by the systems (11.2) and (11.3), (11.30) and (11.31), (11.45) and (11.46), the numerical methods are similar to those used for the Prandtl models and we obtain hysteresis cycles, but we were not able to infer from them relevant information.

Table 11.1 The dimension of the system, the convex function and the invertible matrix used for the above described mechanical models.

System	p	function ϕ	matrix M
(11.2)-(11.3)	2	$\phi(x,y)=\alpha\|y\|$	$\mathrm{diag}(1,\frac{1}{m})$
(11.16)-(11.17), (11.17)-(11.18), (2.79)-(2.80)	3	$\phi(x,y,u)=\psi_{[-\eta,\eta]}(u)$	I
(11.26)-(11.27), (11.27)-(11.28)	$n+2$	$\phi(x,y,u_1,\ldots,u_n)= =\sum_{i=1}^{n}\psi_{[-\eta_i,\eta_i]}(u_i)$	I
(11.37)-(11.38)	$n+2$	$\phi(x,y,z_1,\ldots,z_n)= =\sum_{i=1}^{n}\psi_{[-\alpha_i,\alpha_i]}(z_i)$	$\begin{pmatrix}1&0&0\\0&1&0\\0&0&K\end{pmatrix}$ *K not diagonal*
(11.30)-(11.31)	3n	$\phi(x_1,y_1,u_1,\ldots,x_n,y_n,u_n)= =\sum_{i=1}^{n}\psi_{[-\eta_i,\eta_i]}(u_i)$	I
(11.32)-(11.33)	$2n$	$\phi(x_1,\ldots,x_n,y_1,\ldots,y_n)= =\alpha_1\|y_1\|+\sum_{i=2}^{n}\alpha_i\|y_i-y_{i-1}\|$	$\begin{pmatrix}I&0\\0&D^{-1}\end{pmatrix}$ *D diagonal*
(11.43)-(11.44)	5	$\phi(x_1,y_1,x_2,y_2,u_1)= =\alpha_2\|y_2-y_1\|+\psi_{[-\eta_1,\eta_1]}(u_1)$	$\mathrm{diag}(1,\frac{1}{m_1},1,\frac{1}{m_2},1)$
(11.45)-(11.46)	3	$\phi(x,y,u)= =\alpha_2\|y\|+\psi_{[-\eta_1,\eta_1]}(u)$	$\mathrm{diag}(1,\frac{1}{m_1},1)$

For all these numerical simulations, we solve the differential inclusion (11.54) by using the implicit Euler scheme: let $n \in \mathbb{N}^*$, let $h = T/n$ and for $q \in \{0, \dots, n\}$, let $t_q = qh$. We solve

$$\begin{cases} \forall q \in \{0,\dots,n-1\}, \quad \dfrac{X_{q+1} - X_q}{h} + M\partial\psi(X_{q+1}) \ni G(t_q, X_q), \\ \\ X_0 = \xi. \end{cases} \tag{11.55}$$

The first equality of (11.55) is equivalent to:

$$\forall q \in \{0,\dots,n-1\}, \quad X_{q+1} = (I + hM\partial\psi)^{-1} \left(hG(t_q, X_q) + X_q\right). \tag{11.56}$$

We denote by X^h the linear interpolation of the X_n. The function X^h converges to the solution X of the system (11.54) in $C^0([0,T], \mathbb{R}^p)$. This result is proved in [Crandall and Evans (1975)] which contains a much more general result.

11.1.3.1 *Study of the rheological Prandtl models*

In this section, we study the rheological Prandtl models, which have been seen in section 11.1.1, i.e., the Prandtl model (see the system (11.17) and (11.18)), the Prandl model with linear hardening (see the system (2.79) and (2.80)) and the generalized Prandl model with linear hardening (see the system (11.27) and (11.28)). We choose

$$m = 1. \tag{11.57}$$

We will study a harmonic forcing (section 11.1.3) and other periodic forcings (section 11.1.3).

Harmonic Forcing
For this section, we choose

$$F(t) = f\cos(\omega t). \tag{11.58}$$

We first present essential difference between the Prandtl model and the Prandtl model with linear hardening: we have computed the function x, y and z for the Prandtl model on the interval $[0, T]$ with the following parameters:

$$T = 2000, \quad F(t) = 20\cos(0.1t), \quad \eta = k = 1 \;\; x_0 = y_0 = u_0 = 0, \tag{11.59}$$

and for the Prandtl model with linear hardening on the interval $[0, T]$ with the following parameters:

$$T = 300, \quad F(t) = 20\cos(0.1t), \quad k_0 = 1, \quad \eta = k = 1 \quad x_0 = y_0 = u_0 = 0. \tag{11.60}$$

These functions are presented in the Fig. 11.11 and in the Fig. 11.12.

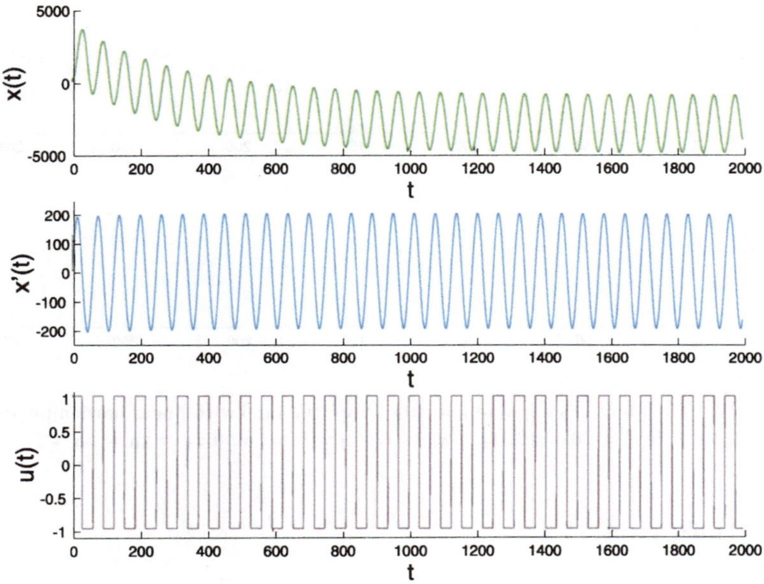

Fig. 11.11 The functions x, y and u for the Prandtl model defined by $T = 2000$, $F(t) = 20\cos(0.1t), \eta = k = 1$ and $x_0 = y_0 = u_0 = 0$.

We can see that the amplitude of the functions x and $\dot{x}$ is larger for the Prandtl model, which corresponds to the system (2.79) with $k_0 = 0$ than for Prandtl model with hardening, which corresponds to the same system with $k_0 \neq 0$. We sketch a physical explanation: if the St-Venant element slips, then we have from (2.79)

$$u \equiv \varepsilon\eta, \tag{11.61}$$

and

$$\ddot{x}(t) + \frac{k_0}{m}x(t) = \frac{f}{m}\cos(\omega t) - \frac{\varepsilon k \eta}{m}, \tag{11.62}$$

where $\varepsilon \in \{-1, 1\}$. If we assume $k_0 = 0$, and if the slip phase starts at t_0,

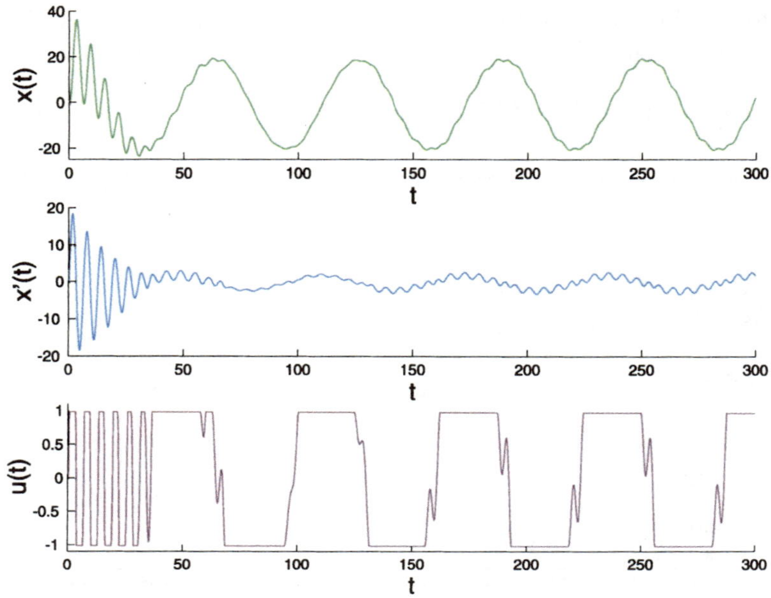

Fig. 11.12 The functions x, y and u for the Prandtl model with linear hardening, defined by $T = 300$, $F(t) = 20\cos(0.1t)$, $k_0 = 1$, $\eta = k = 1$ and $x_0 = y_0 = u_0 = 0$.

then (11.62) gives:

$$\begin{aligned} x(t) =& \frac{2f}{m\omega^2}\sin(\omega(t+t_0))\sin(\omega(t-t_0)) - \frac{\varepsilon k\eta}{2m}(t-t_0)^2 + \\ &+ \left(\dot{x}(t_0) - \frac{f}{m\omega}\sin(\omega t_0)\right)(t-t_0) + x(t_0). \end{aligned} \tag{11.63}$$

For $\omega \ll 1$ and $t \approx t_0$, we have:

$$x(t) \approx \frac{2f}{m\omega^2}\sin(\omega(t-t_0)), \tag{11.64}$$

thus, the approximate amplitude of x is roughly

$$\frac{2f}{m\omega^2} = 4000, \tag{11.65}$$

as can be seen in the Fig. 11.11. If $k_0 = 1$ and if the slip phase starts at t_0, then (11.62) implies:

$$x(t) = D\cos(\omega_0 t + \phi) + \frac{f}{m(\omega_0^2 - \omega^2)}\cos(\omega t) - \frac{\varepsilon k\eta}{m\omega_0^2}, \tag{11.66}$$

where $\omega_0 = \sqrt{k_0/m} \neq \omega$ and (D, ϕ) depends of $(x(t_0), \dot{x}(t_0))$. Here, with $\omega = 0.1$, the amplitude of the sinusoidal component of period ω of the function x is roughly

$$\frac{f}{m(\omega_0^2 - \omega^2)} \approx 20, \tag{11.67}$$

as can be seen in Fig. 11.12.

We will study rheological properties of the Prandtl models. In all cases, the material point is submitted to the external force F and has an abscissa x. The choice of representation is crucial, we have plotted in Fig. 11.13a the curve $\{x(t), F(t)\}_{t\in[3500,8000]}$ for the Prandtl model defined by the following parameters:

$$F(t) = 20\cos(0.1t), \quad \eta = k = 1, \quad \text{and} \quad x_0 = y_0 = u_0 = 0. \tag{11.68}$$

We observe a limit cycle, but it gives no information on the physically relevant parameters of the system; if however we plot as in Fig. 11.13b the curve $\{x(t), F(t) - m\ddot{x}(t)\}_{t\in[3500,8000]}$, we observe that the slope of the oblique parts of the cycle is equal to k.

In Fig. 11.14, we can observe transients ($t \in [0, 1500]$) followed by a periodic regime ($t \geq 1500$), composed of periodic hysteresis cycles. We plot only the periodic regime. In Fig. 11.14, we can see that $x(t)$ is vanishing when $F(t) - m\ddot{x}(t)$ is equal to zero (for $t = 0$), according to the choice of the section 11.1.1.2. Nevertheless, in Fig. 11.15, we can see that the cycle does not contain the point $(0, 0)$ in the plane $(x, F - m\ddot{x})$: we have not plotted transients in this figure.

In Table 11.2, we define the parameters of the simulations for the generalized Prandtl model with linear hardening, which includes as particular cases the Prandtl model and the Prandtl model with hardening: the corresponding system of Eqs. is (11.27) and (11.28).

In Figs. 11.15a, 11.15b, 11.16a, 11.16b, 11.17a, 11.17b and 11.18, we observe hysteresis cycles in the $(x, F - m\ddot{x})$ plane. In the Figs. 11.15a and 11.15b, we can see identical hysteresis cycles up to translation for different initial data. In the Fig. 11.16a ($n = 5$ and $f = 15$), we can see that the hysteresis cycle is composed of 12 line segments. But, in Fig. 11.16b ($n = 5$ and $f = 10$), we can see only 6 line segments. In Figs. 11.17a, 11.17b and 11.18a (large values of n), we can distinguish line segments when magnifying the figure (see Figs. 11.18a and 11.18b).

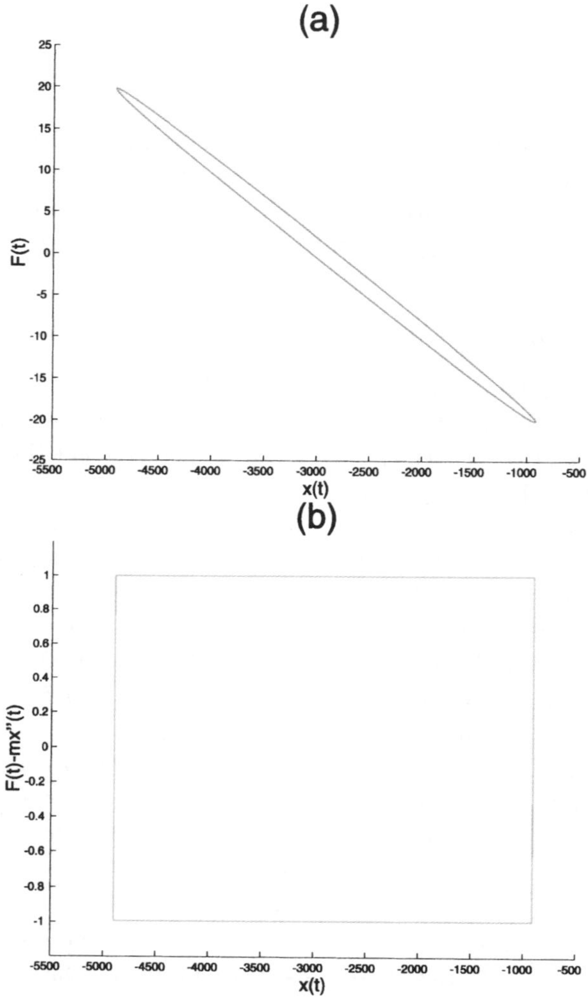

Fig. 11.13 The curves $\{x(t), F(t)\}_{t\in[3500,8000]}$ (a) and $\{x(t), F(t) - m\ddot{x}(t)\}_{t\in[3500,8000]}$ (b) for the Prandtl model defined by $F(t) = 20\cos(0.1t)$, $\eta = k = 1$, and $x_0 = y_0 = u_0 = 0$. These two figures differ only by the choice of coordinates.

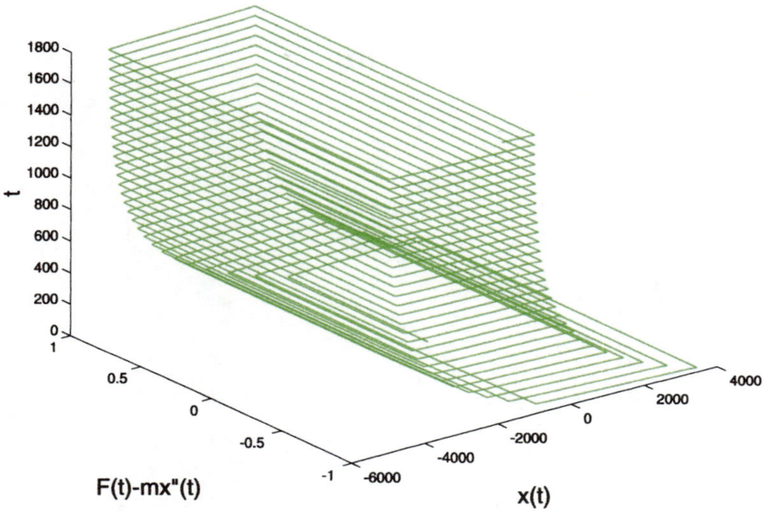

Fig. 11.14 The curve $\{x(t), F(t) - m\ddot{x}(t), t\}_{t\in[0,1800]}$ for the Prandtl model, defined by $F(t) = 20\cos(0.1t)$, $\eta = k = 1$ and $x_0 = y_0 = u_0 = 0$.

Table 11.2 The parameters of the numerical simulations of equations $\dot{x} = y$, $\dot{y} = F/m - (k_0/m)x - (1/m)\sum_{i=1}^{n} k_i u_i$, $\forall i \in \{1, \ldots, n\}$, $\dot{u}_i + \beta(u_i/\eta_i) \ni y$, with the initial conditions $x(0) = x_0$, $y(0) = y_0$, $\forall i \in \{1, ..., n\}$, $u_i(0) = u_{i,0} \in [-\eta_i, \eta_i]$ with forcing $F(t) = f\cos(\omega t)$.

N^o of Fig.	f	ω	a	b	n	k_0	x_0	y_0	η_i	k_i	$u_{0,i}$
11.15a	1	0.5	50	2000	1	0	0	0	1	1	0
11.15b	1	0.5	50	2000	1	0	100	-200	1	1	0.99
11.16a	15	0.5	400	3000	5	0	0	0	i	1	0
11.16b	10	0.5	400	3000	5	0	0	0	i	1	0
11.17a	12	0.5	500	4000	30	0	0	0	i	$1/n$	0
11.17b	12	0.5	500	4000	30	$1/n$	0	0	i	$1/n$	0
11.18	55	0.5	500	4000	100	0	0	0	i	$1/n$	0

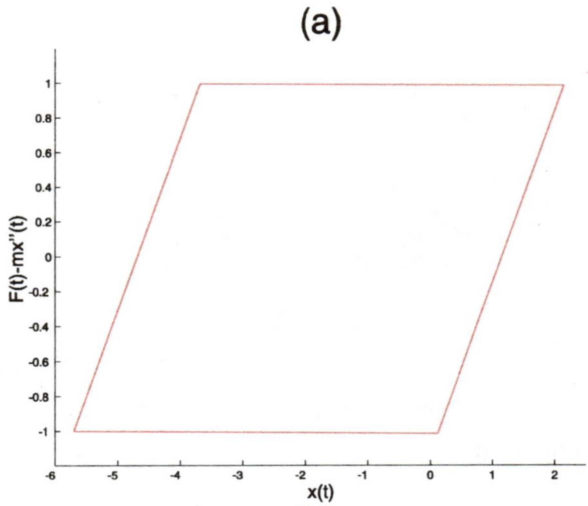

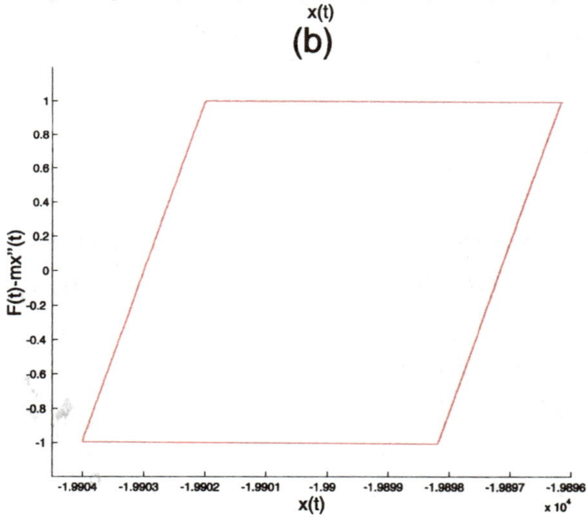

Fig. 11.15 The curve $\{x(t), F(t) - m\ddot{x}(t)\}_{t\in[50,2000]}$ for the generalized Prandtl rheological model with linear hardening, defined by $F(t) = \cos(0.5t)$, $n = 1$, $k_0 = 0$, $\eta_1 = k_1 = 1$ and $(x_0 = y_0 = u_{0,1} = 0)$ (a), $(x_0 = 100,\ y_0 = -200,\ u_{0,1} = 0.99)$ (b). These two figures differ only by the initial conditions.

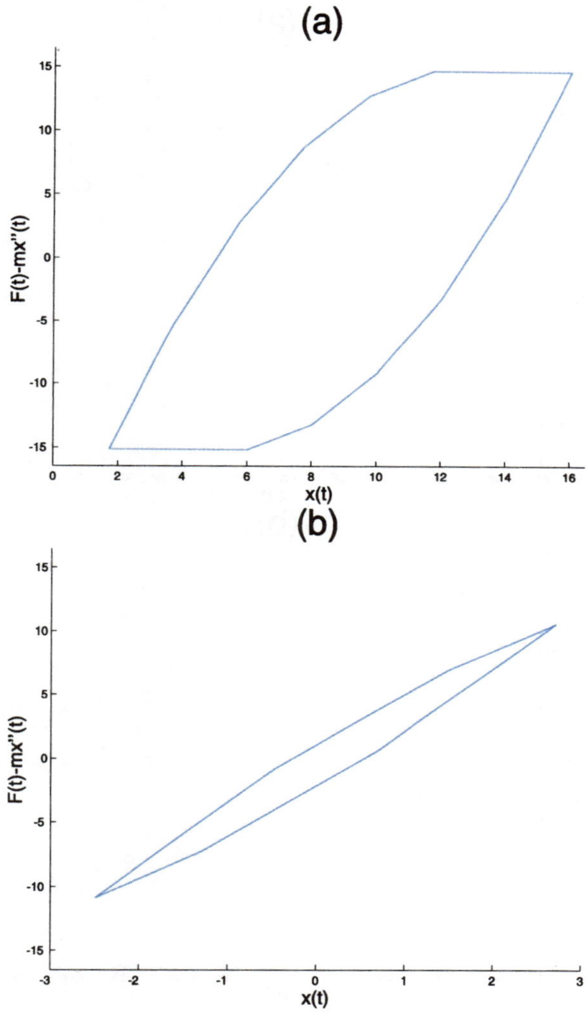

Fig. 11.16 The curve $\{x(t), F(t) - m\ddot{x}(t)\}_{t\in[400,3000]}$ for the generalized Prandtl rheological model with linear hardening, defined by $F(t) = f\cos(0.5t)$, $k_0 = 0$, $n = 5$, $x_0 = y_0 = 0$, $\forall i \in \{1,\ldots,5\}$, $k_i = 1$, $\eta_i = i$, $u_{0,i} = 0$ and $f = 15$ (a) and $f = 10$ (b). These two figures differ only by the amplitude of the forcing.

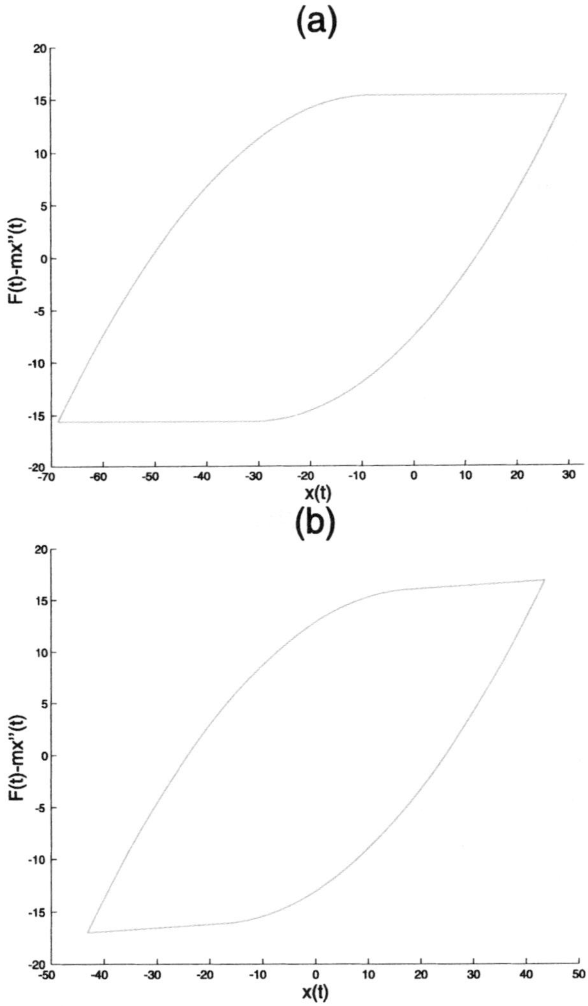

Fig. 11.17 The curve $\{x(t), F(t) - m\ddot{x}(t)\}_{t\in[500,4000]}$ for the generalized Prandtl rheological model with linear hardening, defined by $F(t) = 12\cos(0.5t)$, $n = 30$, $x_0 = y_0 = 0$, and $\forall i \in \{1,\ldots,30\}$, $k_i = 1/30$, $\eta_i = i$, $u_{0,i} = 0$ and $k_0 = 0$ (a) $k_0 = 1/30$ (b). These two figures differ only by the value of k_0.

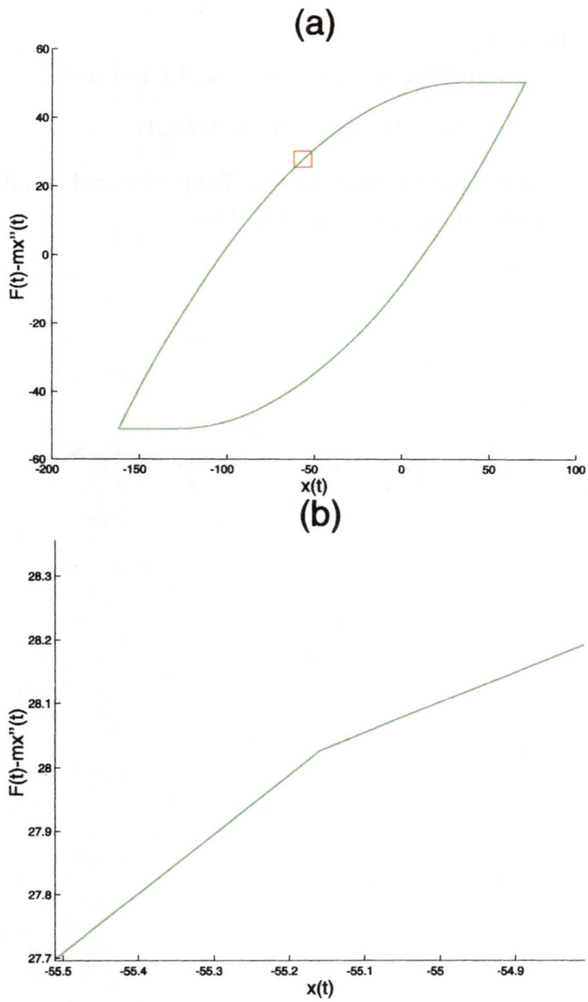

Fig. 11.18 The curve $\{x(t), F(t) - m\ddot{x}(t)\}_{t\in[500,4000]}$ for the generalized Prandtl rheological model with linear hardening, defined by $F(t) = 55\cos(0.5t)$, $n = 100$, $k_0 = 0$, $x_0 = y_0 = 0$, and $\forall i \in \{1, \dots, 100\}$, $k_i = 1/100$, $\eta_i = i$, $u_{0,i} = 0$. Fig. (b) is magnifying and dilating square of Fig. (a).

We have also made simulations with other values of parameters: we obtain hysteresis limit cycles; these cycles are periodic but they are not convex.

Periodic Forcing

We choose now a forcing of pulsation ω and amplitude f:

$$\forall t \in [0,T], \quad F(t) = fH(\omega t), \tag{11.69}$$

where H is 1-periodic and of amplitude 1. We performed simulations using the functions H_0 and H_1 graphed in Fig. 11.19.

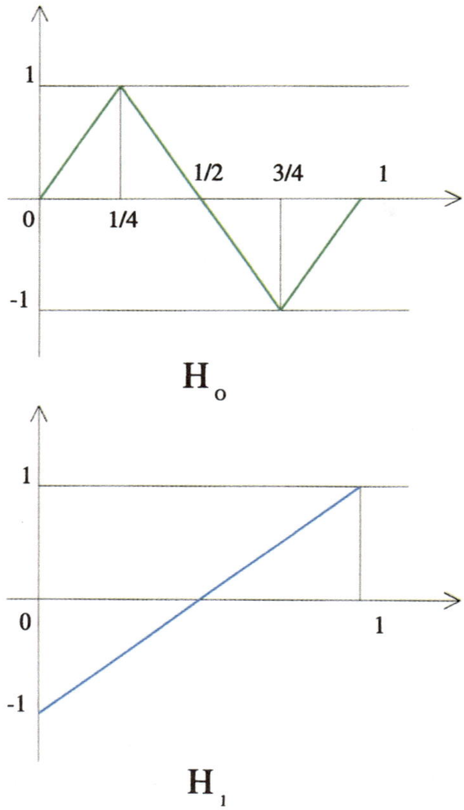

Fig. 11.19 The functions H_0 and H_1.

These simulations are similar to the simulations of section 11.1.3.1. The parameters are defined in Table 11.3.

Table 11.3 The parameters of the numerical simulations of equations $\dot{x} = y$, $\dot{y} = F/m - (k_0/m)x - (1/m)\sum_{i=1}^{n} k_i u_i$, $\forall i \in \{1,...,n\}$, $\dot{u}_i + \beta(u_i/\eta_i) \ni y$, with the initial conditions $x(0) = x_0$, $y(0) = y_0$, $\forall i \in \{1,...,n\}$, $u_i(0) = u_{i,0} \in [-\eta_i, \eta_i]$ with forcing $F(t) = fH(\omega t)$.

N^o of Fig.	H	f	ω	a	b	n	k_0	x_0	y_0	η_i	k_i	$u_{0,i}$
11.20	H_0	15	0.5	5	2000	1	0	0	0	1	1	0
11.21	H_1	80	0.5	400	500	1	0	0	0	1	1	0
11.22	H_1	80	10	400	500	1	0	0	0	1	1	0
11.23a	H_1	80	0.5	400	500	1	1	0	0	1	1	0
11.23b	H_1	80	10	400	500	1	1	0	0	1	1	0
11.24b	H_1	800	0.5	960	2000	30	$1/n$	0	0	i	$1/n$	0

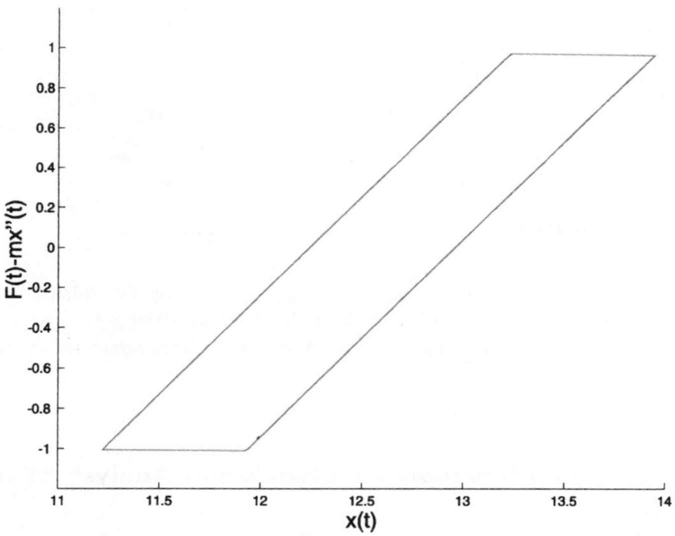

Fig. 11.20 The curve $\{x(t), F(t) - m\ddot{x}(t)\}_{t\in[50,2000]}$ for the generalized Prandtl rheological model with linear hardening, defined by $n = 1, F(t) = 15H_0(0.5t)$, $k_0 = 0, x_0 = y_0 = u_{0,1}0$, $k_1 = \eta_1 = 1$. This curve can be compared to the curve of Fig. 11.15(a) which corresponds to the same values of parameters and to a harmonic forcing.

In Fig. 11.20, we find again the same shape of hysteresis cycle as in Fig. 11.15a. Thus, we can conclude that we obtain hysteresis cycles with a continuous and periodic forcing. On the contrary, we do not obtain necessarily limit periodic hysteresis cycles with a discontinuous forcing, in Fig. 11.21, we observe a resonance, which amplitude decreases when the pulsation increases (see Fig. 11.22). For the static models this resonance phenomenon is known as the ratchet phenomenon. When we choose $k_0 \neq 0$, we obtain limit periodic hysteresis cycles (see Figs. 11.23a, 11.23b and 11.24). In Fig. 11.24, we obtain periodic hysteresis cycles similar to the hysteresis cycle of Fig. 11.17b.

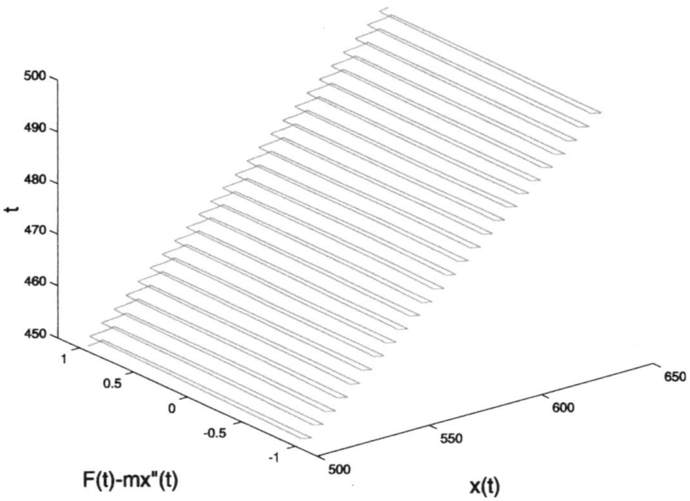

Fig. 11.21 The curve $\{x(t), F(t) - m\ddot{x}(t), t\}_{t\in[450,500]}$ for the generalized Prandtl rheological model with linear hardening, defined by $F(t) = 80H_1(\omega t)$, $n = 1$, $k_0 = 0$, $\eta_1 = k_1 = 1$, $x_0 = y_0 = u_{0,1} = 0$, and $\omega = 0.5$. This curve shows the ratchet phenomenon.

Conclusions on Elastoplastic Models and Analysis of Hysteresis Cycles

Most of the responses of the generalized Prandtl model to a sinusoidal forcing (see Figs. 11.15a, 11.15b, 11.16a, 11.16b, 11.17a, 11.17b or 11.18) or to a periodic (but non harmonic) forcing (see Figs. 11.20, 11.23a, 11.23b and 11.24), tend to hysteresis limit cycles, which we are going to study.

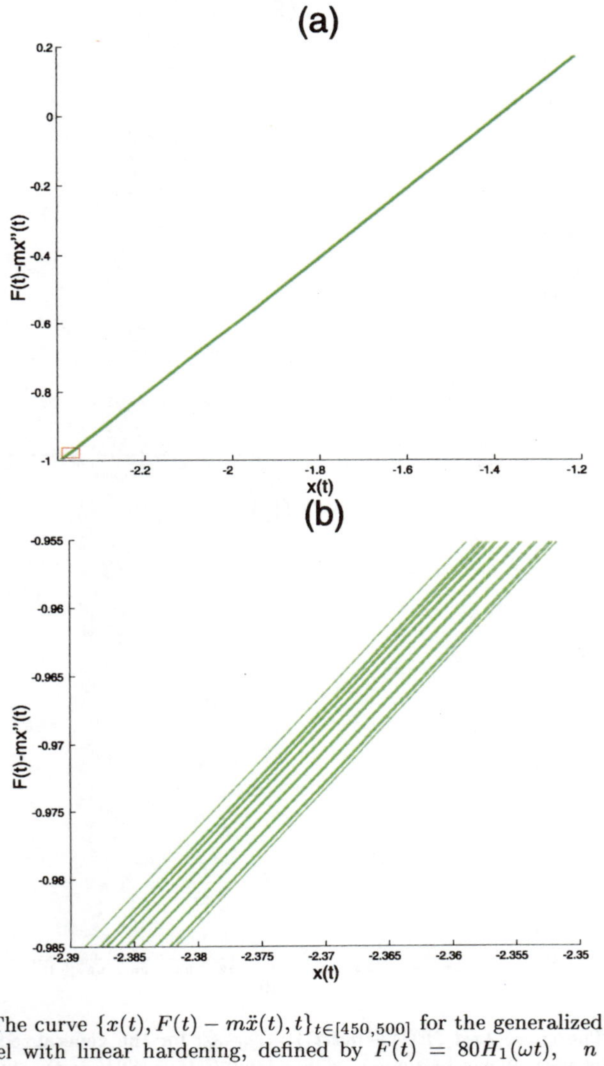

Fig. 11.22 The curve $\{x(t), F(t) - m\ddot{x}(t), t\}_{t\in[450,500]}$ for the generalized Prandtl rheological model with linear hardening, defined by $F(t) = 80H_1(\omega t)$, $n = 1$, $k_0 = 0$, $\eta_1 = k_1 = 1$, $x_0 = y_0 = u_{0,1} = 0$, and $\omega = 10$. Fig. (b) is magnifying square of Fig. (a).

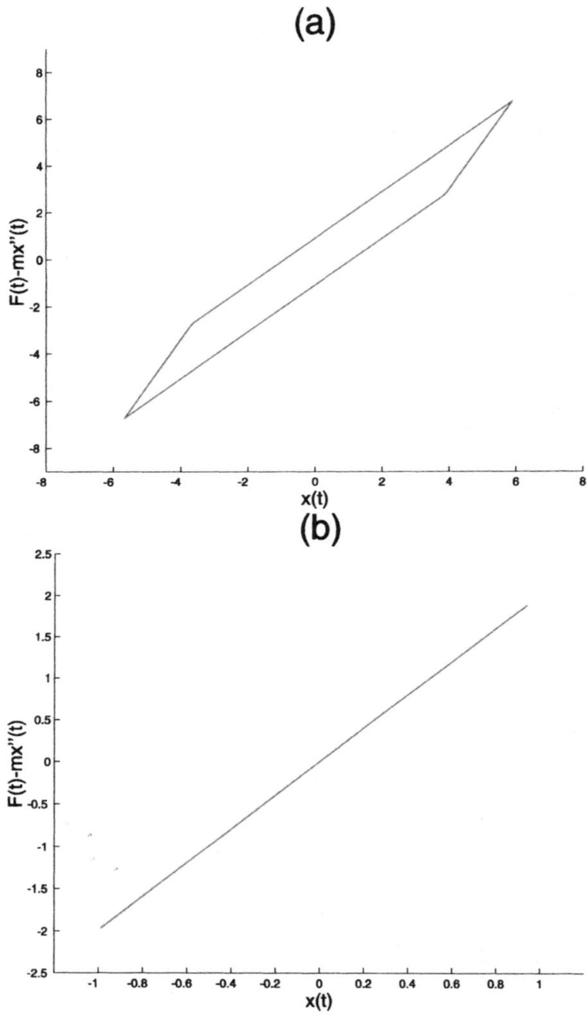

Fig. 11.23 The curve $\{x(t), F(t) - m\ddot{x}(t)\}_{t\in[450,500]}$ for the generalized Prandtl rheological model with linear hardening, defined by $F(t) = 80H_1(\omega t)$, $n = 1$, $k_0 = 1$, $\eta_1 = k_1 = 1$, $x_0 = y_0 = u_{0,1} = 0$, and $\omega = 0.5$ (a) and $\omega = 10$ (b). These two figures differ only by the pulsation of the forcing.

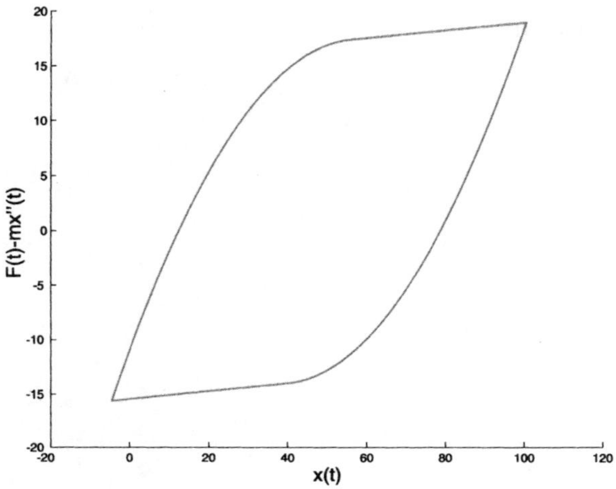

Fig. 11.24 The curve $\{x(t), F(t) - m\ddot{x}(t)\}_{t\in[960,1000]}$ for the generalized Prandtl rheological model with linear hardening, defined by $F(t) = 800H_1(0.5t)$, $n = 30$, $k_0 = 1/n$, $x_0 = y_0 = 0$, and $\forall i \in \{1, ..., 30\}$, $k_i = 1/30$, $\eta_i = i$, $u_{0,i} = 0$.

We observe that these hysteresis cycles have a center of symmetry. Therefore, we study a loading phase corresponding to a halfcycle in the $(x, F - m\ddot{x})$ plan (see Fig. 11.25). Let us denote by A_1, A_2,...,A_{n+1} and A_{n+2} the ends of segments which constitute the hysteresis half-cycle. For $i \in \{1, \dots, n\}$, d_i denotes the difference between abscissa of A_{i+1} and A_i (positive real). For $i \in \{1, \dots, n+1\}$, p_i denotes the slope of the segment $[A_i, A_{i+1}]$.

If the forcing F is not constant on any open not empty subinterval of $[0, T]$, then x has the same property, assume indeed that x is constant on (t_1, t_2), then $\dot{x}$ and $\ddot{x}$ vanish on (t_1, t_2) and by uniqueness of the solution of

$$\begin{cases} \dot{u}_i + \beta(u_i/\eta_i) \ni 0, \\ \quad u_i(t_1) \text{ given}, \end{cases} \tag{11.70}$$

$u_i(t)$ is equal to $u_i(t_1)$ on (t_1, t_2). This implies that the expression

$$F = m\ddot{x} + k_0 x + \sum_{i=1}^{n} k_i u_i, \tag{11.71}$$

is constant over (t_1, t_2), which contradicts our assumption.

We are now able to describe the shape of the representation of a trajectory of the system in the x and $F - m\ddot{x}$ coordinates, in some special case.

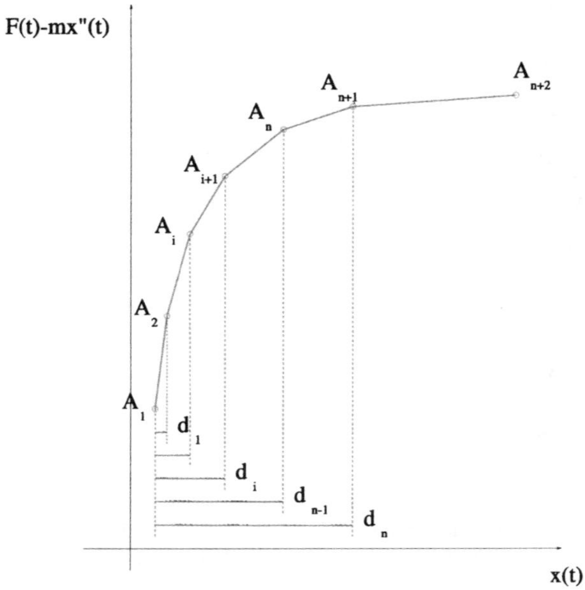

Fig. 11.25 Loading curve for the generalized Prandtl rheological model with linear hardening.

We assume that the $\eta_i = \alpha_i/k_i$ are all distinct, otherwise two elements with identical ratio η_i would enter the plasticity phase simultaneously. We reorder the indices so that:

$$\eta_1 < \eta_2 < ... < \eta_{n-1} < \eta_n. \tag{11.72}$$

Assume that F vanishes on not open, not empty subinterval of $[0, T]$, and the following properties:

1) $u_j(0) = -\eta_j, \quad \forall j = 1, \ldots, n,$
2) x is increasing on $[0, T]$.

Then there exists a increasing sequence $t_1 \leq t_2 \leq \ldots \leq t_n$ such that for some $j \in \{1, \ldots, n\}$:

$$t_1 < t_2 < \ldots < t_j = t_{j+1} = \ldots. = T, \tag{11.73}$$

and we have moreover

$$x(t_j) - x(0) = 2\eta_j, \tag{11.74}$$

$$F - m\ddot{x} - \left(k_0 + \sum_{i=j+1}^{n} k_i\right) x \text{ is constant on } [t_j, t_{j+1}]. \tag{11.75}$$

Observe first that x is strictly increasing on $[0, T]$, thanks to our assumption on F. Let

$$\widehat{u}_j(t) = -\eta_j + x(t) - x(0), \tag{11.76}$$

and denote by ξ the inverse function of x and define:

$$t_j = \begin{cases} \xi\,(2\eta_j + x(0)) & \text{if } \ 2\eta_j + x(0) < x(T), \\ T & \text{if } \ 2\eta_j + x(0) \geq x(T). \end{cases} \tag{11.77}$$

We check immediately that the function

$$u_j(t) = \begin{cases} \widehat{u}_j(t) & \text{if } \ 0 \leq t < t_j, \\ \eta_j & \text{if } \ t \geq t_j, \end{cases} \tag{11.78}$$

is the (unique) solution of

$$\begin{cases} \dot{u}_j + \beta(u_j/\eta_j) \ni \dot{x}, \\ u_j(0) = -\eta_j. \end{cases} \tag{11.79}$$

Therefore, on $[t_j, t_{j+1}]$, relation (11.75) holds. Moreover, the relation (11.77) is equivalent to (11.74). We write Eqs. (11.75) and (11.74) under the form:

$$\forall j \in \{1, \ldots, n+1\}, \quad p_j = k_0 + \sum_{l=j}^{n} k_l, \tag{11.80}$$

$$\forall i \in \{1, \ldots, n\}, \quad d_i = 2\eta_i. \tag{11.81}$$

On the other hand, we remark that the assumptions $k_0 \geq 0$, $k_i > 0$ and the Eq. (11.80) imply:

$$p_{n+1} < p_n < \ldots < p_2 < p_1, \tag{11.82}$$

the Eqs. (11.72) and (11.81) imply:

$$d_1 < d_2 < \ldots < d_{n-1} < d_n. \tag{11.83}$$

From (11.80) and (11.81) we have a one to one correspondence between the parameters k_i and η_i of the generalized Prandtl model and geometrical parameters p_j and d_j of the hysteresis cycle in the $(x, F - m\ddot{x})$ plane.

Thus, a partial identification of the model is possible. But we cannot infer from the values of $(p_i)_{1\leq i\leq n+1}$ and $(d_i)_{1\leq i\leq n}$ the abscissa and the ordinate of the lower left point of the cycle: they depend on the initial conditions and on the forcing: see for example Fig. 11.15a and 11.15b: the two cycles are identical up to translation.

In Figs. 11.15a, 11.15b, 11.16a, 11.17a, 11.17b and 11.18, we can see $(n+1)$ different phases in one half-cycle. On the contrary, in Fig. 11.16b we have just three different phases: the amplitude of forcing is too weak and the five St-Venant elements do not enter the slip phase, two of them are always locked and the generalized Prandtl model works as a model with three S_i-elements, defined by (k_1, η_1), (k_2, η_2) and (k_3, η_3) and one spring of stiffness $K = k_0 + k_4 + k_5$. Let us consider a given hysteresis cycle: we can only calculate the parameters of the Prandtl elements which enter slip phase.

In the case $k_0 = 0$, we can prove that if the amplitude of the forcing is large enough for all the St-Venant elements to be unlocked, the ordinate of the symmetry center of the cycle vanishes. Moreover, the slope p_{n+1} of the last segment is necessary zero.

Numerical Identification of a Model from its Limit Cycles

If we start from the cycles already computed, we find the physical constants of the systems under consideration. For instance, for Fig. 11.16a, we identify five St-Venant elements. We give in Table 11.4 the constants η_i^c and k_i^c that have been obtained and the constants η_i and k_i that characterize the studied Prandtl model (see Table 11.2).

Table 11.4 Constants η_i, k_i and constants η_i^c and k_i^c computed from the cycle of Fig. 11.16(a).

i	0	1	2	3	4	5
η_i^c	–	0.999921	1.999945	3.000250	3.999900	5.000013
η_i	–	1	2	3	4	5
k_i^c	$3.22\,10^{-14}$	0.999892	0.999959	1.000082	0.999971	0.999997
k_i	0	1	1	1	1	1

We find that

$$\max_{1\leq i\leq 5} |\eta_i^c - \eta_i| = 2.501\,10^{-4} \quad \text{and} \quad \max_{0\leq i\leq 5} |k_i^c - k_i| = 1.074\,10^{-4}. \qquad (11.84)$$

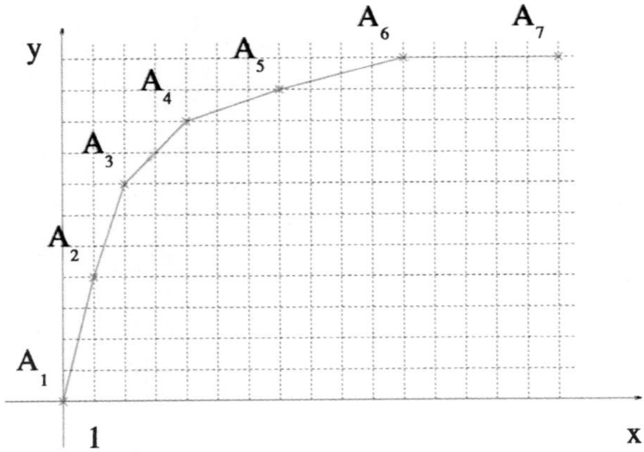

Fig. 11.26 Upper part of an arbitrary limit cycle.

Conversely, let analyse a polygonal convex half-cycle plotted in Fig. 11.26. Formulas (11.80) and (11.81) determine the constant η_i and k_i of a generalized Prandtl model, they are given in Table 11.5.

Table 11.5 Constants η_i and k_i computed from the cycle of Fig. 11.26.

i	0	1	2	3	4	5
η_i	–	1/2	1	2	7/2	11/2
k_i	0	1	2	2/3	1/12	1/4

With these values of η_i and k_i and $n = 5$, we make numerical simulations for the corresponding generalized Prandtl model. The other parameters are defined arbitrarily by

$$F(t) = f\cos(0.5t), \quad m = 1, \quad x_0 = \dot{x}_0 = 0 \forall i \in \{1, \ldots, 5\} \quad u_{0,i} = 0. \tag{11.85}$$

As is natural, the amplitude of the forcing has to be large enough to retrieve the cycle of Fig. 11.26. A forcing amplitude larger than 6.6 increases the size of the last segment of the cycle.

For $f = 6.6$, we obtain a symmetrical hysteresis cycle with 12 segments (see Fig. 11.27). We find

$$\max_{1 \le i \le 5} |\eta_i^c - \eta_i| = 5.450\,10^{-5} \text{ and } \max_{0 \le i \le 5} |k_i^c - k_i| = 1.042\,10^{-4}. \tag{11.86}$$

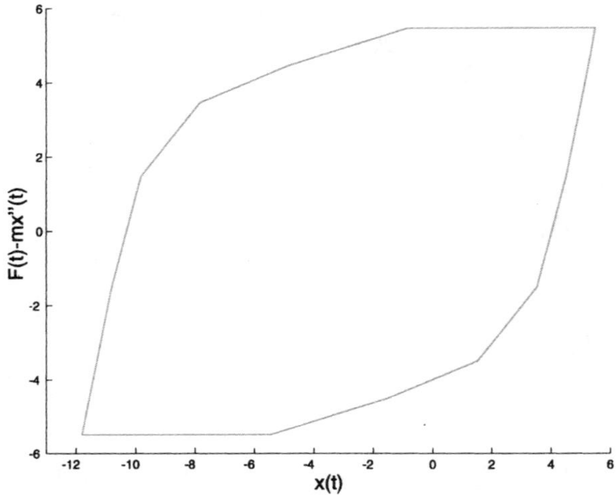

Fig. 11.27 The curve $\{x(t), F(t) - m\ddot{x}(t)_{t\in[750,1000]}\}$ for the generalized Prandtl rheological model with linear hardening, defined by Table 11.5 and $F(t) = f\ \cos(0.5t)$, $n = 5$, $m = 1$, $x_0 = y_0 = 0$, $\forall i \in \{1,\dots,5\}$ $u_{0,i} = 0$ and $f = 6.6$.

11.1.3.2 *A viscoelastoplastic model*

Let us perform some simulations for the systems (11.16) and (11.17). In this section, we choose:

$$F(t) = \cos(0.5t), \tag{11.87}$$

and the parameters have the following values:

$$\eta = k = m = 1 \ \text{ and } \ x_0 = y_0 = u_0 = 0 \ \text{ and } \ c = 100 \ \text{ or } \ c = 1. \tag{11.88}$$

We have plotted the curves $\{\dot{x}(t), F(t) - m\ddot{x}(t)\}_{t\in[400,2000]}$ for $c = 100$ in Fig. 11.28a and for $c = 1$ in Fig. 11.28b. In every case, we obtain hysteresis cycles. For the largest value of c (see Fig. 11.28a), the cycle is similar to those obtained for the Prandtl model with harmonic forcing (see Fig. 11.15a); the behaviour is similar to the elastoplastic case. For smaller value of c, the cycle no longer looks like an elastoplastic cycle (see Fig. 11.28b). The obtained limit cycle does not enable us to identify the parameters of forcing.

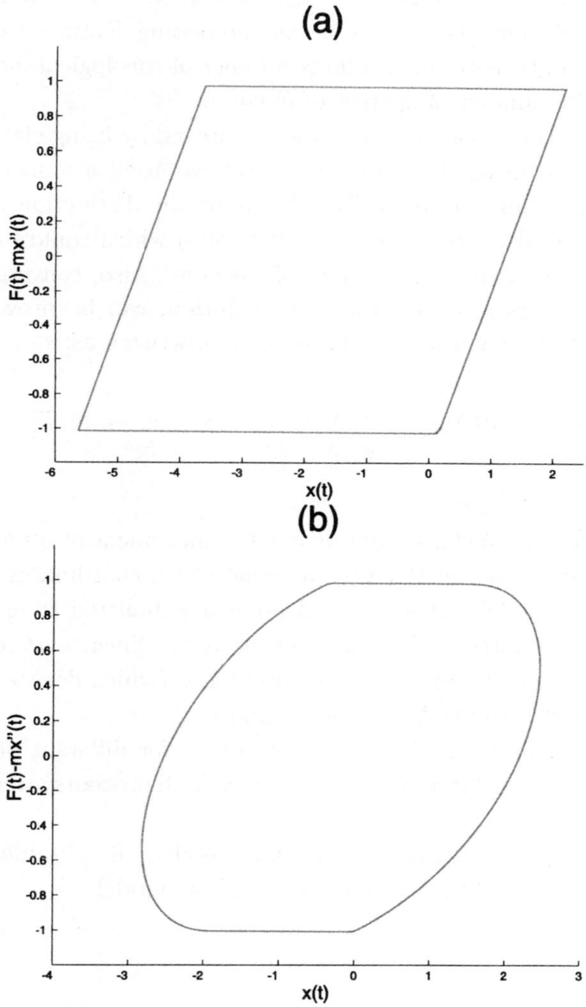

Fig. 11.28 The curves $\{x(t), F(t) - m\ddot{x}(t)\}_{t\in[400,2000]}$ for the viscoelastoplastic model defined by $F(t) = \cos(0.5t)$, $\eta = k = m = 1$, $x_0 = y_0 = u_0 = 0$ and $c = 100$ (a) and $c = 1$ (b). These two figures differ only by value of c.

11.1.4 *Conclusion*

We examine several classes of rheological models constituted of springs, St-Venant elements and dashpots. All these models are governed by the differential inclusion (11.54). Note an interesting feature regarding Eq. (11.54): it permits us to unify a large number of rheological models with a system of finite number of degrees of freedom.

From a mathematical point of view, we proved by using classical results that a unique solution X exists. Moreover, we used a numerical scheme which is simple and easy to realize. It avoids the distinction between the different phases (friction, movement, stick, slip) which could be very laborious for systems with many degrees of freedom. Also, convergence of the approximate solution toward the exact solution, can be proved. From a physical point of view, the Eq. (11.54) can be written as:

$$\begin{cases} \dot{X}(t) + M\partial\phi(X(t)) + C.X(t) \ni H(t) \quad \text{a.e. on } [0,T], \\ X(0) = \xi, \end{cases} \tag{11.89}$$

where C is a matrix.

This differential inclusion points out the movement of a material point with coordinates $X(t)$ in $\mathbb{R}^p$ ($X(t)$ are generalized coordinates of the studied rheological model): this material point is submitted to external force $H(t)$, to repelling force $C.X(t)$ (which denotes the linear part of the model, consisting of springs) and to convex potential ϕ (which denotes the plastic part of the model, the St-Venant elements).

Numerous simulations have been presented for different periodic forcings. All of them exhibit a hysteresis limit cycle that occurs after a transient phase.

In the case of the generalized Prandtl model, the obtained hysteresis cycle determines the physical parameters of the model.

Chapter 12

A Mechanical System with 7 DOF

12.1 Mathematical Model

In order to study the behaviour of a mechanical system consisting of a cable with its shaft, a gear lever and its support (see in Fig. 12.1) we introduce a simplified seven degrees of freedom (DOF) system. An equivalent mass m with displacement X_4 and clearance Δ exhibits impact nonlinearity due to the motor (see Fig. 12.2). Though a real physical external excitation would be more complex, we only consider a simple form of excitation $u(t) = \alpha \sin(\omega t)$ with amplitude α and frequency ω.

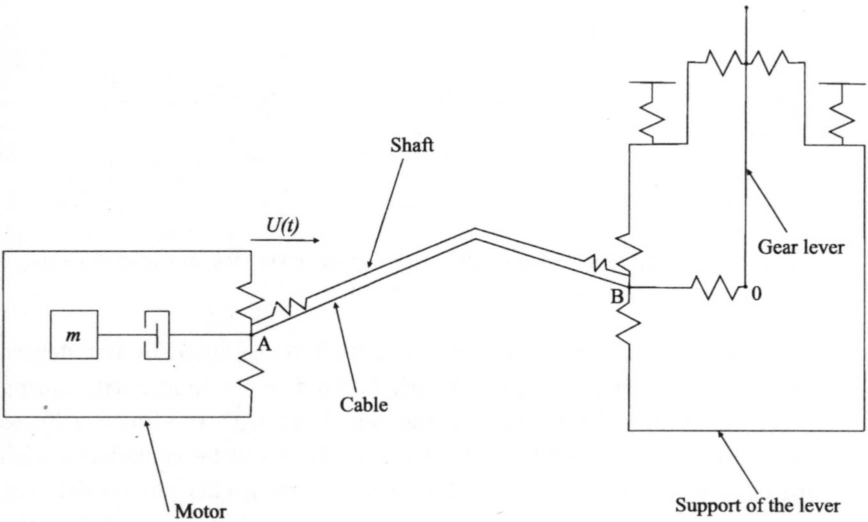

Fig. 12.1 Simplified model of the gear-box.

The cable is represented by a mass M_1 with displacement X_5 and equivalent stiffness k_1 (see Fig. 12.2): this is a one DOF reduction of the continuous system. Impacts between m and M_1 are governed by a classical restitution law with coefficient e. The displacements of masses m and M_1 are governed by the equations:

$$m\ddot{X}_4 = 0 + \text{“impact”} \tag{12.1}$$

$$M_1\ddot{X}_5 = k_1(X_1 - X_5) + \text{“impact”}. \tag{12.2}$$

The functions X_4 and X_5 remain smooth if $X_4 - X_5 \in] - \Delta, \Delta[$. Elastic impacts occur if $X_4 - X_5 = \pm\Delta$. Using some algebra (related to momentum conservation [Brogliato (1996)]) velocities $\dot{X}_4{}^+$ and $\dot{X}_5{}^+$ after an impact are given by the relations:

$$\begin{cases} \dot{X}_4{}^+ - \dot{X}_5{}^+ = -e(\dot{X}_4{}^- - \dot{X}_5{}^-), \\ m\dot{X}_4{}^+ + M_1\dot{X}_5{}^+ = m\dot{X}_4{}^- + M_1\dot{X}_5{}^-, \end{cases} \tag{12.3}$$

where $\dot{X}_4{}^-$ and $\dot{X}_5{}^-$ are velocities before impacts.

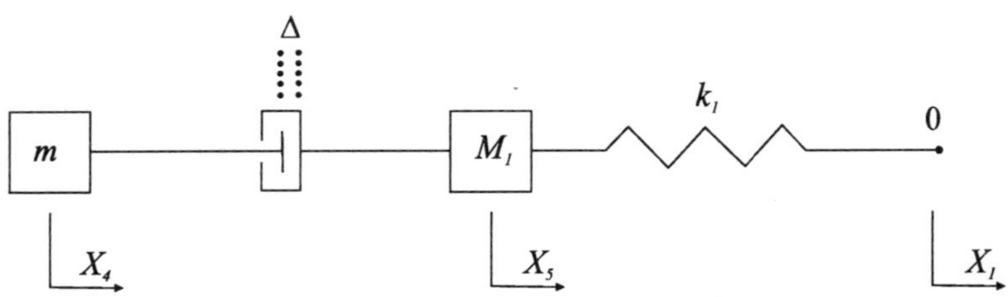

Fig. 12.2 Simplified model of the cable with clearance contact on the motor side.

The cable is assumed to move inside a shaft represented by one degree of freedom with the mass M_2, the length L, fixed to the motor with spring of stiffness k_2 and to the support of the gear lever with the finite stiffness k_3 (see in Figs. 12.3 and 12.5). Another model could be considered with infinite k_3 (see in Figs. 12.4 and 12.5): one has to modify the model that is now presented, in that case, if X_6 and X_7 are the displacement of the junction point between the shaft and the support of the gear lever, respectively, and the displacement of the shaft, $X_6 = X_7$; geometrical relations between X_2, X_7 and other variables and parameters of the problem should be introduced.

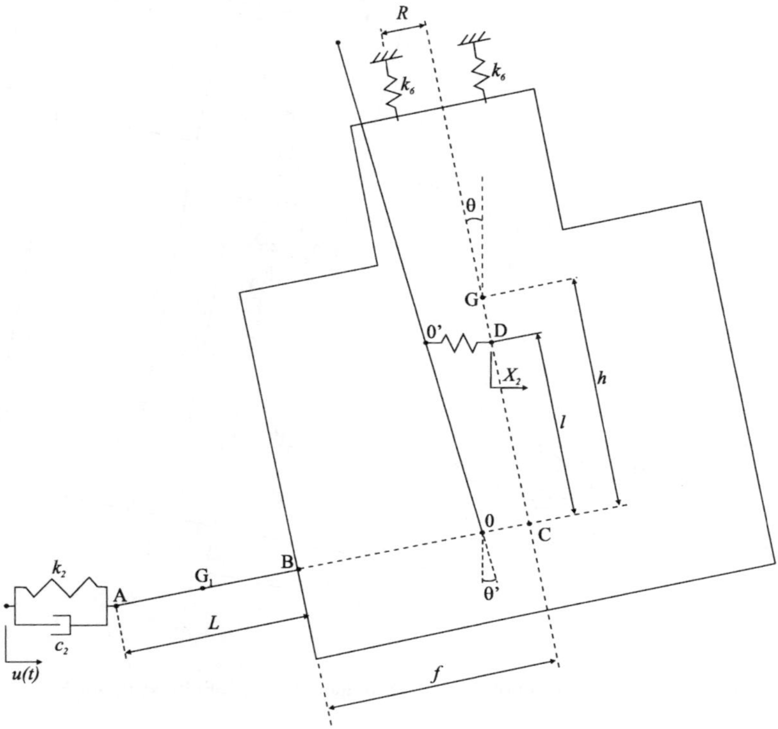

Fig. 12.3 Description of the gear-box model with finite stiffness k_3.

Let us call M_5 the mass of the one DOF system that simplifies the model of a gear lever; let us call θ' the angle between the gear lever and the vertical in the plan where the gear lever is moving. Let us define some parameters: l" is the distance between gravity center G' of the gear lever and rotation point O' that joins the gear lever to its support, l is the length of the gear lever, J'_O is inertia momentum of the gear lever; k_1, k_7 are stiffnesses (see in the Fig. 12.6). The motion of the gear lever is governed by the following equations:

$$M_5\ddot{X}_1 - M_5 l"[\ddot{\theta}'\cos(\theta') - (\dot{\theta}')^2\sin(\theta')] =$$
$$k_7[X_2 - X_1 + l\sin(\theta')] + k_1(X_5 - X_1), \tag{12.4}$$

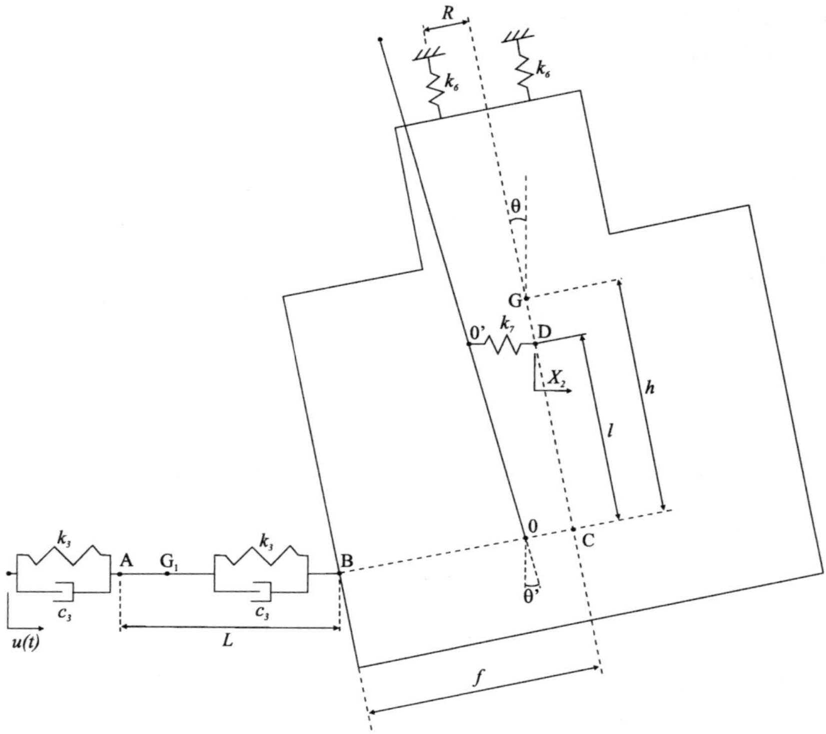

Fig. 12.4 Description of the gear-box model with infinite stiffness k_3.

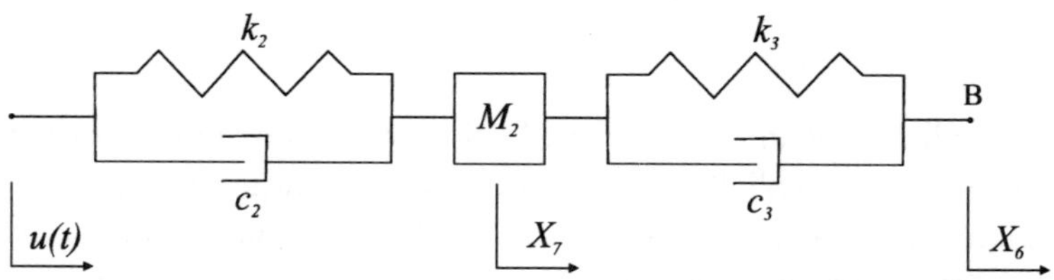

Fig. 12.5 Simplified model of the shaft fixed to the motor and to the support of the gear lever.

and the equation for the momentum is:

$$J'_O\ddot{\theta}' = M_5 g(l" - l)\sin(\theta') - K_1(X_1 - X_5)\cos(\theta') - C\theta', \tag{12.5}$$

with

$$J'_O = J_2 + M_5(l" - l)^2,$$

Some mechanical assumptions are made: There is a reaction force at O', and G' is above O'. Let us call C the momentum that ensures vertical equilibrium of the gear lever. Let θ be the angle between the vertical axis and the support of the gear lever (see in Fig. 12.3 or 12.4). The translation movement of the support of the gear lever is governed by :

$$M_2\ddot{X}_7 = k_2(u(t) - X_7) + k_3(X_6 - X_7). \tag{12.6}$$

The rotation of the support of the gear lever is described by the equation:

$$\begin{cases} J_1\ddot{\theta} = -2k_6R^2\sin(\theta) + k_3(X_7 - X_6)[f\sin(\theta) + h\cos(\theta)] - \\ \qquad k_7[X_2 - X_1 + l\sin(\theta')](h - l)\cos(\theta). \end{cases} \tag{12.7}$$

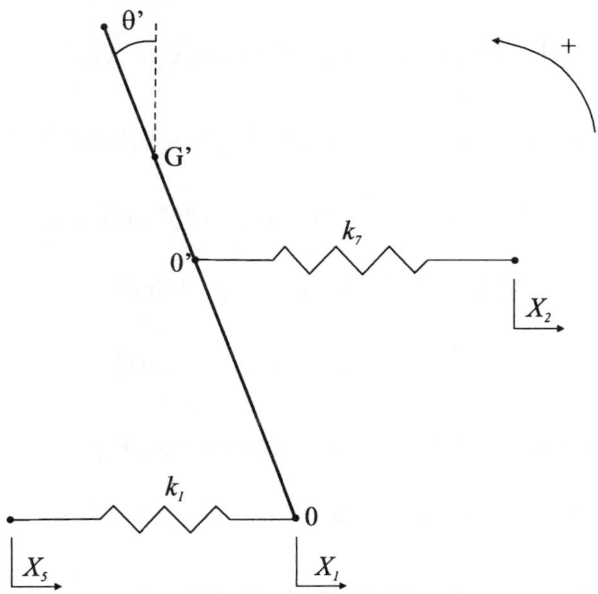

Fig. 12.6 Model of the gear lever.

12.1.1 *Final equations with finite* k_3

Finally, seven differential equations with eight unknown coordinates (X_1, X_2, X_4, X_5, X_6, X_7, θ', θ) are derived. One relation can be added, so that we obtain the following model which is essentially a seven DOF mechanical system:

$$m\ddot{X}_4 = 0 + \text{"impact"}, \tag{12.8}$$

$$M_1\ddot{X}_5 = k_1(X_1 - X_5) + \text{“impact”}, \tag{12.9}$$

$$\begin{cases} \dot{X_4}^+ - \dot{X_5}^+ = -e(\dot{X_4}^- - \dot{X_5}^-), \\ m\dot{X_4}^+ + M_1\dot{X_5}^+ = m\dot{X_4}^- + M_1\dot{X_5}^-, \end{cases} \tag{12.10}$$

$$M_2\ddot{X}_7 = k_2(u(t) - X_7) + k_3(X_6 - X_7), \tag{12.11}$$

$$\begin{cases} M_5\ddot{X}_1 - M_5 l\text{"}[\ddot{\theta'}\cos(\theta') - (\dot{\theta'})^2\sin(\theta')] = \\ k_7[X_2 - X_1 + l\sin(\theta')] + k_1(X_5 - X_1), \end{cases} \tag{12.12}$$

$$[J_2+M_5(l\text{"}-l)^2]\ddot{\theta'} = M_5 g(l\text{"}-l)\sin(\theta')-K_1(X_1-X_5)\cos(\theta')-C\theta', \tag{12.13}$$

$$\begin{aligned} &M_4\ddot{X}_2 - M_4(h-l)[\ddot{\theta'}\cos(\theta') - (\dot{\theta'})^2\sin(\theta')] = \\ &k_3[X_7 - X_6] - k_7[X_2 - X_1 + l\sin(\theta')], \end{aligned} \tag{12.14}$$

$$X_6 = X_2 + f[1 - \cos(\theta)] + l\sin(\theta), \tag{12.15}$$

$$\begin{aligned} J_1\ddot{\theta} = &-2k_6R^2\sin(\theta) + k_3(X_7 - X_6)[f\sin(\theta) + h\cos(\theta)]- \\ &k_7[X_2 - X_1 + l\sin(\theta')](h-l)\cos(\theta). \end{aligned} \tag{12.16}$$

Until now, we did not introduce damping terms.

In a mathematical point of view, it is very easy to add linear viscous damping to each equation: Such terms are proportional to velocities of the different coordinates. They do not modify deeply the question of existence and uniqueness problem. They do not create problem when building adapted numerical schemes.

12.1.2 *Introduction of new coordinates*

Before building a numerical scheme according to the general process described in chapter 2, we change the coordinates in order to adapt the previous differential equations to the convenient mathematical frame that we have presented in chapter 3.

Let us set:

$$\begin{cases} X_8 = X_4 - X_5, \\ X_9 = mX_4 + M_1 X_5. \end{cases} \tag{12.17}$$

The first two equations of the previous subsection become:

$$M_1 \ddot{X}_8 = k_1 \Big[\frac{X_9 - mX_8}{M_1 + m} - X_1 \Big] + \text{“impact”}, \tag{12.18}$$

with impact occurrence if $|X_8| = \Delta$. The main advantage of the new coordinates is that we have only impact according to X_8. The other equation is:

$$\ddot{X}_9 = k_1 \Big[X_1 - \frac{X_9 - mX_8}{m + M_1} \Big]. \tag{12.19}$$

12.1.3 *Numerical scheme*

Let us express the model in the following compact form of a general class of problems. We distinguish between the first equation (with impact) and the other equations (without impact).

The general form of the first equation is now:

$$\ddot{Y}_1 + F_1(Y_1, Y_2, Y_3, Y_4, Y_5, Y_6, Y_7, t) + \text{ “impact” } \ni 0, \tag{12.20}$$

with $Y_1 = X_8$. F_1 is a smooth function of variables $Y_2 = X_9$, $Y_3 = X_7$, $Y_4 = \theta'$, $Y_5 = X_1$, $Y_6 = \theta$, $Y_7 = X_2$ and time t.

The other equations can be written in the form:

$$\begin{cases} \ddot{Y}_2 + F_2(Y_1, Y_2, Y_3, Y_4, Y_5, Y_6, t) = 0, \\ \vdots \\ \ddot{Y}_6 + F_6(Y_1, Y_2, Y_3, Y_4, Y_5, Y_6, t) = 0, \end{cases} \tag{12.21}$$

with smooth functions F_i $(i = 2, \dots, 7)$.

For the six latter equations we can use classical finite difference schemes dealing with smooth nonlinearities included in functions F_i, $i = 2, \dots, 7$ via explicit scheme. Let us choose constant time step h between time $t_n = nh$ and $t_{n+1} = (n+1)h$ in order to simplify. Let us denote Y_{in} approximation (given by the numerical scheme) at time t_n of the exact value $Y_i(nh), i = 1, \dots, 7$ of functions Y_i, $i = 1, \dots, 7$. We write:

$$\frac{Y_{in+1} - 2Y_{in} + Y_{in-1}}{h^2} + F_i(Y_{1n}, Y_{2n}, Y_{3n}, Y_{4n}, Y_{5n}, Y_{6n}, Y_{7n}, nh) = 0. \tag{12.22}$$

The first equation is discretized as follows

$$\begin{aligned} &\frac{Y_{1n+1} - 2Y_{1n} + Y_{1n-1}}{h^2} + F_1(Y_{1n}, \dots, Y_{7n}, nh) + \\ &\qquad \partial\psi_{(1+e)K}(Y_{1n+1} + eY_{1n-1}) = 0. \end{aligned} \tag{12.23}$$

where $(1+e)K$ denotes the convex set $[-(1+e)\Delta, (1+e)\Delta]$, $\psi_{(1+e)K}$ is the characteristic function of the same convex set ($\psi_{(1+e)K}(x) = 0$ if x belongs to $(1+e)K$, else $= +\infty$), and $\partial\psi_{(1+e)K}$ is the subdifferential of the same convex set.

These recursive relations lead easily to any desired values of approximate variables $Y_1, \dots, Y_7$ if values $Y_{10}, Y_{20}, Y_{30}, Y_{40}, Y_{50}, Y_{60}, Y_{70}$ and $Y_{11}, Y_{21}, Y_{31}, Y_{41}, Y_{51}, Y_{61}, Y_{71}$ are known; this is quite natural since these values correspond to the knowledge of initial displacements and initial velocities.

One can see that the case with $6+1$ degrees of freedom (6 without and 1 with impact) that is considered here could be generalized to cases with $N+1$ degrees of freedom.

Let us recall that the main advantage of the scheme that has been presented is its ability to capture impacts without a priori time localization of the impact.

Expressions of the functions F_i, $i = 1, 7$ can be easily obtained from the previous pages.

Remark 12.1 *Numerical scheme could be refined. Indeed the numerical scheme is written for the linear displacement and velocity parts of the system as an implicit form in order to obtain a better behaviour (stability) in practice and simple calculations at each time step. For the nonlinear smooth part, the numerical scheme is written as an explicit form.*

For example, forgetting the impacts for the moment, if a smooth nonlinear differential system is considered as:

$$M\ddot{Z} + C\dot{Z} + KZ + \mathbf{N}(Z, \dot{Z}, t) = 0, \qquad (12.24)$$

with nonlinear terms $\mathbf{N}(Z, \dot{Z}, t)$ depending on Z and $\dot{Z}$ including external solicitations, a numerical scheme of the type

$$\begin{aligned} &M\frac{Z_{n+1} - 2Z_n + Z_{n-1}}{h^2} + C\frac{Z_{n+1} - Z_{n-1}}{2h} + \\ &K[Z_n \text{ or } Z_{n+1}] + \mathbf{N}(Z_n, \frac{Z_n - Z_{n-1}}{h}, t_n) = 0. \end{aligned} \qquad (12.25)$$

can be build. The latter expression is a linear system (of small size here) with unknown Z_{n+1}. This linear system has to be solved. The computational cost is slightly increased, but the numerical behaviour is improved.

This form of the numerical scheme can be used directly in the case of the models that have been presented here before modified by diagonal damping terms. These numerical schemes have been implemented and provide the numerical results given hereafter.

12.2 Numerical Results and Comments for Finite k_3

We choose some parameters to be used by the model: $k_6 = 1000000 \ Nm^{-1}$, $\Delta = 0.0001 \ m$, $u(t) = 0.002\sin(\omega t)$, $e = 0.9$, $g = 9.81 \ ms^{-2}$, $m = 0.9 \ kg$, $M_4 = 0.3 \ kg$, $C = 2M_5 g(l" - l)$, $\omega = 1200 \ rd\,s^{-1}$ and $L = 1 \ m$. The mass of the cable is $M_1 = 0.12 \ kg$, the stiffness of the cable $k_1 = 120000 \ Nm^{-1}$, the mass of the shaft $M_2 = 0.32 \ kg$, the stiffness of the shaft (motor side))$k_2 = 102000 \ Nm^{-1}$, the stiffness of the joint between the shaft and the support $k_3 = 1000000 \ Nm^{-1}$, the stiffness of the joint between gear lever and support of the gear lever $k_7 = 18000 \ Nm^{-1}$, the distance between the basis O of the gear lever and the rotation point $l = 0.071 \ m$, the length of the gear lever $l' = 0.333 \ m$, the distance between the gravity center of the gear lever and O $l" = 0.150 \ m$, the location of the gravity

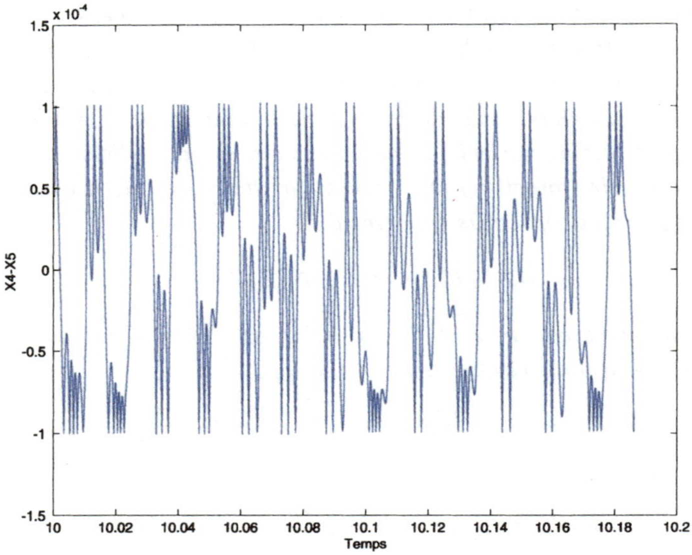

Fig. 12.7 $k_3 = 1000000 \quad Nm^{-1}$. Response of X_4X_5 versus time t between 6.5 s and 6.7 s.

center G of the support vs the fixation point B of the shaft $h = 0.033 \quad m$, the mass of the gear lever $M_5 = 0.530 \quad kg$, the inertia momentum of the gear lever versus its gravity center $J_2 = 3.32 \times 10^{-6} \quad kgm^2$, the distance between the ordinate of G and the one of B $f = 0.126 \quad m$, the inertia momentum of the support around G $J_1 = 2.65 \times 10^{-6} \quad kgm^2$, distance between the fixation point of the gear lever and the vertical line crossing through G $R = 0.056 \quad m$. Details of the numerical implementation are not given. They are of the same kind than for finite stiffness k_3 case. Damping does not create any problem. So only some results are given and described. They are similar to those that could be obtained for the infinite k_3. They are summed up in Figs. 12.7 and 12.8. The former describes the behaviour of $X_4 - X_5$ versus time.

Accumulation of impacts can be seen justifying the care in building the numerical schemes. The response is a complex one; it is harmonically a very rich one. The same kind of behaviour could be observed in the case of infinite k_3.

But the time response of θ' is almost periodic (see in Fig. 12.8) as it can be shown by the analysis of the frequency spectrum. So the complexity of the response due to the impacts is not easily carried to the response of the

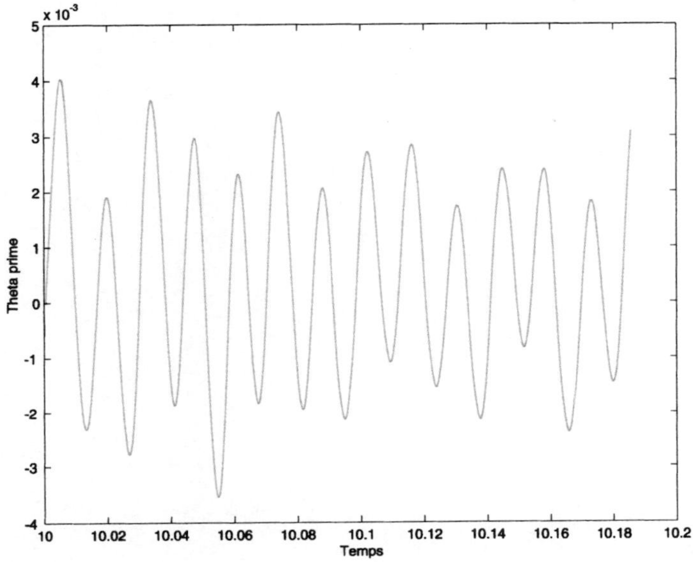

Fig. 12.8 $k_3 = 1000000 \quad Nm^{-1}$. Response of θ' versus time t between 6.5 s and 6.7 s.

gear lever via such a simple model.

As a conclusion, the simple modal models of the cable and the shaft should be enriched.

Chapter 13

Stability of Singular Periodic Motions in Single Degree of Freedom Vibro-Impact Oscillators and Grazing Bifurcations

13.1 Introduction

Impacts in mechanical systems (such as in [Abraham (1993)], [Brogliato (1996)], [Chatterjee and Mallik (1996)], [Lenci and Rega (1998b)], [Peterka and Vacik (1992)]), lead to the occurrence of particular periodic solutions often denoted as grazing solutions (see [Ivanov (1994)], [Nordmark (1991)], [Nordmark (1992)], [Nordmark (1997)]), which are characterized by a non-differentiability of the Poincaré map at the corresponding fixed point. The aim of this chapter is to investigate the stability of a grazing periodic solution in a general single degree of freedom mechanical system with impacts. The starting point is Nordmark's method and the results given in [Molenaar *et. al.* (1999)], [Nordmark (1991)] are completed.

In Section 13.2, the mechanical system considered is introduced and a change of coordinates that simplifies the writing of the Poincaré map is proposed. A local expansion of this map around the nondifferentiable fixed point is then obtained in Section 13.3, with a distinction between two cases depending on the derivatives of the flow at the fixed point. This expansion enables to investigate the stability of the grazing periodic solution in Section 13.4. Finally, the results of the preceding sections are applied to two mechanical systems in Section 13.5.

13.2 Mechanical System and Change of Coordinates

The general model of a single degree of freedom vibro-impact mechanical system is considered in this paper. As long as there are no impacts, the equations of motion are given by:

$$\ddot{x} = f(x, \dot{x}, t), \tag{13.1}$$

where f is a continuously differentiable function such that:

$$f(x, y, t) = f(x, y, t+T), \quad \forall (x, y, t) \in \mathbb{R}^3. \tag{13.2}$$

The motion is constrained to remain in an half plane of the space defined by $x \leq x_{max}$ because of an obstacle. Impacts against this obstacle are taken into account as instantaneous processes that lead to a jump in the velocity according to the rule:

$$x(t) = x_{max} \Rightarrow \dot{x}(t^+) = G(\dot{x}(t^-)), \tag{13.3}$$

where G is a continuously differentiable function verifying $G(0) = 0$. Define $r = -G'(0)$ and $s = -G''(0)$.

This chapter focuses on grazing periodic solutions, namely T-periodic solutions of the system given by Eqs. (13.1) and (13.3) such that one impact with zero velocity occurs at $t_1^0 \in [0, T]$. After a re-scaling of time, it is assumed that the impact occurs at T: The solution x starting from $(x_{max}, 0)$ at $t_0 = 0$ verifies $x(T) = x_{max}$ and $\dot{x}(T) = 0$. This periodic solution can be seen as a fixed point of the Poincaré map defined in $\mathbf{S}^1 \times \mathbb{R}^2$ by:

$$(t_0, x_0, \dot{x}_0) \mapsto (t_0 + T, \phi_1^{\text{impact}}(t_0, x_0, \dot{x}_0, T), \phi_2^{\text{impact}}(t_0, x_0, \dot{x}_0, T)), \tag{13.4}$$

where ϕ^{impact} is the flow of the system governed by the Eqs. (13.1) and (13.3): $\phi_1^{\text{impact}}(t_0, x_0, \dot{x}_0, t)$ is the displacement at $t_0 + t$ of the solution starting from $(x_0, \dot{x}_0)$ at t_0, and $\phi_2^{\text{impact}}(t_0, x_0, \dot{x}_0, t)$ is the corresponding velocity.

By a translation of coordinates, the fixed point $(x_{max}, 0)$ can be set to $(0, 0)$ in the new coordinates (λ, μ). In these coordinates, the flow of the Eq. (13.1) (without impacts) is referred to as $P = (P_1, P_2)$, and the admissible area is defined by $\lambda \leq 0$.

For any initial condition close to $(0, 0, 0)$, the solution will either impact once on $[0, T]$ (near to T), or miss the obstacle. Therefore, a local condition for an impact to occur needs to be determined. To express the required condition in a convenient way, a nonlinear change of variables is defined by:

$$\begin{aligned} u &= P_1(0, \lambda, \mu, T), \\ v &= P_2(0, \lambda, \mu, T). \end{aligned} \tag{13.5}$$

The previous equations define an invertible change of variables if and only if the eigenvalues of P at $(0,0,0,T)$ are nonzero. Let us denote by λ_1 and λ_2 the eigenvalues of the Jacobian matrix of P (or – it is equivalent – of the flow without impacts ϕ) at $(0,0,0,T)$: In the following, it is assumed that λ_1 and λ_2 are nonzero.

As for $(u,v)=(0,0)$ the displacement is maximum at $t=0$ and $t=T$, when u and v are small some maxima are likely to occur at $\tau_1(u,v)$ and $T-\tau_2(u,v)$. The occurrence of a maximum at $T-\tau_2(u,v)$ implies:

$$P_2(0,\lambda,\mu,T-\tau_2(u,v))=0. \tag{13.6}$$

Taking into account the change of variables previously defined yields:

$$\begin{aligned}\frac{\partial P_2}{\partial t}(0,\lambda,\mu,T) &= f(u+x_{max},v,t_1^0)\\ &= -\nu-\beta u-\gamma v\ +\text{h.o.t.},\end{aligned} \tag{13.7}$$

and

$$\frac{\partial^2 P_2}{\partial t^2}(0,\lambda,\mu,T)=\gamma\nu-\delta\ +\text{h.o.t.}, \tag{13.8}$$

where

$$\nu=-f(x_{max},0,t_1^0), \tag{13.9}$$

$$\beta=-\frac{\partial f}{\partial x}(x_{max},0,t_1^0), \tag{13.10}$$

$$\gamma=-\frac{\partial f}{\partial y}(x_{max},0,t_1^0), \tag{13.11}$$

$$\delta=-\frac{\partial f}{\partial t}(x_{max},0,t_1^0), \tag{13.12}$$

and "h.o.t" denotes higher order terms. Let us notice that ν is the opposite of the acceleration of the grazing solution at the impact, which implies that $\nu\geq 0$; it is moreover assumed that $\nu\neq 0$.

By using a second order Taylor expansion in the Eq. (13.6), $\tau_2(u,v)$ is obtained as a solution of the second order equation

$$v-\frac{\partial P_2}{\partial t}(0,\lambda,\mu,T)\tau_2+\frac{1}{2}\frac{\partial^2 P_2}{\partial t^2}(0,\lambda,\mu,T)\tau_2^2=0, \tag{13.13}$$

and since $\tau_2(u,v)$ must tend to 0 as (u,v) approaches $(0,0)$, it is given by:

$$\tau_2(u,v)=-\frac{1}{\nu}v+\frac{\beta}{\nu^2}uv+\frac{\gamma\nu+\delta}{2\nu^3}v^2\ +\text{h.o.t.} \tag{13.14}$$

Two cases arise then depending on the sign of $\tau_2(u,v)$: If $\tau_2(u,v) > 0$ (namely if the maximum of displacement occurs before T), an impact has occurred on $[0,T]$ if and only if $P_1(0,\lambda,\mu,T-\tau_2(u,v)) > 0$, whereas when $\tau_2(u,v) < 0$ (namely when the maximum of displacement occurs after T), the system impacts on $[0,T]$ if and only if $P_1(0,\lambda,\mu,T) > 0$. A second order expansion yields the condition for an impact to occur, which is $u+\frac{1}{2\nu}v^2 > 0$ in the first case and $u > 0$ in the latter one. Therefore, for any local initial conditions (u,v) leading to an impact the quantity

$$\chi = -K\left(u + \frac{1}{2\nu}v^2\right), \tag{13.15}$$

is negative (where $K > 0$ is a constant that will be set later).

At this point, a new change of variables seems appropriate to express the impact time – and hence the Poincaré map – under a convenient form. The first coordinate is given by χ, and the second coordinate ψ is found by requiring the axis $\chi = 0$ to be mapped onto $\psi = 0$ by the Poincaré map following the procedure given in the reference [Nordmark (1991)]. To achieve this, first of all the flow Π in the (u,v) variables needs to be inverted, i.e., (u,v) has to be expressed as a function of (Π_1,Π_2). A restriction to the second order gives:

$$\begin{aligned}
u = {} & \frac{\beta_2}{\lambda_1\lambda_2}\Pi_1(0,u,v,T) - \frac{\beta_1}{\lambda_1\lambda_2}\Pi_2(0,u,v,T) + y_1\Pi_1(0,u,v,T)^2 + \\
& y_2\Pi_1(0,u,v,T)\Pi_2(0,u,v,T) + y_3\Pi_2(0,u,v,T)^2, \qquad (13.16)\\
v = {} & -\frac{\alpha_2}{\lambda_1\lambda_2}\Pi_1(0,u,v,T) + \frac{\alpha_1}{\lambda_1\lambda_2}\Pi_2(0,u,v,T) + z_1\Pi_1(0,u,v,T)^2 + \\
& z_2\Pi_1(0,u,v,T)\Pi_2(0,u,v,T) + z_3\Pi_2(0,u,v,T)^2. \qquad (13.17)
\end{aligned}$$

By expanding $\Pi_1(0,u,v,T)$ and $\Pi_2(0,u,v,T)$ to the second order and by considering both relationships in the basis (u,v), the ten unknown coefficients are obtained, and the main four ones are the elements of the Jacobian matrix of ϕ at $(0,0,0,T)$, given by:

$$\alpha_1 = \frac{\partial \phi_1}{\partial x_0}(0,0,0,T), \tag{13.18}$$

$$\alpha_2 = \frac{\partial \phi_2}{\partial x_0}(0,0,0,T), \tag{13.19}$$

$$\beta_1 = \frac{\partial \phi_1}{\partial \dot{x}_0}(0,0,0,T), \tag{13.20}$$

$$\beta_2 = \frac{\partial \phi_2}{\partial \dot{x}_0}(0,0,0,T). \tag{13.21}$$

Let us notice that these coefficients verify $\alpha_1 + \beta_2 = \lambda_1 + \lambda_2$ and $\alpha_1\beta_2 - \alpha_2\beta_1 = \lambda_1\lambda_2$. Eqs. (13.16) and (13.17) can be inserted into the equation $\chi = 0$ in order to get an equation connecting Π_1 and Π_2, and hence to obtain the coordinate ψ:

$$\psi = LK(-\beta_2 u + \beta_1 v) + a_3 u^2 + a_4 uv + a_5 v^2, \tag{13.22}$$

where a_3, a_4 and a_5 can be expressed in terms of α_1, α_2, β_1, β_2 and the elements of the Hessian matrix of ϕ at $(0,0,0,T)$, and $L \neq 0$ is a constant that will be set later. The change of variables $(u,v) \mapsto (\chi,\psi)$ occurs to be valid if and only if $\beta_1 \neq 0$ (because it was assumed that $\nu \neq 0$).

13.3 Local Expansion of the Poincaré Map

13.3.1 *General case ($\beta_1 \neq 0$)*

The Poincaré map expressed in the (χ,ψ) variables will be referred to as Ξ. The first order relationship connecting the coordinates (u,v) to (χ,ψ) is given by:

$$u = -\frac{1}{K}\chi, \tag{13.23}$$

$$v = -\frac{\beta_2}{K\beta_1}\chi + \frac{1}{LK\beta_1}\psi. \tag{13.24}$$

Restricting to the first order in (χ,ψ), the border between impact and non impact side is given by $\chi = 0$. Furthermore, the admissible area is locally defined in the (χ,ψ) variables by:

$$L\lambda_1\lambda_2\psi \geq 0. \tag{13.25}$$

13.3.1.1 *Nonimpact side ($\chi \geq 0$)*

In that case, it is sufficient to restrict the analysis to the first order. By using a first order Taylor expansion in the Eq. (13.5) the Poincaré map is obtained in the (u, v) variables:

$$\begin{aligned} \Pi_1(0, u, v, T) &= \alpha_1 u + \beta_1 v, \\ \Pi_2(0, u, v, T) &= \alpha_2 u + \beta_2 v, \end{aligned} \tag{13.26}$$

which, by applying the change of variables defined by Eqs. (13.15), (13.22), (13.23) and (13.24), leads to:

$$\begin{aligned} \Xi_1(\chi, \psi) &= (\lambda_1 + \lambda_2)\chi - \frac{1}{L}\psi, \\ \Xi_2(\chi, \psi) &= L\lambda_1\lambda_2\chi. \end{aligned}$$

In order to obtain a Poincaré map similar to the expression given in the reference [Nordmark (1991)], L has to be set to -1, which yields:

$$\begin{aligned} \Xi_1(\chi, \psi) &= (\lambda_1 + \lambda_2)\chi + \psi, \\ \Xi_2(\chi, \psi) &= -\lambda_1\lambda_2\chi. \end{aligned} \tag{13.27}$$

13.3.1.2 *Impact side ($\chi < 0$)*

As pointed out in the Section 13.2, a response starting from (u, v) at $t_0 = 0$ can exhibit a maximum at $\tau_1(u, v)$ and at $T - \tau_2(u, v)$. In the first case, the occurrence of a maximum implies $P_2(0, \lambda, \mu, \tau_1) = 0$, and hence with a second order expansion:

$$\mu + \frac{\partial P_2}{\partial t}(0, \lambda, \mu, 0)\tau_1 + \frac{1}{2}\frac{\partial^2 P_2}{\partial t^2}(0, \lambda, \mu, 0)\tau_1^2 = 0. \tag{13.28}$$

Taking into account the fact that $\tau_1(u, v)$ tends to 0 as (u, v) approaches $(0, 0)$, the following local expansion of τ_1 is obtained:

$$\tau_1(u, v) = -\frac{1}{\nu\lambda_1\lambda_2}(\alpha_2 u - \alpha_1 v). \tag{13.29}$$

The maximum of the response is then found to be in (χ, ψ) variables:

$$P_1(0, \lambda, \mu, \tau_1) = \frac{1}{K\lambda_1\lambda_2}\psi, \tag{13.30}$$

which by the relation (13.25) is always negative for initial conditions lying in the admissible area. Therefore an impact cannot occur in the neighborhood of $t = 0$, but only in the neighborhood of $t = T$.

Assume that starting from the initial condition $(0, \lambda, \mu)$ the solution impacts once at $t_1 = T - \tau \in [0, T]$. The occurrence of an impact at t_1 implies:

$$P_1(0, \lambda, \mu, T - \tau) = 0. \tag{13.31}$$

By using a third order Taylor expansion in the Eq. (13.31), τ is obtained as a solution of the third order equation

$$u - v\tau + \frac{1}{2}\frac{\partial^2 P_1}{\partial t^2}(0, \lambda, \mu, T)\tau^2 - \frac{1}{6}\frac{\partial^3 P_1}{\partial t^3}(0, \lambda, \mu, T)\tau^3 = 0. \tag{13.32}$$

Since on the impact side $\chi < 0$, τ can be locally sought under the form:

$$\tau(\chi, \psi) = A_1\sqrt{-\chi} + B_1\chi + C_1\psi, \tag{13.33}$$

where A_1, B_1 and C_1 are unknown coefficients. By inserting (13.33) into (13.32) and by restricting to the first order in (χ, ψ), the unknown coefficients are found to be:

$$A_1 = \sqrt{\frac{2}{\nu K}}, \tag{13.34}$$

$$B_1 = \frac{\gamma\nu - \delta + 3\nu\beta_2/\beta_1}{3K\nu^2}, \tag{13.35}$$

$$C_1 = \frac{1}{K\nu\beta_1}. \tag{13.36}$$

Once the expansion of the impact time has been obtained, the local Poincaré map can be sought by analytical means. The impact at $t_1 = T - \tau$ implies:

$$P_1^{\text{impact}}(0, \lambda, \mu, T) = P_1(T - \tau, P_1(0, \lambda, \mu, t_1^+), P_2(0, \lambda, \mu, t_1^+), \tau). \tag{13.37}$$

Hence, because $P_1(0, \lambda, \mu, t_1^+) = 0$ and $P_1(T, 0, 0, 0) = 0$, a second order

Taylor expansion yields:

$$\begin{aligned}
P_1^{\text{impact}}(0,\lambda,\mu,T) = &\left[\frac{\partial P_1}{\partial t}(T,0,0,0) - \frac{\partial P_1}{\partial t_0}(T,0,0,0)\right]\tau + \\
&+\frac{\partial P_1}{\partial \mu}(T,0,0,0)P_2(0,\lambda,\mu,t_1^+) + \\
&+\left[\frac{1}{2}\frac{\partial^2 P_1}{\partial t_0^2}(T,0,0,0) + \frac{1}{2}\frac{\partial^2 P_1}{\partial t^2}(T,0,0,0) + \right. \\
&\left.\frac{\partial^2 P_1}{\partial t_0 \partial t}(T,0,0,0)\ \right]\tau^2 + \\
&+\left[\frac{\partial^2 P_1}{\partial \mu \partial t_0}(T,0,0,0) - \frac{\partial^2 P_1}{\partial \mu \partial t}(T,0,0,0)\right]\tau P_2(0,\lambda,\mu,t_1^+) + \\
&+\frac{1}{2}\frac{\partial^2 P_1}{\partial \mu^2}(T,0,0,0)P_2(0,\lambda,\mu,t_1^+)^2.
\end{aligned} \tag{13.38}$$

The impact law given in (13.3) locally yields:

$$P_2(0,\lambda,\mu,t_1^+) = -r\left(v + \nu\tau + \frac{\gamma\nu-\delta}{2}\tau^2\right) - \frac{1}{2}s\nu^2\tau^2, \tag{13.39}$$

because $\frac{\partial^2 P_2}{\partial t^2}(0,0,0,T) = \gamma\nu - \delta$. All the quantities appearing in the Eq. (13.38) can then be reduced to the first order in (χ,ψ):

$$P_2(0,\lambda,\mu,t_1^+) = -r\sqrt{\frac{2\nu}{K}}\sqrt{-\chi} + \frac{2r(\gamma\nu-\delta)+3s\nu^2}{3K\nu}\chi, \tag{13.40}$$

$$\tau P_2(0,\lambda,\mu,t_1^+) = \frac{2r}{K}\chi, \tag{13.41}$$

$$\tau^2 = -\frac{2}{K\nu}\chi, \tag{13.42}$$

$$P_2(0,\lambda,\mu,t_1^+)^2 = -2r^2\frac{\nu}{K}\chi. \tag{13.43}$$

Moreover, the following relationship holds:

$$\forall t \in [0,T], \quad P_1(t, P_1(0,\lambda,\mu,t), P_2(0,\lambda,\mu,t), T-t) = P_1(0,\lambda,\mu,T), \tag{13.44}$$

for all (λ, μ) such that no impacts occur on $[0, T]$. Differentiating this equation with respect to t, and setting $(\lambda, \mu, t) = (0, 0, T)$ leads to:

$$\frac{\partial P_1}{\partial t_0}(T, 0, 0, 0) - \frac{\partial P_1}{\partial t}(T, 0, 0, 0) = \nu \frac{\partial P_1}{\partial \mu}(T, 0, 0, 0). \tag{13.45}$$

The same procedure is applied by differentiating twice Eq. (13.44) with respect to t in order to get:

$$\begin{aligned}\frac{1}{2}\frac{\partial^2 P_1}{\partial t_0^2}(T, 0, 0, 0) + \frac{1}{2}\frac{\partial^2 P_1}{\partial t^2}(T, 0, 0, 0) - \frac{\partial^2 P_1}{\partial t_0 \partial t}(T, 0, 0, 0) \\ = \nu \left[\frac{\partial^2 P_1}{\partial \mu \partial t_0}(T, 0, 0, 0) - \frac{\partial^2 P_1}{\partial \mu \partial t}(T, 0, 0, 0) \right] + \\ \frac{1}{2}\left[\nu \frac{\partial P_1}{\partial \lambda}(T, 0, 0, 0) - (\gamma\nu - \delta)\frac{\partial P_1}{\partial \mu}(T, 0, 0, 0) \right] - \\ \frac{\nu^2}{2}\frac{\partial^2 P_1}{\partial \mu^2}(T, 0, 0, 0).\end{aligned} \tag{13.46}$$

Furthermore, if Eq. (13.44) is differentiated with respect to λ, then (λ, μ, t) is set to $(0, 0, T)$, and the same procedure starting with a differentiation with respect to μ is applied, a 2×2 linear system is obtained which yields:

$$\frac{\partial P_1}{\partial \mu}(T, 0, 0, 0) = 0, \tag{13.47}$$

$$\frac{\partial P_1}{\partial \lambda}(T, 0, 0, 0) = 1. \tag{13.48}$$

By iterating the method one step beyond, namely by differentiating Eq. (13.44) twice with respect to (λ, λ), (λ, μ), (μ, μ) and then setting $(\lambda, \mu, t) = (0, 0, T)$, a 3×3 linear system is obtained that gives:

$$\begin{aligned}\frac{\partial^2 P_1}{\partial \mu^2}(T, 0, 0, 0) = 0, \\ \frac{\partial^2 P_1}{\partial \mu \partial \lambda}(T, 0, 0, 0) = 0, \\ \frac{\partial^2 P_1}{\partial \lambda^2}(T, 0, 0, 0) = 0.\end{aligned} \tag{13.49}$$

Finally, Eq. (13.44) can be differentiated with respect to t, next on one hand with respect to λ and on the other hand with respect to μ, and then both relations can be applied at $(\lambda, \mu, t) = (0, 0, T)$. A linear combination of the two relations and the results of Eqs. (13.47) and (13.48) yield:

$$\frac{\partial^2 P_1}{\partial\mu\partial t_0}(T,0,0,0) - \frac{\partial^2 P_1}{\partial\mu\partial t}(T,0,0,0) - \nu\frac{\partial^2 P_1}{\partial\mu^2}(T,0,0,0)$$
$$= -\frac{\partial P_1}{\partial\lambda}(T,0,0,0) + \gamma\frac{\partial P_1}{\partial\mu}(T,0,0,0)$$
$$= -1. \tag{13.50}$$

Insertion of the results of Eqs. (13.45), (13.46), (13.47), (13.48), (13.49) and (13.50) into Eq. (13.38) gives:

$$P_1^{\text{impact}}(0,\lambda,\mu,T) = \frac{1+2r}{K}\chi, \tag{13.51}$$

and with similar calculations:

$$P_2^{\text{impact}}(0,\lambda,\mu,T) = -(1+r)\sqrt{\frac{2\nu}{K}}\sqrt{-\chi} - \frac{1}{K\beta_1}\psi +$$
$$\Big\{2\nu(1+r)\beta_1(\beta\beta_1 - \gamma\alpha_1) - \nu\beta_2\lambda_1\lambda_2 + \tag{13.52}$$
$$\left[2\gamma(1+\nu) + \frac{2}{3}(1+r)(\gamma\nu - \delta) + s\nu^2\right]\beta_1\lambda_1\lambda_2\Big\}\chi.$$

Eq. (13.51) shows that the Poincaré map lies in the admissible area since $\chi < 0$ implies $P_1^{\text{impact}}(0,\lambda,\mu,T) \leq 0$. Moreover, from these expressions in the coordinates (λ,μ), the Poincaré map Π in the (u,v) coordinates can be obtained by using a second order Taylor expansion of (13.5) which gives, with Eqs. (13.18)-(13.21):

$$u = \alpha_1\lambda + \beta_1\mu + \frac{1}{2}\frac{\partial^2\phi_1}{\partial x_0^2}(0,0,0,T)\lambda^2 + \frac{\partial^2\phi_1}{\partial x_0\partial\dot{x}_0}(0,0,0,T)\lambda\mu +$$
$$\frac{1}{2}\frac{\partial^2\phi_1}{\partial\dot{x}_0^2}(0,0,0,T)\mu^2, \tag{13.53}$$
$$v = \alpha_2\lambda + \beta_2\mu + \frac{1}{2}\frac{\partial^2\phi_2}{\partial x_0^2}(0,0,0,T)\lambda^2 + \frac{\partial^2\phi_2}{\partial x_0\partial\dot{x}_0}(0,0,0,T)\lambda\mu +$$
$$\frac{1}{2}\frac{\partial^2\phi_2}{\partial\dot{x}_0^2}(0,0,0,T)\mu^2. \tag{13.54}$$

Once the Poincaré map Π has been obtained, it only remains to apply the change of variables defined in (13.15) and (13.22) to obtain the local

Poincaré map:

$$\begin{aligned}\Xi_1(\chi,\psi) &= (1+r)\beta_1\sqrt{2\nu K}\sqrt{-\chi}+\alpha\chi+\psi,\\ \Xi_2(\chi,\psi) &= -\lambda_1\lambda_2 r^2\chi.\end{aligned} \tag{13.55}$$

Setting $K = 1/\left(2\nu(1+r)^2\right)$ finally yields

$$\begin{aligned}\Xi_1(\chi,\psi) &= \beta_1\sqrt{-\chi}+\alpha\chi+\psi,\\ \Xi_2(\chi,\psi) &= -\lambda_1\lambda_2 r^2\chi,\end{aligned} \tag{13.56}$$

where:

$$\begin{aligned}\alpha = \lambda_1+\lambda_2-\nu\beta_1 s-2(1+r)\alpha_1+(1+r)^2\left(\beta_2^2+\nu\frac{\partial^2\phi_1}{\partial\dot{x}_0^2}(0,0,0,T)\right)+\\ 2(1+r)(\gamma\alpha_1-\beta\beta_1)\frac{\beta_1}{\lambda_1\lambda_2}-\frac{2}{3\nu}\left[3\gamma(1+\nu)+(1+r)(\gamma\nu-\delta)\right]\beta_1.\end{aligned} \tag{13.57}$$

The results of (13.56) are similar to what was given in [Nordmark (1991)], except the term $\alpha\chi$ in the first component of Ξ, which has no reason to be zero in the general case. Moreover, the result given by (13.56) includes the case when $\beta_1 < 0$ which has not been discussed in detail in [Nordmark (1991)]. Finally, the local expansion of the Poincaré map given in (13.56) shows that this map is not differentiable at the fixed point $(0,0)$.

13.3.2 *Particular case* ($\beta_1 = 0$)

As underlined in Section 13.2, the change of variables $(u,v)\mapsto(\chi,\psi)$ is not valid when $\beta_1 = 0$. In that case, $\alpha_1+\beta_2 = \lambda_1+\lambda_2$ and $\alpha_1\beta_2 = \lambda_1\lambda_2$ hence $\{\alpha_1,\beta_2\} = \{\lambda_1,\lambda_2\}$: for example, set $\lambda_1 = \beta_2$ and $\lambda_2 = \alpha_1$. Note that the preceding relationships show that in that case the eigenvalues λ_1 and λ_2 are necessarily real.

Another change of coordinates can be defined to give a convenient local form to the Poincaré map. Let us define:

$$\chi = -\tilde{K}\left(u+\frac{1}{2\nu}v^2\right), \tag{13.58}$$

$$\psi = \tilde{L}v. \tag{13.59}$$

The local boundary between the impact and nonimpact side is still given by $\chi = 0$. Moreover, the admissible area delimited by the obstacle is defined, to the first order in the (χ,ψ) variables, by:

$$\lambda_2\chi \geq 0. \tag{13.60}$$

Therefore, if $\lambda_2 < 0$, for any initial condition close to $(0,0)$ inside the admissible area, the system impacts once on $[0,T]$, and if $\lambda_2 > 0$, for any initial condition close to $(0,0)$ inside the admissible area, the system never impacts on $[0,T]$.

The inverse transformation restricted to the first order in (χ, ψ) is given by:

$$u = -\frac{1}{\tilde{K}}\chi, \tag{13.61}$$

$$v = \frac{1}{\tilde{L}}\psi. \tag{13.62}$$

Following the same procedure as in Section 13.3.1, the first order local Poincaré map can be obtained:
if $\chi \geq 0$:

$$\Xi_1(\chi, \psi) = \lambda_2 \chi,$$
$$\Xi_2(\chi, \psi) = -\alpha_2 \frac{\tilde{L}}{\tilde{K}}\chi + \lambda_1 \psi,$$

and if $\chi < 0$:

$$\Xi_1(\chi, \psi) = (r^2\lambda_2 - k_1)\chi,$$
$$\Xi_2(\chi, \psi) = -(1+r)\lambda_1 \tilde{L}\sqrt{\frac{2\nu}{\tilde{K}}}\sqrt{-\chi} + \alpha'\chi + \lambda_1\psi,$$

where

$$k_1 = (1+r)^2\left(\lambda_2 - \lambda_1^2 - \nu\frac{\partial^2\phi_1}{\partial \dot{x}_0^2}(0,0,0,T)\right). \tag{13.63}$$

By setting $\tilde{L} = -1$ and $\tilde{K} = 2\nu(1+r)^2$, the local map becomes:
if $\chi \geq 0$:

$$\Xi_1(\chi, \psi) = \lambda_2 \chi, \tag{13.64}$$

$$\Xi_2(\chi, \psi) = \frac{\alpha_2}{2\nu(1+r)^2}\chi + \lambda_1\psi, \tag{13.65}$$

and if $\chi < 0$:

$$\Xi_1(\chi, \psi) = (r^2\lambda_2 - k_1)\chi, \tag{13.66}$$

$$\Xi_2(\chi, \psi) = \lambda_1\sqrt{-\chi} + \alpha'\chi + \lambda_1\psi, \tag{13.67}$$

where

$$\alpha' = \frac{-1}{2\nu(1+r)^2}\Big\{\frac{\lambda_1}{\nu}\left[\frac{2}{3}(1+r)(\gamma\nu-\delta)+2\gamma(1+\nu)+s\nu^2\right]+$$
$$(1+2r)\alpha_2-\nu(1+r)^2\frac{\partial^2\phi_2}{\partial\dot{x}_0^2}(0,0,0,T)-2\gamma(1+r)\Big\}. \qquad (13.68)$$

13.4 Stability of the Nondifferentiable Fixed Point

13.4.1 *General case ($\beta_1 \neq 0$)*

13.4.1.1 *Complex eigenvalues*

On the nonimpact side, the local Poincaré map is given by (13.27). By another change of coordinates defined by $x = \chi$ and $y = (\Re(\lambda_1)\chi + \psi)/\Im(\lambda_1)$, for which the impact boundary is given by $x = 0$, the map becomes:

$$\Upsilon(x,y) = \rho\mathbf{R}(\theta)\begin{pmatrix} x \\ y \end{pmatrix}, \qquad (13.69)$$

where $\lambda_1 = \rho e^{i\theta}$ and $\mathbf{R}(\theta)$ is the rotation of angle θ. Therefore, any point entering the nonimpact side will eventually leave it after a finite number of iterations. Moreover, on the impact side, the square root term in (13.56) leads to an infinite stretching in the neighborhood of $(0,0)$. If $|\lambda_1| \geq 1$ (namely if the periodic motion without impact is unstable), then the mapping is repulsive on both sides. Otherwise, the mapping is stable on the nonimpact side, but the finite contraction competes with the infinite stretching of the impact side, and because the iterations cannot all remain on the nonimpact side, it can be inferred that the local map is repulsive.

Thus, when the eigenvalues λ_1 and λ_2 are complex the nondifferentiable fixed point is unstable.

13.4.1.2 *Real eigenvalues*

The stability of the nondifferentiable fixed point depends on the eigenvalues λ_1 and λ_2. The results of the following study are summarized in Table 13.4.1.2. The case when λ_1 and λ_2 have opposite signs have been included in the study because it is interesting from the theoretical point of view, but it has not been mentioned in Table 13.4.1.2 since it cannot occur for the physical system governed by Eq. (13.1).

If $|\lambda_1| \geq 1$ or $|\lambda_2| \geq 1$: In that case, the grazing periodic solution without

Table 13.1 Stability of the nondifferentiable fixed point when $\beta_1 \neq 0$ and λ_1 and λ_2 are real.

$(\lambda_1 < 0$ and $\lambda_2 < 0)$ or $\lambda_1 \geq 1$ or $\lambda_2 \geq 1$	$0 < \lambda_1 < 1$ and $0 < \lambda_2 < 1$	
	$\beta_1 > 0$	$\beta_1 < 0$
UNSTABLE	STABLE	STABLE $(\max\{\lambda_1, \lambda_2\}\chi + \psi \geq 0$ and $\psi \leq 0)$ UNSTABLE (otherwise)

impacts is unstable. Because of the square root term, the fixed point is also repulsive on the impact side. Therefore, the nondifferentiable fixed point is unstable.

If $|\lambda_1| < 1$ and $|\lambda_2| < 1$ and $\lambda_1 \neq \lambda_2$: Let us consider an initial condition (χ, ψ) lying in the nonimpact side. As λ_1 and λ_2 are not equal, it can be assumed that $\lambda_1 > \lambda_2$. A new change of variables can be defined to give a convenient form to the local mapping on the nonimpact side. By setting,

$$x = \frac{\lambda_1 \chi + \psi}{\lambda_1 - \lambda_2}, \tag{13.70}$$

$$y = \frac{-\lambda_2 \chi - \psi}{\lambda_1 - \lambda_2}, \tag{13.71}$$

the impact condition becomes $x + y < 0$ and when $x + y \geq 0$ the Poincaré map is locally given by

$$\Upsilon(x, y) = \begin{pmatrix} \lambda_1 x \\ \lambda_2 y \end{pmatrix}. \tag{13.72}$$

The point (χ, ψ) such that $\chi \geq 0$ is represented by a point (x, y) such that $x + y \geq 0$. Several situations then occur depending on the sign of λ_1 and λ_2.

- If $\lambda_1 < 0$ and $\lambda_2 < 0$:
  the admissible area is defined by $\psi \leq 0$, and the Poincaré map is given by $\Xi_1(\chi, \psi) = (\lambda_1 + \lambda_2)\chi + \psi \leq 0$. Hence the first iterate of Ξ lies on the impact side.
- If $\lambda_1 > 0$ and $\lambda_2 > 0$:
  the admissible area is defined by $\psi \leq 0$, which is equivalent to

$$\lambda_2 x + \lambda_1 y \geq 0. \tag{13.73}$$

The position of $\Upsilon^{on}(x,y)$ is determined by the sign of

$$q_n = \lambda_1^n x + \lambda_2^n y. \tag{13.74}$$

If $x \geq 0$, $\lambda_1 > \lambda_2$ implies that $q_n \geq 0$ for all n, and thus the nonimpact side is stable. If $x < 0$, then Eq. (13.73) implies $y \geq 0$, and q_n has the same sign as $x + \left(\frac{\lambda_2}{\lambda_1}\right)^n y$ which decreases from $x + y \geq 0$ to $x < 0$ as n increases from 0 to $+\infty$. Therefore for n sufficiently large $\Upsilon^{on}(x,y)$ passes into the impact side.

- If $\lambda_1 > 0$ and $\lambda_2 < 0$:
  the admissible area is defined by $\psi \geq 0$, which is equivalent to $\lambda_2 x + \lambda_1 y \leq 0$. The conditions $x + y \geq 0$ and $\lambda_2 x + \lambda_1 y \leq 0$ imply $x \geq 0$.
  - If $\lambda_1 < -\lambda_2$: q_n has the same sign as $x + (-1)^n \left|\frac{\lambda_2}{\lambda_1}\right|^n y$. If $y \neq 0$, the sign of q_n is determined by $(-1)^n y$ if n is sufficiently large: whatever the sign of y, an integer n can be found such that $\Upsilon^{on}(x,y)$ lies in the impact side. Moreover, if $y = 0$ then q_n is always positive: $\Upsilon^{on}(x,0)$ remains in the nonimpact side for all n, and the line $y = 0$ is stable under iterations of Υ.
  - If $\lambda_1 > -\lambda_2$: q_n has the same sign as $(-1)^n y + \left|\frac{\lambda_1}{\lambda_2}\right|^n x$. If n is even, $(-1)^n y + \left|\frac{\lambda_1}{\lambda_2}\right|^n x$ increases from $x + y \geq 0$ to $+\infty$ as n goes from 0 to $+\infty$. If n is odd, $(-1)^n y + \left|\frac{\lambda_1}{\lambda_2}\right|^n x$ increases from $-y - \frac{\lambda_1}{\lambda_2}x$ to $+\infty$ as n goes from 1 to $+\infty$. But $(-y - \frac{\lambda_1}{\lambda_2}x)$ has the same sign as $\lambda_1 x + \lambda_2 y$ which is equal to $(\lambda_1 + \lambda_2)(x + y) - (\lambda_2 x + \lambda_1 y) \geq 0$. Therefore, $q_n \geq 0$ for all n, which means that the nonimpact side is stable under iterations of Υ.
  - If $\lambda_1 = -\lambda_2$: q_n has the same sign as $x + (-1)^n y$. If n is even, this quantity is clearly positive, whereas if n is odd then $x - y$ has the same sign as $-\lambda_2 x - \lambda_1 y \geq 0$. The nonimpact side is thus stable under iterations of Υ, as it was the case when $\lambda_1 > -\lambda_2$.

Let us now consider an initial condition (χ, ψ) close to $(0,0)$ lying in the impact side. According to Eq. (13.56), $\Xi_1(\chi,\psi)$ has locally the same sign as $\beta_1\sqrt{-\chi}$. Hence if $\beta_1 > 0$ then $\Xi(\chi,\psi)$ lies on the nonimpact side, and if $\beta_1 < 0$ then the impact side is stable by iteration of Ξ.

The preceding study can be summarized as follows:

- If $\lambda_1 < 0$ and $\lambda_2 < 0$: the fixed point is unstable.
- If $\lambda_1 > 0$ and $\lambda_2 > 0$:
  - If $\beta_1 > 0$: the fixed point is stable.
  - If $\beta_1 < 0$: for all initial conditions (χ, ψ) close to $(0,0)$ such that $\lambda_1\chi + \psi \geq 0$ and $\psi \leq 0$, the iterates of the Poincaré map converge to $(0,0)$ and remain inside this domain. For any other initial condition the iterates of the Poincaré map move away from the fixed point $(0,0)$.
- If $\lambda_1 > 0$ and $\lambda_2 < 0$:
  - If $\lambda_1 < -\lambda_2$: for all initial conditions (χ, ψ) close to $(0,0)$ such that $\lambda_2\chi + \psi = 0$ and $\psi \geq 0$, the iterates of the Poincaré map converge to $(0,0)$ and remain inside this line. For any other initial condition the iterates of the Poincaré map move away from the fixed point $(0,0)$.
  - If $\lambda_1 \geq -\lambda_2$:
    - If $\beta_1 > 0$: the fixed point is stable.
    - If $\beta_1 < 0$: for all initial conditions (χ, ψ) close to $(0,0)$ such that $\chi \geq 0$ and $\psi \geq 0$, the iterates of the Poincaré map converge to $(0,0)$ from the nonimpact side. For any other initial condition the iterates of the Poincaré map move away from the fixed point $(0,0)$.

If $\lambda_2 = \lambda_1$ and $|\lambda_1| < 1$: The matrix defining the local Poincaré map on the nonimpact side can no longer be transformed into diagonal form. Nevertheless, the nth iterate of Ξ can easily be expressed under the form:

$$\Xi^{on}(\chi, \psi) = \begin{bmatrix} (n+1)\lambda_1^n & n\lambda_1^{n-1} \\ -n\lambda_1^{n+1} & -(n-1)\lambda_1^n \end{bmatrix} \begin{pmatrix} \chi \\ \psi \end{pmatrix}. \tag{13.75}$$

Therefore, $\Xi_1^{on}(\chi, \psi)$ has the same sign as $\lambda_1^{n-1}\left[\left(1 + \dfrac{1}{n}\right)\lambda_1\chi + \psi\right]$. Moreover, the admissible area is defined by $\psi \leq 0$.

- If $\lambda_1 > 0$: $\Xi_1^{on}(\chi, \psi)$ has the same sign as $\left(1 + \frac{1}{n}\right)\lambda_1\chi + \psi$, which decreases from $2\lambda_1\chi + \psi$ to $\lambda_1\chi + \psi$ as n increases from 1 to $+\infty$. This property implies that if $\lambda_1\chi + \psi < 0$ then $\Xi^{on}(\chi, \psi)$ lies on the impact side for n sufficiently large, and that $\Xi^{on}(\chi, \psi)$ remains on the nonimpact side if $\lambda_1\chi + \psi \geq 0$ (the area $\lambda_1\chi + \psi \geq 0$ being stable under iterations of Ξ).

- If $\lambda_1 < 0$: $\Xi_1^{on}(\chi, \psi)$ has the same sign as $(-1)^{n-1}\left[\left(1+\frac{1}{n}\right)\lambda_1\chi+\psi\right]$.

  Similarly to the preceding case, if $\lambda_1\chi+\psi \leq 0$ then $\left(1+\frac{1}{n}\right)\lambda_1\chi+\psi \leq 0$ for all n, and hence $\Xi^{o2}(\chi,\psi)$ lies on the impact side. If $\lambda_1\chi+\psi<0$ then $\left(1+\frac{1}{n}\right)\lambda_1\chi+\psi \geq 0$ is either always positive, or negative when n is small and positive afterwards. It is therefore clear that some n can be found so that $\Xi_1^{on}(\chi,\psi)<0$, namely so that the nth iterate of Ξ lies on the impact side.

The results of stability are thus the same as what would have been obtained by setting $\lambda_1 = \lambda_2$ in the previous case.

13.4.2 *Particular case ($\beta_1 = 0$)*

13.4.2.1 *Positive eigenvalues*

In this subsection, the case when both eigenvalues λ_1 and λ_2 are positive is studied. As can be seen from the inequality (13.60) and since λ_2 is positive, the admissible area delimited by the obstacle is locally defined by $\chi \geq 0$: any admissible initial condition lies in the nonimpact side. The local map is in that case merely linear so the study of stability can be achieved by examining the eigenvalues of Ξ, which are λ_1 and λ_2. Moreover, the nonimpact side is stable under iterations of Ξ since $\lambda_2 > 0$. Therefore the nondifferentiable fixed point is stable if and only if $0 < \lambda_2 < 1$ and $0 < \lambda_1 < 1$.

13.4.2.2 *Negative eigenvalues*

In this subsection, the case when both eigenvalues λ_1 and λ_2 are negative is studied. As can be seen from the inequality (13.60) and since λ_2 is negative, the admissible area delimited by the stop is locally defined by $\chi \leq 0$. Therefore, any admissible initial condition lies in the impact side.

Let us consider an initial condition (χ_0, ψ_0) close to $(0,0)$, χ_0 and ψ_0 being of the same order with $\chi_0 < 0$. As can be seen from Eqs. (13.66)-(13.67), after the first iteration χ_1 is of the same order as ψ_1^2: one has $\chi_1 \approx (k_1 - r^2\lambda_2)\psi_1^2/\lambda_1^2$, and ψ_1 has locally the same sign as λ_1. Therefore, in order to study the stability of the nondifferentiable fixed point from the impact side, it is necessary to start from (χ_0, ψ_0) such that $\chi_0 = x\psi_0^2$ with $x < 0$ and $\psi_0 < 0$.

Because of such a relationship connecting χ_0 and ψ_0, higher order terms need to be considered in the local expansion of the Poincaré map given in Eq. (13.66):

$$\Xi_1(\chi,\psi) = (r^2\lambda_2 - k_1)\chi + 2k_1\sqrt{-\chi}\psi + k_1\psi^2. \tag{13.76}$$

As long as $\chi_n = \Xi_1^{on}(\chi,\psi)$ remains negative, the local expression of the nth iterate of Ξ is given by:

$$\begin{aligned} \chi_n &= x_n\psi_n^2, \\ \psi_n &= y_n\psi_0, \end{aligned} \tag{13.77}$$

where

$$y_n = \prod_{i=0}^{n-1} \lambda_1\left(1 - \sqrt{-x_i}\right), \tag{13.78}$$

and $(x_n)_{n\geq 0}$ is iteratively defined by $x_0 = x$, $x_{n+1} = \varphi(x_n)$ with:

$$\varphi(x) = \frac{k_1}{\lambda_1^2} + \frac{r^2\lambda_2}{\lambda_1^2}\frac{x}{(1-\sqrt{-x})^2}. \tag{13.79}$$

If (χ_n,ψ_n) remains on the impact side for n sufficiently large, it is easy to show that if $(x_n)_{n\geq 0}$ converges to $l < 0$ then the nondifferentiable fixed point is stable from the impact side if and only if $y_n \to 0$, which is equivalent to:

$$\left|\lambda_1\left(1 - \sqrt{-l}\right)\right| < 1. \tag{13.80}$$

Because the impact side leads to $\chi = x\psi^2$ with $\psi < 0$, the boundary between the impact and nonimpact sides must be expressed in terms of x. The time τ_2 defined in Section 13.2 can be locally written as $\tau_2 = \psi/\nu$ which is negative: the impact condition is then given by $u > 0$. A local expansion implies that an impact occurs on $[0, T]$ when $\chi = x\psi^2$ and $\psi < 0$ if and only if $x < -(1+r)^2$. Moreover, the admissible area (defined by $\lambda \leq 0$) is given by $x \leq x_l$, where

$$x_l = \frac{k_1 - (1+r)^2\lambda_2}{\lambda_1^2}. \tag{13.81}$$

Let us consider an initial condition (χ_0,ψ_0) such that $\chi_0 < 0$. After one iteration the local map gives $\chi_1 = x_0\psi_1^2$, with $x_0 \approx \dfrac{k_1 - r^2\lambda_2}{\lambda_1^2}$ and $\psi_1 < 0$. Two cases have to be considered depending on the position of x_0.

First, if $x_0 > -(1+r)^2$ (locally equivalent to $k_1 > r^2\lambda_2 + (1+r)^2\lambda_1^2$) then (χ_1, ψ_1) lies in the nonimpact side. Since $\chi_1 = x_0\psi_1^2$, higher order terms need to be considered in the expansion given in Eq. (13.64):

$$\Xi_1(\chi,\psi) = \lambda_2\chi + k_1\psi^2. \tag{13.82}$$

Therefore the next iteration gives $\psi_2 = \lambda_1^2\psi_0$ and $\chi_2 = \varphi(x_0)\psi_2^2$ where:

$$\varphi(x) = \frac{\lambda_2 x - k_1}{\lambda_1^2}. \tag{13.83}$$

The previous equation extends the definition of φ to $]-(1+r)^2, x_l[$. The condition $k_1 > r^2\lambda_2 + (1+r)^2\lambda_1^2$ implies that $x_l > -(1+r)^2$: x_l is on the nonimpact side, so the interval of study is restricted to $]-(1+r)^2, x_l]$. Since $\varphi(x_l) + (1+r)^2$ has the same sign as $\lambda_2 + \lambda_1^2$, two cases arise again:

- If $\lambda_2 + \lambda_1^2 > 0$: the nonimpact side is stable under iterations of Ξ. For all $x_0 \in [-(1+r)^2, x_l]$ it can be shown that the sequence $(x_n)_{n\geq 0}$ converges to the attractive fixed point $x_f = \dfrac{k_1}{\lambda_1^2 - \lambda_2}$ of φ. The nth iterate of Ξ is given by:

$$\chi_n = x_{n-1}\psi_n^2, \tag{13.84}$$
$$\psi_n = \lambda_1^{n-1}\psi_1. \tag{13.85}$$

  Therefore, the nondifferentiable fixed point is stable if and only if $-1 < \lambda_1 < 0$.
- If $\lambda_2 + \lambda_1^2 < 0$: it can be shown that for all $x_0 \in [-(1+r)^2, x_l] - \{x_f\}$ the nth iterate of Ξ passes into the impact side for n sufficiently large.

Let us now investigate the case when $x_0 < -(1+r)^2$, which is locally equivalent to $k_1 < r^2\lambda_2 + (1+r)^2\lambda_1^2$, corresponding to the impact side. Two cases have to be examined depending on the position of x_l.

- If $x_l \leq -(1+r)^2$ (equivalent to $k_1 \leq (1+r)^2(\lambda_2 - \lambda_1^2)$): The interval of study on the impact side is restricted to $]-\infty, x_l]$ and it can be shown that the impact side is stable under iterations of Ξ. The fixed points of φ are given by $x = -X^2$ where $X > 0$ are the real roots of the fourth order polynomial:

$$P(X) = \lambda_1^2 X^4 - 2\lambda_1^2 X^3 + (k_1 + \lambda_1^2 - r^2\lambda_2)X^2 - 2k_1 X + k_1. \tag{13.86}$$

  The real roots of P are sought in the interval $]X_l, +\infty[$ where $X_l = \sqrt{-x_l}$. After noticing that $P(X_l) < 0$ and $\lim_{X\to+\infty} P(X) = +\infty$, it can be shown that P has either one real root X_{f_1} or three real roots $X_{f_3} < X_{f_2} < X_{f_1}$ in the interval of study, which define the possible fixed points of φ: $x_{f_1} < x_{f_2} < x_{f_3}$. In the first case, x_{f_1} is attractive and all initial conditions $x_0 \in]-\infty, x_l]$ are attracted by x_{f_1}. According to Eq. (13.80), the nondifferentiable fixed point is stable if and only if $\lambda_1(1 - X_{f_1}) < 1$ which is equivalent to $X_{f_1} < X_s$ where $X_s = 1 - 1/\lambda_1$. By considering the monotony of P it follows that the nondifferentiable fixed point is stable if and only if $X_s > X_l$ and $P(X_s) > 0$ which is equivalent to:

$$k_1 > \lambda_2(1+r)^2 - (1-\lambda_1)^2, \tag{13.87}$$

$$k_1 > -(1-r^2\lambda_2)(1-\lambda_1)^2. \tag{13.88}$$

In the case where φ admits three fixed points, x_{f_1} and x_{f_3} are attractive while x_{f_2} is repulsive. Moreover, all initial conditions $x_0 \in]-\infty, x_{f_2}[$ are attracted by x_{f_1} and all initial conditions $x_0 \in]x_{f_2}, x_l[$ are attracted by x_{f_3}. Since $X_{f_3} < X_{f_1}$, the nondifferentiable fixed point is stable if and only if $X_{f_1} < X_s$ which is equivalent to $X_s > X_l$, $P(X_s) > 0$, $P'(X_s) > 0$ and $P''(X_s) > 0$. Therefore in that case the stability conditions are given by (13.87) and (13.88) and by the two other conditions:

$$k_1 > (1-\lambda_1)(\lambda_1 + r^2\lambda_2 - 2), \tag{13.89}$$

$$k_1 > -\lambda_1^2 + 6\lambda_1 - 6 + r^2\lambda_2. \tag{13.90}$$

- If $x_l > -(1+r)^2$ (equivalent to $k_1 > (1+r)^2(\lambda_2 - \lambda_1^2)$): The interval of study for φ is $]-\infty, -(1+r)^2[$. By considering the different cases of monotony for the polynomial P defined in (13.86), it can be shown that φ has either two fixed points $x_{f_1} < x_{f_2}$ (corresponding to the roots X_{f_1} and X_{f_2} of P) or none. In the first case, x_{f_1} is attractive while x_{f_2} is repulsive, and for all $x_0 \in]-\infty, x_{f_2}[$ the sequence $(x_n)_{n\geq 0}$ converges to x_{f_1} while for all $x_0 \in]x_{f_2}, -(1+r)^2[$ an iterate of Ξ passes into the nonimpact side for n sufficiently large. The nondifferentiable fixed point is thus stable from the impact side if and only if $X_s > 1+r$, $P(X_s) > 0$, $P'(X_s) > 0$ and $P''(X_s) > 0$, which is equivalent to $\lambda_1 > -1/r$ and conditions (13.88), (13.89) and (13.90). In the case when φ admits no fixed points, the sequence $(x_n)_{n\geq 0}$ is strictly increasing and eventually an iterate of Ξ passes into the nonimpact side.

The cases when the iterates of Ξ alternatively lie in the impact and nonimpact side remain to be examined. First of all, when $k_1 > r^2\lambda_2 + (1+$

$r)^2\lambda_1^2$ and $\lambda_2 + \lambda_1^2 < 0$, then (χ_1, ψ_1) is on the nonimpact side and (χ_n, ψ_n) passes into the impact side for n sufficiently large, and we are in the case when $x_l > -(1+r)^2$. Because $k_1 > r^2\lambda_2 + (1+r)^2\lambda_1^2$, the impact side is mapped into the nonimpact side by the local mapping Ξ. The sequence $(x_n)_{n\geq 0}$ is not convergent but for all $n \geq 0$ one has $x_n \in]x_m, x_l[$ where:

$$x_m = \frac{k_1(\lambda_1^2 + \lambda_2) - (1+r)^2\lambda_2^2}{\lambda_1^4}. \tag{13.91}$$

Let n be a given integer and assume that k iterates of Ξ lie on the nonimpact side, and $n-k$ lie in the impact side. The nth iterate of Ξ is given by:

$$\chi_n = x_n \psi_n^2, \tag{13.92}$$

$$\psi_n = \lambda_1^k \prod_{i=1}^{n-k} \lambda_1 \left(1 - \sqrt{-x_i}\right) \psi_0. \tag{13.93}$$

Since $(x_n)_{n\geq 0}$ is bounded, the nondifferentiable fixed point is stable if and only if $\psi_n \to 0$ as n tends to $+\infty$ and because $-r\lambda_1 \leq \lambda_1 \left(1 - \sqrt{-x_i}\right) \leq \lambda_1(1 - X_m)$, from Eq. (13.93) a sufficient condition for stability is given by $\lambda_1 > -1/r$ and $X_m < X_s$ where $X_m = \sqrt{-x_m}$. Such a condition is equivalent to $\lambda_1 > -1/r$ and:

$$k_1 < -\frac{\lambda_1^2(1-\lambda_1)^2 - (1+r)^2\lambda_2^2}{\lambda_1^2 + \lambda_2}. \tag{13.94}$$

The final case left to study is when $(1+r)^2(\lambda_2 - \lambda_1^2) < k_1 < r^2\lambda_2 + (1+r)^2\lambda_1^2$, which corresponds to $x_0 < -(1+r)^2 < x_l$. In that case φ has either two fixed points or none on the impact side. The behaviour on the nonimpact side leads to distinguish between two cases:

- If $\lambda_2 + \lambda_1^2 > 0$: The nonimpact side is stable under iterations of Ξ and the nondifferentiable fixed point is stable on this side if and only if $-1 < \lambda_1 < 0$. If φ has no fixed point on the impact side, it has been shown that every time an iterate of Ξ has entered the impact side, it eventually leaves it after a finite number of iterations. There is hence no extra condition for stability in that case. If φ has two fixed points on the impact side, according to the preceding study the conditions (13.88), (13.89) and (13.90) must be verified to obtain stability of the nondifferentiable fixed point also on the impact side.
- If $\lambda_2 + \lambda_1^2 < 0$: as it can be seen from the preceding study, two cases may occur:

– If φ has two fixed points on the impact side: the sequence $(x_n)_{n\geq 0}$ is attracted by x_{f_1} if $x_n < x_{f_2}$ for any $n \geq 1$. The nondifferentiable fixed point is then stable when the conditions $\lambda_1 > -1/r$, (13.88), (13.89) and (13.90) are fulfilled. When $(x_n)_{n\geq 0}$ does not converge to any fixed point of φ but oscillates in the range $]x_{f_2}, x_l[$, a sufficient condition for stability can be derived from the preceding study holding for $k_1 > r^2\lambda_2 + (1+r)^2\lambda_1^2$ and $\lambda_2 + \lambda_1^2 < 0$. It leads to $\lambda_1 > -1/r$ and $X_{f_2} < X_s$, which is already verified to ensure stability in the case when $(x_n)_{n\geq 0}$ converges to x_{f_1}.
– If φ has no fixed points on the impact side: this case is similar to the case when $k_1 > r^2\lambda_2 + (1+r)^2\lambda_1^2$ and $\lambda_2 + \lambda_1^2 < 0$ studied above. A sufficient condition for stability of the nondifferentiable fixed point is then given by $\lambda_1 > -1/r$ and the inequality (13.94).

Finally, let us sum up the conditions for stability of the nondifferentiable fixed point with the assumption that $-1/r < \lambda_1 < 0$ and $-1/r < \lambda_2 < 0$ (this condition is fulfilled in particular when the grazing solution is stable without the presence of an obstacle). Let us define the following conditions:

$$
\begin{aligned}
&\mathrm{A}: \ -(1-r^2\lambda_2)(1-\lambda_1)^2 < k_1 \leq (1+r)^2(\lambda_2 - \lambda_1^2),\\
&\mathrm{B}: \ k_1 > (1+r)^2(\lambda_2 - \lambda_1^2),\\
&\mathrm{C}: \ 0 < -\lambda_2 < \lambda_1^2 < 1,\\
&\mathrm{D}: \ -1/r < \lambda_2 < -\lambda_1^2 < 0,\\
&\mathrm{E}: \ P \text{ defined in (13.86) has at least one real root greater than } 1+r,\\
&\mathrm{F}: \ P \text{ has no real roots greater than } 1+r,\\
&\mathrm{G}: \ k_1 < -\frac{\lambda_1^2(1-\lambda_1)^2 - (1+r)^2\lambda_2^2}{\lambda_1^2 + \lambda_2}.
\end{aligned}
$$

The nondifferentiable fixed point is stable if the following proposition is true:

$$\text{A or (B and (C or (D and (E or (F and G))))).}$$

The preceding study of stability shows that the same kind of behaviour as the case when $\beta_1 \neq 0$ can be obtained: for some sets of parameters, the grazing solution can be stable from the impact side and unstable from the nonimpact side or conversely. Moreover, it is possible for an unstable periodic solution to be stabilized by the presence of a stop leading to zero velocity impacts.

13.5 Applications

13.5.1 *Linear harmonic oscillator*

In the case of a linear harmonic oscillator with damping, f is given by:

$$f(x, y, t) = \rho \cos(\omega t) - 2\varepsilon\omega_1 y - \omega_1^2 x, \tag{13.95}$$

where $\varepsilon \geq 0$. If the impact law is chosen to be the classical restitution law, G is given by:

$$G(x) = -rx, \tag{13.96}$$

where $r \in [0, 1]$ is the coefficient of restitution. Moreover, a grazing T-periodic solution (where $T = 2\pi/\omega$) exists if only if:

$$\rho^2 = x_{max}^2 \left[(\omega_1^2 - \omega^2)^2 + 4\varepsilon^2\omega_1^2\omega^2\right]. \tag{13.97}$$

In that case, the constants introduced in the Section 13.2 are given by:

$$t_1^0 = \frac{1}{\omega} \arctan\left(\frac{\omega_1^2 - \omega^2}{2\varepsilon\omega_1\omega}\right), \tag{13.98}$$

$$\nu = \omega^2 x_{max}, \tag{13.99}$$

$$\beta = \omega_1^2, \tag{13.100}$$

$$\gamma = 2\varepsilon\omega_1, \tag{13.101}$$

$$\delta = \nu\gamma, \tag{13.102}$$

$$s = 0. \tag{13.103}$$

13.5.1.1 *Underdamped oscillator* $(0 \leq \varepsilon < 1)$

The flow can be analytically expressed to obtain:

$$\alpha_1 = e^{-\varepsilon\omega_1 T} \left[\cos(\tilde{\omega}_1 T) + \eta \sin(\tilde{\omega}_1 T)\right], \tag{13.104}$$

$$\beta_1 = e^{-\varepsilon\omega_1 T} \frac{\sin(\tilde{\omega}_1 T)}{\tilde{\omega}_1}, \tag{13.105}$$

$$\alpha_2 = -\tilde{\omega}_1 e^{-\varepsilon\omega_1 T}(1 + \eta^2) \sin(\tilde{\omega}_1 T), \tag{13.106}$$

$$\beta_2 = e^{-\varepsilon\omega_1 T} \left[\cos(\tilde{\omega}_1 T) - \eta \sin(\tilde{\omega}_1 T)\right], \tag{13.107}$$

where $\tilde{\omega}_1 = \omega_1\sqrt{1-\varepsilon^2}$ and $\eta = \dfrac{\varepsilon}{\sqrt{1-\varepsilon^2}}$. The eigenvalues λ_1 and λ_2 are given by $\lambda_1 = \bar{\lambda_2} = e^{-\varepsilon\omega_1 T}e^{i\tilde{\omega}_1 T}$. The study of stability conducted in Section 13.4 yields:

- If $\sin(\tilde{\omega}_1 T) \neq 0$: $\beta_1 \neq 0$ and λ_1 and λ_2 are complex. The nondifferentiable fixed point is hence unstable.
- If $\sin(\tilde{\omega}_1 T) = 0$: $\beta_1 = 0$, so the results of Section 13.4.2 hold. Since $\sin(\tilde{\omega}_1 T) = 0$, there exists $k \in \mathbb{N}$ such that $\tilde{\omega}_1 T = k\pi$ and $\lambda_1 = \lambda_2$ verifies $|\lambda_1| < 1$.
  - If k is even: then λ_1 and λ_2 are both positive. Since $\lambda_1 < 1$ and $\lambda_2 < 1$, the results established in Section 13.4.2.1 show that the grazing solution is stable.
  - If k is odd: then λ_1 and λ_2 are both negative. Moreover, the constant k_1 defined in Eq. (13.63) is given by $k_1 = (1+r)^2\lambda_1(1-\lambda_1)$. Thus the parameters verify:

$$-(1-r^2\lambda_2)(1-\lambda_1)^2 < k_1 = (1+r)^2(\lambda_2 - \lambda_1^2).$$

According to the conditions given in the Section 13.4.2.2, the grazing solution is stable.

13.5.1.2 *Critical damping* ($\varepsilon = \mathbf{1}$)

The flow can be analytically expressed to obtain:

$$\alpha_1 = (1+\omega_1 T)e^{-\omega_1 T}, \tag{13.108}$$

$$\beta_1 = Te^{-\omega_1 T}, \tag{13.109}$$

$$\alpha_2 = -\omega_1^2 Te^{-\omega_1 T}, \tag{13.110}$$

$$\beta_2 = (1-\omega_1 T)e^{-\omega_1 T}. \tag{13.111}$$

The eigenvalues are then given by $\lambda_1 = \lambda_2 = e^{-\omega_1 T}$, and because $\beta_1 > 0$, Table 13.1 shows that the nondifferentiable fixed point is stable.

13.5.1.3 *Overdamped oscillator ($\varepsilon > 1$)*

The flow can be analytically expressed to obtain:

$$\alpha_1 = e^{-\omega_1 T}\left[\cosh(\tilde{\omega}_1 T) + \eta \sinh(\tilde{\omega}_1 T)\right], \tag{13.112}$$

$$\beta_1 = e^{-\omega_1 T}\frac{\sinh(\tilde{\omega}_1 T)}{\tilde{\omega}_1}, \tag{13.113}$$

$$\alpha_2 = \tilde{\omega}_1 e^{-\omega_1 T}(1-\eta^2)\sinh(\tilde{\omega}_1 T), \tag{13.114}$$

$$\beta_2 = e^{-\omega_1 T}\left[\cosh(\tilde{\omega}_1 T) - \eta \sinh(\tilde{\omega}_1 T)\right], \tag{13.115}$$

where $\tilde{\omega}_1 = \omega_1\sqrt{\varepsilon^2 - 1}$ and $\eta = \dfrac{\varepsilon}{\sqrt{\varepsilon^2-1}}$. The eigenvalues λ_1 and λ_2 are given by $\lambda_1 = e^{(1-\eta)\tilde{\omega}_1 T}$ and $\lambda_2 = e^{-(1+\eta)\tilde{\omega}_1 T}$, hence $0 < \lambda_1 < 1$ and $0 < \lambda_2 < 1$. Since $\beta_1 > 0$, the results of Table 13.4.1.2 show that the nondifferentiable fixed point is stable.

13.5.2 *Forced damped pendulum*

Let us consider the mathematical case of a harmonically forced damped pendulum, where f is given by:

$$f(x, y, t) = \rho\cos(\omega t) - c_1 y - c_2\sin(x). \tag{13.116}$$

The impact law is chosen to be Newton's restitution law, so G is given by:

$$G(x) = -rx, \tag{13.117}$$

where $r \in [0, 1]$ is the coefficient of restitution. A grazing T-periodic solution can be found when $c_1 = 0.2$, $c_2 = 1$, $\omega = 1$, $\rho = 1.7$ and $x_{max} = 2.40165433$. The constants introduced in the Section 13.2 can then be estimated by numerical procedures to obtain $t_1^0 = 2.83336472$, $\alpha_1 = 0.8644$, $\beta_1 = 2.919$, $\alpha_2 = -0.0065$, $\beta_2 = 0.3056$, $\delta_1 = 1.02$, $\eta_1 = -6.3469$, $\delta_2 = 0.4433$, $\eta_2 = 4.7358$, $\gamma_1 = -2.59$ and $\gamma_2 = 1.0868$. The eigenvalues of the Jacobian matrix of the Poincaré map are $\lambda_1 = 0.8281$ and $\lambda_2 = 0.3419$.

13.6 Conclusion

In this chapter, the dynamics of grazing periodic solutions with one impact at zero velocity per cycle have been investigated: The method and the results of Nordmark have been extended, the analytical calculations have been given in detail, and new situations have been pointed out. Cases of stability of a grazing solution, not predicted by the previous works on the subject, have been exhibited and illustrated on two examples. Some other peculiarities have also been exhibited, such as the partitioning of the neighborhood of the fixed point into an area of stability and an area of instability. Finally, numerical examples have outlined that a new situation analysed in this chapter can occur.

Chapter 14

Triple Pendulum with Impacts

In this chapter the triple pendulum with damping, external forcing and with impacts is investigated. We integrate the obtained system of governing equations between two successive impacts, and the discontinuity points are detected (by halving time step until obtaining required precision). At each impact time, the state of the system is transformed using the extended restitution coefficient rule. The theory of Aizerman and Gantmakher is used to calculate the fundamental solution matrices in the analysed system exhibiting discontinuities. The fundamental matrices are used during calculation of Lyapunov exponents, in stability analysis of periodic solutions (Floquet multipliers) and in shooting method for finding periodic orbits. Then some examples for three coupled identical rods with horizontal barrier are presented.

14.1 Introduction

A pendulum plays a very important role in mechanics since many interesting nonlinear dynamical behaviour can be illustrated and analysed using this simple system [Acheson and Mullin (1993)], [Bishop and Clifford (1996)], [Skeledon (1994)], [Yagasaki (1994)]. Even an externally excited simple pendulum can exhibit periodic, quasi-periodic and chaotic dynamics. It is obvious that coupled pendulums with obstacles can serve as models even for more complicated behaviour including that of energy pumped to the model, various resonances, jumps between different system states, various continuous and discontinuous bifurcations, symmetry breaking and crisis bifurcations, pools of attractions, oscillatory-rotational attractors, etc. We observe recently also an interest focused on experimental investigation of either simple or coupled mechanical and electronic setup of pendulums

[Blackburn *et. al.* (1987)], [Doerner *et. al.* (1994)], [Heng *et. al.* (1994)]. A driven pendulum can serve as a paradigm for to exploring the influence of nonlinearity on a model with a smal number of variables, and to model nonlinear systems arising from different fields like charge density naves [Grüner *et. al.* (1981)], Josephson junctions [Steward (1968)] and others [Baker and Gollub (1990)], [Di Lieto *et. al.* (1991)]. Three different chaotic motions, i.e. via the period doubling bifurcation, rotating chaotic solution and the so called tumbling chaos exhibited by the parametrically excited simple pendulum have been reported in [Bishop and Clifford (1996)]. The pendulum oscillations caused by random vibrations of its suspension axis has been studied in the reference [Landa and Zaikin (1996)]. Singularity theory yielding a integrable approximation of the Poincaré map has been applied to investigate resonances in a spring-pendulum (see [Broer *et. al.* (1998)]). The criteria of existence of chaos using Melnikov's method in a pendulum with feedback control has been investigated in [Yagasaki (1994)], [Yagasaki (1996)]. The works of Szemplińska-Stupnicka and her co-workers [Szemplińska-Stupnicka (1990)], [Szemplińska-Stupnicka *et. al.* (2000)], [Szemplińska-Stupnicka *et. al.* (2001)], [Tyrkiel *et. al.* (2000)] can be treated as a bridge between classical approaches to nonlinear dynamical systems and modern dynamics, using a mathematical pendulum as a simplest object exhibiting rich spectrum of nonlinear phenomena. Szemplińska-Stupnicka and Tyrkiel have also proposed a link between theoretical formulations of global bifurcations and a fractal-like structure of attraction boundaries, as well as a unification of theories of chaotic oscillations and fractals by analyzing the fractal structure of a strange chaotic attractor. Unfortunately, a single degree-of-freedom models are only the first step to understand a real behaviour of either natural or engineering systems. Many physical objects are modelled by a few degrees of freedom, i.e., by a system of second order ordinary differential equations. This is well known in mechanics, but even in physics an attempt to investigate coupled pendulums is recently observed. For instance, in references [Acheson and Mullin (1993)], [Skeledon and Mullin (1992)] both the theoretical and experimental (laboratory) investigations have been carried out on two and three coupled pendulums. In addition, we would like to emphasize that one can also approximate a continuous nonlinear system via the discrete ones in many branches of science (for example, earth-quake caused vibrations of a high building can be modelled as a triple inverted pendulum). On the other hand, it is well known that impact and friction accompany almost all real behaviour. Those problem appear during study of joints [Paoli *et. al.* (1992)], models of beams [Dowell and Schwartz (1983a)], [Dowell and Schwartz (1983b)], [Whiteman and Ferri (1996)], rotor-casing [Li and Paidoussis (1994)] or impact in gearboxes. This branch of research includes extremely many works and possesses a very long history, and therefore we only refer readers to the works oriented on numerical investigations, (see [Awrejcewicz and Delfs (1990a)], [Awrejcewicz and Delfs (1990b)], [Cone and Zadoks (1995)], [Foale and Bishop (1994)], [Lamarque and Bastien (2000)], [Popp and Stelter (1990)], [Shaw and Shaw (1989)], [Shaw (1986)], [Whiston (1987)]). The nonsmooth dynamical systems are analysed in both pure [Kunze (2000)] and applied sciences [Brogliato (1999)] (see also the long lists of the cited bibliography therein). This chapter matches all the mentioned research directions by illustrating some chosen nonlinear phenomena of three coupled pendulums with impacts. In spite of the mentioned possible applications of our investigations of the system to model other systems, we would like to point out briefly one more example. Consider the piston-connecting rod-crankshaft system shown schematically in Fig. 14.1.

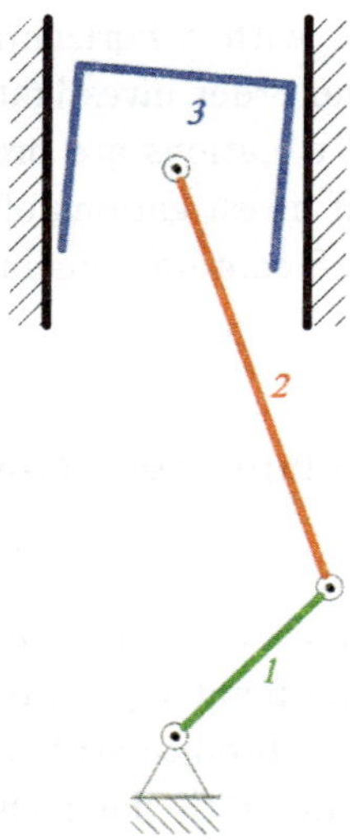

Fig. 14.1 The piston-connecting rod-crankshaft system.

This can be modelled very roughly also by our system, since we have three links and since between the piston and cylinder we have clearance, and an impact can occur. Observe, however, that this modelling can be only treated as a first step, because the real dynamics exhibited by the piston-connecting rod-crankshaft system is extremely complicated (see, for example, [Sygniewicz (1991)]).

This chapter is organized in the following manner (for more details see [Awrejcewicz *et. al.* (2002)]). First, in section 14.2 we describe the general model of triple physical pendulum without impacts, and the governing equations. In section 14.3 we provide the obstacles described by the set of algebraic inequalities and we present the generalized (for multi-degree-of-freedom systems) impact law based on restitution coefficient law. Section 14.4 contains description of the fundamental solution matrices and their usefulness in Lyapunov exponents calculation and in the investigation of periodic solutions. Particularly, we briefly describe the Aizerman-Gantmakher theory, which is used to calculate 'jumps' in the time evolution of the fundamental solution matrix in the discontinuity points. In section 14.5 we describe the model of a special case of the triple pendulum with impacts: three identical rods with a horizontal barrier. In section 14.6 the results of the accuracy and order investigations of the used method of integrating of the differential equations are presented. In section 14.7 we show the results of numerical investigations of three coupled rods with a horizontal barrier. Section 14.8 contains concluding remarks.

14.2 Investigated Pendulum and Governing Equations (Without Impacts)

Three joined physical pendulums moving on the plane of a global coordinate system x, y (with origin in the point O_i) are presented in Fig. 14.2. It is assumed that the links are absolute stiff bodies moving in a vacuum and coupled by viscous damping with the coefficients $\bar{c}_i$ $(i = 1, 2, 3)$. The first body is harmonically excited by $\bar{q}_1 \cos \bar{\omega}_1 \tau$, where τ denotes time. In addition, the following assumptions have been introduced:

- The mass centers of the links lie on the lines including the joints O_i.
- One of the principal central inertia axes of each link (z_{Ci}) is perpendicular to a movement pendulum plane $x - y$.

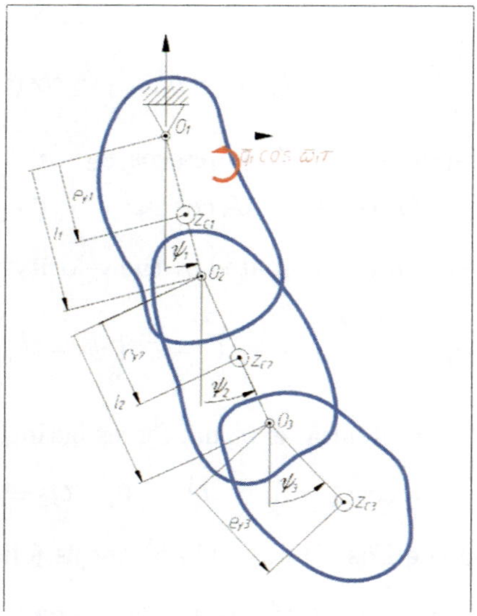

Fig. 14.2 The investigated triple pendulum.

Since the system position is defined by angles ψ_i, the following generalized-coordinate vector is used:

$$\boldsymbol{\psi} = \begin{Bmatrix} \psi_1 \\ \psi_2 \\ \psi_3 \end{Bmatrix}. \tag{14.1}$$

The equations of motion of the investigated triple pendulum can be written using the following Lagrangian formulation:

$$\frac{d}{d\tau}\left(\frac{\partial \bar{T}}{\partial \dot{\psi}_j}\right) + \frac{\partial \bar{F}_t}{\partial \dot{\psi}_j} - \frac{\partial \bar{T}}{\partial \psi_j} + \frac{\partial V}{\partial \psi_j} = \bar{Q}_j, \qquad j = 1, 2, 3, \tag{14.2}$$

where symbol (.!.) means derivative with respect to real time τ, $\bar{T}$ and $\bar{V}$ are correspondingly real kinetic and potential energy,

$$
\begin{aligned}
\bar{T} &= \frac{1}{2}B_1\dot{\psi}_1^{\,2} + \frac{1}{2}B_2\dot{\psi}_2^{\,2} + \frac{1}{2}B_3\dot{\psi}_3^{\,2} + N_{12}\dot{\psi}_1\dot{\psi}_2\cos(\psi_1 - \psi_2) + \\
&+ N_{13}\dot{\psi}_1\dot{\psi}_3\cos(\psi_1 - \psi_3) + N_{23}\dot{\psi}_2\dot{\psi}_3\cos(\psi_2 - \psi_3), \\
\bar{V} &= -M_1\cos\psi_1 - M_2\cos\psi_2 - M_3\cos\psi_3.
\end{aligned}
\tag{14.3}
$$

$\bar{F}_t$ is related to real energy dissipated by the velocity-proportional forces:

$$
\bar{F}_t = \frac{1}{2}\left(\bar{c}_1\dot{\psi}_1^{\,2} + \bar{c}_2\left(\dot{\psi}_1 - \dot{\psi}_2\right)^2 + \bar{c}_3\left(\dot{\psi}_2 - \dot{\psi}_3\right)^2\right), \tag{14.4}
$$

and Q_j are the real nonpotential external forces acting on the system

$$
\bar{Q}_1 = \bar{q}_1\cos(\bar{\omega}_1\tau), \qquad \bar{Q}_2 = 0, \quad \bar{Q}_3 = 0. \tag{14.5}
$$

The symbols used in the Eqs. (14.3) - (14.5) are as follows

$$
\begin{aligned}
B_1 &= J_{z1} + e_{y1}^2 m_1 + l_1^2(m_2 + m_3), \\
B_2 &= J_{z2} + e_{y2}^2 m_2 + l_2^2 m_3, \\
B_3 &= J_{z3} + e_{y3}^2 m_3, \\
N_{12} &= m_2 e_{y2} l_1 + m_3 l_1 l_2, \\
N_{13} &= m_3 e_{y3} l_1, \\
N_{23} &= m_3 e_{y3} l_2, \\
M_1 &= m_1 g e_{y1} + (m_2 + m_3) g l_1, \\
M_2 &= m_2 g e_{y2} + m_3 g l_2, \\
M_3 &= m_3 g e_{y3}.
\end{aligned}
\tag{14.6}
$$

Above J_{zi} $(i = 1,2,3)$ denote the appropriate principal central moments of inertia, m_i $(i = 1,2,3)$ denote masses of respective links and g is the gravitation acceleration. The other symbols are clarified in Fig. 14.2. In order to obtain the nondimensional form of the governing equations, we introduce first the nondimensional time

$$
t = \alpha_1\tau, \tag{14.7}
$$

where: $\alpha_1 = \left(M_1 B_1^{-1}\right)^{\frac{1}{2}}$ is the frequency of small free vibrations without damping of a single pendulum formed by adding a point mass equal to the sum of the masses of the second and the third pendulum to the first link in the point O_2.

The relations between derivatives with respect to real time and nondimensional one are

$$\frac{d(\ldots)}{d\tau} = \frac{d(\ldots)}{dt} \cdot \frac{dt(\tau)}{d\tau} = \alpha_1 \frac{d(\ldots)}{dt}, \tag{14.8}$$

and therefore

$$\overset{|}{\psi}_j = \alpha_1 \dot{\psi}_j \quad j = 1, 2, 3. \tag{14.9}$$

Moreover

$$\frac{\partial(\ldots)}{d\overset{|}{\psi}_j} = \frac{\partial(\ldots)}{d\dot{\psi}_j} \frac{\partial \dot{\psi}_j(\overset{|}{\psi}_j)}{\partial \overset{|}{\psi}_j} = \frac{1}{\alpha_1} \frac{\partial(\ldots)}{\partial \dot{\psi}_j}, \qquad j = 1, 2, 3. \tag{14.10}$$

Now we divide both sides of Eq. (14.2) by $2E_1$, where

$$E_1 = \frac{1}{2}\alpha_1^2 B_1 = \frac{1}{2} M_1, \tag{14.11}$$

and we obtain the following nondimensional form of equations of motion of the investigated triple pendulum in the Lagrangian formulation:

$$\frac{d}{dt}\left(\frac{\partial T}{\partial \dot{\psi}_j}\right) + \frac{\partial F_t}{\partial \dot{\psi}_j} - \frac{\partial T}{\partial \psi_j} + \frac{\partial V}{\partial \psi_j} = Q_j, \qquad j = 1, 2, 3, \tag{14.12}$$

where T and V are correspondingly nondimensional kinetic and potential energy:

$$\begin{aligned} T = \frac{\bar{T}}{2E_1} &= \frac{1}{2}\left(\dot{\psi}_1^2 + \beta_2 \dot{\psi}_2^2 + \beta_3 \dot{\psi}_3^2\right) + \nu_{12}\dot{\psi}_1\dot{\psi}_2 \cos(\psi_1 - \psi_2) + \\ &+ \nu_{13}\dot{\psi}_1\dot{\psi}_3 \cos(\psi_1 - \psi_3) + \nu_{23}\dot{\psi}_2\dot{\psi}_3 \cos(\psi_2 - \psi_3), \\ V = \frac{\bar{V}}{2E_1} &= -\cos\psi_1 - \mu_2 \cos\psi_2 - \mu_3 \cos\psi_3. \end{aligned} \tag{14.13}$$

F_t is the nondimensional equivalent of $\bar{F}_t$

$$F_t = \frac{\bar{F}_t}{2\alpha_1 E_1} = \frac{1}{2}\left(c_1 \dot{\psi}_1^2 + c_2\left(\dot{\psi}_2 - \dot{\psi}_1\right)^2 + c_3\left(\dot{\psi}_3 - \dot{\psi}_2\right)^2\right), \tag{14.14}$$

and Q_j are the nondimensional external forces acting on the system

$$Q_1 = \frac{\bar{Q}_1}{2E_1} = q_1 \cos(\omega_1 t), \qquad Q_2 = 0, \qquad Q_3 = 0. \tag{14.15}$$

The following nondimensional parameters have been introduced and used in Eqs. (14.13)-(14.15):

$$\beta_2 = \frac{B_2}{B_1}, \qquad \beta_3 = \frac{B_3}{B_1},$$
$$\nu_{12} = \frac{N_{12}}{B_1}, \qquad \nu_{13} = \frac{N_{13}}{B_1}, \qquad \nu_{23} = \frac{N_{23}}{B_1},$$
$$\mu_2 = \frac{M_2}{M_1}, \qquad \mu_3 = \frac{M_3}{M_1}, \tag{14.16}$$
$$c_1 = \frac{\bar{c}_1}{\sqrt{M_1 B_1}}, \; c_2 = \frac{\bar{c}_2}{\sqrt{M_1 B_1}}, \; c_3 = \frac{\bar{c}_3}{\sqrt{M_1 B_1}},$$
$$q_1 = \frac{\bar{q}_1}{M_1}, \qquad \omega_1 = \frac{1}{\alpha_1}\bar{\omega}_1.$$

The Lagrange Eqs. (14.12) lead to following nondimensional form of the differential equations of motion of the system without impacts (they are presented using matrix notation):

$$\boldsymbol{M}(\boldsymbol{\psi})\ddot{\boldsymbol{\psi}} + \boldsymbol{B}(\boldsymbol{\psi})\dot{\boldsymbol{\psi}}^{\boldsymbol{2}} + \boldsymbol{C}\dot{\boldsymbol{\psi}} + \boldsymbol{D}(\boldsymbol{\psi}) = \boldsymbol{F}(\boldsymbol{t}), \tag{14.17}$$

where:

$$\ddot{\boldsymbol{\psi}} = \begin{Bmatrix} \ddot{\psi}_1 \\ \ddot{\psi}_2 \\ \ddot{\psi}_3 \end{Bmatrix}, \quad \dot{\boldsymbol{\psi}}^{\boldsymbol{2}} = \begin{Bmatrix} \dot{\psi}_1^2 \\ \dot{\psi}_2^2 \\ \ddot{\psi}_3^2 \end{Bmatrix}, \quad \dot{\boldsymbol{\psi}} = \begin{Bmatrix} \dot{\psi}_1 \\ \dot{\psi}_2 \\ \dot{\psi}_3 \end{Bmatrix}, \tag{14.18}$$

$\boldsymbol{M}(\boldsymbol{\psi})$ is the 3x3 inertia matrix:

$$\boldsymbol{M}(\boldsymbol{\psi}) = \begin{bmatrix} 1 & \nu_{12}\cos(\psi_1 - \psi_2) & \nu_{13}\cos(\psi_1 - \psi_3) \\ \nu_{12}\cos(\psi_1 - \psi_2) & \beta_2 & \nu_{23}\cos(\psi_2 - \psi_3) \\ \nu_{13}\cos(\psi_1 - \psi_3) & \nu_{23}\cos(\psi_2 - \psi_3) & \beta_3 \end{bmatrix}, \tag{14.19}$$

$\boldsymbol{B}(\boldsymbol{\psi})$ is the 3x3 skew-symmetric matrix:

$$\boldsymbol{B}(\boldsymbol{\psi}) = \begin{bmatrix} 0 & \nu_{12}\sin(\psi_1 - \psi_2) & \nu_{13}\sin(\psi_1 - \psi_3) \\ -\nu_{12}\sin(\psi_1 - \psi_2) & 0 & \nu_{23}\sin(\psi_2 - \psi_3) \\ -\nu_{13}\sin(\psi_1 - \psi_3) & -\nu_{23}\sin(\psi_2 - \psi_3) & 0 \end{bmatrix}, \tag{14.20}$$

C is the 3x3 dissipative matrix:

$$C = \begin{bmatrix} c_1 + c_2 & -c_2 & 0 \\ -c_2 & c_2 + c_3 & -c_3 \\ 0 & -c_3 & c_3 \end{bmatrix}, \tag{14.21}$$

$$D(\psi) = \begin{pmatrix} \sin\psi_1 \\ \mu_2 \sin\psi_2 \\ \mu_3 \sin\psi_3 \end{pmatrix}, \tag{14.22}$$

and $F(t)$ is the vector of time-depending external excitations, which reads:

$$F(t) = \begin{pmatrix} q_1 \cos\omega_1 t \\ 0 \\ 0 \end{pmatrix}. \tag{14.23}$$

Introduction of the nondimensional form of the governing equations has allowed to decrease the number of parameters from 16 real ones to 12 nondimensional ones. Moreover, these nondimensional equations may govern not only this system of triple pendulum, but also others systems, not necessary mechanical ones [Bayly and Virgin (1992)], [Heng *et. al.* (1994)]. The equations of motion can also be written in the following standard form as a set of n first-order differential equations:

$$\dot{x} = f(x(t)), \qquad x(t_0) = x_0, \tag{14.24}$$

where the state vector in the case of the investigated triple pendulum is chosen as:

$$x = \left[\psi_1,\ \psi_2,\ \psi_3,\ \dot{\psi}_1,\ \dot{\psi}_2,\ \dot{\psi}_3,\ \omega_1 t\right]^T, \tag{14.25}$$

where $x_7 = \omega_1$ is the phase of the external excitation (it is assumed that $x_7 \in \langle 0, 2\pi)$) and the vector function $f(x)$ is derived from (14.17) and has the following form:

$$f(x) = \left\{ \begin{matrix} x_4 \\ x_5 \\ x_6 \\ M(x)^{-1}\left(F(t) - D(x) - C\begin{Bmatrix} x_4 \\ x_5 \\ x_6 \end{Bmatrix} - B(x)\begin{Bmatrix} x_4^2 \\ x_5^2 \\ x_6^2 \end{Bmatrix}\right) \\ \omega_1 \end{matrix} \right\}. \tag{14.26}$$

Observe, that now we have formally 7th-order ($n = 7$) autonomous system.

14.3 Introduction of the Obstacles

When in the investigated system the physical rigid obstacles appear, the governing equations can be written as follows:

$$\boldsymbol{M}(\boldsymbol{\psi})\ddot{\boldsymbol{\psi}} + \boldsymbol{B}(\boldsymbol{\psi})\dot{\boldsymbol{\psi}}^2 + \boldsymbol{C}\dot{\boldsymbol{\psi}} + \boldsymbol{D}(\boldsymbol{\psi}) = \boldsymbol{F}(t), \quad h_i(\boldsymbol{\psi}) \geq 0, \quad (i = 1, \ldots, m), \tag{14.27}$$

where the inequality $h_i(\boldsymbol{\psi}) \geq 0$ represents an unilateral constraint that is imposed on the position of the system, and m denotes the number of constraints (obstacles). Since the investigated system (14.27) is a Lagrangian one, it can be represented by a point moving in its configuration space $\boldsymbol{\psi}$. The unilateral constraints define domains of this space, and the point representing the system strikes the boundaries of these domains. At the time instants of these "generalized" collisions, the velocities of the system undergo jumps. Here we present the manner of calculation of the post-impact velocities for the system with many degrees of freedom according to the description given in reference [Brogliato (1999)]. During the impacts, the following shock dynamics equation is valid:

$$\boldsymbol{M}(\boldsymbol{\psi})\left(\dot{\boldsymbol{\psi}}^{+} - \dot{\boldsymbol{\psi}}^{-}\right) = \boldsymbol{p}_{\boldsymbol{\psi}}, \tag{14.28}$$

where $\boldsymbol{M}(\boldsymbol{\psi})$ denotes the inertia matrix (14.19), $\boldsymbol{p}_{\boldsymbol{\psi}}$ is the generalized percussion vector for coordinates $\boldsymbol{\psi}$ (the force impulse vector due to the impact), and $\dot{\boldsymbol{\psi}}^{-}$ is the vector of velocities immediately before impact (in the time t_k^-) and $\dot{\boldsymbol{\psi}}_i^{+}$ is the vector of the post-impact velocities (in the time t_k^+), where t_k denotes the impact time.

For frictionless constraints (as we now assume for further consideration), $\boldsymbol{p}_{\boldsymbol{\psi}}$ occurs along $\nabla_{\boldsymbol{\psi}} h_i(\boldsymbol{\psi})$ (because the interaction force due to the impact occurs along an Euclidean normal to the surface $h_i(\boldsymbol{\psi}) = 0$, which results from the virtual work principle). Now from (14.28) we can obtain the following two algebraic equations versus three unknowns

$$\boldsymbol{t}_{\boldsymbol{\psi},i}^T \boldsymbol{M}(\boldsymbol{\psi})\left(\dot{\boldsymbol{\psi}}^{+} - \dot{\boldsymbol{\psi}}^{-}\right) = 0, \tag{14.29}$$

where: $\boldsymbol{t}_{\psi,1}$ and $\boldsymbol{t}_{\psi,2}$ are tangent unitary vectors, chosen mutually independent, i.e. $\boldsymbol{t}_{\boldsymbol{\psi},1}^T \nabla_{\boldsymbol{\psi}} h_i(\boldsymbol{\psi}) = 0$, $\boldsymbol{t}_{\boldsymbol{\psi},2}^T \nabla_{\boldsymbol{\psi}} h_i(\boldsymbol{\psi}) = 0$ and $\boldsymbol{t}_{\boldsymbol{\psi},1}^T \boldsymbol{t}_{\boldsymbol{\psi},2}^T = 0$.

The last lacking equation is the extended Newton's (restitution coefficient) rule applied to the component of $\dot{\boldsymbol{\psi}}$ along $\nabla_{\boldsymbol{\psi}} h_i(\boldsymbol{\psi})$:

$$\boldsymbol{n}_{\boldsymbol{\psi}}^T \dot{\boldsymbol{\psi}}^+ = -e\boldsymbol{n}_{\boldsymbol{\psi}}^T \dot{\boldsymbol{\psi}}^-, \tag{14.30}$$

where: $\boldsymbol{n}_{\boldsymbol{\psi}} = \dfrac{\nabla_{\boldsymbol{\psi}} h_i(\boldsymbol{\psi})}{\|\nabla_{\boldsymbol{\psi}} h_i(\boldsymbol{\psi})\|}$. From Eq. (14.29) one realizes that there is in general a discontinuity in the tangential velocity due to inertia coupling.

The change of the kinetic energy at impact is defined by the formula

$$\Delta T(t_k) = \frac{1}{2}\left(\dot{\boldsymbol{\psi}}^+\right)^T \boldsymbol{M}(\boldsymbol{\psi})\dot{\boldsymbol{\psi}}^+ - \frac{1}{2}\left(\dot{\boldsymbol{\psi}}^-\right)^T \boldsymbol{M}(\boldsymbol{\psi})\dot{\boldsymbol{\psi}}^-. \tag{14.31}$$

From Eq. (14.31), using Eqs. (14.29)-(14.30), one obtains:

$$\Delta T(t_k) = \frac{1}{2}(e^2 - 1)\left(\frac{\boldsymbol{M}(\boldsymbol{\psi})^{-1}\boldsymbol{n}_{\boldsymbol{\psi}}}{\sqrt{\boldsymbol{n}_{\boldsymbol{\psi}}^T \boldsymbol{M}(\boldsymbol{\psi})^{-1}\boldsymbol{n}_{\boldsymbol{\psi}}}}\boldsymbol{M}(\boldsymbol{\psi})\dot{\boldsymbol{\psi}}^-\right)^2. \tag{14.32}$$

As we can see, for $0 \le e \le 1$, the kinetic energy change due to impact is $\Delta T(t_k) \le 0$, for $e = 1$ it is $\Delta T(t_k) = 0$ and for $e = 0$, $\Delta T(t_k)$ reaches the lowest possible value.

In summary, when the investigated nonlinear system (14.24) achieves at the time $t = t_k$ a discontinuity point indicated by $h(\boldsymbol{x}(t_k^-)) = 0$, the state vector is transformed by

$$\boldsymbol{x}(t_k^+) = \boldsymbol{g}(\boldsymbol{x}(t_k^-)), \tag{14.33}$$

where the vector function $\boldsymbol{g}(\boldsymbol{x})$ is derived from Eqs. (14.29)-(14.30), and has the following form

$$\boldsymbol{g}(\boldsymbol{x}) = \left\{\begin{matrix} x_1 \\ x_2 \\ x_3 \\ \begin{bmatrix} \boldsymbol{n}_{\boldsymbol{\psi}}^T \\ \begin{bmatrix} \boldsymbol{t}_{\boldsymbol{\psi},1}^T \\ \boldsymbol{t}_{\boldsymbol{\psi},2}^T \end{bmatrix} \cdot \boldsymbol{M}(\boldsymbol{x}) \end{bmatrix}^{-1} \begin{bmatrix} -e\boldsymbol{n}_{\boldsymbol{\psi}}^T \\ \begin{bmatrix} \boldsymbol{t}_{\boldsymbol{\psi},1}^T \\ \boldsymbol{t}_{\boldsymbol{\psi},2}^T \end{bmatrix} \cdot \boldsymbol{M}(\boldsymbol{x}) \end{bmatrix} \cdot \begin{Bmatrix} x_4 \\ x_5 \\ x_6 \end{Bmatrix} \\ x_7 \end{matrix}\right\}. \tag{14.34}$$

14.4 Calculation of the Fundamental Solution Matrices for Dynamical Systems with Impacts

A disturbance $\boldsymbol{\delta x_0}$ on the initial condition will cause a disturbance $\boldsymbol{\delta x}(t)$ on the state $\boldsymbol{x}(t)$ of the investigated nonlinear system (14.24). The time evaluation of the vector $\boldsymbol{\delta x}(t)$ is represented by linearized Eq. (14.24):

$$\boldsymbol{\delta \dot{x}} = \boldsymbol{F}(t)\boldsymbol{\delta x}(t), \qquad \boldsymbol{\delta x}(t_0) = \boldsymbol{\delta x_0}, \tag{14.35}$$

where:

$$\boldsymbol{F}(t) = \left.\frac{\partial \boldsymbol{f}(\boldsymbol{x})}{\partial \boldsymbol{x}^T}\right|_{\boldsymbol{x}=\boldsymbol{x}(t)}. \tag{14.36}$$

The matrix consisting of n linear independent solutions $\boldsymbol{\delta x}(t)$ of the n-dimensional linear system (14.35) is called the fundamental solution matrix

$$\boldsymbol{\delta X}(t) = [\boldsymbol{\delta x_1}, \ldots, \boldsymbol{\delta x_n}]. \tag{14.37}$$

Let us consider a particular fundamental matrix $\boldsymbol{\Phi}(t, t_0)$ obtained from the following initial value problem

$$\dot{\boldsymbol{\Phi}}(t, t_0) = \boldsymbol{F}(t)\boldsymbol{\Phi}(t, t_0), \qquad \dot{\boldsymbol{\Phi}}(t, t_0) = \boldsymbol{\Phi}_0 = \boldsymbol{I}, \tag{14.38}$$

where $\boldsymbol{I}$ is the unitary matrix. Now, each disturbance $\boldsymbol{\delta x}(t)$ can be found on the basis of the fundamental solution matrix $\boldsymbol{\Phi}(t, t_0)$, which simply maps $\boldsymbol{\delta x}(t)$ into $\boldsymbol{\delta x}_0$ as follows

$$\boldsymbol{\delta x}(t) = \boldsymbol{\Phi}(t, t_0)\boldsymbol{\delta x}_0. \tag{14.39}$$

The concept of a fundamental solution matrix is directly used during the calculation of the Lyapunov exponents. The spectrum of the Lyapunov exponents λ_j $(j = 1, \ldots, n)$ of an attractor of dynamical system (14.24) can be calculated as follows:

$$\lambda_j = \lim_{t\to\infty} \frac{1}{t - t_0} \ln \frac{||\boldsymbol{\delta x}_j(t)||}{||\boldsymbol{\delta x}_j(t_0)||}, \qquad \boldsymbol{\delta x}_j(t_0) = \boldsymbol{i}_j, \tag{14.40}$$

where $\boldsymbol{i}_j$ is the j-th unitary vector. A spectrum of the Lyapunov exponents characterizes an attractor in the state space. It measures the average expansion of a small volume element in orthogonal directions along trajectories in the state space. A positive Lyapunov exponent measures the average exponential divergence of nearby trajectories, and the negative exponent measures the exponential convergence of trajectories into the attractor in some directions. Any attractor of continuously differentiable dynamical system

(14.24), except of a fixed point, has at least one zero exponent corresponding to the direction along the flow. The sum of the Lyapunov exponents is the average volume contraction rate and in dissipative dynamical systems it is negative, and there is at least one negative exponent. If the maximal exponent is positive, some adjacent trajectories will exponentially diverge from $\boldsymbol{x}(t)$, a sensitive dependence on initial conditions is observed and a chaotic motion can appear. In summary, the spectrum of Lyapunov exponents allows us to identify the attractor of a dynamical system according to Table 14.1.

Table 14.1 Identification of attractors by means of their Lyapunov exponents.

Attractor	Lyapunov exponents
Stable fixed point	$0 > \lambda_1 \geq \ldots \geq \lambda_n$
Limit cycle	$\lambda_1 = 0 > \lambda_2 \geq \ldots \geq \lambda_n$
2-periodic (torus)	$\lambda_1 = \lambda_2 = 0 > \lambda_3 \geq \ldots \geq \lambda_n$
k-periodic (k-torus)	$\lambda_1 = \ldots \lambda_k = 0 > \lambda_{k+1} \geq \ldots \geq \lambda_n$
Chaotic	$\lambda_1 > \lambda_2 = 0 > \lambda_3 \geq \ldots \geq \lambda_n$
Hyperchaotic	$\lambda_1 \geq \lambda_2 > \lambda_3 = 0 > \lambda_4 \geq \ldots \geq \lambda_n$
Other cases ($k \geq 3$)	$\lambda_1 \geq \lambda_2 \geq \ldots \lambda_k > \lambda_{k+1} = 0 > \lambda_{k+2} \geq \ldots \geq \lambda_n$

An algorithm of computing a Lyapunov exponent is based on Eq. (14.40) [Wolf *et. al.* (1985)]. In order to obtain the Jacobian matrix (14.36), the trajectory $\boldsymbol{x}(t)$ is built by integrating the nonlinear Eq. (14.24) for initial condition on the attractor. In addition, to avoid the overflow problem of exponentially diversing solutions of the Eq. (14.35), after some integration time, the Gram-Schmidt reorthonormalization procedure is applied in order to define a new set of initial conditions, and then the integration is continued. The factors of the normalization are summed up over a long time in order to estimate the Lyapunov exponents (14.40).

Fundamental solution matrices are also important in the stability analysis of periodic solutions of dynamical systems. As the Lyapunov exponents are useful in identifing the nonperiodic attractors, the concept of Floquet multipliers gives us the information about stability of a periodic solution. The Floquet multipliers are the eigenvalues of a fundamental solution matrix. Each Floquet multiplier, over one period of a closed orbit, provides a measure of the local orbital divergence or convergence along a particular direction. Periodic orbit is stable if all n Floquet multipliers (for n-dimensional system) lie within the unit circle on the complex plane. If at least one multiplier (of the highest module) crosses the unit circle, the orbit changes stability and a bifurcation of periodic solution appears. In addition, the fundamental solution matrices are used in shooting method for finding periodic solutions and in continuation methods to follow its branches [Awrejcewicz (1990c); Awrejcewicz (1991d); Awrejcewicz (1991e)], [Awrejcewicz and Someya (1991)]. In continuous systems the fundamental solution matrix can be easily obtained by integrating Eq. (14.38). In discontinuous systems (with impacts, dry friction), the time evolution of the fundamental matrix undergoes discontinuities (called also 'jumps' or 'saltations'). The jumps in the fundamental solution matrix (saltation matrices) can be, however, calculated using the Aizerman-Gantmakher theory. The theory of Aizerman and Gantmakher was used to calculate Lyapunov exponents in systems with impacts in [Müller (1995)] and to analyse bifurcations of periodic solutions in systems with dry friction [Leine *et. al.* (1999)]. Here, this theory is used to calculate Lyapunov exponents and to analyse the bifurcations of periodic solutions numerically, in our dynamical system with impacts.

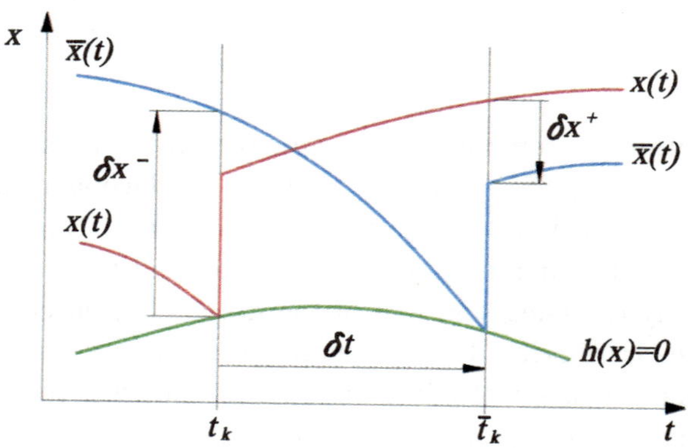

Fig. 14.3 Undisturbed and disturbed motion.

When the nonlinear system of the investigated triple pendulum at the time instant $t = t_k$ reaches the discontinuity point (see Fig. 14.3) indicated by $h(\boldsymbol{x}(t_k^-)) = 0$, the state vector $\boldsymbol{x}(t)$ is transformed using Eq. (14.33).

The disturbed motion $\bar{\boldsymbol{x}}(t)$ given by

$$\bar{\boldsymbol{x}}(t) = \boldsymbol{x}(t) + \boldsymbol{\delta} x(t), \tag{14.41}$$

encounters this discontinuity at time

$$\bar{t}_k = t_k + \delta t. \tag{14.42}$$

The transition condition for perturbation $\boldsymbol{\delta x}(t)$ due to the Aizerman-Gantmakher theory is given in [Müller (1995)] and has the following form

$$\boldsymbol{\delta x}(t_k^+) = \boldsymbol{p}\left(\boldsymbol{x}(t_k^-),\ \boldsymbol{\delta x}(t_k^-)\right), \tag{14.43}$$

where the function $\boldsymbol{p}(\boldsymbol{x}, \boldsymbol{\delta x})$ reads

$$\boldsymbol{p}(\boldsymbol{x}, \boldsymbol{\delta x}) = \boldsymbol{G}(\boldsymbol{x})\boldsymbol{\delta x} + \left[\boldsymbol{G}(\boldsymbol{x})\boldsymbol{f}(\boldsymbol{x}) - \boldsymbol{f}(\boldsymbol{g}(\boldsymbol{x}))\right]\delta t(\boldsymbol{x}, \boldsymbol{\delta x}), \tag{14.44}$$

where

$$\delta t(\boldsymbol{x}, \boldsymbol{\delta x}) = -\frac{\boldsymbol{H}(\boldsymbol{x})\boldsymbol{\delta x}}{\boldsymbol{H}(\boldsymbol{x})\boldsymbol{f}(\boldsymbol{x})}, \tag{14.45}$$

$$\boldsymbol{H}(\boldsymbol{x}) = \frac{\partial \boldsymbol{h}(\boldsymbol{x})}{\partial \boldsymbol{x}^T}, \tag{14.46}$$

$$\boldsymbol{G}(\boldsymbol{x}) = \frac{\partial \boldsymbol{g}(\boldsymbol{x})}{\partial \boldsymbol{x}^T}, \tag{14.47}$$

The same function can be of course used to calculate the complete post-impact fundamental solution matrix:

$$\boldsymbol{\Phi}(t_k^+, t_0) = \boldsymbol{p}(\boldsymbol{x}(t_k^-),\ \boldsymbol{\Phi}(t_k^-, t_0)). \tag{14.48}$$

14.5 Simplification of the System

As an example, a special simple case of the introduced system of coupled physical pendulums will be analysed further. We consider only three identical rods with damping, external excitation q_1 and with the obstacle in the form of a horizontal wall (Fig. 14.4).

The physical obstacle can be expressed in the following nondimensional form:

$$\begin{aligned} h_1(\boldsymbol{\psi}) &= \eta - \cos\psi_1 \geq 0, \\ h_2(\boldsymbol{\psi}) &= \eta - (\cos\psi_1 + \cos\psi_2) \geq 0, \\ h_3(\boldsymbol{\psi}) &= \eta - (\cos\psi_1 + \cos\psi_2 + \cos\psi_3) \geq 0, \end{aligned} \tag{14.49}$$

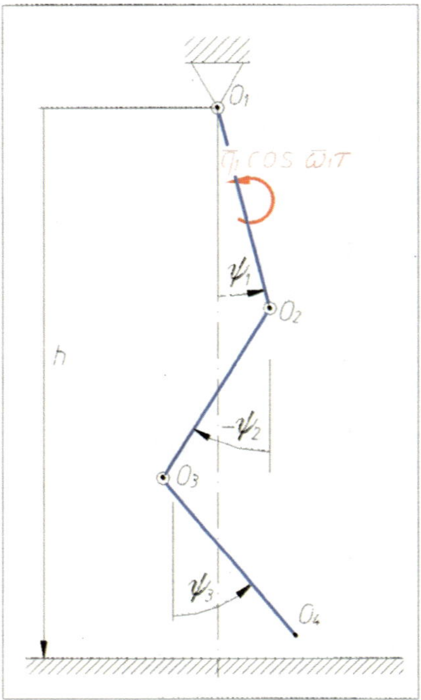

Fig. 14.4 Three coupled rods with the obstacle.

where:

$$\eta = \frac{h}{l}. \tag{14.50}$$

Above h is the real position of the horizontal wall, and l is the length of the rod.

For three identical rods the nondimensional parameters have the form:

$$\begin{aligned}
&\beta_2 = \frac{4}{7}, \quad \beta_3 = \frac{1}{7}, \\
&\mu_2 = \frac{3}{5}, \quad \mu_3 = \frac{1}{5}, \\
&\nu_{12} = \frac{9}{14}, \nu_{13} = \frac{3}{14}, \nu_{23} = \frac{3}{14}.
\end{aligned} \tag{14.51}$$

During presentation of the obtained results in Section 14.7, we will often use the following new variables as the functions of the state of the system

$$\begin{aligned} x_{O4}(\boldsymbol{\psi}) &= \sin\psi_1 + \sin\psi_2 + \sin\psi_3, \\ y_{O4}(\boldsymbol{\psi}) &= -(\cos\psi_1 + \cos\psi_2 + \cos\psi_3), \end{aligned} \tag{14.52}$$

where x_{O4} and y_{O4} are simply the position of the third rod's end (O_4 point) in the coordinate system x, y.

14.6 The Method Used for Integration of the System and its Accuracy

The fourth order Runge-Kutta method is used for integration of the system of differential equations between two successive impacts and the impact times are detected with an arbitrary accuracy. In each discontinuity point a state vector of the system is transformed due to the extended Newton's law using Eqs. (14.33)-(14.34). For all examples shown in Section 14.7 the Runge-Kutta integration time step is equal to (if it is not additionally clarified) $h = 2\pi/400$ and the precision of detection of the discontinuity point $\boldsymbol{x}_k$ is $SP = 10^{-10}$.

In Fig. 14.5 we have diagrams of the maximum difference between two trajectories obtained with the integration time steps $h(\boldsymbol{x}_h(t))$ and $\frac{h}{2}(\boldsymbol{x}_{\frac{h}{2}}(t))$ starting from the same point of an attractor, compared during the calculation time $t_{cal} = 10 \cdot 2\pi$, versus the time step h (using the logarithmic scales). Since the trajectory $\boldsymbol{x}_{\frac{h}{2}}(t)$ is assumed to be exact (in relation to $\boldsymbol{x}_h(t)$), the max $\left|x_h(t) - \boldsymbol{x}_{\frac{h}{2}}(t)\right|$ is the precision of obtained solution $\boldsymbol{x}_h(t)$ and the gradient of the line on each diagram in Fig. 14.5 is the order of the used method. We investigate the accuracy of the integration method for three different cases of motion with impacts: periodic, quasi-periodic and chaotic motion and we compare them with results for periodic motion without impacts. As we can see, in all three cases the order of the method is approximately four, and the chosen discontinuity detection accuracy as well as the integration time step are sufficient.

In the accuracy investigation, the following (see next section for more details) attractors have been used: Fig. 14.8a - periodic solution without impacts, Fig. 14.8b - periodic solution with impacts, Figs. 14.9–14.10 - quasi-periodic solution with impacts, Figs. 14.15b/c - chaotic solution with impacts.

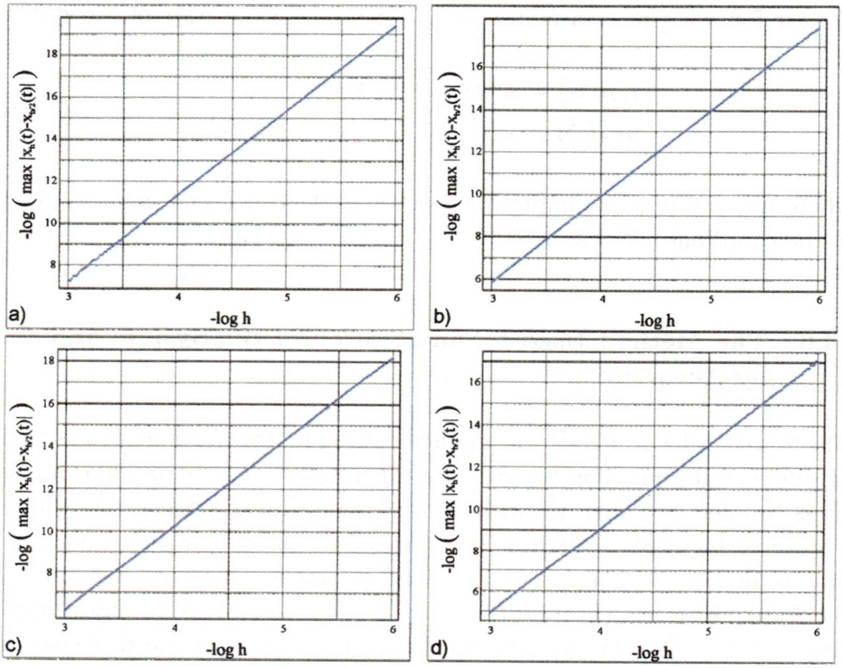

Fig. 14.5 Maximum error of the used method for different time steps: a) periodic solution without impacts; b) periodic solution with impacts; c) quasi-periodic solution with impacts; d) hyperchaotic solution with impacts.

14.7 Numerical Examples

We present here some examples of numerical investigation of three identical coupled rods with physical obstacle in the form of a horizontal barrier, shown in Fig. 14.4 (see section 14.5). The following nondimensional parameters are fixed: $\eta = 2.5$, $e = 1$ (restitution coefficient) and $c_1 = c_2 = c_3 = 0.1$. The q_1 parameter is varied within the interval $(0.75, 0.7885)$, the ω_1 is varied within the interval $(1, 1.005)$ and the initial conditions of the system are varied, too. First we present basins of attraction of three different coexisting stable periodic solutions for $q_1 = 0.7885$ and $\omega_1 = 1$ (Fig. 14.6). This diagram is obtained by integrating the Eqs. (14.24) for different initial conditions, and after omitting an initial transitional process, the attained attractor is recognized using the Lyapunov exponents. The corresponding three trajectories are shown in Fig. 14.8. One of them (green) avoids the obstacle, whereas two of them (blue and red) are with impacts.

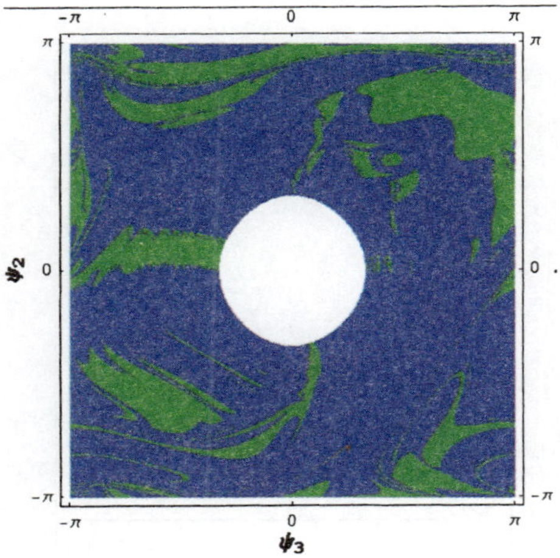

Fig. 14.6 Basins of attraction of three coexisting stable periodic solutions (blue, green and red colors); the initial conditions: $\psi_1(t=0)=0$, $\psi_2(0)=(-\pi,\pi)$, $\psi_3(0)=(-\pi,\pi)$, $\dot{\psi}_1(0)=\dot{\psi}_2(0)=\dot{\psi}_3(0)=0$; the parameters $q=0.7885$ and $\omega_1=1$.

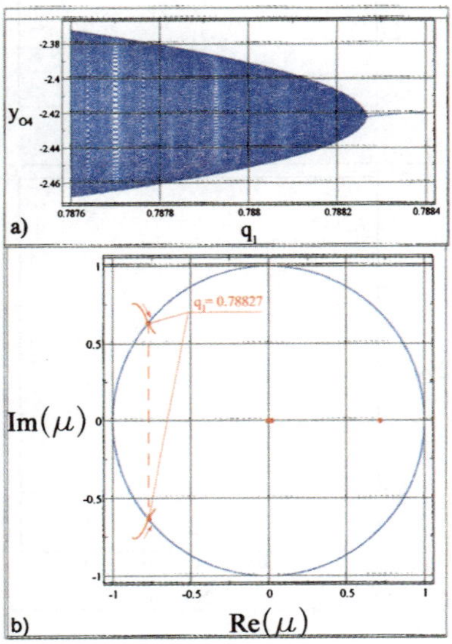

Fig. 14.7 Neimark-Sacker bifurcation for $q_1=q_1^*=0.78827$: bifurcational diagram of Poincaré sections (a) and the complex plane of the Floquet multipliers (b).

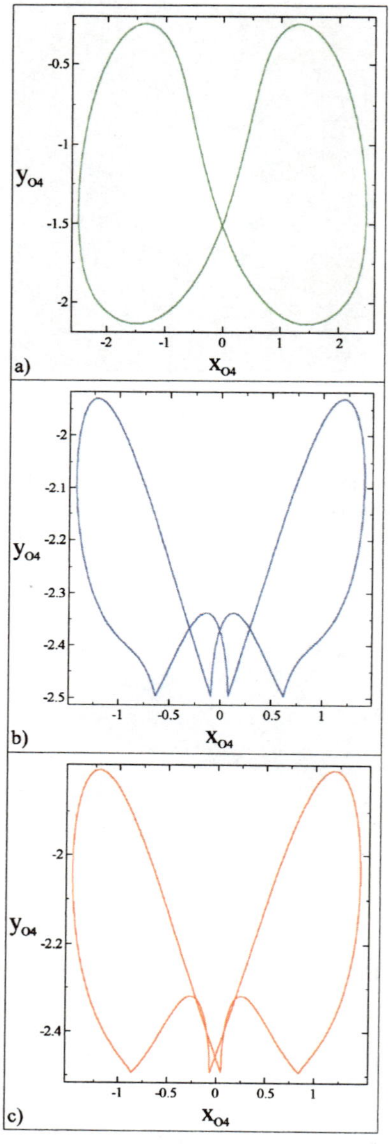

Fig. 14.8 Three periodic attractors corresponding to Fig. 14.6.

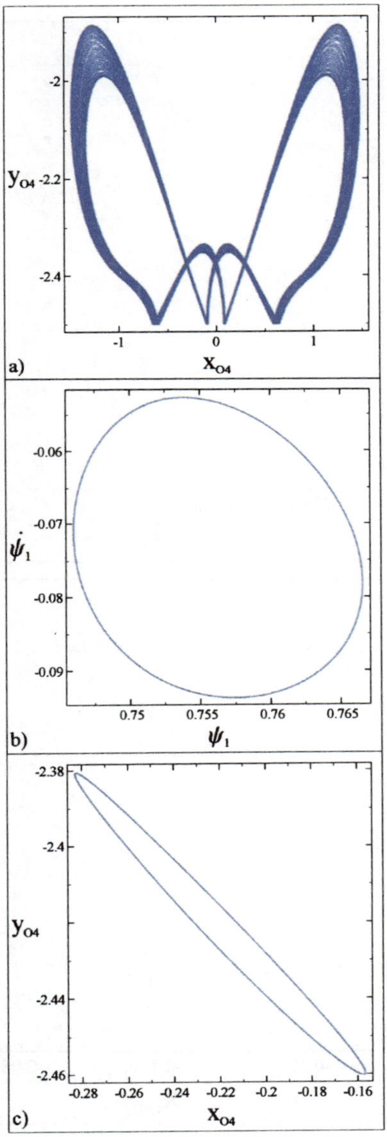

Fig. 14.9 Quasi-periodic attractor for $q_1 = 0.7878$: the trajectory (a) and Poincaré section (b,c).

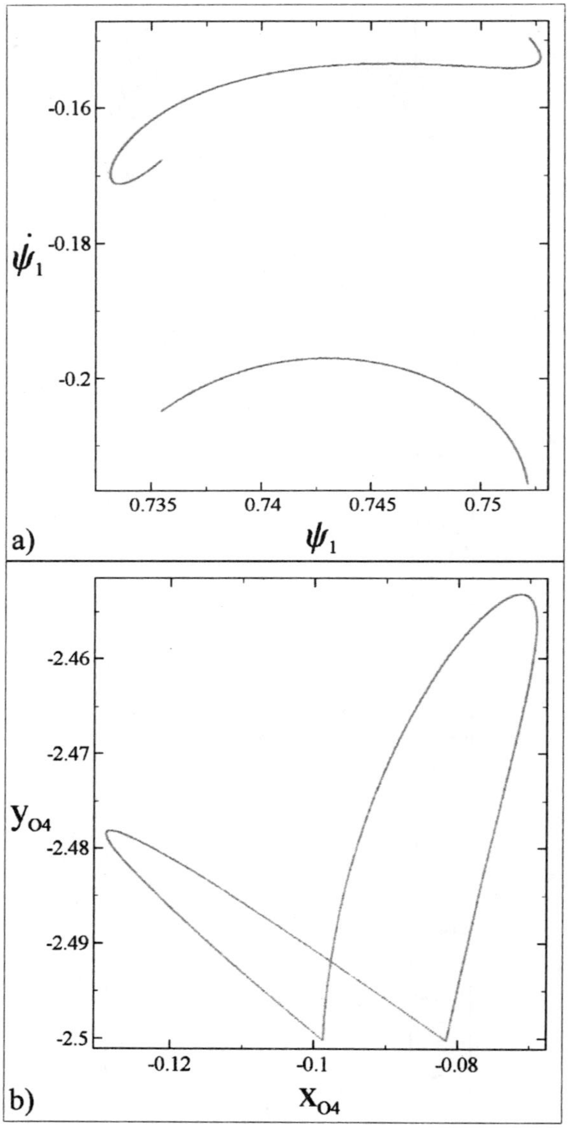

Fig. 14.10 Poincaré section of quasi-periodic attractor from Fig. 14.9, but for different chosen section.

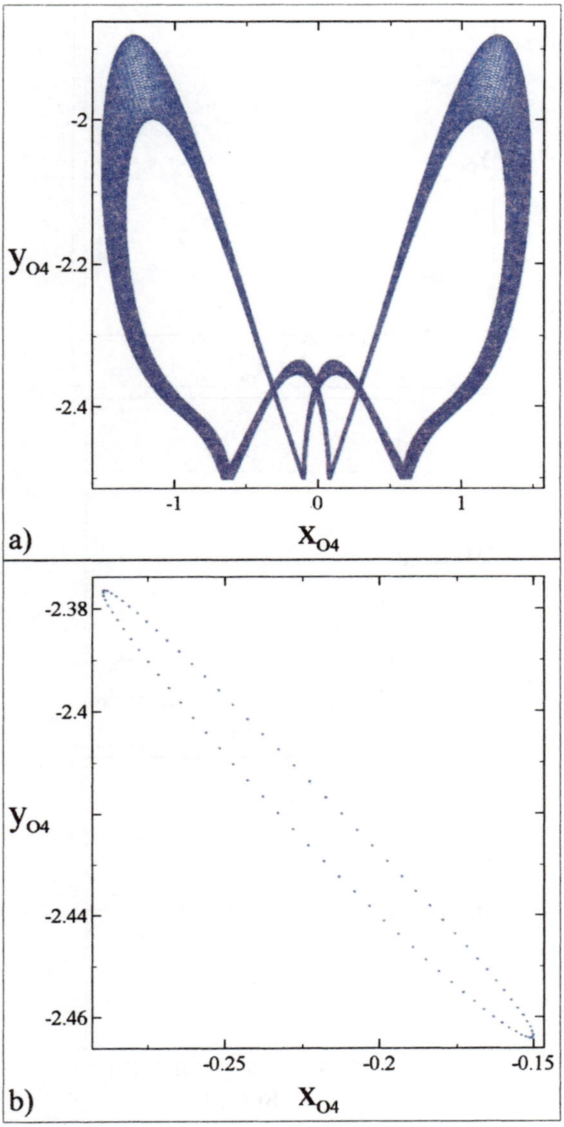

Fig. 14.11 Multi-periodic attractor on torus for $q_1 = 0.7877$: trajectory (a) and Poincaré section (b).

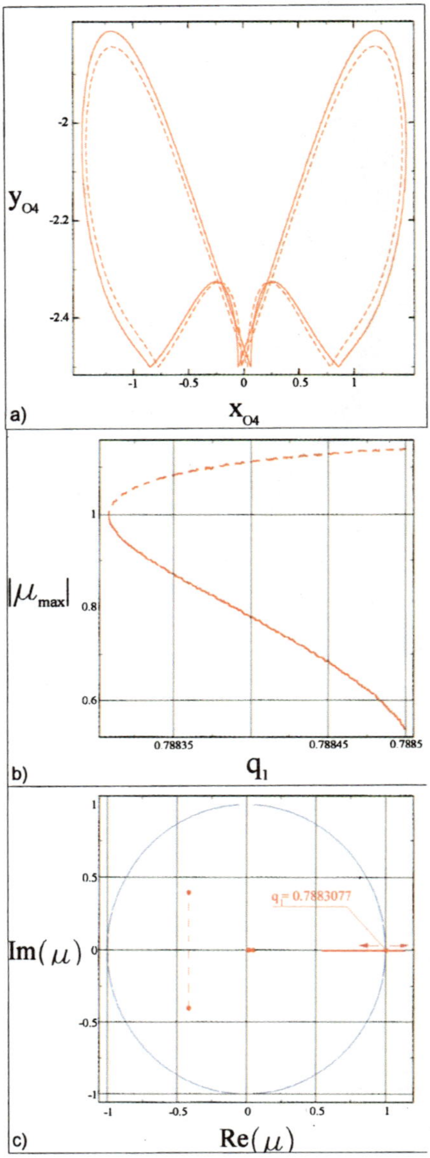

Fig. 14.12 Saddle-node bifurcation of periodic solutions for $q_1 = q_1^* = 0.7883077$: two coexisting periodic solutions (stable and unstable) for $q_1 = 0.7885$ (a); bifurcational diagram of the largest Floquet multipliers' module (b); the complex plane of the Floquet multipliers (c).

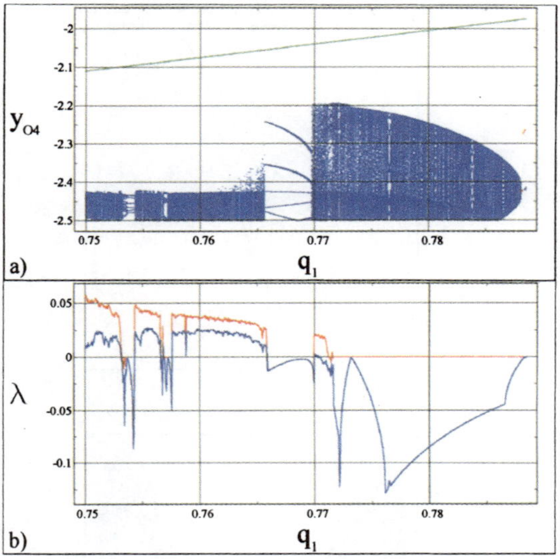

Fig. 14.13 Bifurcational diagram for varied parameter $q_1 \in (0.7500,\ 0.7885)$ for two coexisting solutions (blue and green) (a); and the corresponding diagram of two largest Lyapunov exponents for blue the solution (b).

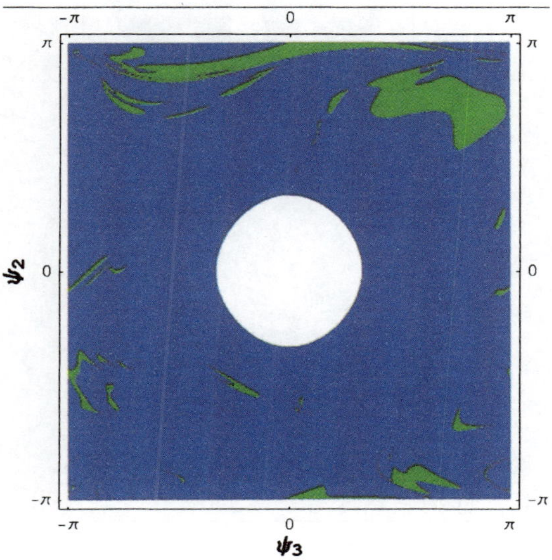

Fig. 14.14 Basins of attraction of two coexisting attractors (blue and green); the initial conditions are the same as in Fig. 14.6; the parameters $q_1 = 0.75$ and $\omega_1 = 1$.

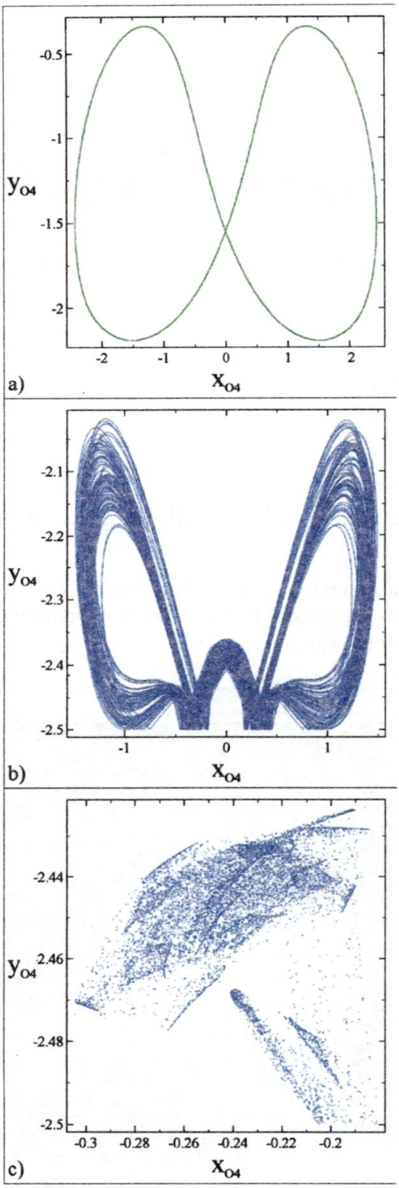

Fig. 14.15 Two attractors corresponding to Fig. 14.14: the stable periodic orbit (green) (a) and the hyperchaotic attractor (blue) - the trajectory (b) and their Poincaré section (c).

Now we quasi-statically decrease the bifurcational parameter q_1 and observe the blue solution from Fig. 14.8b. For the critical value of $q_1 = q_1^* = 0.78827$, the pair of two conjugated Floquet multipliers (of the highest module) crosses the unit circle on the complex plane (Fig. 14.7b), the periodic orbit loses its stability and the invariant torus is generated (Neimark-Sacker bifurcation). The unstable periodic solution exists now within the torus. On the left side of q_1^*, the region of quasi-periodicity begins, with many periodic windows - see bifurcation diagram of Poincaré sections' projection on axis y_{O4} (Fig. 14.7a).

We present an example of quasi-periodic attractor (from the mentioned above region) for $q_1 = 0.7878$. In Fig. 14.9 we have the trajectory and two different projections of its Poincaré map. The Poincaré section has been chosen here and in all places of this report (with exception of Fig. 14.10, as $x_7 = \pi/2$, using notation (14.24)-(14.26)). This Poincaré section can be also understood as the stroboscopic sampling of the state vector in times $t_i = \dfrac{1}{\omega_1}\left(\dfrac{\pi}{2} + 2\pi i\right)$, where $i = 0, 1, 2, \ldots$. Note that the Poincaré section of the same quasi-periodic attractor with impacts may have form of the closed curve or form of discontinuous one: see the section $x_7 = 1.67$ in Fig. 14.10 of the same attractor as in Fig. 14.9.

In Fig. 14.11 we present $87T$-periodic solution (when T is period of external exciation) existing on the investigated torus. This solution corresponds to one of the mentioned earlier periodic windows ($q_1 = 0.7877$). Let us trace the red solution (Fig. 14.8c) with a decrease of the parameter q_1. It has been found that for $q_1 = 0.7885$, this stable periodic solution coexists with another unstable periodic orbit (Fig. 14.12a). For the critical value of the parameter $q_1 = q_1^* = 0.7883077$, one of the Floquet multipliers (of the highest module) of both solutions crosses the unit circle on the complex plane in the point $(1, 0)$ (Fig. 14.12b,c). These two periodic solutions join each other and vanish via saddle-node bifurcation.

Let us trace the further behaviour of both blue (Fig. 14.8b) and green (Fig. 14.8a) solutions with further changes of q_1 (see Fig. 14.13): the bifurcational diagram of Poincaré sections for both solutions (a) and the corresponding bifurcational diagram of two largest Lyapunov exponents for the blue solution (b), with the interval $q_1 \in (0.7500, 0.7885)$. The green solution is always periodic and it does not undergo any bifurcation. In the case of blue solution a transition to quasiperiodicity by Neimark-Sacker bifurcation is detected. A further decrease of q_1 causes a transition to chaos for $q_1 = 0.7715$ with a short periodic window. For $q_1 = 0.7700$

again a relatively large periodic orbit domain occurs, which disappears for $q_1 = 0.7659$. Then hyperchaos occurs, which is observed until the end of the considered interval of q_1. The hyperchaos is interrupted by narrow periodic windows for $q_1 \in (0.7530, 0.7543)$ and $q_1 \in (0.7566, 0.7576)$ which with a high probability include also quasi-periodic and chaotic orbits. However, the latter ones are difficult to distinguish in Fig. 14.13b. In Fig. 14.14 we present basins of attraction of two coexisting attractors: periodic (green) and hyperchaotic (blue) for $q_1 = 0.75$ and their phase portraits are shown in Fig. 14.15.

Table 14.2 The Lyapunov exponents λ_i $(i = 1, \dots, 6)$ for the presented solutions.

Figure	λ_1	λ_2	λ_3	λ_4	λ_5	λ_6	Attractor
14.8a	-0.11	-0.11	-0.26	-0.26	-1.71	-2.29	Limit cycle
14.8b	-0.0017	-0.0017	-0.0660	-1.2626	-1.7733	-1.7733	Limit cycle
14.8c	-0.14	-0.28	-0.28	-0.28	-1.39	-2.56	Limit cycle
14.9/14.10	0	-0.006	-0.074	-1.264	-1.764	-1.778	Quasi-periodic
14.15a	-0.01	-0.01	-0.25	-0.25	-1.69	-2.40	Limit cycle
14.15b/c	0.06	0.01	-0.67	-0.94	-1.48	-2.10	Hyperchaotic
14.17	0	-0.042	-0.042	-1.325	-1.702	-1.772	Quasi-periodic

Finally, let us trace behaviour of the blue periodic solution for fixed $q_1 = 0.7885$ (Fig. 14.8b), when the frequency of external excitation is varied in the interval $\omega_1 \in (1, 1.005)$. The obtained results are shown in Fig. 14.16. As we can see, it also undergoes bifurcation leading to quasi-periodicity. The quasi-periodic attractor is shown in Fig. 14.17.

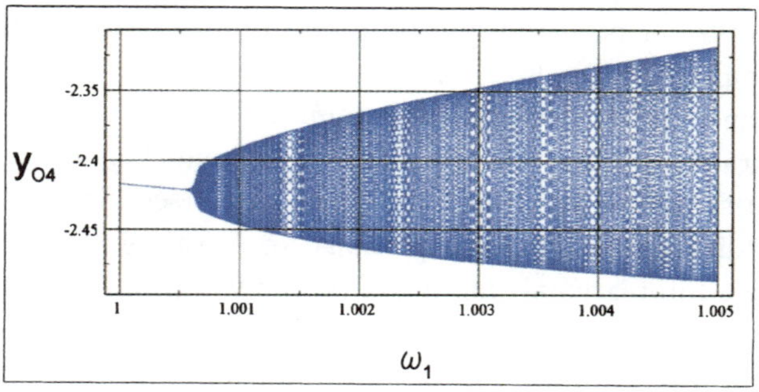

Fig. 14.16 Bifurcational diagram for $\omega_1 \in (1,\ 1.005)$ and $q_1 = 0.7885$ fixed.

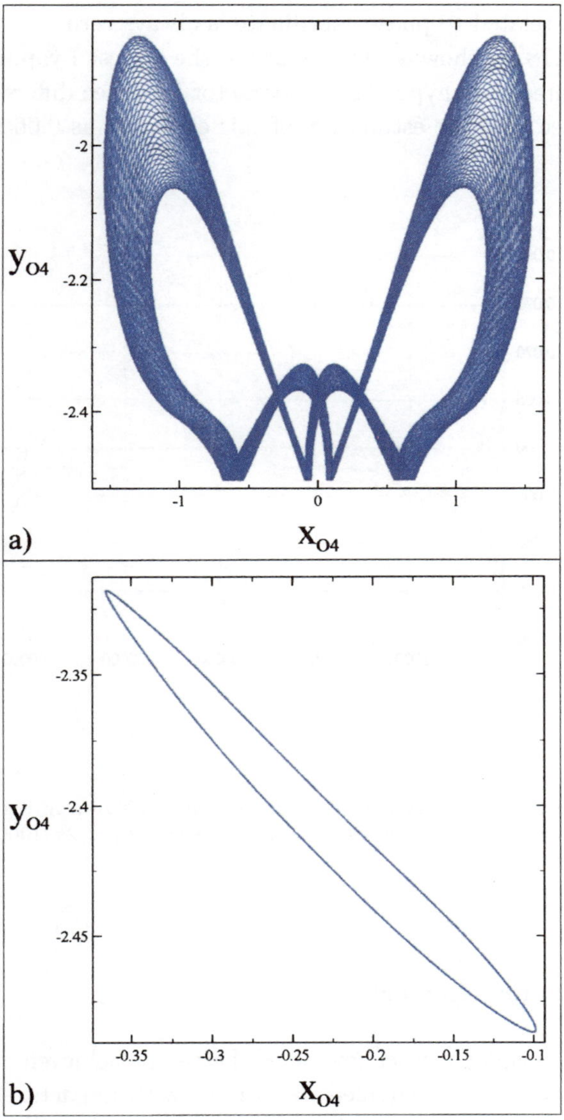

Fig. 14.17 Quasi-periodic attractor for $q_1 = 0.7885$ and $\omega_1 = 1.005$: trajectory (a) and Poincaré section (b).

The Lyapunov exponents for the presented attractors are listed up in Table 14.7. Note, that only six Lyapunov exponents are needed, because one exponent related to phase coordinate is always zero.

In Fig. 14.18 we show a convergence of the largest Lyapunov exponent for the presented here hyperchaotic attractor for three different time steps and we can see that the estimation of this exponent as 0.06 is the correct one.

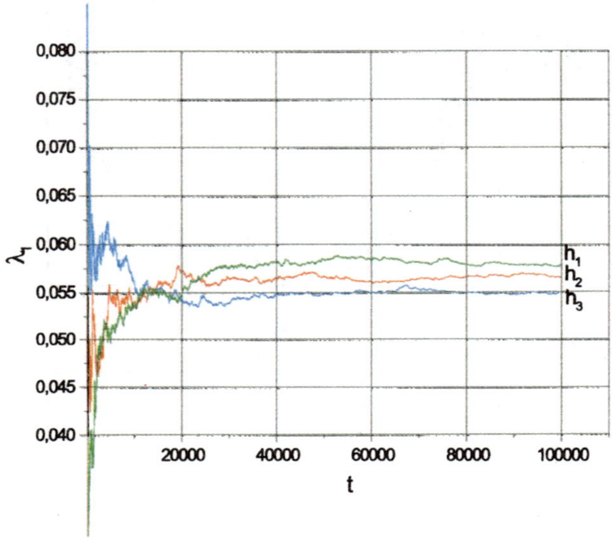

Fig. 14.18 The first Lyapunov exponent for the solution shown in Fig. 14.15b/c (hyperchaotic motion) for three different integral time steps: $h_1 = 2\pi/1000$, $h_2 = 2\pi/400$ and $h_3 = 2\pi/200$.

14.8 Concluding Remarks

This chapter includes a brief review of the recent achievements in a field of nonlinear dynamics of coupled pendulums with impacts. First, the investigated system of three coupled physical pendulums are described and the governing equations are derived from the Lagrangian formulation. After introducing of the obstacles a calculation of the fundamental solution matrices for dynamical systems with impacts is outlined with an emphasis on the Aizerman-Gantmakher theory, saltation matrices and the Lyapunov exponents. Then the general considerations are simplified to a case of only three identical rods with damping, external excitation and with the obstacle in the form of a horizontal wall. In Section 14.6 the method used for numerical integration of the governing equations and its accuracy is discussed and illustrated. Some arbitrarily chosen numerical examples are presented in Section 14.7. It has been shown, among others, that the coexisting periodic solutions (stable and unstable), quasi-periodic, chaotic and hyperchaotic attractor are typical for our investigated system. Some examples of basins of attraction of coexisting attractors have been shown. In addition, the continuous Neimark-Sacker and the saddle-node bifurcations of periodic solutions have been reported. We have also shown, using bifurcation diagrams, a transition from periodicity, through quasiperiodicity and chaotic motions to a hyperchaotic behaviour of the system. For the attractors presented in this chapter the associated Lyapunov exponents have been calculated.

Chapter 15

Analytical Prediction of Stick-Slip Chaos

15.1 Introduction

The simplest description of self-excited vibrations is related to the energy flow from one form to another. In general for an arbitrarily taken physical system, an influence of the surroundings causes the dissipation of vibrations. However, the loss of energy is compensated by another source of energy, then an equilibrium of exchange between the lost and supplied energies is obtained and we usually observe periodic behaviour (a stable periodic orbit). The self-excited process is characterized by an energy source (it does not vibrate), a vibrating system and a device opening or closing the energy flow from the source. Self-excited oscillations are often observed in nature and belong to one of the oldest problems of nonlinear mechanics and physics.

Writing with chalk on the blackboard with accompanying squeak, tram squeak effects during turnings, squeak of the door hinges, bridge self-excited oscillations caused by wind forces (e.g. Tacoma Narrows bridge), a submerged pipeline in a tidal estuary oscillating transversely to the flow, shimming of the landing plane wheels, oscillations of a turning-lathe knife during cutting process certainly do not show all possible unfavorable self-excited mechanical vibrations [Andronov *et. al.* (1966)], [Den Hartog (1956)], [Kauderer (1958)], [Minorsky (1962)], [Pfeiffer (1992)], [Stoker (1950)]. In spite of the mentioned detrimental vibrations, they are sometimes desirable for both technical or entertainment reasons. Here self-sustained violin string oscillations explained by Lord Rayleigh, electric bells, pendulum watches or some electronic components should be mentioned [Feely and Chua (1992)], [McIntyre *et. al.* (1981)], [Schelleng (1971)].

Our attention will be focused on the self-sustained oscillations induced by friction. A decreasing part of the friction coefficient versus a relative velocity between sliding bodies is responsible for an occurrence of self-excited oscillations. When this relative velocity equals zero, then static friction can occur (stick), otherwise a slip mode corresponding to dynamic friction occurs. There are many papers devoted to that subject, which is also described in classical books already mentioned [Awrejcewicz (1988); Awrejcewicz (1991c)], [Awrejcewicz and Delfs (1990a); Awrejcewicz and Delfs (1990b)], [Awrejcewicz and Mrozowski (1989)], [Feeny and Moon (1993)], [Irretier (1983)], [Nayfeh and Mook (1979)], [Pfeiffer and Hajek (1992)], [Popp and Stelter (1990)], [Pratt and Williams (1981)].

Up to now, a classical approach to systems with dry friction has been briefly described. On the other hand, there exists some more detailed modelling of frictional behaviour, including microshapes of the contact surfaces or even a geometry of atomic meshes. Special attention is paid to an accurate modelling of constitutive relations for both normal and tangential body deformations and the normal motion [Feder and Feder (1991)], [Oden and Martins (1985)], [Tworzydlo *et. al.* (1992)].

It is also well known, that systems with friction are nonsmooth ones and cause some difficulties in both the theoretical and numerical analyses. The direct calculations of Lyapunov exponents cannot be used in that case and a special approach during the analysis of slip-slip, stick-slip and slip-stick transition should be developed. Also some progress in the theoretical approach of the analysis of nonsmooth dynamical systems has been achieved, showing a possibility of the concept generalization of the monodromy matrix, stability, etc. for both smooth and nonsmooth dynamical systems.

As it has been mentioned, although there are some research results showing stick-slip behaviour during chaotic dynamics of simple dynamical systems [Galvanetto *et. al.* (1993); Galvanetto *et. al.* (1995)], [Popp and Stelter (1990)], they are rather oriented towards a numerical analysis, and an analytical prediction of stick-slip chaos is very rarely given. In this chapter the Melnikov's method has been applied to that nonsmooth dynamical system and explicitly given formula corresponding to the stick behaviour has been successfully obtained. Therefore, the obtained analytical results allow one to analyse the stick-slip chaotic dynamics, or even to control it.

Another goal of our considerations is to predict stick-slip chaos in very weakly forced (quasi-autonomous) dynamical systems. There are many papers illustrating chaos in nonautonomous dynamical systems, however, with a reasonably large number of external excitations. In this chapter, due to an analytical approach, we are able to predict stick-slip chaos using extremely small external forcing.

We predict stick-slip chaotic dynamics in a one degree of freedom very weakly forced (quasi-autonomous) oscillator using the Melnikov's technique. Among others, we have found critical chaotic threshold points, where infinitely small external periodic perturbations applied to an autonomous system cause an occurrence of chaos.

15.2 The Melnikov's Method

The Melnikov's method is widely described in classical books and papers (see e.g. [Melnikov (1963)], [Yagasaki (1992)]. Here this method will be applied to the weakly forced nonlinear mechanical oscillators.

The dynamics of a one-degree-of-freedom nonlinear oscillator can be governed by the equation

$$\ddot{x} + S(x) = \varepsilon G(x, \dot{x}, t), \tag{15.1}$$

where: x is the displacement, $S(x)$ represents the nonlinear stiffness, $G(x, \dot{x}, t)$ is the time T-periodical function depending on the displacement and velocity, and $\varepsilon \geq 0$ is the small parameter. Eq. (15.1) can be rewritten in the form of two first-order differential equations

$$\begin{aligned} \dot{v} &= -S(x) + \varepsilon G(x, v, t), \\ \dot{x} &= v, \end{aligned} \tag{15.2}$$

where v is the velocity. For $\varepsilon = 0$ we get an unperturbed system of the following form

$$\begin{aligned} \dot{v} &= -S(x), \\ \dot{x} &= v. \end{aligned} \tag{15.3}$$

Let us suppose that the unperturbed system described by Eq. (15.3) has the equilibrium point p_0 at the origin of the phase plane. If the point p_0 is of the saddle type, then it is possible to find a homoclinic orbit q_0 associated to that point. In the case when $G(x, v, t)$ is T-periodic in time, the perturbed system (15.2) can be written as an autonomous one

$$\begin{aligned}\dot{v} &= -S(x) + \varepsilon g(x, v, \eta),\\ \dot{x} &= v,\\ \dot{\eta} &= \omega,\end{aligned} \tag{15.4}$$

where the frequency $\omega = 2\pi/T$. The phase space of the system (15.4) for $\varepsilon = 0$ has the cycle structure caused by η and the hyperbolic orbit. Therefore, we can define a Poincaré map $P_\varepsilon : \Sigma^{\eta_0} \to \Sigma^{\eta_0}$ which transforms $\Sigma^{\eta_0} = \{(x, v, \eta) | \eta = \eta_0 \in [0, T)\}$ to itself and which has a saddle fixed point with a homoclinic orbit Σ^{η_0}, corresponding to the phase portrait of the system (15.3). For $\varepsilon > 0$ the homoclinic orbit splits and it yields the stable $W^s(p_\varepsilon)$ and unstable $W^n(p_\varepsilon)$ manifolds of the hyperbolic point p_ε lying near p_0, of the form

$$\begin{aligned}W^s(p_\varepsilon) &= \left\{(x, v) \in \Sigma^{\eta_0} \Big| \lim_{n \to +\infty} P_\varepsilon^n = p_\varepsilon\right\},\\ W^n(p_\varepsilon) &= \left\{(x, v) \in \Sigma^{\eta_0} \Big| \lim_{n \to +\infty} P_\varepsilon^{-n} = p_\varepsilon\right\}.\end{aligned} \tag{15.5}$$

The projection $\bar{d}(t_0)$ on the normal to the homoclinic orbit $q_0(t - t_0)$ of the distance $d(t_0)$ between $W^s(p_\varepsilon)$ and $W^n(p_\varepsilon)$ can be obtained (see e.g. [Yagasaki (1995)], [Lamarque and Malasoma (1995)], where applications to mechanical systems are given)

$$\bar{d}(t_0) = -\varepsilon \frac{M(t_0)}{|f|} + O(\varepsilon^2), \tag{15.6}$$

where: $|f| = v^2 + S^2(x)$ and

$$M(t_0) = \int_{-\infty}^{\infty} \{v[q_0(t - t_0)] G[q_0(t - t_0), t]\}\, dt, \tag{15.7}$$

is called the Melnikov function. For $M(t_0) = 0$ the stable $W^s(p_\varepsilon)$ and unstable $W^n(p_\varepsilon)$ manifolds of the hyperbolic point p_ε intersect and when $M(t_0)$ has simple zeros (additionally $\partial M(t_0)/\partial t_0 \neq 0$), then by the Smale-Birkhoff homoclinic theorem [Wiggins (1990)], the power of the set of the intersection points equals to the power of the set of the integer numbers ($\aleph$) and the chaotic motions can appear.

15.3 Analyzed System

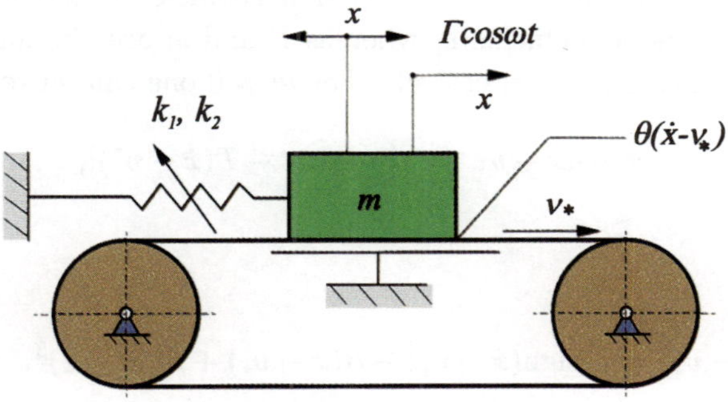

Fig. 15.1 Analysed system.

The analysed mechanical system with a small periodic external excitation and the Duffing type stiffness is shown in Fig. 15.1. One degree-of-freedom stick-slip oscillations are governed by the following second-order differential equation

$$m\ddot{x} - k_1 x + k_2 x^3 = \varepsilon[\Gamma \cos \omega t - \theta(\dot{x} - v^*)], \tag{15.8}$$

where:

$$\theta(\dot{x} - v_*) = \theta_0 \text{sign}(\dot{x} - v_*) - A(\dot{x} - v_*) + B(\dot{x} - v_*)^3, \tag{15.9}$$

and it corresponds to the friction characteristic versus the relative velocity.

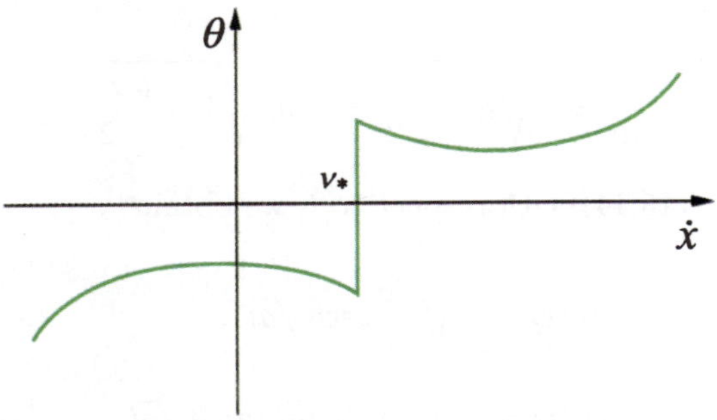

Fig. 15.2 Friction versus velocity.

In Eqs. (15.8) and (15.9) $\varepsilon > 0$ is the perturbation parameter and v_* corresponds to the velocity of the belt on which lies the considered mechanical system. θ_0, A, B are friction coefficients (see Fig. 15.2), k_1 and k_2 are stiffness coefficients, whereas Γ and ω are the amplitude and frequency of excitation, respectively. For $m \neq 0$ one can write

$$\ddot{x} - ax + bx^3 = \varepsilon[\gamma \cos \omega t - T(\dot{x} - v^*)], \tag{15.10}$$

where:

$$T(\dot{x} - v_*) = T_0 \text{sign}(\dot{x} - v_*) - \alpha(\dot{x} - v_*) + \beta(\dot{x} - v_*)^3, \tag{15.11}$$

and $a = k_1/m$, $b = k_2/m$, $\gamma = \Gamma/m$, $T = \theta/m$, $T_0 = \theta_0/m$, $\alpha = A/m$, $\beta = B/m$. Eq. (15.10) can be written as the following set of first-order ODE's

$$\begin{aligned} \dot{v} &= ax - bx^3 + \varepsilon[\gamma \cos \omega t - T(v - v_*)], \\ \dot{x} &= v. \end{aligned} \tag{15.12}$$

For $\varepsilon = 0$ we have the unperturbed system with the following Hamiltonian

$$H = \frac{1}{2}(v^2 - ax^2 + \frac{1}{2}bx^4), \tag{15.13}$$

and there are three critical points for $a, b > 0$: two centres at $(\pm\sqrt{a/b}, 0)$ and a hyperbolic saddle at the origin with the homoclinic orbit satisfying the equation

$$\frac{dx}{dt} = x\sqrt{a - b\frac{x^2}{2}} = \sqrt{a}x\sqrt{1 - \frac{b}{a}\frac{x^2}{2}}. \tag{15.14}$$

Integrating Eq. (15.14) with respect to t, we obtain

$$\begin{aligned} x_0(t) &= \pm\sqrt{\frac{2a}{b}} \text{sech}\sqrt{a}t, \\ v_0(t) &= a\sqrt{\frac{2}{b}} \text{sech}\sqrt{a}t \tan h\sqrt{a}t. \end{aligned} \tag{15.15}$$

The homoclinic orbit of Eq. (15.15) is shown in Fig. 15.3.

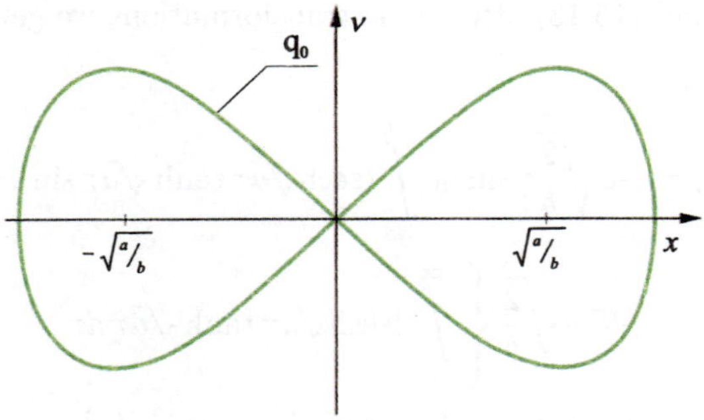

Fig. 15.3 Homoclinic orbit q_0 for $\varepsilon = 0$.

15.4 Analytical Results

From Eq. (15.7) one gets the following Melnikov function

$$M(t_0) = \int_{-\infty}^{\infty} \{v[q_0(t - t_0)]G[q_0(t - t_0), t]\}\, dt, \qquad (15.16)$$

where for our system we have

$$G(x, v, t) = \gamma \cos \omega t - T(v - v_*), \qquad (15.17)$$

and $q_0(x_0(t), v_0(t))$ is defined by Eq. (15.15). Substituting (15.15) and (15.17) in (15.16) we obtain

$$M(t_0) = \int_{-\infty}^{\infty} \Big\{ \gamma v_0(t - t_0) \cos \omega t - v_0(t - t_0) \Big[T_0 \mathrm{sign}[v_0(t - t_0) - v_*] + -\alpha(v_0(t - t_0) v_* + \beta(v_0(t - t_0) - v_*)^3 \Big] \Big\}\, dt. \qquad (15.18)$$

Using (15.15) and (15.18) after some transformations we get

$$\begin{aligned} M(t_0) = &-a\sqrt{\frac{2}{b}}\gamma\sin t_0 \int_{-\infty}^{\infty} (\mathrm{sech}\sqrt{a}\tau \tanh\sqrt{a}\tau \sin\omega\tau)\, d\tau \\ &-T_0 a\sqrt{\frac{2}{b}}\Bigg\{ \int_{-\infty}^{\infty} \Big[\mathrm{sech}\sqrt{a}\tau \tanh\sqrt{a}\tau\, d\tau \\ &\times \mathrm{sign}(\mathrm{sech}\sqrt{a}\tau \tanh\sqrt{a}\tau - v_*)\Big]\, d\tau\Bigg\} \\ &-4\beta\frac{a^4}{b^2}\int_{-\infty}^{\infty} (\mathrm{sech}\sqrt{a}\tau \tanh\sqrt{a}\tau)^4\, d\tau \\ &+3\beta v_* a^3 \left(\frac{2}{b}\right)^{\frac{3}{2}} \int_{-\infty}^{\infty} \mathrm{sech}^3\sqrt{a}\tau \tanh^3\sqrt{a}\tau\, d\tau \\ &+2(\alpha - 3\beta v_*^2)\frac{a^2}{b}\int_{-\infty}^{\infty} \mathrm{sech}^2\sqrt{a}\tau \tanh^2\sqrt{a}\tau\, d\tau \\ &+(\alpha - \beta v_*^2) v_* a\sqrt{\frac{2}{b}}\int_{-\infty}^{\infty} \mathrm{sech}\sqrt{a}\tau \tanh\sqrt{a}\tau\, d\tau, \end{aligned} \tag{15.19}$$

where: $\tau = t - t_0$. The first integral of Eq. (15.19) can be calculated using the method of residues, which yields

$$\begin{aligned} &a\sqrt{\frac{2}{b}}\gamma\sin t_0 \int_{-\infty}^{\infty} (\mathrm{sech}\sqrt{a}\tau \tanh\sqrt{a}\tau \sin\omega\tau)\, d\tau = \\ &\qquad \pi\omega\sqrt{\frac{2}{b}}\gamma \sin t_0 \mathrm{sech}\left(\frac{\pi\omega}{2\sqrt{a}}\right). \end{aligned} \tag{15.20}$$

To calculate the second integral we use

$$v_0(\tau) sgn[v_0(\tau) - v_*] = \begin{cases} -v_0(\tau), & v_0 < v_*, \\ v_0(\tau), & v_0 > v_*. \end{cases} \tag{15.21}$$

If τ_{1*}, τ_{2*} are the solutions of $v_* = v_0(\tau) = a\sqrt{2/b}\operatorname{sech}\sqrt{a}\tau \tan h\sqrt{a}\tau$, then we have

$$\begin{aligned} \operatorname{sech}\sqrt{a}\tau_{1*} &= \sqrt{\frac{1}{2} - \sqrt{\frac{1}{4} - \frac{b}{2a^2}v_*^2}}, \\ \text{and } \operatorname{sech}\sqrt{a}\tau_{2*} &= \sqrt{\frac{1}{2} + \sqrt{\frac{1}{4} - \frac{b}{2a^2}v_*^2}} \text{ for } v_* \le \frac{a}{\sqrt{2b}}. \end{aligned} \tag{15.22}$$

Therefore, we have

$$\begin{aligned} &T_0 a\sqrt{\frac{2}{b}} \int_{-\infty}^{\infty} \operatorname{sech}\sqrt{a}\tau \tanh\sqrt{a}\tau sgn \left[\operatorname{sech}\sqrt{a}\tau \tanh\sqrt{a}\tau - v_*\right] d\tau \\ &\quad = T_0 a\sqrt{\frac{2}{b}} \left[- \int_{-\infty}^{\tau_{1*}} \operatorname{sech}\sqrt{a}\tau \tanh\sqrt{a}\tau\, d\tau + \right. \\ &\quad\quad + \int_{\tau_{1*}}^{\tau_{2*}} \operatorname{sech}\sqrt{a}\tau \tanh\sqrt{a}\tau\, d\tau \\ &\quad\quad \left. - \int_{\tau_{2*}}^{\infty} \operatorname{sech}\sqrt{a}\tau \tanh\sqrt{a}\tau\, d\tau \right] = \\ &= 2T_0\sqrt{\frac{2a}{b}} \left[\sqrt{\frac{1}{2} + \sqrt{\frac{1}{4} - \frac{b}{2a^2}v_*^2}} - \sqrt{\frac{1}{2} - \sqrt{\frac{1}{4} - \frac{b}{2a^2}v_*^2}} \right]. \end{aligned} \tag{15.23}$$

If $v_* > (a/\sqrt{2b})$, then $\tau_* \notin \mathbf{R}$ and

$$\begin{aligned} &T_0 a\sqrt{\frac{2}{b}} \int_{-\infty}^{\infty} \operatorname{sech}\sqrt{a}\tau \tanh\sqrt{a}\tau \operatorname{sign}\left[\operatorname{sech}\sqrt{a}\tau \tanh\sqrt{a}\tau - v_*\right] d\tau = \\ &\quad = T_0 a\sqrt{\frac{2}{b}} \int_{-\infty}^{\infty} \operatorname{sech}\sqrt{a}\tau \tanh\sqrt{a}\tau\, d\tau = 0. \end{aligned} \tag{15.24}$$

Thus,

$$T_0 a\sqrt{\frac{2}{b}} \int_{-\infty}^{\infty} \left[\operatorname{sech}\sqrt{a}\tau \tanh\sqrt{a}\tau \operatorname{sign}(\operatorname{sech}\sqrt{a}\tau \tanh\sqrt{a}\tau - v_*)\right] d\tau \quad (15.25)$$

$$= \begin{cases} -2T_0\sqrt{\frac{2a}{b}}\left[\sqrt{\frac{1}{2}+\sqrt{\frac{1}{4}-\frac{b}{2a^2}v_*^2}} - \sqrt{\frac{1}{2}-\sqrt{\frac{1}{4}-\frac{b}{2a^2}v_*^2}}\right] & \text{for } v_* < \dfrac{a}{\sqrt{2b}} \\ 0 & \text{for } v_* \geq \dfrac{a}{\sqrt{2b}} \end{cases}$$

The third integral is given by

$$4\beta\frac{a^4}{b^2}\int_{-\infty}^{\infty} (\operatorname{sech}\sqrt{a}\tau \tanh\sqrt{a}\tau)^4\, d\tau =$$

$$= 4\beta\frac{a^4}{\sqrt{a}b^2}\left.\frac{\operatorname{sech}^3\tau(105\sinh\tau - 49\sinh 3\tau + 7\sinh 5\tau + \sinh 7\tau)}{1120}\right|_{-\infty}^{\infty} =$$

$$= \frac{16}{35}\beta\frac{a^4}{\sqrt{a}b^2}. \quad (15.26)$$

The fourth one disappears

$$3\beta v_* a^3\left(\frac{2}{b}\right)^{\frac{3}{2}}\int_{-\infty}^{\infty}(\operatorname{sech}\sqrt{a}\tau \tanh\sqrt{a}\tau)^3\, d\tau = 0, \quad (15.27)$$

and the fifth one is elementary

$$2(\alpha + 3\beta v_*^2)\frac{a^2}{b}\int_{-\infty}^{\infty}(\operatorname{sech}\sqrt{a}\tau \tanh\sqrt{a}\tau)^2\, d\tau = \quad (15.28)$$

$$= \frac{2}{3}(\alpha + 3\beta v_*^2)\frac{a^2}{\sqrt{a}b}\tanh^3\tau\Big|_{-\infty}^{\infty} = \frac{4}{3}(\alpha + 3\beta v_*^2)\frac{a^2}{\sqrt{a}b}.$$

The last integral is equal to zero

$$(\alpha + \beta v_*^2)v_* a\sqrt{\frac{2}{b}}\int_{-\infty}^{\infty}\operatorname{sech}\sqrt{a}\tau \tanh\sqrt{a}\tau\, d\tau = 0. \quad (15.29)$$

Substituting Eqs. (15.20), (15.26) - (15.29) in Eq. (15.19) we finally obtain

$$M(t_0) = -\pi\gamma\omega\sqrt{\frac{2}{b}}\text{sech}\left(\frac{\pi\omega}{2\sqrt{a}}\right)\sin\omega t_0 - \frac{16}{35}\beta\frac{a^4}{\sqrt{ab^2}} + \frac{4}{3}(\alpha - 3\beta v_*^2)\frac{a^2}{\sqrt{ab}} \tag{15.30}$$

$$-\begin{cases} -2T_0\sqrt{\frac{2a}{b}}\left[\sqrt{\frac{1}{2}+\sqrt{\frac{1}{4}-\frac{b}{2a^2}v_*^2}} - \sqrt{\frac{1}{2}-\sqrt{\frac{1}{4}-\frac{b}{2a^2}v_*^2}}\right] & \text{for } v_* < \frac{a}{\sqrt{2b}} \\ 0 & \text{for } v_* \geq \frac{a}{\sqrt{2b}}. \end{cases}$$

The obtained formula is applicable in both the cases, i.e. for stick-slip and slip-slip motions. The latter one corresponds to a zero additional value given in the brackets. Then, the Melnikov criterion is given by

$$\pi\gamma\omega\text{sech}\left(\frac{\pi\omega}{2}\right) > \frac{16}{35}\beta\frac{a^4}{b\sqrt{2ab}} - \frac{4}{3}(\alpha - 3\beta v_*^2)\frac{a^2}{\sqrt{2ab}} \tag{15.31}$$

$$+\begin{cases} 2T_0 a\left[\sqrt{\frac{1}{2}+\sqrt{\frac{1}{4}-\frac{b}{2a^2}v_*^2}} - \sqrt{\frac{1}{2}-\sqrt{\frac{1}{4}-\frac{b}{2a^2}v_*^2}}\right] & \text{for } v_* < \frac{a}{\sqrt{2b}} \\ 0 & \text{for } v_* \geq \frac{a}{\sqrt{2b}}. \end{cases}$$

Eq. (15.31) gives the possibility of chaotic threshold estimation in our weakly forced stick-slip oscillator.

15.4.1 *Numerical results*

In order to check the validity of our results, we have taken $a = b = 1$, $\alpha = \beta = T_0 = 0.3$, $\omega = 2$, $v_* = 0.4$, and we obtained $\gamma_c = 0.65$. The *Mathematica* program and the MATLAB-SIMULINK package have been used to simulate the analysed system. The MATLAB-SIMULINK scheme for the analysed system is shown in Fig. 15.4. There is a Duffing block representing the Duffing stiffness. The friction block simulates the friction shown in Fig. 15.2, and v_*-block is the constant value of the belt velocity. The phase planes and Poincaré maps are generated by the phase plane block and the Poincaré map block. The obtained results for $a = b = 1$, $\alpha = \beta = T_0 = 0.3$, $\omega = 2$ are shown in the (γ, v_*) plane in Fig. 15.5. The $\gamma(v_*)$ curve presented in Fig. 15.5 is obtained using an analytical prediction and separates the graph into two parts. Above this curve, chaos can appear, because near the line, the stable and unstable manifolds intersect transversally. For $v_* = a/\sqrt{2b} = \sqrt{2}/2 \approx 0.71$ there is a characteristic edge.

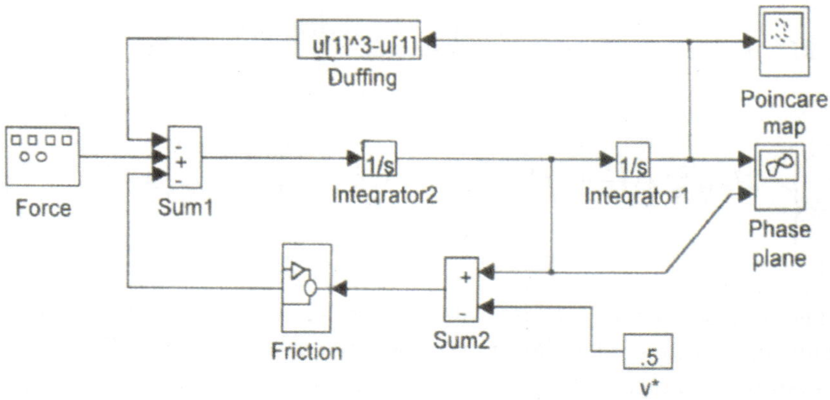

Fig. 15.4 The MATLAB-SIMULINK scheme of the analysed system.

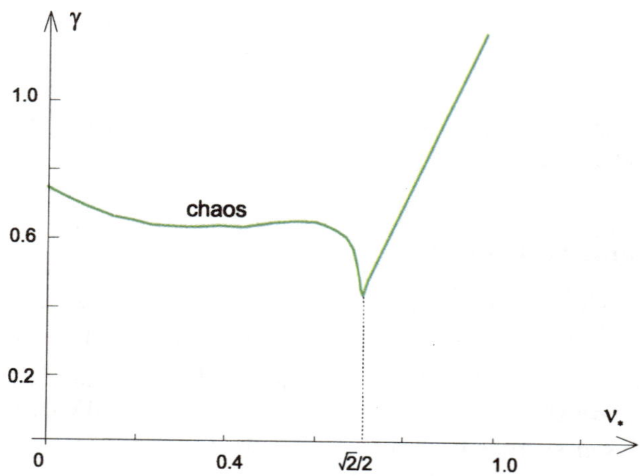

Fig. 15.5 Chaotic threshold in the (γ, v_*) plane ($a = b = 1$, $\alpha = \beta = T_0 = 0.3$, $\omega = 2$).

The slip-slip motions are dominated for $v_* < 0.71$, otherwise the stick-slip ones step forward.

We are going now to check numerically the validity of the obtained results. For this aim we have fixed (apart from the mentioned parameters) also $v_* = 0.4$ and we have used γ as the control parameter. The results,

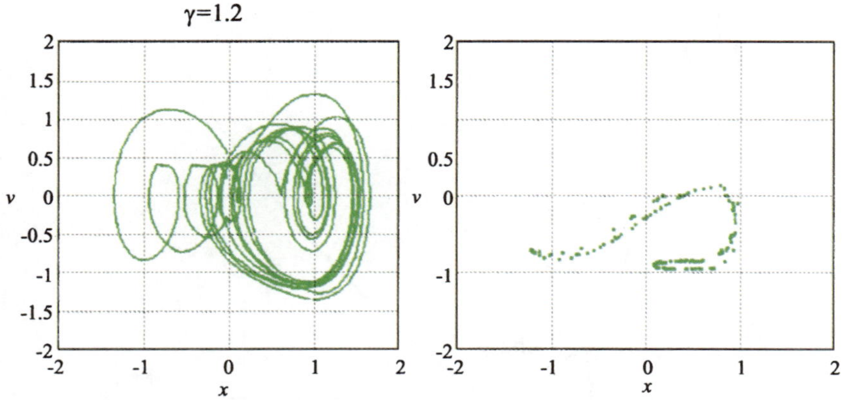

Fig. 15.6 Phase portrait and Poincaré map of the analysed system for $\gamma = 1.2$.

in the form of the phase portraits and the Poincaré maps, are presented in Fig. 15.6. For $\gamma = 0.6$ we have obtained a periodic orbit, which with an increase in γ doubles its period. Close to $\gamma_c = 0.65$ (analytical prediction) a very "slight" chaotic behaviour has been observed. For $\gamma = 1.2$ two-well potential chaos is shown. Note in all the phase portraits the cusps correspond to a sign change of the relative velocity, as well as short horizontal parts correspond to the sticks during motions. The obtained results illustrate practically a good agreement with the analytical method used. Since we have found the cusp of the curve $\gamma(v_*)$ (shown in Fig. 15.5), we have tried to manipulate the parameters in such a way that the cusp can touch the horizontal coordinate.

This corresponds to a special threshold for which we have $\gamma = 0$, i.e. an autonomous system. Then, an infinitely small periodic perturbation will lead the system to chaos. To achieve that we have used the parameter b. For example, for $b = 5$ there exist two following values of v_* (two cusps), for which $\gamma = 0$: $v_* = 0.310429$ and $v_* = 0.557204$. The obtained results are shown in Fig. 15.7. The γ parameter represents the forcing amplitude and $\gamma = 0$ corresponds to a one-degree-of-freedom autonomous stick-slip system. Because according to the Poincaré-Bendixon theorem we cannot get chaos in an autonomous one-degree-of-freedom system, we have found the autonomous system lying on the border of chaos. The simulation results are given for the root $v_* = 0.557204$ and for $\gamma = 1.2$ in Fig. 15.6. We have obtained for $\gamma = 0$ a periodic motion with the fundamental frequency corresponding to that introduced by self-excitation due to friction. Also

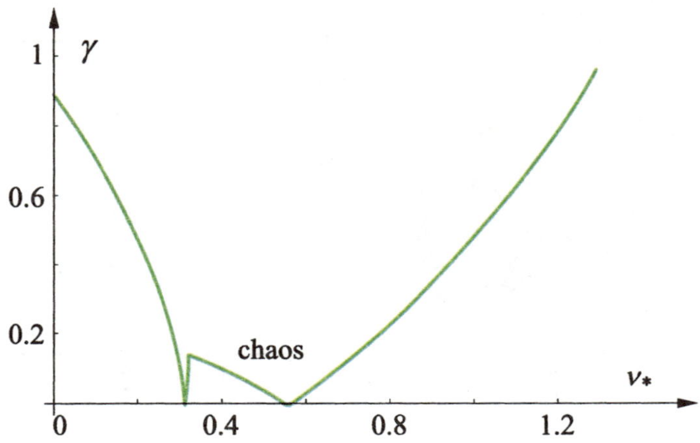

Fig. 15.7 Chaotic threshold in the (γ, v_*) plane ($a = 1$, $b = 5$, $\alpha = \beta = T_0 = 0.3$, $\omega = 2$).

slip-stick phenomena are visible. While increasing γ, a quasiperiodic stick-slip attractor is observed ($\gamma = 0.05$). A self-locking phenomenon has been further observed, leading to a period-2ω-periodic window. Beginning with $\gamma = 0.54$ stick-slip chaos is found. Although chaotic behaviour has been tested for $\gamma = 0.54$ (see also $\gamma = 0.543$) we suspect that an extremely low order of chaos can also occur for $\gamma \approx 0$, which was extremely difficult to detect numerically. We have also calculated the Lyapunov exponent which again was located very close to zero.

Roughly speaking, for the second root $v_* = 0.31043$, we have used similar numerical quantification and we have got qualitatively similar results (Fig. 15.7). We have started with the periodic orbit ($\gamma = 0$), then we have found the stick-slip quasi-periodic one ($\gamma = 0.05$), a periodic window ($\gamma = 0.2$), and two-well potential stick-slip chaos ($\gamma = 0.3875$).

The next numerical example is given for a narrow distance between two roots of v_*, i.e. $v_* = 0.4996$ and $v_* = 0.5255$ and $b = 2$ (see Fig. 15.8). There is an interval between the roots of the small value of γ. Again, we have used numerical simulations and beginning with $\gamma = 0$ and increasing this parameter to the value of $\gamma = 0.55$, we have observed a gradual route from quasi-periodicity to periodicity and then period doubling leading to stick-slip chaos. The latter case is presented in Fig. 15.9 for $\gamma = 0.692$. The last step in our consideration is to predict the so-called chaotic threshold for a self-excited autonomous oscillator. Substituting $\gamma = \omega = 0$ in Eq.

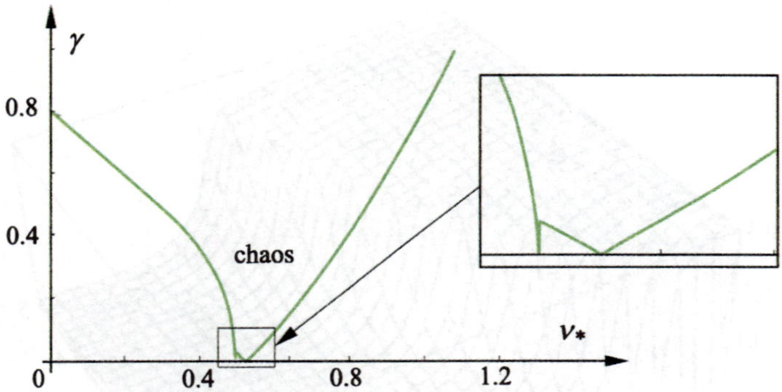

Fig. 15.8 Chaotic threshold in the (γ, v_*) plane ($a = 1$, $b = 2$, $\alpha = \beta = T_0 = 0.3$, $\omega = 2$).

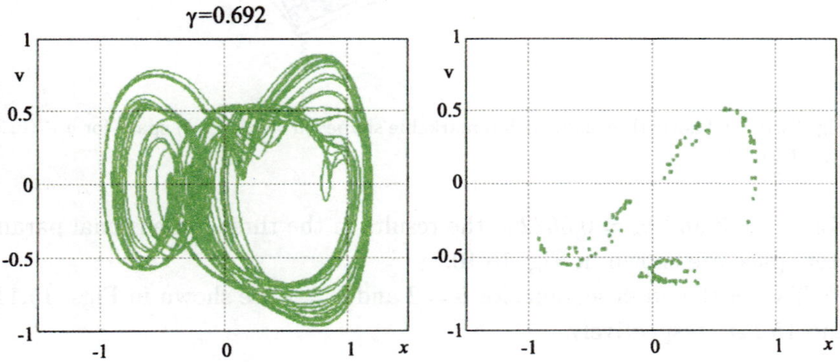

Fig. 15.9 Phase portrait and Poincaré map of the analysed system for $\gamma = 0.692$ and $v_* = 0.5$.

(15.31) and solving it with respect to β, we obtain the following analytical formula

$$\beta = \begin{cases} \dfrac{35}{6} \cdot \dfrac{b}{a(4a^2 + 35bv_*^2)} \Bigg[2a\alpha + \\ + 3T_0\sqrt{b} \left(\sqrt{1 - \sqrt{1 - \dfrac{bv_*^2}{a^2}}} - \sqrt{1 + \sqrt{1 - \dfrac{bv_*^2}{2a^2}}} \right) \Bigg] & \text{for } v_* < \dfrac{a}{\sqrt{2b}}, \\ \dfrac{35}{3} \cdot \dfrac{\alpha b}{4a^2 + 35bv_*^2} & \text{for } v_* \geq \dfrac{a}{\sqrt{2b}}. \end{cases}$$

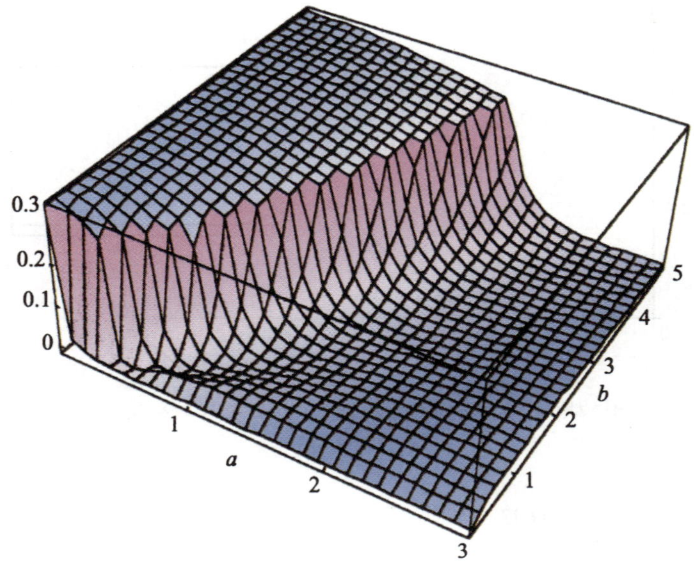

Fig. 15.10 Chaotic threshold with remarkable shapes in the (β, a, b) space for $a \in (0;3)$, $b \in (0;5)$.

For $\alpha = 0.3$ and $v_* = 0.557204$ the results in the three-dimensional parameter space are shown in Fig. 15.10.

Two of the cross-sections for $b = 2$ and $a = 1$ are shown in Figs. 15.11a and 15.11b, respectively.

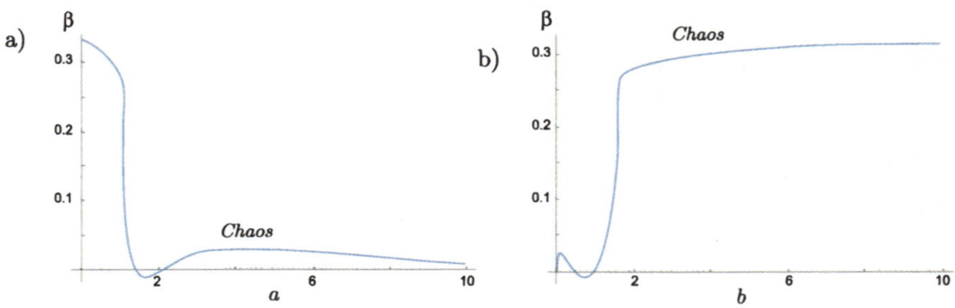

Fig. 15.11 Chaotic threshold in the (β, a, b) plane for $b = 2$ (a), and $a = 1$ (b).

These results show that we can either get chaos or regular motions threshold depending on the parameters. We can predict stick-slip or slip chaos on the basis of the Melnikov's technique. An analysis based on the observation of the Lyapunov exponent versus a parameter is widely used in literature. Here, we have proposed similar results in the sense of distinguishing between regular and chotic motions but in an analytical form. These results can also be obtained using the standard numerical methods, but they are time-consuming.

Chapter 16

Thermoelasticity, Wear and Stick-Slip Movements of a Rotating Shaft with a Rigid Bush

16.1 Introduction

Friction, wear, heat generation and temperature deformation are complex processes which influence each other and integrate into a sole diverse process of friction. During the transient friction process the variation of the parameters of the problem are mutually connected and conditioned. During small sliding velocities rapid un-uniform movements can be observed very often. These movements take place intermittently with periodical disruptions and stops. In this chapter we have made an attempt to investigate this mutual connection on the simplest model that remains in conditions of such relative movements as stick-slip oscillations [Awrejcewicz and Pyryev (2002)].

Uneven movement during friction (stick-slip motion) is the phenomenon of alternate relative sliding and relative rest that arises due to self-oscillations during decreasing of friction coefficient while sliding velocity increases. The main reason of mechanical oscillations arising is presence of positive difference between friction force of rest and sliding. This difference is caused by the decreasing of sliding friction force with the increasing of relative velocity. Self-oscillations are undamped oscillations supported by external energy sources in the nonlinear dissipative system. They principally differ from other oscillatory processes in dissipative systems in that way that the supporting periodical influence from outside is not necessary. Under wear process we will understand the process of material separation from the surface of the body that leads to gradual change of dimensions. Wear is the result of wear process that can be measured in the length units.

To the present time the different models of nonlinear mechanical systems frictional self-oscillations and the methods of their solving were considered in references [Andronov *et. al.* (1966)], [Awrejcewicz *et. al.* (1998)], [Ibrahim (1994)], [Krahelskyy and Hittis (1987)], [Martins *et. al.* (1990)]. Investigation of dry friction is presented and the approaches for the evaluation of this kind of friction are developed in references [Chichinadze *et. al.* (1979)], [Krahelskyy and Hittis (1987)], [Martins *et. al.* (1990)]. Friction units wear investigation is performed and corresponding methods of wear calculation are discussed in reference [Chichinadze *et. al.* (1979)].

However, less investigated remains question concerning frictional oscillations in conditions of heat generation and wear. Investigation of such oscillations arising conditions and behaviour of contact characteristics (contact temperature, contact pressure, wear) could be useful for the explanation of different phenomena in disc brakes, grinding machines, high accuracy mechanisms (when it is necessary to assure displacements with small velocities) and other machines with friction couples.

The problem related to oscillations of a spring-fixed on the steady base bush with respect to uniformly rotating shaft in conditions of high friction is investigated in reference [Andronov *et. al.* (1966)]. This problem can be treated as the draft sketch of ordinary braking pad or Prony's clamp. For small rotating velocity of a shaft a stick-slip motion of a braking pad is observed. For large velocities the pad undergoes the damped periodical oscillations. A similar like behaviour is also observed in many various models which are described and illustrated in references [Ibrahim (1994)], [Krahelskyy and Hittis (1987)], [Martins *et. al.* (1990)]. As it is well known a decreasing part of the kinetic friction F_t^k versus (small) velocity is responsible for an occurrence of stick-slip oscillations [Awrejcewicz (1983); Awrejcewicz (1987); Awrejcewicz (1988); Awrejcewicz and Delfs (1990a)].

Thermoelastic contact of a rotating shaft with an uninertial steady bush with wear processes is considered in [Pyryev (2000); Pyryev and Hrylitskyy (1996)]. Observe that during shaft rotation its volume is extended due to occurrence of a heat process. On the other hand the process of material wear from the bush surface occurs. All of the contact characteristics of two sliding bodies are coupled. In addition, when so called critical rotation velocity is achieved a surrounding medium does not keep up with heat collection and the system is overheated (the contact characteristics increase exponentially), and thermoelastic instability occurs. The mentioned bush wear process leads to increase of the critical velocity, and finally, for enough high wearing, the critical velocity do not appear. Dynamical models of thermoelastic contact in conditions of frictional heating are also presented in references [Alexandrov and Annakulova (1990); Alexandrov and Annakulova (1992); Olesiak and Pyryev (1999); Olesiak and Pyryev (2000); Olesiak *et. al.* (1997); Pyryev (2000)].

In this chapter more general plane axially-symmetric problem about thermoelastic contact of the rotating shaft with the rigid bush fixed elastically to the steady base by springs in conditions of frictional self-oscillations and wear is investigated. Observe that in our considered system a shaft temperature, the contact pressure, wear of the bush, and the bush movement of our system are coupled with each other. This chapter is mainly focused on analysis of the mentioned self-coupling processes.

16.1.1 *Statement of the problem*

Consider thermoelastic thermoelastic contact of solid isotropic circular shaft (cylinder) of radius R_1 with a cylindrical tube-like rigid bush (solid liner, rigid ring) which is fitted to the cylinder according to the expression $U_0 h(t)$ (Fig. 16.1).

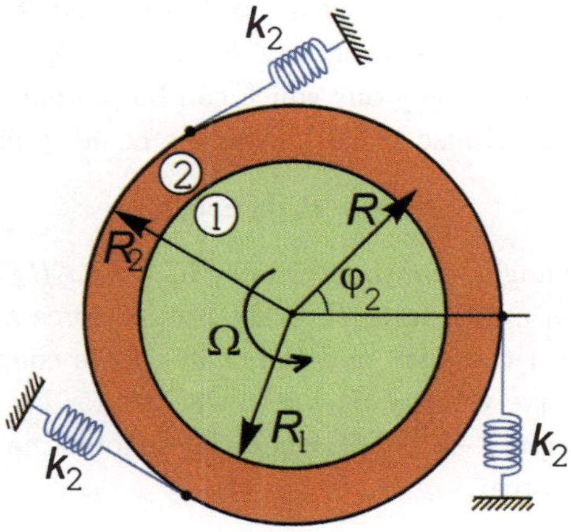

Fig. 16.1 A model of the problem.

Calculated per unit of the length inertia moment of the bush is equal to B_2. Bush is fixed to the steady-state base through the spring with reduced coefficient of rigidity k_2. Shaft rotates with angular velocity $\Omega(t) = \Omega_* \omega_1(t)$. On the contact surface between the cylinder and the bush friction force F_t arises and as the consequence heat generation and wear U_z take place. The work of friction force is transformed into heat energy. The temperature of the shaft at the initial instant of time is equal to T_0. Assume that between shaft and the bush Newton's heat exchange law takes place and that the bush has constant temperature T_0.

Both thermal and stress-strain state of the shaft is considered using the cylinder coordinates R, ϕ, Z rotating with the angular velocity Ω with its origin situated in the center of the rotating elastic body. The governing equations of motion of uncoupled thermoelastic problem along Z axis have the following form [Kovalenko (1975)], [Ulitko (1990)]:

$$\mu\nabla^2\mathbf{u} + (\lambda+\mu)\operatorname{grad}\operatorname{div}\mathbf{u} + \rho\Omega^2 R\mathbf{e}_R = (3\lambda+2\mu)\alpha_1 \operatorname{grad} T + \rho\frac{\partial^2\mathbf{u}}{\partial t^2},$$

$$\nabla^2 T = \frac{1}{a_1}\frac{\partial T}{\partial t}, \tag{16.1}$$

where ∇^2 - Laplace operator, $\mathbf{u} = U\mathbf{e}_R + V\mathbf{e}_\phi + W\mathbf{e}_Z$ - vector of relative displacement, T - shaft temperature, λ, μ - Lamé coefficients, ρ - bush density, α_1 - coefficient of linear thermal expansion, a_1 - thermal diffusivity.

We consider the bush as an ideal rigid body and its movement is described using the cylinder coordinates R, φ_2, Z. A set of three Euler's equations are reduced to only one which can be presented as the following differential inclusion [Kunze (2000)], [Martins *et. al.* (1990)],

$$B_2\ddot{\varphi}_2 + F_s R_2 \in F_t R_1, \tag{16.2}$$

where $\varphi_2(t)$ is the angle of bush deviation, $F_s = k_2\varphi_2 R_2$ is the spring force related to the bush unit length, F_t is the friction force acting on the bush unit length between the bush and the shaft. F_t is equal to kinetic force $F_t^k(V_w)$ if the relative velocity $|V_w| > 0$, whereas for $V_w = 0$ its values are located in the interval $[-F_t^s,\ F_t^s]$, where F_t^s denotes the maximal value of the static friction force:

$$F_t = \operatorname{sign}(V_w)\begin{cases} F_t^k(V_w) & \text{if } |V_w| > 0, \\ F_t^s & \text{if } |V_w| = 0, \end{cases}$$

$$\operatorname{sign}(V_w) = \begin{cases} 1 & \text{if } V_w > 0, \\ [-1,1] & \text{if } V_w = 0, \\ -1 & \text{if } V_w < 0. \end{cases} \tag{16.3}$$

We consider a one-dimensional model of thermal friction and wear during stick-slip motion taking into account the following assumptions:

(1) The external excitation of the system allows for neglection of the term $\rho\partial^2\mathbf{u}/\partial t^2$ in the Lamé Eq. (16.1). Shaft rotates with small angular velocity $\Omega(t)$ so that centrifugal forces $\rho\Omega^2 R$ could be neglected [Timoshenko and Goodier (1951)].

(2) The vector components related to displacements as well as the shaft temperature do not depend on ϕ, Z, and the unequal zero components $U(R,t)$, $V(R,t)$ and $T(R,t)$ depend only on the radial coordinate and time.

(3) We also assume that according to Coulomb's law friction force is proportional to normal part of reaction $F_t = f(V_w)N(t)$, with the following friction coefficient

$$f(V_w) = \text{sign}\,(V_w) \begin{cases} f_k(V_w) & \text{if } |V_w| > 0, \\ f_s & \text{if } |V_w| = 0, \end{cases} \tag{16.4}$$

where f_s denotes maximal value of the static friction coefficient.

(4) The coefficient of kinetic friction $f_k(V_w)$ is dependent from relative velocity of shaft and bush in a way governed by equation $V_w = \Omega R_1 - \dot{\varphi}_2(t)R_1$ (see the Fig. 16.2). The so called Harsy-Stribeck's curve [Ibrahim (1994)], [Krahelskyy and Hittis (1987)], has a minimum when $V_w = V_{\min}$, whereas for $V_w < V_{\min}$ friction coefficient decreases $f_k'(V_w) < 0$. Observe that rather various functions are taken in a literature to approximate the experimentally found values. It leads to different approximation of both friction force or momentum. The most popular is that governed by simple equation $f(V_w) = \text{sign}\,(V_w)f_s$. A review of various friction models can be found in already mentioned references [Andronov *et. al.* (1966)], [Ibrahim (1994)], [Krahelskyy and Hittis (1987)], [Martins *et. al.* (1990)]. As a criterion choice of a friction model usually a minimal set of the parameters together with a good agreement with the experiment are taken. The dependence of kinetic friction coefficient versus sliding velocity (Fig. 16.2) can be presented in the form

$$f_k(V_w) = \begin{cases} f_{\min} + (f_s - f_{\min})\exp(-b_1|V_w|), & \text{if} \quad |V_w| \le V_{\min}, \\ f_{\min} + (f_s - f_{\min})\exp(-b_1|V_{\min}|) + \\ \quad + \dfrac{b_2 b_3(|V_w| - V_{\min})^2}{1 + b_2(|V_w| - V_{\min})}, & \text{if} \quad |V_w| \le V_{\min}, \end{cases} \tag{16.5}$$

where: f_s, $f_{\min}$, b_1, b_2, b_3, $V_{\min}$ can be defined via experiment, and the function $f_k(V_w)$ is smooth for $|V_w| > 0$, $f_k(0+) = f_s$,

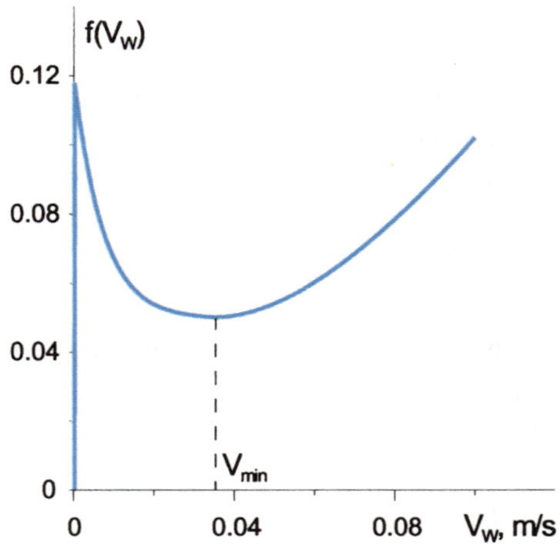

Fig. 16.2 Dependence of kinetic friction coefficient versus relative velocity.

$f_k(V_w)/V_w|_{V_w \to \infty} = b_3$.
For computation purposes, the multi-valued relation $\mathrm{sgn}\,(x)$ is approximated by the function $\mathrm{sgn}_{\varepsilon_0}(x)$ defined in the reference [Martins *et. al.* (1990)]

$$\mathrm{sgn}_{\varepsilon_0}(x) = \begin{cases} 1 & \text{if } x > \varepsilon_0, \\ (2 - \frac{|x|}{\varepsilon_0})\frac{x}{\varepsilon_0} & \text{if } |x| < \varepsilon_0, \\ -1 & \text{if } x < -\varepsilon_0, \end{cases} \tag{16.6}$$

where $\varepsilon_0 = 0.0001$.

(5) The heat flows q_1 and q_2 are generated on the contact surface due to Ling rule [Ling (1959)] and governed by the equation $q_1 + q_2 = (1-\eta)F_{fr}V_w$, where $\eta \in [0,1]$ denotes the part of heat energy which goes on the wear [Alexandrov and Annakulova (1990)]. Both flows q_1 and q_2 go into shaft and bush, respectively:

$$q_1 = \lambda_1 \frac{\partial T(R_1,t)}{\partial R}, \quad q_2 = -\alpha_t \left(T_0 - T(R_1,t)\right), \tag{16.7}$$

where λ_1 is the heat conduction coefficient, α_T is the heat exchange coefficient between shaft and a bush.

(6) One of the most popular wear model is governed by the equation

$$\dot{U}_z(t) = K_z|V_w(t)|^m P(t)^n, \tag{16.8}$$

where m, n - exponents, K_z - wear coefficient of the bush, $P(t) = N(t)/2\pi R_1$ - contact pressure. We assume Archard's law of wear [Archard (1959)], [Horyacheva and Dobychin (1988)] in the form of (16.8) where $m = n = 1$. The chosen rule is typical for an abrasive wear.

After taking in to account the above assumptions our problem (from a mathematical viewpoint) is governed by the Eqs. (16.1) and it can be reduced to that of finding a solution to the quasi-static thermoelasticity equations [Kovalenko (1975)] described in the cylinder coordinates of the form

$$\frac{\partial^2 U(R,t)}{\partial R^2} + \frac{1}{R}\frac{\partial U(R,t)}{\partial R} - \frac{1}{R^2}U(R,t) = \alpha_1 \frac{1+\nu}{1-\nu}\frac{\partial T(R,t)}{\partial R}, \tag{16.9}$$

$$\frac{\partial^2 T(R,t)}{\partial R^2} + \frac{1}{R}\frac{\partial T(R,t)}{\partial R} = \frac{1}{a_1}\frac{\partial T(R,t)}{\partial t}, \quad 0 < R < R_1, \quad 0 < t < t_c. \tag{16.10}$$

The differential inclusion (16.2) is then approximated by the equation of bush movement as a solid body

$$B_2\ddot{\varphi}_2(t) + k_2 R_2^2 \varphi_2(t) = f(V_w) 2\pi R_1^2 P(t), \quad 0 < t < t_c. \tag{16.11}$$

The following mechanical boundary conditions are attached to Eqs. (16.9)-(16.11):

$$U(0,t) = 0, \quad U(R_1,t) = -U_0 h(t) + U_z(t), \quad 0 < t < t_c, \tag{16.12}$$

as well as both the heat boundary conditions

$$\lambda_1 \frac{\partial T(R_1,t)}{\partial R} + \alpha_T (T(R_1,t) - T_0) = (1-\eta) f(V_w) V_w P(t), \tag{16.13}$$

$$R\left.\frac{\partial T(R,t)}{\partial R}\right|_{R\to 0} = 0, \quad 0 < t < t_c, \tag{16.14}$$

and the initial conditions

$$T(R,0) = T_0, \quad 0 < R < R_1, \quad \varphi_2(0) = \varphi_2^\circ, \quad \dot{\varphi}_2(0) = \omega_2^\circ. \tag{16.15}$$

A speed of bush or shaft wear is proportional to the power of friction force [Archard (1959)], [Horyacheva and Dobychin (1988)] in the following manner:

$$\dot{U}_z(t) = K_z|V_w(t)|P(t). \tag{16.16}$$

Radial stress $\sigma_R(R,t)$ in the cylinder can be determined after radial displacements $U(R,t)$ and temperature $T(R,t)$ in the cylinder as follows

$$\sigma_R(R,t) = \frac{E_1}{1-2\nu}\left[\frac{1-\nu}{1+\nu}\frac{\partial U(R,t)}{\partial R} + \frac{\nu}{1-\nu}\frac{U(R,t)}{R} - \alpha_1(T(R,t)-T_0)\right]. \tag{16.17}$$

In the above expressions the following definitions are used: $P(t) = -\sigma_R(R_1,t)$, E_1 is the elasticity modulus, ν is Poisson's ratio. We find temperature, contact pressure, radial displacements, the velocity and the magnitude of wear at the moment of time $t \in [0, t_c]$, where t_c is time of contact ($0 < t < t_c$, $P(t) > 0$).

Integrating (16.9) and taking into account (16.12), (16.17) we can write down contact pressure as follows:

$$P(t) = \frac{2E_1\alpha_1}{1-2\nu}\frac{1}{R_1^2}\int_0^{R_1}[T(\xi,t)-T_0]\,\xi\,d\xi + \frac{E_1}{(1-2\nu)(1+\nu)R_1}\left[U_0h(t)-U_z\right].$$

At the above expressions the following dimensionless values are introduced

$$r = \frac{R}{R_1}, \quad \tau = \frac{t}{t_*}, \quad u = \frac{U}{U_0}, \quad u_z = \frac{U_z}{U_0}, \quad p = \frac{P}{P_*}, \quad \theta = \frac{T-T_0}{T_*},$$

$$Bi = \frac{\alpha_t R_1}{\lambda_1}, \quad k_z = \frac{\Omega_* P_* t_* K_z R_1}{U_0}, \quad \gamma = \frac{2(1-\eta)E_1\alpha_1 R_1^2\Omega_*}{\lambda_1(1-2\nu)}, \quad \varepsilon = \frac{P_* t_* 2\pi R_1^2}{B_2\Omega_*},$$

$$\tilde{\omega} = \frac{t_* a_1}{R_1^2}, \quad \varphi^\circ = \frac{\varphi_2^\circ}{\Omega_* t_*}, \quad \omega^\circ = \frac{\omega_2^\circ}{\Omega_*}, \quad \varphi(\tau) = \frac{\varphi_2(t_*\tau)}{\Omega_* t_*}, \quad \tau_c = \frac{t_c}{t_*},$$

$$F(\omega_1 - \dot{\varphi}) = f(R_1\Omega_*(\omega_1 - \dot{\varphi})), \quad h(\tau) = h(t_*\tau),$$

where the corresponding characteristic parameters read:

$$t_* = \sqrt{\frac{B_2}{k_2R_2^2}}, \quad P_* = \frac{E_1U_0}{(1+\nu)(1-2\nu)R_1}, \quad T_* = \frac{U_0}{2\alpha_1(1+\nu)R_1}.$$

In the dimensionless form the considered problem has the form:

$$\ddot{\varphi}(\tau) + \varphi(\tau) = \varepsilon F(\omega_1 - \dot{\varphi})p(\tau), \quad 0 < \tau < \tau_c, \tag{16.18}$$

$$\varphi(0) = \varphi^\circ, \quad \dot{\varphi}(0) = \omega^\circ; \tag{16.19}$$

$$p(\tau) = h(\tau) - u_z(\tau) + \int\limits_0^1 \theta(\xi,\tau)\xi\, d\xi, \quad 0 < \tau < \tau_c; \tag{16.20}$$

$$u_z(\tau) = k_z \int\limits_0^\tau |\omega_1 - \dot{\varphi}(\tau)| p(\tau)\, d\tau, \quad 0 < \tau < \tau_c; \tag{16.21}$$

$$\frac{\partial^2\theta(r,\tau)}{\partial r^2} + \frac{1}{R}\frac{\partial\theta(r,\tau)}{\partial r} = \frac{1}{\tilde{\omega}}\frac{\partial\theta(r,\tau)}{\partial\tau}, \quad 0 < \tau < \tau_c, \quad 0 < r < 1; \tag{16.22}$$

$$\frac{\partial\theta(1,\tau)}{\partial r} + Bi\theta(1,\tau) = \gamma F(\omega_1 - \dot{\varphi}(\tau))(\omega_1 - \dot{\varphi}(\tau))p(\tau), \quad 0 < \tau < \tau_c; \tag{16.23}$$

$$r\left.\frac{\partial\theta(r,\tau)}{\partial r}\right|_{r\to 0} = 0, \quad 0 < \tau < \tau_c; \quad \theta(r,0) = 0, \quad 0 < r < 1. \tag{16.24}$$

16.1.2 *Solution of the problem*

Applying the Laplace transform to nonlinear problem (16.20)-(16.24) (s - the transformation parameter)

$$\{\bar{\theta}(r,s),\ \bar{p}(s),\ \bar{u}_z(s),\ \bar{h}(s),\ \bar{q}(s)\} = \\ = \int\limits_0^\infty \{\theta(r,\tau),\ p(\tau),\ u_z(\tau),\ h(\tau),\ q(\tau)\}\, e^{-s\tau}\, d\tau,$$

where nonlinear part has the form

$$q(\tau) = \gamma F(\omega_1 - \dot{\varphi})(\omega_1 - \dot{\varphi})p(\tau), \tag{16.25}$$

we obtain the solution in the form of Laplace transforms

$$\bar{\theta}(r,s) = G_\theta(r,s)\bar{q}(s),\, \bar{p}(s) = \bar{h}(s) + G_p(s)\bar{q}(s),$$

$$G_\theta(r,s) = \frac{I_0(s_\omega r)}{\tilde{\omega}\Delta_1(s)}, \quad G_u(s) = \frac{\Delta_1(s)}{\tilde{\omega}\Delta_1(s)}, \tag{16.26}$$

$$\Delta_2(s) = \frac{I_1(s_\omega)}{s_\omega}, \ \Delta_1(s) = s_\omega^2\Delta_2(s) + BiI_0(s_\omega), \ s_\omega = \sqrt{\frac{s}{\tilde{\omega}}},$$

where $I_0(x)$, $I_1(x)$ are the modified Bessel functions.

Performing the inverse Laplace transformation with the help of residua's and convolution theorems [Carslaw and Jaeger (1959)] one can find the following relations

$$p(\tau) = h(\tau) - u_z(\tau) + \gamma\tilde{\omega}\int_0^\tau G_u(\tau-\xi)F(\omega_1-\dot{\varphi})p(\xi)(\omega_1-\dot{\varphi})\,d\xi, \tag{16.27}$$

$$\theta(r,\tau) = \gamma\tilde{\omega}\int_0^\tau G_\theta(r,\tau-\xi)F(\omega_1-\dot{\varphi})p(\xi)(\omega_1-\dot{\varphi})\,d\xi, \tag{16.28}$$

where:

$$\{G_u(\tau),\ G_\theta(1,\tau)\} = \sum_{m=1}^{\infty}\frac{\{2Bi,\ 2\mu_m^2\}}{Bi^2+\mu_m^2}e^{-\mu_m^2\tilde{\omega}\tau}, \tag{16.29}$$

and μ_m are the characteristic equation roots ($m = 1, 2, 3, \ldots$) of the equation

$$BiJ_0(\mu) - \mu J_1(\mu) = 0. \tag{16.30}$$

Thus, the stated problem is reduced to the solution of the system of nonlinear differential (16.18) and integral (16.27) equations taking into account (16.21) for the contact pressure $p(\tau)$ and the velocity $\dot{\varphi}(\tau)$. Temperature and wear could be found using expressions (16.28) and (16.21), respectively.

16.1.3 *Steady-state solution analysis*

16.1.3.1 *Analysis of steady-state solution in case of wear absence ($k_z = 0$)*

Consider the case $\omega_1(\tau) = \omega_1^\circ H(\tau)$, $h(\tau) = H(\tau)$; $H(\tau) = 1$, $\tau > 0$, $H(\tau) = 0$, $\tau < 0$. The steady-state solution is found for these input data when wear is absent (in Eqs. (16.18), (16.22) terms consisting time derivatives are neglected)

$$p_{st} = \frac{1}{1-\upsilon},\ \theta_{st} = \frac{2\upsilon}{1-\upsilon},\ \varphi_{st} = \frac{\varepsilon F(\omega_1^\circ)}{1-\upsilon},$$
$$\gamma_1 = \frac{\varepsilon F(\omega_1^\circ)}{\omega_1^\circ}, \quad \upsilon = \frac{\gamma\omega_1^\circ F(\omega_1^\circ)}{2Bi}. \tag{16.31}$$

For the determination of the solution's behaviour the linearization of the problem was performed in the vicinity of steady-state point (16.31). Assume perturbations as follows: $h(\tau) = 1 + h^*(\tau)$, $\omega_1(\tau) = \omega_1^\circ + \omega_1^*(\tau)$, $|h^*| \ll 1$, $\omega_1^* \ll 1$. Then a solution can be represented in the form

$$\varphi(r,\tau) = \varphi_{st} + \varphi^*(\tau), \quad \theta(r,\tau) = \theta_{st}(r) + \theta^*(r,\tau), \quad p(\tau) = p_{st} + p^*(\tau), \tag{16.32}$$

where $|\varphi^*| \ll 1$, $|\theta^*| \ll 1$, $|p^*| \ll 1$. The right hand sides of the equation (16.18) and the boundary condition (16.23) were linearized. For perturbations we obtain the linear problem

$$\ddot{\varphi}^*(t) + \varphi^*(t) = \varepsilon F(\omega_1^\circ)p^*(\tau) + \varepsilon F'(\omega_1^\circ)p_{st}(\omega_1^* - \dot{\varphi}^*), \quad 0 < \tau < \infty, \tag{16.33}$$

$$\varphi^*(0) = 0, \quad \dot{\varphi}^*(0) = 0; \tag{16.34}$$

$$p^*(\tau) = h^*(\tau) + \int_0^1 \theta^*(\xi,\tau)\xi\, d\xi, \quad 0 < \tau < \infty; \tag{16.35}$$

$$\frac{\partial^2\theta^*(r,\tau)}{\partial r^2} + \frac{1}{r}\frac{\partial\theta^*(r,\tau)}{\partial r} = \frac{1}{\tilde{\omega}}\frac{\partial\theta^*(r,\tau)}{\partial\tau}, \quad 0 < \tau < \infty, \quad 0 < r < 1; \tag{16.36}$$

$$\frac{\partial\theta^*(1,\tau)}{\partial r} + Bi\theta^*(1,\tau) = \gamma\left[\omega_1^\circ F(\omega_1^\circ)p^* + p_{st}(\omega_1^* - \dot{\varphi}^*)(F(\omega_1^\circ) + \omega_1^\circ F'(\omega_1^\circ))\right],$$

$$0 < \tau < \infty, \tag{16.37}$$

$$r\left.\frac{\partial\theta^*(r,\tau)}{\partial r}\right|_{r\to 0} = 0, \quad 0 < \tau < \infty; \quad \theta^*(r,0) = 0, \quad 0 < r < 1. \tag{16.38}$$

Applying Laplace transformation to the linear problem (16.33)-(16.38) (s - parameter of transformation) of the form

$$\{\bar{\theta}^*(r,s),\ \bar{p}^*(s),\ \bar{\varphi}^*(s),\ \bar{h}^*(s),\ \bar{\omega}_1^*(s)\} = \int_0^\infty \{\theta^*,\ p^*,\ \varphi^*,\ h^*,\ \omega_1^*\}\, e^{-s\tau}\, d\tau,$$

we obtain a solution in Laplace transforms

$$\bar{\varphi}^*(s) = G_{\varphi h}(s)\bar{h}^*(s) + G_{\varphi\omega}(s)\bar{\omega}_1^*(s), \tag{16.39}$$

$$\bar{p}^*(s) = G_{ph}(s)\bar{h}^*(s) + G_{p\omega}(s)\bar{\omega}_1^*(s), \tag{16.40}$$

$$\bar{\theta}^*(r,s) = G_{\theta h}(r,s)\bar{h}^*(s) + G_{\theta\omega}(s)\bar{\omega}_1^*(s), \tag{16.41}$$

where:

$$\{G_{\varphi h}(s),\ G_{\varphi\omega}(s),\ G_{ph}(s),\ G_{p\omega}(s),\ G_{\theta h}(s),\ G_{\theta\omega}(s)\} = \frac{\{\cdot\}}{\Delta^*(s)},$$

$$\{\cdot\} = \{F(\omega_1^\circ)\Delta_1(s),\ p_{st}\left[\gamma F^2(\omega_1^\circ)\Delta_2(s) + F'(\omega_1^\circ)\Delta_1(s)\right],\ \Omega_2(s)\Delta_1(s),$$

$$\gamma\varepsilon(s^2+1)\Delta_2(s)\left(F(\omega_1^\circ) + \omega_1^\circ F'(\omega_1^\circ)\right)p_{st},\ I_0(s_\omega)\gamma\omega_1^\circ F(\omega_1^\circ)\Omega_1(s),$$

$$\gamma\varepsilon I_0(s_\omega)(s^2+1)p_{st}(F(\omega_1^\circ) + \omega_1^\circ F'(\omega_1^\circ))\}. \tag{16.42}$$

The characteristic equation of the steady-state problem has the form

$$\Delta^*(s) = 0, \qquad \Delta^*(s) = \Delta_1(s)\Omega_2(s) - 2Biv\Delta_2(s)\Omega_1(s),$$
$$\Omega_1(s) = s^2 - \frac{\gamma_1}{1-v}s + 1, \quad \Omega_2(s) = s^2 + \frac{\gamma_2}{1-v}s + 1, \quad \gamma_2 = \varepsilon F'(\omega_1^\circ). \tag{16.43}$$

The roots s_m ($\text{Re}s_1 > \text{Re}s_2 > \ldots \text{Re}s_m > \ldots, m = 1,2,3,\ldots$) of the characteristic Eq. (16.43) can be located in the left-hand side $\text{Re}s < 0$ (steady solution is stable) or the right-hand side $\text{Re}s > 0$ (steady solution is unstable) of the complex plane depending to the input data of the problem. The parameters of the problem for which the roots transition takes place will be called further critical. The analysis of the parameter's regions of steady-state solution stability is carried out. The characteristic function is presented in the form

$$\Delta^*(s) = \sum_{m=0}^{\infty} \left(\frac{s}{\tilde{\omega}}\right)^m d_m, \tag{16.44}$$

$$d_m = (1-\delta_{m1})\tilde{\omega}^2 \left(d_{m-2}^{(1)} - 2Bivd_{m-2}^{(2)}\right) + \frac{\tilde{\omega}}{1-v}\left(\gamma_2 d_{m-1}^{(1)} - 2Biv\gamma_1 d_{m-1}^{(2)}\right) +$$
$$+ d_m^{(1)} - 2Bivd_m^{(2)}, \quad m = 1, 2, \ldots; \quad \delta_{mn} = 1, \ m = n; \quad \delta_{mn} = 0, \ m \neq n;$$

$$d_0 = Bi(1-v), \quad d_m^{(1)} = \frac{Bi + 2m}{2^{2m}(m!)^2}, \quad d_m^{(2)} = \frac{1}{2^{2m+1}m!(1+m)!}.$$

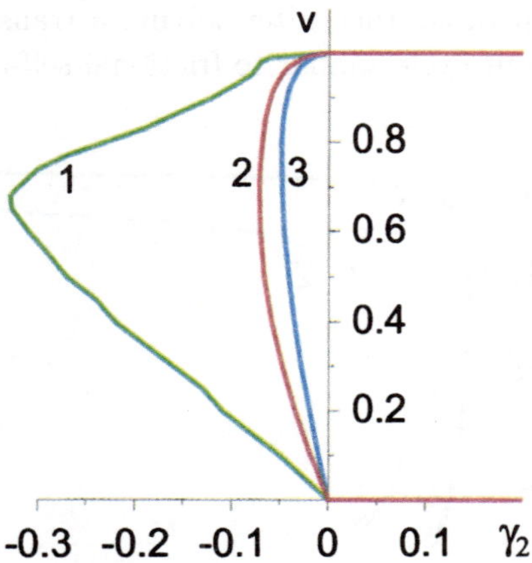

Fig. 16.3 Dependence of critical value of v versus critical value of γ_2 for different values of the parameter $\tilde{\omega}$ ($k_z = 0$). Curve 1: $\tilde{\omega} = 0.005$, 2: $\tilde{\omega} = 0.1$, 3: $\tilde{\omega} = 1$. The parameter regions inside the curves correspond to the area of steady-state solution stability.

The dependence of critical value v versus parameter γ_2 for different $\tilde{\omega} = 0.05; 0.1; 1$ is presented in Fig. 16.3 (curves 1-3, correspondingly). Fig. 16.4 contains the dependence of critical value v versus parameter $\tilde{\omega}$ for different values $\gamma_2 = -0.05; -0.08$ (curves 1-2), and the other parameters are: $B_i =$

$10, \gamma_1 = 0.586$. For the parameters that are situated inside curves, a steady-state solution is stable. The detailed analysis shows that decreasing of γ_2 leads to narrowing the stability area according to parameter v and to increasing of critical values $\tilde{\omega}$. When $v = 0$ and neglecting heat expansion of the cylinder we obtain a model of self-oscillations [Andronov *et. al.* (1966)], with the characteristic equation of the linearized problem $\Omega_2(s) = 0$. In this case a steady-state solution is stable, when $\gamma_2 > 0$. When $\tilde{\omega} = 0$ (bush is immovable) we obtain a model [Pyryev and Hrylitskyy (1996)] with the $\Delta_1(s) - 2Biv\Delta_2 = 0$. In this case, when $v > 1$ a steady solution is unstable (root $s_1 > 0$). The so called thermal instability [Pyryev and Hrylitskyy (1996)] or thermal explosion [Alexandrov and Annakulova (1990)], [Alexandrov and Annakulova (1992)] takes place. The analysis of the particular cases of the considered model shows that from one side steady-state solution is stable, when $\gamma_2 > 0$. This corresponds to $V_w > V_{min}$ (see Fig. 16.2). From the other side a steady solution is stable, when $v < 1$. The specific parameters of the system, the view of Stribeck's curve and analysis of the characteristic equation roots (16.43) can help definitely to answer the question about the stability of steady-state solution. If steady solution (16.31) is unstable, then after solving a transient problem it can behave as a stable limit cycle similar to frictional self-oscillations or it can increase with time.

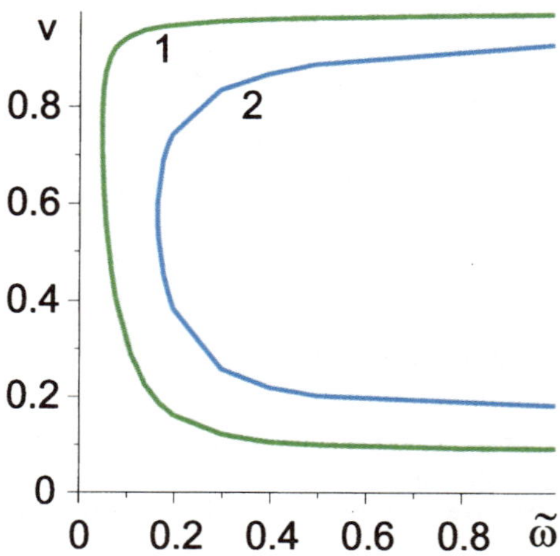

Fig. 16.4 Critical value of v dependence versus critical value of $\tilde{\omega}$ for different γ_2 ($k_z = 0$). Curve 1: $\gamma_2 = -0.05$, 2: $\gamma_2 = -0.08$. Regions of steady-state solution stability are inside the corresponding curves.

16.1.4 *Analysis of steady-state solution in the presence of wear* ($k_z \neq 0$)

The steady state solution of the problem in the case of presence of wear is found (in Eqs. (16.18), (16.22) the terms with time derivatives and $\dot{u}_z(\tau) = 0$ are neglected):

$$p_{st} = 0, \quad \theta_{st} = 0, \quad \varphi_{st} = 0, \quad u_z^{st} = 1. \tag{16.45}$$

Performing similar to the previous case procedure the characteristic equation of linearized problem is obtained in the vicinity of steady solution (16.45):

$$\Delta^*(s) = 0, \quad \Delta^*(s) = (s + k_z\omega_1^\circ)\Delta_1(s) - 2Bivs\Delta_2(s). \tag{16.46}$$

The analysis of the parameter's regions in which steady-state solution (16.45) is stable is carried out. For this purpose the characteristic function is sought in the form

$$\Delta^*(s) = \sum_{m=0}^{\infty} \left(\frac{s}{\tilde{\omega}}\right)^m b_m, \tag{16.47}$$

$$b_0 = kd_0^{(1)}, \quad k = k_z\omega_1^\circ, \quad b_m = kd_m^{(1)} + \tilde{\omega}\left(d_{m-1}^{(1)} - 2Bivd_{m-1}^{(2)}\right), \; m = 1, 2, \ldots \tag{16.48}$$

First, a root of the characteristic equation for small wear $k \ll 1$ can be presented in the form

$$s_1 = -\frac{k}{1-v}. \tag{16.49}$$

As it can be seen from (16.49) for low wear, when $v < 1$, the steady solution (16.45) is stable and in opposite case - unstable. In the conditions of wear presence system's contact time t_c is limited. The material of the bush will wear with time. That is why we should accurately state conditions of stability. Parameters of the problem under which roots s_1, s_2 of the characteristic equation have positive real part and they are complex conjugate are related to intensive wear. For these parameters the amplitude of the oscillations increases according to exponential law but system's contact time is limited. System will go out from the contact faster than contact characteristics will reach critical values (initial assumptions will loose sense). Parameters of the problem under which the roots of the characteristic equation s_1, s_2 have

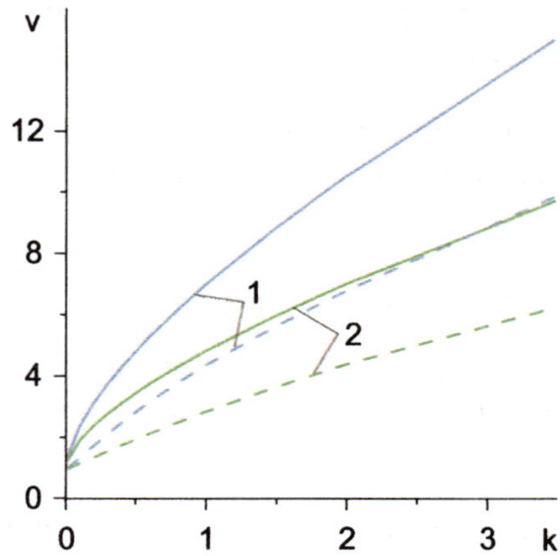

Fig. 16.5 Critical values v versus k for different $\tilde{\omega}$ ($k_z \neq 0$). Curves 1: $\tilde{\omega} = 0.05$, 2: $\tilde{\omega} = 0.1$. Regions under solid curves are stable. Parameter regions between solid and dashed curves correspond to intensive wear.

only positive real part are referred as the critical parameters. Under these parameters there will be no oscillations but contact characteristics will rise according to an exponential law. The system will be overheated. The rate of heat expansion is bigger than wear rate. In Fig. 16.5 the dependence of critical value v (solid curves) versus parameter k for different values of $\tilde{\omega} = 0.05,\ 0.1$ (curves 1-2 respectively) is presented. Taking into account three terms in the decomposition (16.47) the formula for critical values is derived (roots coincide on real axis Res_1 =Re$s_2 > 0$, Ims_1 =Ims_2=0)

$$v_{cr} = 1 + \frac{k + \sqrt{2Bik\tilde{\omega}(4 + Bi)}}{2Bi\tilde{\omega}}.$$

This formula is the more accurate the less wear parameter k is. For the parameters that are situated under the solid curve $v < v_{cr}$ the system is stable. Moreover, between solid and dashed curves $v_0 < v < v_{cr}$ time, when the system is in contact is limited and intensive wear takes place. The formula for the dashed curve is also found (roots are transiting through imaginary axis Res_1 =Re$s_2 = 0$, Im$s_1 = -$Im$s_2 \neq 0$)

$$v_0 = 1 + \frac{k(2 + Bi)}{4Bi}.$$

This formula is the more accurate the less wear parameter k is.

16.1.5 *Numerical analysis of the transient solution*

The numerical analysis of the problem is performed using Runge-Kutta method by taking into account the following asymptotes

$$G_\theta(1,\tau) \approx \sqrt{1/\pi\tau\tilde{\omega}},\ G_u(\tau) \approx 1,\ \tau \to 0.$$

The following kinetic friction parameters have been fixed: $f_s = 0.12$, $f_{min} = 0.05$, $b_1 = 140\ \text{sm}^{-1}$, $b_2 = 10\ \text{sm}^{-1}$, $b_3 = 2\ \text{sm}^{-1}$, $V_{min} = 0.035\ \text{sm}^{-1}$. The results of calculations for different values of the parameter $\gamma = 0,\ 200,\ 400$ for the steel cylinder ($\alpha_1 = 14 \times 10^{-6}\ {}^\circ C^{-1}$, $\lambda_1 = 21\ \text{W}/(\text{m}^\circ\text{C}^{-1})$, $\nu = 0.3$, $a_1 = 5.9 \times 10^{-6}\ \text{m}^2/\text{s}$, $E_1 = 19 \times 10^{10}$ Pa) and $R_1 = 0.03$ m, $\Omega = 1$ rad/s, $\varepsilon = 10$, $\tilde{\omega} = 0.1$, $Bi = 10$, $\varphi^\circ = 0$, $\omega^\circ = 0$ are shown in Figs. 16.6–16.14. Solid curves correspond to the case of wear absence $k_z = 0$, dashed - to the case of wear with the dimensionless wear coefficient $k_z = 0.1$. In this case $t_* = 0.25$ s, $P_* = 1.22 \times 10^4$ Pa, $\gamma_1 = 0.51$, $\gamma_2 = -196$. In the case of heat expansion absence time evolutions of the dimensionless speed $\dot{\varphi}(\tau)$ (curve 1), dimensionless displacement $\varphi(\tau)$ (curve 2) of the bush and dimensionless friction force $\varepsilon F(\omega_1 - \dot{\varphi})p(\tau)$ (dashed curve 3) for $\gamma = 0$ ($v = 0$, $s_1 = 0.0051$ the steady-state solution is unstable) are shown in Fig. 16.6.

Fig. 16.7 illustrates the behaviour of the phase trajectory in the phase plane without taking into account heat expansion of the cylinder $\gamma = 0$ and wear of the bush $k_z = 0$ (solid curve) and with wear (dashed curve). In conditions of wear absence a transient solution leads with time to stable limit cycle with the period 6.608. Stick-slip phenomenon takes place. Let us point out that (as it can be seen from this figure) the friction force has a jump at the instant of time when bush speed reaches speed of the cylinder $V_w = 0$. At that moment the rest friction force changes the sign. Then dimensionless friction force becomes equal to dimensionless displacement and this lasts to the moment when it reaches maximal value of static friction force. From that instant of time friction force decreases. After that it reaches local minimum when bush acceleration is equal to zero. Then friction force increases till it reaches cylinder's velocity $V_w = 0$. Process continues cyclically. In conditions of wear presence friction force will tend with time to zero and bush will perform its own oscillations with the period of 2π. Time evolutions of the dimensionless speed $\dot{\varphi}(\tau)$ (curve 1), dimensionless displacement $\varphi(\tau)$ (curve 2) of the bush and dimensionless kinetic friction force $\varepsilon F(\omega_1 - \dot{\varphi})p(\tau)$ (dashed curve 3) are shown in Fig. 16.8 for $\gamma = 200$ ($v = 0.51, s_1 = 0.0012$ steady-state solution is unstable). Fig. 16.9

illustrates the behaviour of the phase trajectory in phase plane without taking into account a heat expansion of the cylinder $\gamma = 200$ and $k_z = 0$ (solid curve) and with wear (dashed line). As it can be seen in conditions of bush wear absence the contact characteristics tends with time to the limit cycle with the period equal to 5.5.

Fig. 16.10 illustrates changing with time of the dimensionless speed $\dot{\varphi}(\tau)$ (curve 1), dimensionless displacement $\varphi(\tau)$ (curve 2) of the bush and dimensionless kinetic friction force $\varepsilon F(\omega_1 - \dot{\varphi})p(\tau)$ (dashed curve 3) for $\gamma = 400$ ($v = 1.02$, $s_1 = 2 \times 10^{-6}$). Fig. 16.11 illustrates the behaviour of phase trajectory on phase plane without taking into account the heat expansion of the cylinder $\gamma = 400$ and $k_z = 0$ (solid curve) and with wear (dashed curve). In the case of wear absence thermoelastic instability takes place. System's characteristics do not tend with time to stable limit cycle but exponentially increase. In that case cylinder does not have enough time to become cool. Wear presence (dashed curves) leads to thermoelastic instability disappearing.

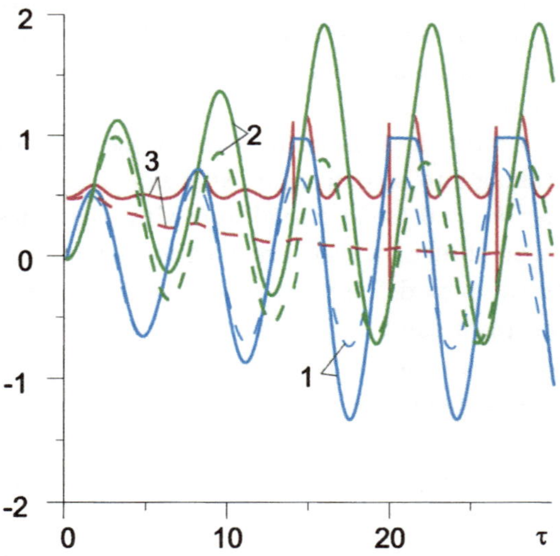

Fig. 16.6 Dependence of bush movement dimensionless velocity $\dot{\varphi}$ (curve 1), dimensionless displacements φ (curves 2) and dynamic friction force $\varepsilon F(\omega_1 - \dot{\varphi})$ (curve 3) versus dimensionless time τ in conditions of heat expansion absence $\gamma = 0$. Solid curves: $k_z = 0$, dashed curves: $k_z = 0.1$.

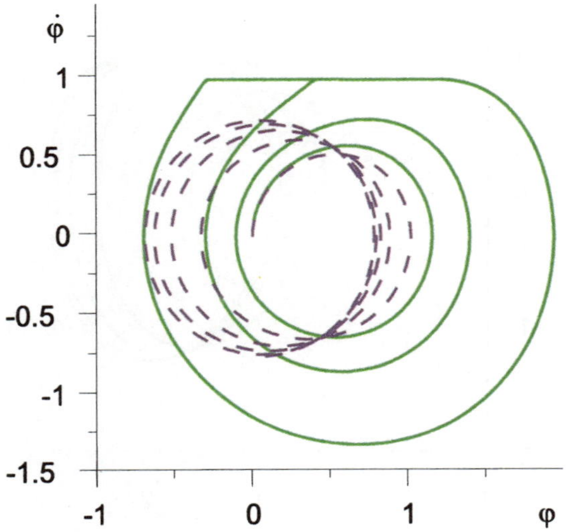

Fig. 16.7 Phase trajectory of bush movement in conditions of heat expansion absence $\gamma = 0$. Solid curves: $k_z = 0$, dashed curves: $k_z = 0.1$.

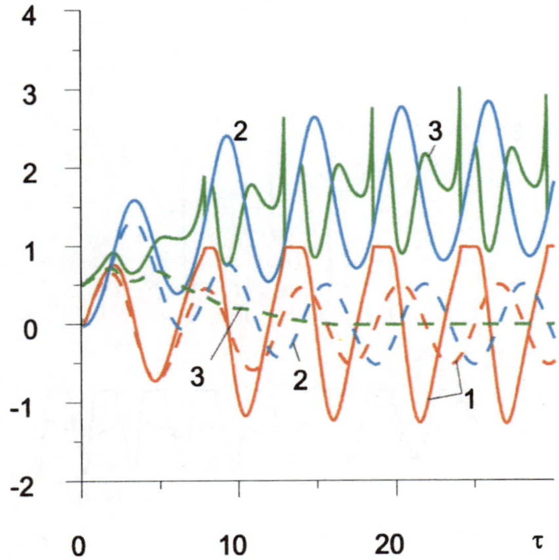

Fig. 16.8 Dependence of bush dimensionless velocity $\dot{\varphi}$ (curves 1), dimensionless displacements φ (curve 2) and dynamic friction force $\varepsilon F(\omega_1 - \dot{\varphi})$ (curve 3) versus dimensionless time τ, when $\gamma = 200$. Solid curves: $k_z = 0$, dashed curves $k_z = 0.1$.

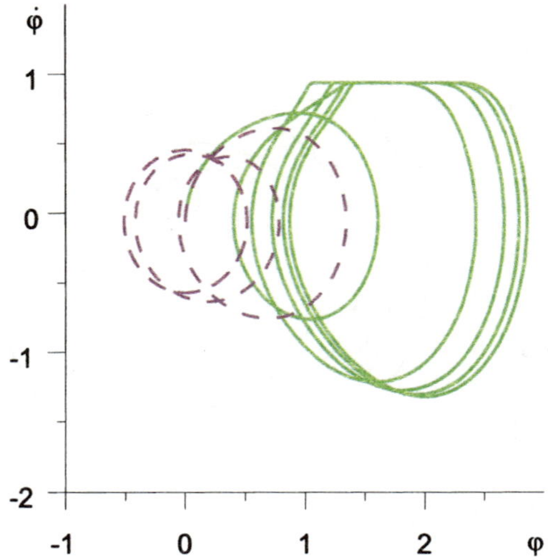

Fig. 16.9 Phase trajectory of bush movement in conditions, when $\gamma = 200$. Solid curves: $k_z = 0$, dashed curves $k_z = 0.1$.

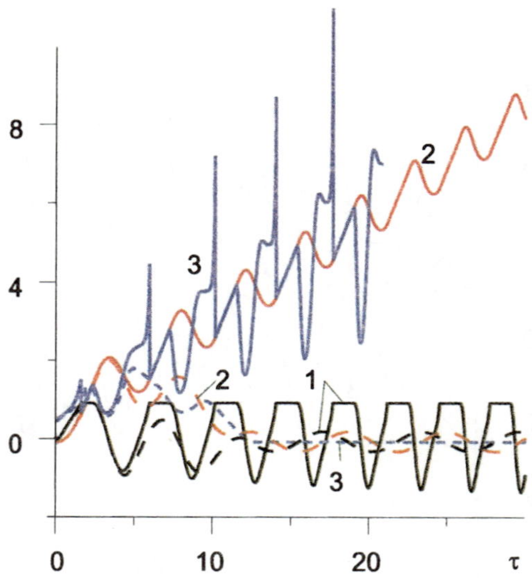

Fig. 16.10 Dependence of bush dimensionless velocity $\dot{\varphi}$ (curves 1), dimensionless displacements φ (curve 2) and dynamic friction force $\varepsilon F(\omega_1 - \dot{\varphi})$ (curve 3) versus dimensionless time τ, when $\gamma = 400$. Solid curves: $k_z = 0$, dashed curves $k_z = 0.1$.

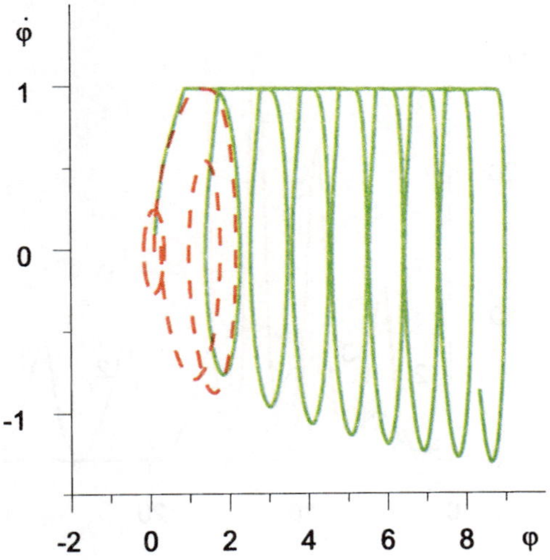

Fig. 16.11 Phase trajectory of bush movement for $\gamma = 400$. Solid curves: $k_z = 0$, dashed curves $k_z = 0.1$.

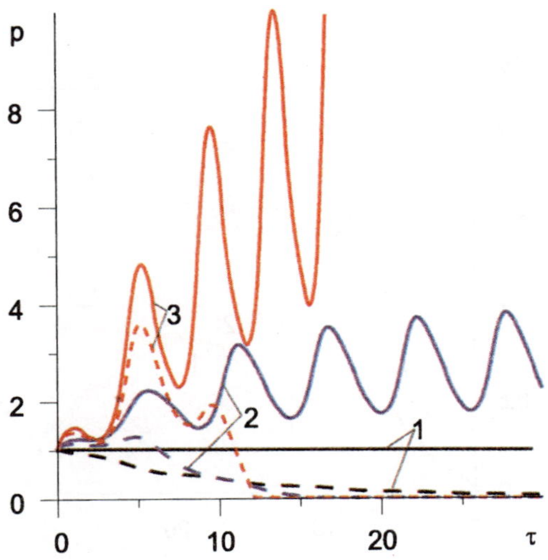

Fig. 16.12 Behaviour of dimensionless contact pressure p versus dimensionless time τ for different values of γ. Curve 1: $\gamma = 0$, 2: $\gamma = 200$, 3: $\gamma = 400$. Solid curves: $k_z = 0$, dashed curves $k_z = 0.1$.

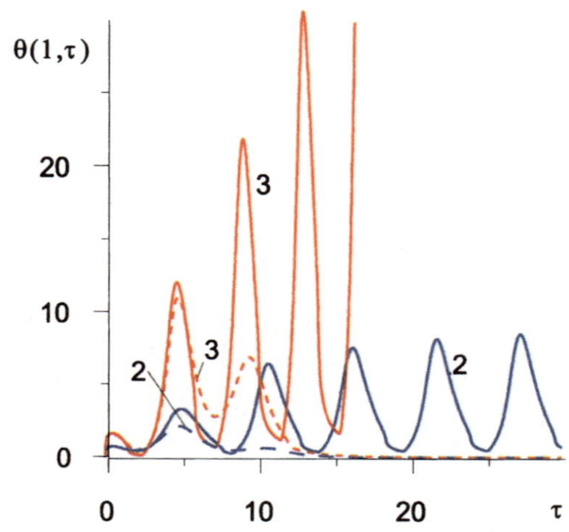

Fig. 16.13 Behaviour of dimensionless contact temperature $\theta(1,\tau)$ versus dimensionless time τ for different values of γ. Curve 1: $\gamma = 0$, 2: $\gamma = 200$, 3: $\gamma = 400$. Solid curves: $k_z = 0$, dashed curves $k_z = 0.1$.

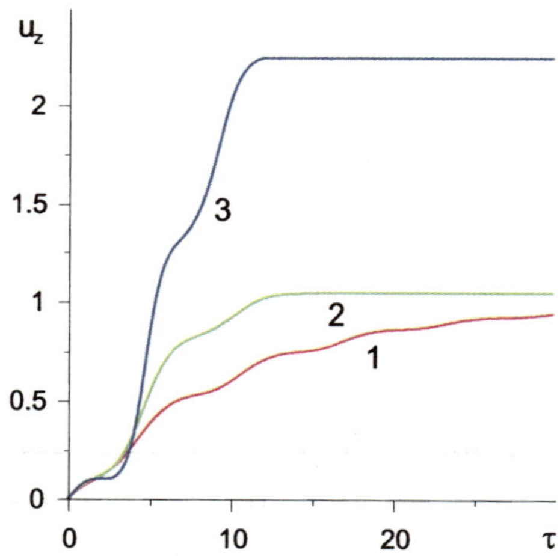

Fig. 16.14 Time evolution of dimensionless wear u_z for different values of parameter γ. Curve 1: $\gamma = 0$, 2: $\gamma = 200$, 3: $\gamma = 400$.

Time changing laws for the contact pressure, temperature and wear are shown in Figs. 16.12–16.14 by solid (dashed) curves for the case of wear absence (presence), correspondingly. Curve 1 corresponds to the case of absence of heat expansion $\gamma = 0$, curve 2 - to $\gamma = 200$, curve 3 - to $\gamma = 400$. In the last case contact characteristics increase and cylinder does not have time for cooling. While parameter $\tilde{\omega}$ decreases and $v < 1$ time necessary for leading of contact characteristics on limit cycle increases. Wear presence leads to decreasing of contact characteristics values and to limiting of system's contact time (curves 2 and 3 in Fig. 16.12). With the increasing of the parameter γ system's contact time decreases and, correspondingly, bush wear increases and its value becomes bigger than value of initial deformation caused by an initial stress state of the cylinder.

Chapter 17

Control for Discrete Models of Buildings Including Elastoplastic Terms

17.1 Introduction

In order to analyze dynamical behavior of buildings, and to take into account soil-structure coupling, simple models have been introduced both for buildings (see [Filliatrault (1996)] for example) and soil-structure and foundation (see [Wolf (1994)] for example).

Translational cone and rotational cone models provide spring-dashpot mass models (see [Wolf (1994)]). Here we intend to introduce an extension of the model resulting from translational cone model for soil-foundation interaction: this extension includes elastoplastic terms. Such terms have been introduced in previous works ([Bastien (2000)], [Schatzman *et. al.* (1999)]) and have been described from experiments and identified via nonparametric models [Benedettini *et. al.* (1995)]. More general models are also considered. We are interested indeed in control procedure for such an extended model. Control procedures have already been extensively studied. In the literature, many papers are concerned with mechanical engineering or civil engineering applications of control. First results and applications involve linear modelling and different strategies of control based on minimization of quadratic energies (optimal, active, modal control etc. see [Filliatrault (1996)]). Recently, control strategies have been introduced for smooth nonlinear systems [Zhu *et. al.* (2001)]. For further literature on the topic, see in [Housner *et. al.* (1997)] where a survey on structural control is given. In recent works, Yamada and Kobori [Yamada and Kobori (2001)] present a survey of strategies for a seismic structural control. In particular, they describe nonlinear velocity feedback laws. Many other recent references can be found in this paper. In [Wojtkiewicz and Bergman (2001)], authors introduce an algorithm for control of nonlinear systems: Smooth systems and non smooth systems are taken into account, but in the last case, the non smooth constitutive law is known via the hysteretic loop. Applications of control to structures have been given in [Roberti *et. al.* (1995)], [Roberti (1994)] (e.g.) for linear model of buildings.

Here we intend to apply the results of active control method for linear systems to mechanical systems including elastoplastic terms: These terms represent nonsmooth nonlinearities involved in constitutive laws for example. But we assume that only these laws are known and not any hysteretic loop. The control procedure is based on the calculation of command from a linear system extracted from the nonlinear model.

This chapter is organized as follows. Section 17.2 provides basic description of Prandtl rheological model. Section 17.3 describes the general models we intend to study. Section 17.4 recalls some mathematical results. In Section 17.5, we describe the adopted numerical scheme that will be used to deal with nonsmooth nonlinearities (elastoplastic terms).

In Section 17.6, control procedure is introduced. The algorithm of control and Riccati equation ([Inman (1989)], [Roberti (1994)]) are presented in Section 17.7. Numerical results are presented in Section 17.8 for a three-degrees-of-freedom mechanical system under deterministic or stochastic solicitations. Extension to nonlinear cases are presented via an example with three-degrees-of-freedom of Section 17.9. Conclusions are given in Section 17.10.

17.2 Reminder about Prandtl Rheological Model

The Prandtl rheological model corresponds to serial association of one spring of stiffness $k > 0$ and one St-Venant element of threshold $\alpha > 0$ (see in Fig. 17.1a).

The constitutive law of this rheological model is mathematically described by

$$f = -kw \in -\alpha\sigma(\dot{v}),$$

where w and v define properly coordinates of displacement of spring and of de St-Venant element. The graph σ is defined in 11.1.1 with its inverse graph β. Thus, (17.2) is equivalent to

$$\dot{w} + \beta(\frac{kw}{\alpha}) \ni \dot{x}.$$

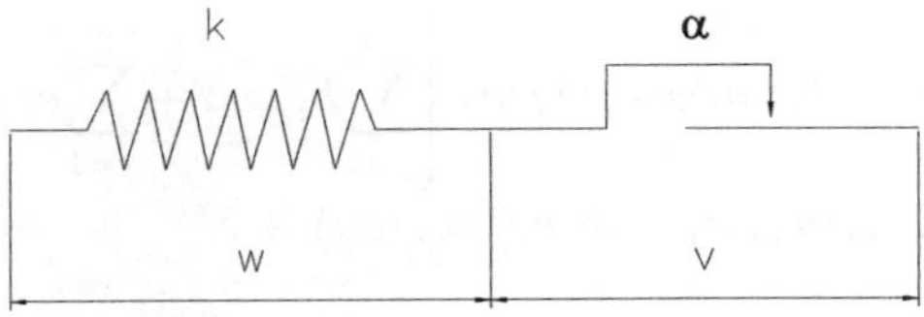

Fig. 17.1 *Prandtl* rheological model.

Details could be found in the references [Bastien (2000)], [Bastien *et. al.* (2000)].

17.3 The Studied Models with n DOF

In this section, we first consider models described in Fig. 17.2. It consists of parallel association of m-Prandtl rheological models and m-dampers with linear hardening.

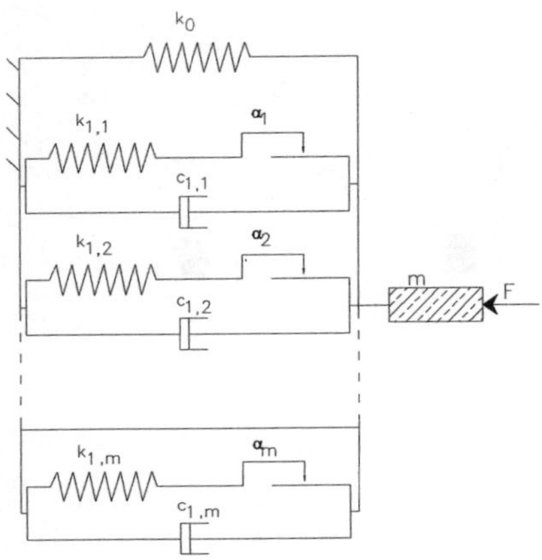

Fig. 17.2 *G* model.

This mechanical system is described by the differential inclusion

$$\begin{cases} \ddot{x} = F/m - (k_0/m)x - (1/m)\left[\sum_{j=1}^{m} k_{1,j}w_j + \sum_{j=1}^{m} c_{1,j}y\right], \\ \forall j \in \{1,\dots,m\}, \dot{w}_j + \beta(\, k_{1,j}\, w_j/\alpha_j) \ni \dot{x}. \end{cases} \tag{17.1}$$

Setting

$$y = \dot{x} \text{ and } \forall j \in \{1,\dots,m\}, \quad \eta_j = \alpha_j/k_{1,j},$$

we write Eq. (17.1) under the form

$$\begin{cases} \dot{x} = y, \\ \dot{y} = F/m - (k_0/m)x - (1/m)\left[\sum_{j=1}^{m} k_{1,j}w_j + \sum_{j=1}^{m} c_{1,j}y\right], \\ \forall j \in \{1,\dots,m\}, \dot{w}_j + \beta(w_j/\eta_j) \ni y. \end{cases} \tag{17.2}$$

Initial data are:

$$\begin{cases} x(0) = x_0, \quad y(0) = y_0, \\ \forall j \in \{1,\dots,m\}, \quad w_j(0) = w_{j,0} \in [-\eta_j, \eta_j]. \end{cases}$$

We also consider association in series of one G-model, $n-1$ Kelvin-Voigt

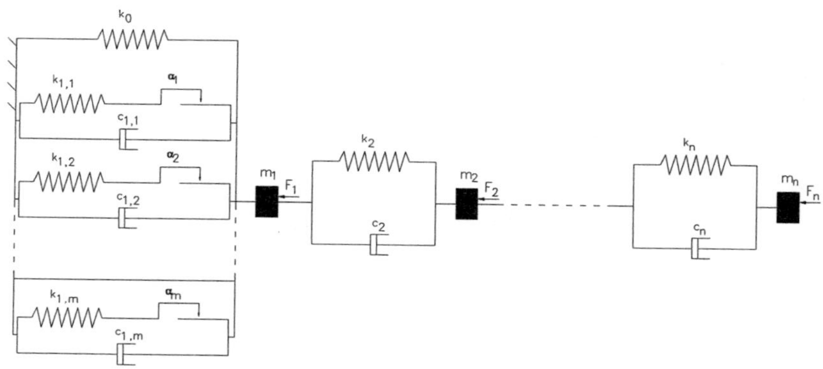

Fig. 17.3 Association in series of one G-model, $n-1$ Kelvin-Voigt models and n material points.

models and n material points (see in the Fig. 17.3); for $i = 1, \ldots, n$ x_i denotes the abscissa of material point of mass m_i. By setting

$$\forall i \in \{1, \ldots, n\} \quad y_i = \dot{x}_i \text{ and } \forall j \in \{1, \ldots, m\} \quad \eta_j = \alpha_j / k_{1,j},$$

we obtain the following system:

$$\begin{cases} \begin{cases} \dot{x}_1 = y_1, \\ \dot{y}_1 = \left(F_1 - (k_0 + k_2)x_1 + k_2 x_2 - \displaystyle\sum_{j=1}^{m} k_{1,j} w_j - \right. \\ \qquad \left. \left(\displaystyle\sum_{j=1}^{m} c_{1,j} + c_2 \right) y_1 + c_2 y_2 \right) / m_1, \\ \forall j \in \{1, \ldots, m\}, \quad \dot{w}_j + \beta(w_j/\eta_j) \ni y_1, \end{cases} \\ \forall i \in \{2, \ldots, n\}, \begin{cases} \dot{\tilde{X}} = \tilde{Y}, \\ \dot{\tilde{Y}} = D^{-1}H - D^{-1}C\tilde{Y} - D^{-1}K\tilde{X}, \end{cases} \end{cases} \tag{17.3}$$

with initial data:

$$\begin{cases} \forall i \in \{1, \ldots, n\}, \quad x_i(0) = x_{i,0}, \quad y_i(0) = y_{i,0}, \\ \forall j \in \{1, \ldots, m\}, \quad w_j(0) = w_{j,0} \in [-\eta_j, \eta_j]. \end{cases}$$

We have set

$$\tilde{X} = \begin{bmatrix} x_2 \ldots x_n \end{bmatrix}^T, \quad \tilde{Y} = \begin{bmatrix} y_2 \ldots y_n \end{bmatrix}^T,$$

$$D = diag \begin{bmatrix} m_1 \ldots m_n \end{bmatrix},$$

$$K = \begin{bmatrix} -k_2 & k_2+k_3 & -k_3 & 0 & \dots & 0 & 0 \\ 0 & -k_3 & k_3+k_4 & -k_4 & \dots & 0 & 0 \\ & \ddots & \ddots & \ddots & & & \\ 0 & 0 & \dots & -k_i & k_i+k_{i+1} & -k_{i+1} & \dots \\ & & & \ddots & \ddots & \ddots & \\ 0 & 0 & & \dots & 0 & -k_n & k_n \end{bmatrix},$$

$$H = \begin{bmatrix} F_2 \\ \vdots \\ F_n \end{bmatrix},$$

$$C = \begin{bmatrix} -c_2 & c_2+c_3 & -c_3 & 0 & \dots & 0 & 0 \\ 0 & -c_3 & c_3+c_4 & -c_4 & \dots & 0 & 0 \\ & \ddots & \ddots & \ddots & & & \\ 0 & 0 & \dots & -c_i & c_i+c_{i+1} & -c_{i+1} & \dots \\ & & & \ddots & \ddots & \ddots & \\ 0 & 0 & & \dots & 0 & -c_n & c_n \end{bmatrix}.$$

The model presented in Fig. 17.3 can be used to describe dynamical behaviour of buildings with soil-structure coupling. First material point and G-model represent soil and foundation coupled. Association of $n-1$ Kelvin-Voigt models and $n-1$ masses corresponds to $(n-1)$ floors of building.

Nonlinear extensions of the later model will be also considered: Prandtl models could be added in parallel to the springs and dashpots $k_3, \dots, k_n$ (see 17.18).

17.4 Existence and Uniqueness Results

We observe that the introduced mechanical model (17.2) can be subsumed under form conveniently described in the language of maximal monotone operators.

17.4.1 *Reminder about maximal monotone graphs σ and β*

The reader is referred to references ([Bastien (2000)], [Bastien *et. al.* (2000)], [Brezis (1973)]) or to the Chapters 2 and 3 of this book for details. Let $<,>$ be scalar product on $\mathbb{R}^n$. If ϕ is a convex proper and lower semi-continuous function from $\mathbb{R}^n$ to $[-\infty, \infty]$, we can define its sub-differential $\partial\phi$ by

$$\begin{cases} y \in \partial\phi(x) \Longleftrightarrow \forall h \in \mathbb{R}^n, \quad \phi(x+h) - \phi(x) \geq < y, h >, \\ \\ D(\partial\phi) = \{x : \partial\phi(x) \neq \emptyset\}. \end{cases}$$

Moreover, $\partial\phi$ is a maximal monotone graph in $\mathbb{R}^n \times \mathbb{R}^n$. The maximal monotone graphs σ and β are sub-differentials of proper semi-continuous convex functions $|x|$ and $\psi_{[-1,1]}$ defined by

$$\forall x \in \mathbb{R}, \quad \psi_{[-1,1]}(x) = \begin{cases} 0 & \text{if } x \in [-1,1], \\ +\infty & \text{if } x \notin [-1,1], \end{cases}$$

and the choice of the canonical scalar product in $\mathbb{R}$. Therefore

$$\forall x \in \mathbb{R}, \quad \sigma(x) = \partial|x| \text{ and } \beta(x) = \partial\psi_{[-1,1]}(x).$$

We observe that if $\mathbb{R}^n$ is equipped with its canonical scalar product and with another scalar product

$$< x, y >_M = x^T M^{-1} y,$$

where M is symmetric positive definite, then we can relate the sub-differential $\partial\phi$ of ϕ relatively to the canonical scalar product and the sub-differential $\partial_M\phi$ relatively to $<,>_M$ by:

$$\partial_M\phi(x) = M\partial\phi(x).$$

17.4.2 *Mathematical study of a differential system*

We give now the general mathematical formulation of our problem. We assume that T is strictly positive and that G is function from $[0,T] \times \mathbb{R}^n$ to $\mathbb{R}^n$ which is Lipschitz continuous with respect to its second argument, i.e. there exist $\omega \geq 0$ such that:

$$\forall t \in [0,T], \forall X_1, X_2 \in \mathbb{R}^n, \quad ||G(t, X_1) - G(t, X_2)|| \leq ||X_1 - X_2||.$$

Moreover, we assume that

$$\forall Y \in \mathbb{R}^n, \quad G(.,Y) \in L^\infty(0,T,\mathbb{R}^n).$$

The matrix M is symmetric positive definite (and consequently is invertible), and ϕ is convex proper and lower semi-continuous on $\mathbb{R}^n$.
Under the previous assumptions, for all $\xi \in D(\partial\phi)$, there exists a unique function X in $W^{1,1}(0,T,\mathbb{R}^n)$ such that

$$\begin{cases} \dot{X}(t) + M\partial\phi(X(t)) \ni G(t, X(t)) & \text{a.e on }]0,T[, \\ X(0) = \xi, \end{cases}$$

because we observed above that $\partial_M\phi(x) = M\partial\phi(x)$; therefore proposition 3.13 page 107 of [Brezis (1973)] applies.
Thus, the system (17.2) can be written under the form (17.4.2) and possesses a unique solution. The convex function ϕ for the above described mechanical model (17.2) is a sum of functions $\psi_{[-\eta_j,\eta_j]}$

$$\phi(x_1, y_1, u_1, \ldots, u_m) = \sum_{j=1}^{m} \psi_{[-\eta_j,\eta_j]}(u_j),$$

and the invertible matrix used M is equal to the identity matrix I.

Results are also valid for nonlinear extensions if the control force is depending on displacements and velocities that are solutions of a linear differential system: this point will be clarified in Section 17.9.
Such mathematical results give existence and uniqueness results for systems involving nonlinear coupling terms of elastoplastic type between the terms of Fig. 17.3 that are linearly coupled.

If we add a command to such models so that the command is depending upon displacements and Riccati or Lyapunov matrices obtained for the classical control theory of a linear system, the previous mathematical results apply. This is because the command can simply split into a stiffness matrix and external solicitation.

17.5 Numerical Scheme

We solve the differential inclusion (17.4.2) by using the following implicit Euler scheme. Let $n \in \mathbb{N}$. Let $h = T/n$ and for $q \in \{0, \ldots, n\}$, let $t_q = qh$.

We solve:

$$\begin{cases} \forall q \in \{0,\ldots,n-1\}, \ \dfrac{X_{q+1}-X_q}{h} + M\partial\psi(X_{q+1}) \ni G(t_q,X_q), \\ X_0 = \xi. \end{cases}$$

The first equality of (17.5) is equivalent to:

$$\forall q \in \{0,\ldots,n-1\}, \ X_{q+1} = (I+hM\partial\phi)^{-1}(hG(t_q,X_q)+X_q).$$

In this last equation $(I+hM\partial\phi)^{-1}$ is defined for $h>0$ because M is symetric positive definite and $\partial\phi$ maximal monotone ([Bastien (2000)], [Brezis (1973)]).
We denote by X^h the linear interpolation of the $X_n^{\prime}s$. The function X^h converges to the solution X of system (17.4.2) in $C^0([0,T],\mathbb{R}^n)$. This result is proved in [Crandall and Evans (1975)] which contains a much more general result.
Thus, system (17.3) for numerical simulations by using implicit Euler scheme will have the following form

$$\forall q \in \{0,\ldots,n-1\},$$

$$\begin{cases} \begin{cases} x_{1,q+1} = hy_{1,q} + x_{1,q}, \\ y_{1,q+1} = h(F_1 - k_0x_{1,q} - \sum_{j=1}^m k_{1,j}w_j - \sum_{j=1}^m c_{1,j}y_{1,q})/m_1 + y_{1,q}, \\ \forall j \in \{1,\ldots,m\} \quad w_{j,q+1} = \begin{cases} w_{j,q}+hy_{1,q} & \text{if } |w_{j,q}+hy_{j,q}| < \eta_j, \\ +\eta_j & \text{if } w_{j,q}+hy_{j,q} \geq \eta_j, \\ -\eta_j & \text{if } w_{j,q}+hy_{j,q} \leq -\eta_j. \end{cases} \end{cases} \\ \forall i \in \{2,\ldots,n\}, \begin{cases} \tilde{X}_{q+1} = h\tilde{Y} + \tilde{X}_q, \\ \tilde{Y}_{q+1} = h(D^{-1}H - D^{-1}C\tilde{X}_q - D^{-1}K\tilde{X}_q) + \tilde{Y}_q, \end{cases} \end{cases} \tag{17.4}$$

with initial data:

$$\begin{cases} \forall i \in \{1,\ldots,n\}, \quad x_i(0) = x_{i,0}, \quad y_i(0) = y_{i,0}, \\ \forall j \in \{1,\ldots,m\}, \quad w_j(0) = w_{j,0} \in [-\eta_j,\eta_j]. \end{cases}$$

17.6 Control Procedure

The reader is referred to [Roberti (1994)] or [Roberti *et. al.* (1995)] for some technical details. Let us consider a mechanical model including soil-structure coupling where a building is modeled by a $n-1$ DOF system, and the soil by a system with a 1 DOF system. This discrete system with n DOF is governed by the following second order differential equations (ODEs):

$$M\ddot{y}(t) + C\dot{y}(t) + Ky(t) + \hat{K}w(t) + \hat{C}y_1(t) = D_1 f(t) + B_1 u(t). \quad (17.5)$$

The $n \times 1$ vector $y(t)$ represents displacements of soil and the building floors in a previously established co-ordinate system and $y_1(t)$ is the first term of $y(t)$. M, C and K are $n \times n$ order mass, damping and stiffness matrices, correspondingly. The matrix B_1 informs us about the control force positions in structure. The matrix D_1 represents the influence of external excitation f(t). The $i \times 1$ vector $w(t)$ describes displacements of i elastic elements being component of Prandtl's models and the $n \times i$ matrices $\hat{K}$ and $\hat{C}$ represent corresponding stiffness and damping matrix:

$$\hat{K} = \begin{bmatrix} k_{1,1} & k_{1,2} & \dots & k_{1,i} \\ 0 & 0 & \dots & 0 \\ \vdots & \vdots & & \vdots \\ 0 & 0 & \dots & 0 \end{bmatrix}, \hat{C} = \begin{bmatrix} c_{1,1} & c_{1,2} & \dots & c_{1,i} \\ 0 & 0 & \dots & 0 \\ \vdots & \vdots & & \vdots \\ 0 & 0 & \dots & 0 \end{bmatrix}. \quad (17.6)$$

The introduced control $u(t)$ is assumed in the following form:

$$u(t) = F_1 y(t) + F_2 \dot{y}(t) + F_3 f(t), \quad (17.7)$$

where F_i , $i = 1, 2, 3$, are the amplification matrices. Substituting (17.7) into (17.5) we get:

$$\begin{cases} M\ddot{y}(t) + (C - B_1F_2)\dot{y}(t) + (K - B_1F_1)y(t) + \hat{K}w(t) + \\ \qquad \hat{C}y_1(t) = (D_1 + B_1F_3)f(t). \end{cases} \quad (17.8)$$

It is seen from (17.8) that the control modifies parameters of the structure, i.e. its stiffness and damping in order to achieve a suitable response of the building to the excitation $f(t)$. One can notice that the form of the control provides a differential inclusion so that existence and uniqueness results are governed by the recalled mathematical results.

17.7 Algorithm of Control

Eqs. (17.5) are transformed into a set of first order ODEs

$$\dot{x}(t) = A(t)x(t) - \tilde{K}w(t) - \tilde{C}y_1(t) + B(t)u(t) + Dz(t),$$

where

$$x(t) = \left[y(t)\ \dot{y}(t) \right]^T$$

is the $2n$ vector of a structure state, and

$$z(t) = \left[f(t)\ \dot{f}(t) \right]$$

is the $2n$ vector of an external excitation.
The new $2n \times i$ matrices $\tilde{K}$ and $\tilde{C}$ are writen in the following way:

$$\tilde{K} = \begin{bmatrix} 0_{(n\times i)} \\ \hat{K} \end{bmatrix}, \tilde{C} = \begin{bmatrix} 0_{(n\times i)} \\ \hat{C} \end{bmatrix},$$

where $0_{(n\times i)}$ is the $n \times i$ zero matrix. The $2n \times 2n$ matrix A represents the structure parameters, and it reads:

$$A = \begin{pmatrix} 0_n & I_n \\ -M^{-1}K & -M^{-1}C \end{pmatrix}.$$

The $2n \times n$ matrix

$$B^T = \left(0\ M^{-1}B_1 \right)$$

defines control positions, whereas the $2n \times n$ matrix D

$$D^T = \left(0\ M^{-1}D_1 \right)$$

defines an influence of the external excitation on the system behaviour. Finally, 0_n and I_n are the $n \times n$ zero and unit matrices, respectively.

17.7.1 *Improvement of the control*

In order to achieve an efficient control, $u(t)$ is found through a minimization condition applied to the following functional:

$$J = \int_{t_0}^{t_f} \left[x^T(t)Qx(t) + u^T(t)Ru(t) \right] dt.$$

By a proper choice of the equilibrium matrices Q and R during the action of excitations $[t_o, t_f]$ one can find a suitable compromise between the efficiency of control and the energy needed to apply the control law.

17.7.2 *Riccati equation*

If seismic considered is considered as deterministic, law of optimum control for given construction can be counted by minimizing function of performance of second degree

$$\begin{cases} J(x,z,u) = \frac{1}{2}x^T(t_f)\theta(t_f)x(t_f) + \\ \frac{1}{2}\int_{t_0}^{t_f} \begin{pmatrix} x(t) \\ z(t) \\ u(t) \end{pmatrix}^T \begin{pmatrix} Q(t) & S(t) & T(t) \\ S^T(t) & N(t) & L(t) \\ T^T(t) & L^T(t) & R(t) \end{pmatrix} \begin{pmatrix} x(t) \\ z(t) \\ u(t) \end{pmatrix} dt. \end{cases}$$

where θ, Q, S, T, N, L, and R are symmetric equilibrium matrices. Under the differential condition Hamiltonian of system (17.7.2) can be written in the form

$$\begin{cases} H(x,u,\lambda,t) = \frac{1}{2}\begin{pmatrix} x(t) \\ z(t) \\ u(t) \end{pmatrix}^T \begin{pmatrix} Q(t) & S(t) & T(t) \\ S^T(t) & N(t) & L(t) \\ T^T(t) & L^T(t) & R(t) \end{pmatrix} \begin{pmatrix} x(t) \\ z(t) \\ u(t) \end{pmatrix} + \\ \lambda^T(Ax(t) + Bu(t)) + Dz(t). \end{cases}$$

Let us differentiate (17.7.2) with respect to λ:

$$\frac{\partial H}{\partial \lambda} = \dot{x}(t) = A(t)x(t) + B(t)u(t)) + D(t)z(t).$$

The Euler-Lagrange equation provides:

$$\frac{d\lambda}{dt} = -\frac{\partial H}{\partial x} = -Q(t)x(t) - A^T(t)\lambda(t) - S(t)z(t) - T(t)u(t).$$

If H is constant then $\dfrac{\partial H}{\partial u} = 0$, so we have:

$$\frac{\partial H}{\partial u} = R(t)u(t) + B^T(t)\lambda(t) + T^T(t)x(t) + L^T(t)z(t) = 0.$$

Then (17.7.2) can be solved:

$$u(t) = -R^{-1}(t)[(B^T(t)\lambda(t) + T^T(t))x(t) + L^T(t)z(t)].$$

By setting (17.7.2) to (17.7.2) and (17.7.2) we obtain a system of $2n$ ODE's

$$\begin{pmatrix} \dot{x}(t) \\ \dot{\lambda}(t) \end{pmatrix} = \begin{pmatrix} A_n(t) & -B(t)R^{-1}B^T(t) \\ -Q_n(t) & -A_n^T(t) \end{pmatrix} \begin{pmatrix} x(t) \\ \lambda(t) \end{pmatrix} + \begin{pmatrix} D_n(t) \\ -S_n(t) \end{pmatrix} z(t),$$

where:

$$A_n = A - BR^{-1}T^T, \quad Q_n = Q - TR^{-1}T^T,$$

$$D_n = D - BR^{-1}L^T, \quad S_n = S - TR^{-1}L^T.$$

The solution of the Eq. (17.7.2) can be represented by using the final conditions and the transition matrix ϕ:

$$\begin{pmatrix} x(t_f) \\ \lambda(t_f) \end{pmatrix} = \phi(t_f, t) \begin{pmatrix} x(t) \\ \lambda(t) \end{pmatrix} + \begin{pmatrix} r_1(t) \\ r_2(t) \end{pmatrix},$$

where the second term of sum is a particular solution of the Eq. (17.7.2). Referring the latter expression (17.7.2) to the final condition $\lambda(t_f) = \theta(t_f)x(t_f)$ we obtain:

$$x(t_f) = \phi_{11}(t_f, t)x(t) + \phi_{12}(t_f, t)\lambda(t) + r_1(t),$$

$$\theta(t_f)x(t_f) = \phi_{21}(t_f, t)x(t) + \phi_{22}(t_f, t)\lambda(t) + r_2(t).$$

By setting the expression of $x(t_f)$ into the second equation we have

$$\theta\phi_{11}(t_f, t)x(t) + \theta\phi_{12}(t_f, t)\lambda(t) = \phi_{21}(t_f, t)x(t) + \phi_{22}(t_f, t)\lambda(t) + r_2(t) - \theta r_1(t),$$

and then

$$\lambda(t) = \{\phi_{22}(t_f, t) - \theta\phi_{12}(t_f, t)\}^{-1} \left(\{\theta\phi_{11}(t_f, t) - \phi_{21}(t_f, t)\}x(t) + r_2(t) - \theta r_1(t) \right).$$

Now, we write λ in the following form

$$\lambda(t) = K_x(t)x(t) + r(t),$$

where K_x is $n \times n$ Riccati matrix and $r(t)$ represents filter of external excitation $z(t)$.
Using Eqs. (17.7.1) and (17.7.2) we obtain:

$$u(t) = -R^{-1}(t)[(B^T(t)K_x(t) + T^T(t))x(t) + B^T(t)r(t) + L^T(t)z(t)].$$

Differentiating (17.7.2) versus time t gives

$$\dot{\lambda}(t) = K_x(t)\dot{x}(t) + \dot{K}_x(t)x(t) + \dot{r}(t),$$

and then using (17.7.2) in order to eliminate $\dot{x}$ we obtain:

$$\dot{\lambda} = (\dot{K}_x + K_x A_n - K_x B R^{-1} B^T K_x) x - K_x B R^{-1} B^T r + K_x D_n z + \dot{r}.$$

The comparison of last expression with (17.7.2) gives

$$\begin{cases} (\dot{K}_x + K_x A_n + A_n^T K_x - K_x B R^{-1} B^T K_x) x = -\dot{r} + (K_x B R^{-1} B^T - \\ \qquad\qquad A_n^T) r - (K_x D_n + S_n) z. \end{cases}$$

Because $x(t)$ and $z(t)$ are arbitrary vectors, Eq. (17.7.2) is verified only if K_x satisfies condition

$$\dot{K}_x = -K_x A_n - A_n^T K_x + K_x B R^{-1} B^T K_x - Q_n, \qquad K_x(t_f) = \theta(t_f),$$

and r verifies condition

$$\dot{r} = (K_x B R^{-1} B^T - A_n^T) r - (K_x D_n + S_n) z, \qquad r(t_f) = 0.$$

If the applied force is a deterministic force, the element r of (17.7.2) can be represented as $K_z z(t)$. Wanted Lagrange multiplier will accept the following form

$$\lambda(t) = K_x(t) x(t) + K_z(t) z(t),$$

with final condition

$$\lambda(t_f) = \theta(t_f) x(t_f).$$

Let us rewrite(17.7.2) and (17.7.2) in the form

$$\begin{cases} K_x A_n + A_n^T K_x + Q_n - K_x B R^{-1} B^T K_x = -\dot{K}_x(t), \\ K_x(t_f) = \theta(t_f), \\ K_x D_n + K_z A_z + A_n^T K_z + S_n - K_x B R^{-1} B^T K_z = -\dot{K}_z(t), \\ K_z(t_f) = 0, \end{cases}$$

where matrix K_x is solution of a Riccati equation and K_z is solution of a Lyapunov type equation. In order to estimate the matrices K_x and K_z, let us integrate numerically the system (17.7.2). Numerical integration starts at instant $t = t_f$ and ends at instant $t = t_0$. We run this operation by using the classical fourth order Runge-Kutta method RK34.

17.8 Numerical Results for a System with 3 DOF

In order to illustrate the control of a structure, which consists of a building and its foundation coupled to the soil, a three degrees of freedom approximation will be used. In order to control the buildings floors (modelled as the material points and represented by m_2 and m_3) a force generator situated between two first masses is applied (see Fig. 17.4).

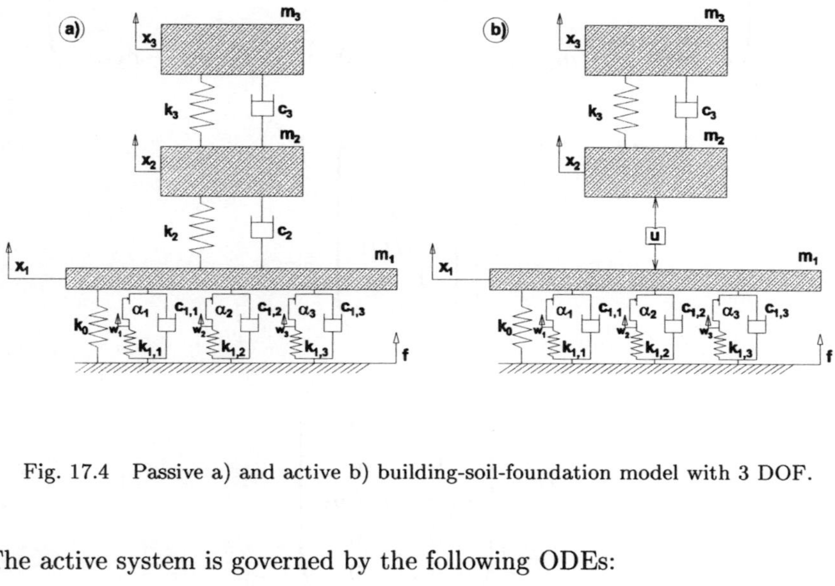

Fig. 17.4 Passive a) and active b) building-soil-foundation model with 3 DOF.

The active system is governed by the following ODEs:

$$\begin{cases} m_1\ddot{x}_1 = -\sum_{j=1}^{3} k_{1,j}w_j + \sum_{j=1}^{3} c_{1,j}(\dot{f} - \dot{x}_1) + k_0(f - x_1) - u, \\ m_2\ddot{x}_2 = k_3(x_3 - x_2) + c_3(\dot{x}_3 - \dot{x}_2) + u, \\ m_3\ddot{x}_3 = k_3(x_3 - x_2) - c_3(\dot{x}_3 - \dot{x}_2). \end{cases} \tag{17.9}$$

The control law is governed by the equation:

$$u(t) = -R^{-1}(t)B^T(t)\left[K_x(t) + K_z(t)\right] = F_x(t)x(t) + F_z(t)z(t),$$

where the matrix K_x is the solution to a Riccati equation, and K_z is the solution to a Lyapunov type equation.

The matrices F_x and F_z are called the amplification matrices.

For t_f large enough ($t_f \longrightarrow \infty$), the matrices A, B and D can be treated as the constant ones:

$$A = \begin{pmatrix} 0 & 0 & 0 & 1 & 0 & 0 \\ 0 & 0 & 0 & 0 & 1 & 0 \\ 0 & 0 & 0 & 0 & 0 & 1 \\ -k_1/m_1 & 0 & 0 & -\sum_{j=1}^{3} c_{1,j}/m_1 & 0 & 0 \\ 0 & -k_3/m_2 & k_3/m_2 & 0 & -c_3/m_2 & c_3/m_2 \\ 0 & k_3/m_3 & -k_3/m_3 & 0 & c_3/m_3 & -c_3/m_3 \end{pmatrix},$$

$$B = \begin{pmatrix} 0 \\ 0 \\ 0 \\ -1/m_1 \\ 1/m_2 \\ 0 \end{pmatrix},$$

$$D = \begin{pmatrix} 0 & 0 \\ 0 & 0 \\ 0 & 0 \\ k_1/m_1 & c_1/m_1 \\ 0 & 0 \\ 0 & 0 \end{pmatrix}.$$

The equilibrium matrices Q, S and R are written as:

$$Q = \begin{pmatrix} q_1 + q_2 & -q_2 & 0 & 0 & 0 & 0 \\ -q_2 & q_2 + q_3 & -q_3 & 0 & 0 & 0 \\ 0 & -q_3 & q_3 & 0 & 0 & 0 \\ 0 & 0 & 0 & 0 & 0 & 0 \\ 0 & 0 & 0 & 0 & 0 & 0 \end{pmatrix},$$

$$S^T = \begin{pmatrix} -q_1 & 0 & 0 & 0 & 0 & 0 \\ 0 & 0 & 0 & 0 & 0 & 0 \end{pmatrix},$$

$$R = r.$$

If we replace the matrices (17.8, 17.8, 17.8) in (17.7.1), the cost function J can be written as:

$$J = \frac{1}{2}\int_{t_0}^{t_f} (ru^2 + q_1(x_1 - f)^2 + q_2(x_2 - x_1)^2 + q_3(x_3 - x_2)^2\, dt.$$

The form of the matrix A_z depends on the kind of applied loading. Therefore the matrices K_x and K_z (solutions of the system of Eqs. (17.7.2)) are approached by constant values $\overline{K}_x$ and $\overline{K}_z$. The mentioned situation corresponds to the stationary case:

$$\begin{cases} \overline{K}_x A_n + A_n^T \overline{K}_x + Q_n - \overline{K}_x B R^{-1} B^T \overline{K}_x = 0, \\ \overline{K}_x D_n + \overline{K}_z A_z + A_n^T \overline{K}_z + S_n - \overline{K}_x B R^{-1} B^T \overline{K}_z = 0. \end{cases}$$

In order to test the stationary property of the system (17.8) we apply the deterministic force $f = f_o \cos(\omega t) e^{-\frac{\lambda}{2}t}$ and then stochastic process. One can design seismic excitation by means of filtered white noise. Let seismic state result from:

$$x_s(t) = G_s(t) z_a(t),$$

where z_a represents the state of seismic excitation, recorded with second order differential linear equation:

$$\ddot{z}_a(t) + 2\xi_d \omega_d \dot{z}_a(t) + \omega_d^2 z_a(t) = \varpi(t).$$

The parameters ω_d and ξ_d represent local soil conditions. Eq. (17.8) can also be written in the following first ODE form:

$$\dot{z}_a(t) = A_z z_a(t) + B_z \varpi(t),$$

where matrices A_z and B_z are respectively

$$A_z = \begin{pmatrix} 0 & 1 \\ -\omega_d^2 & -2\xi_d \omega_d \end{pmatrix}, \quad \text{i } B_z = \begin{pmatrix} 0 \\ 1 \end{pmatrix},$$

and ϖ is stationary white noise defined as limit of Ornstein-Uhlenbeck's process. $G_s(t)$ determines instead envelope z_a defined by

$$G_s(t) = \begin{cases} 0 & t < 0, \\ (t/t_1)^2 & 0 \le t \le t_1, \\ 1 & t_1 \le t \le t_2, \\ \exp(-c[t - t_2]) & t > t_2. \end{cases} \tag{17.10}$$

In order to compare the passive and active (with the control forces) control the value $u(t)$ has been computed using (17.9). Let us present numerical results for displacements and accelerations of the masses.

17.8.1 *System with 3 DOF under stochastic loading*

The parameters of the system are:
$m_1 = 1500$ kg, $m_2 = 11000$ kg, $m_3 = 11000$ kg, $k_0 = 575000$ N/m, $k_{1,1} = 575000$ N/m, $k_{1,2} = 575000$ N/m, $k_{1,3} = 575000$ N/m, $k_2 = 53000$ N/m, $k_2 = 53000$ N/m, $\eta_1 = 0.1$ m, $\eta_2 = 0.15$ m, $\eta_3 = 0.2$ m, $c_{1,1} = 5000$ Ns/m, $c_{1,2} = 5000$ Ns/m, $c_{1,3} = 5000$ Ns/m, $c_2 = 15000$ Ns/m, $c_3 = 15000$ Ns/m, $q_1 = 1, q_2 = 1, q_3 = 10$, $\omega_d = 18.87$ rad/s, $\xi_d = 0.65$, $t_1 = 3$ s, $t_2 = 10$ s and $c = 0.76\ s^{-1}$.

Initial conditions are:
$x_{1,0} = x_{2,0} = 0.$, $w_1 = 0$, $\dot{x}_{1,0} = \dot{x}_{2,0} = 0$, $x_{3,0} = 0. = \dot{x}_{3,0}$.

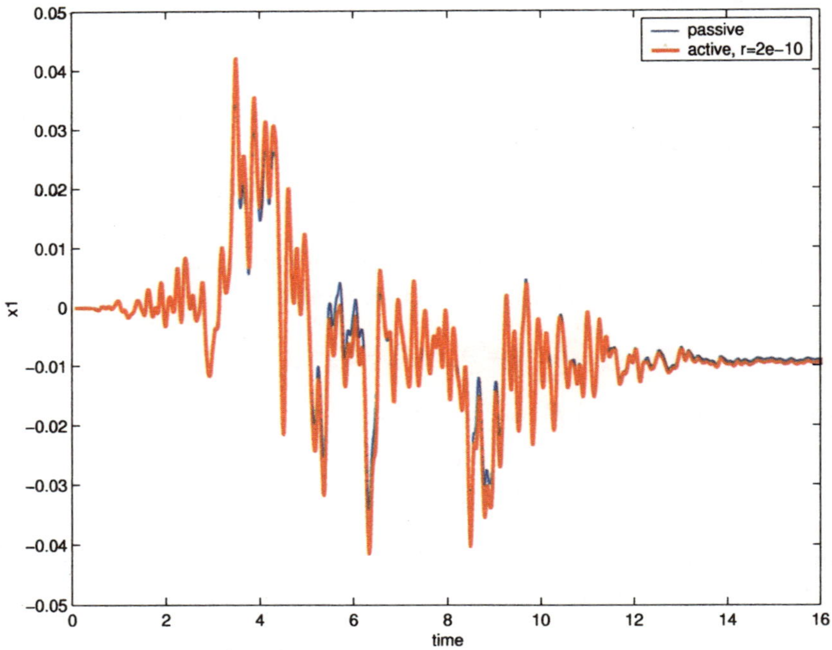

Fig. 17.5 Displacement x_1 versus time t for the passive and active control.

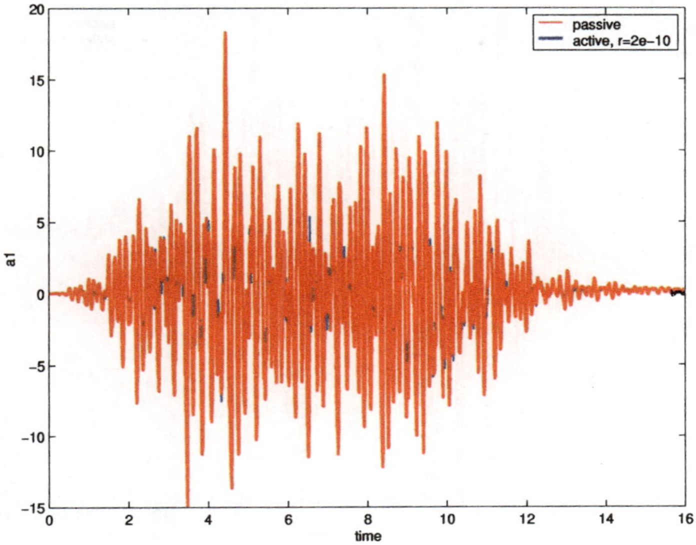

Fig. 17.6 Acceleration $\ddot{x}_1$ (denoted a_1 on y label axis) versus time t for the passive and active control.

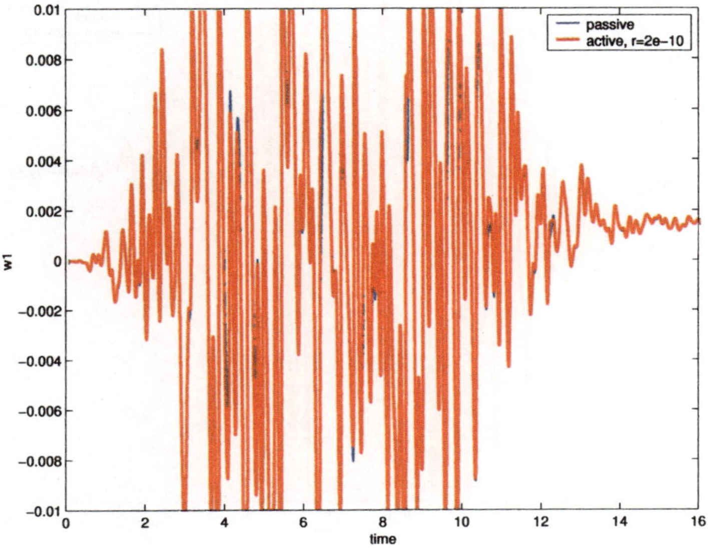

Fig. 17.7 Displacement w_1 versus time t for the passive and active control.

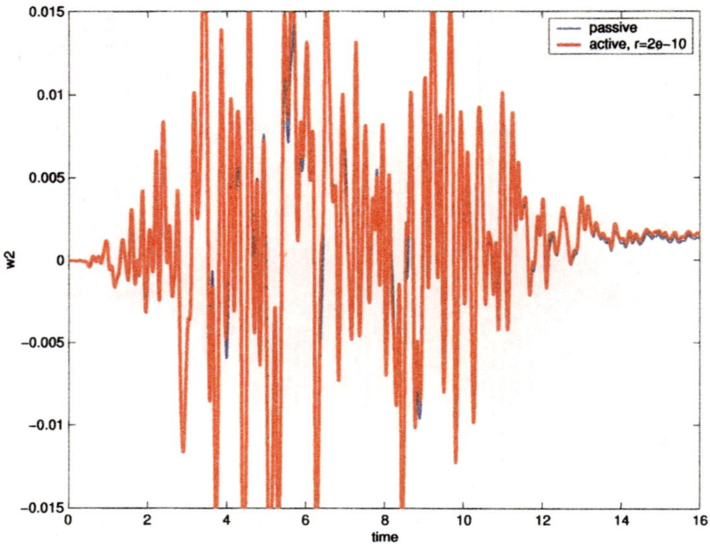

Fig. 17.8 Displacement w_2 versus time t for the passive and active control.

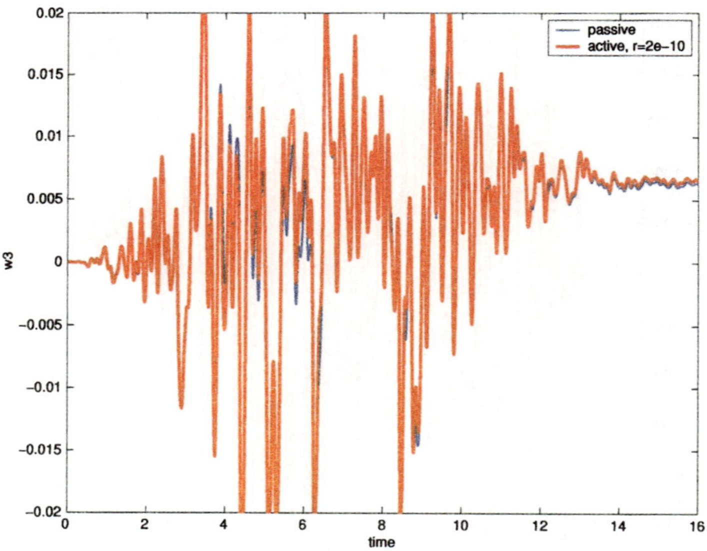

Fig. 17.9 Displacement w_3 versus time t for the passive and active control.

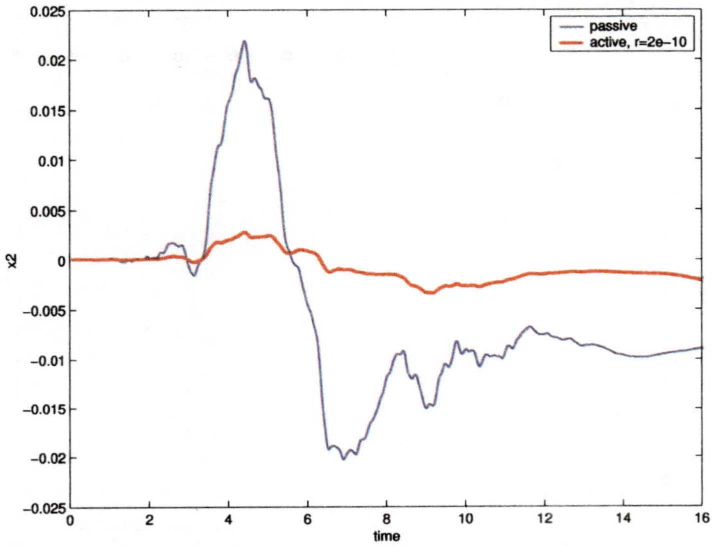

Fig. 17.10 Displacement x_2 versus time t for the passive and active control.

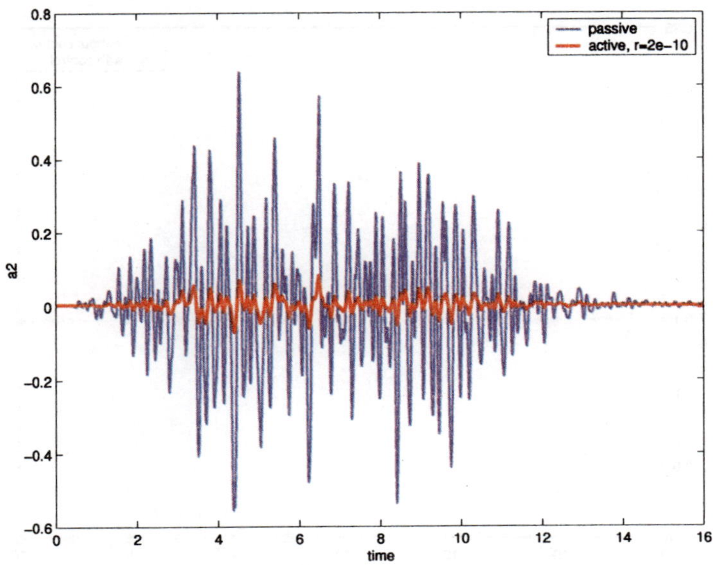

Fig. 17.11 Acceleration $\ddot{x}_2$ (denoted a_2 on y label axis) versus time t for the passive and active control.

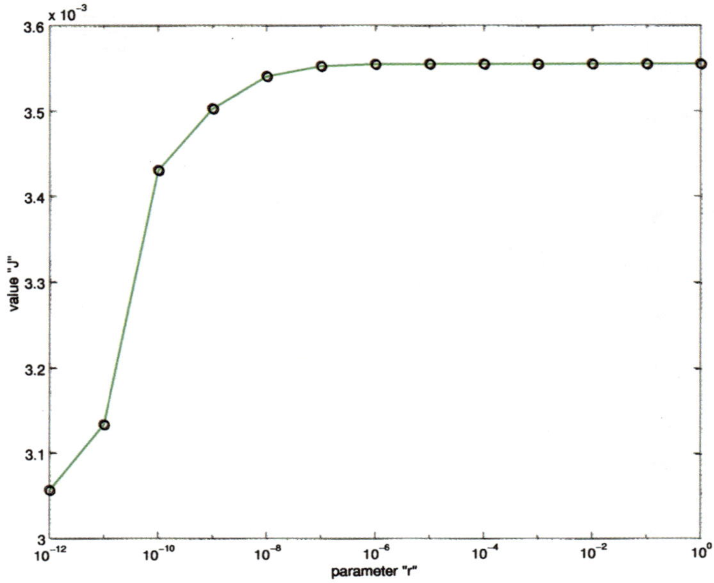

Fig. 17.12 Influence of r on values J.

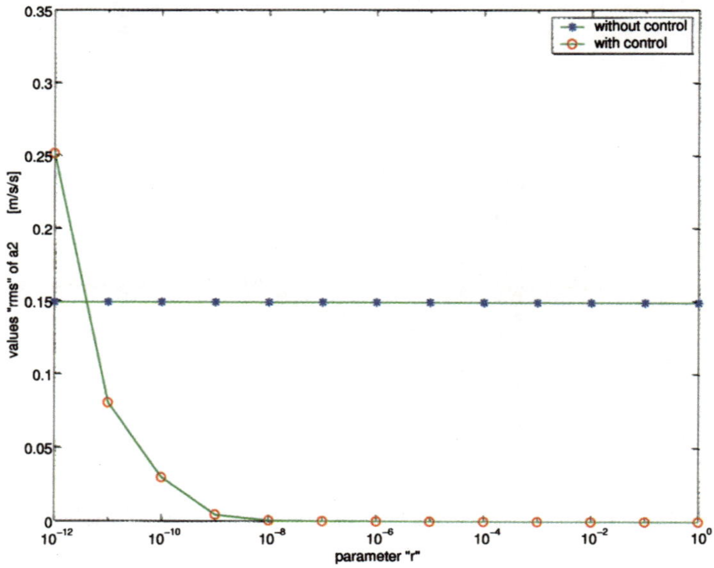

Fig. 17.13 Influence of r on values rms of $\ddot{x}_2$ (denoted a_2).

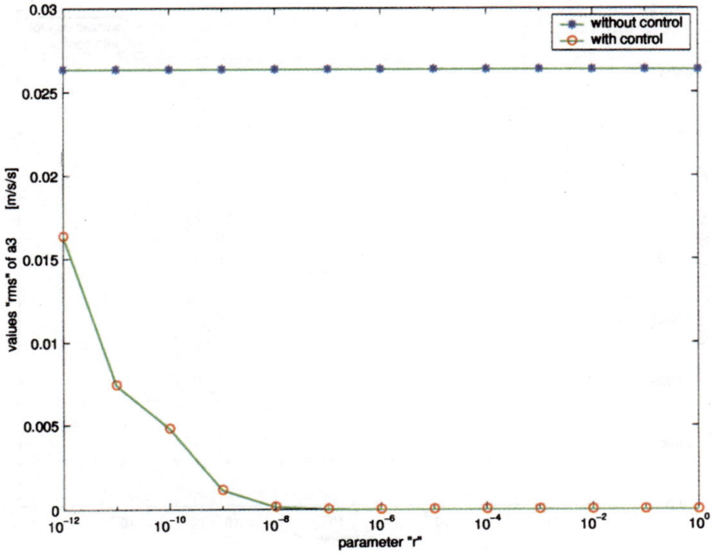

Fig. 17.14 Influence of r on values rms of $\ddot{x}_3$ (denoted a_3).

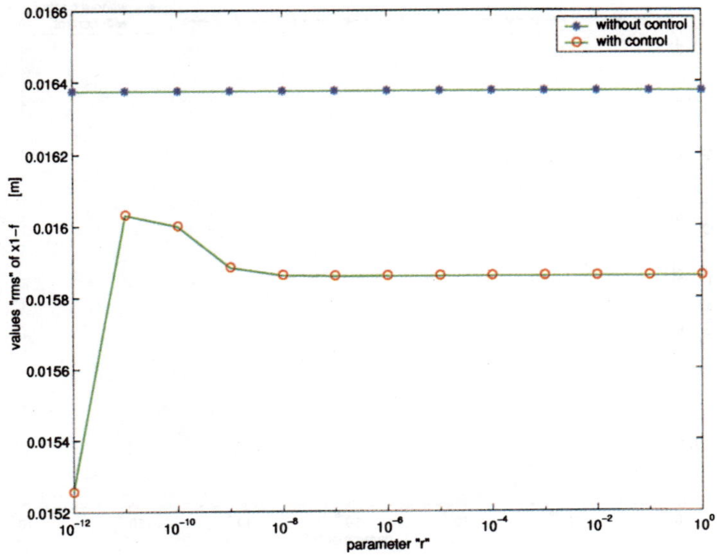

Fig. 17.15 Influence of r on values rms of $x_1 - f$.

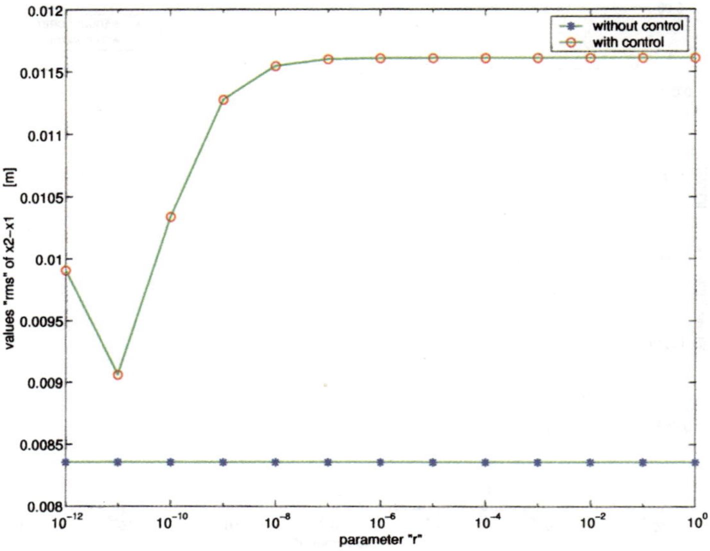

Fig. 17.16 Influence r on values rms of $x_2 - x_1$.

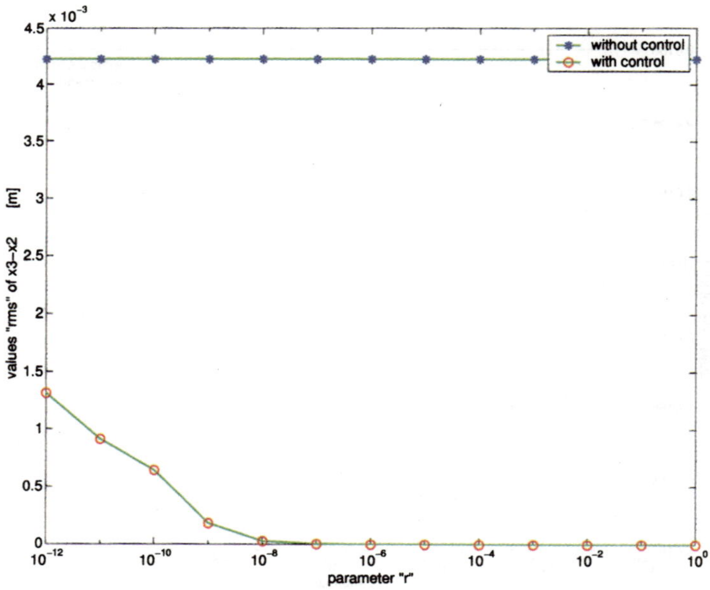

Fig. 17.17 Influence r on values rms of $x_3 - x_2$.

In Fig. 17.5, the displacement x_1 is presented versus time without control and with active control. In Fig. 17.6, the acceleration of the first mass, $\ddot{x}_1$ is plotted versus time without and with active control. In Figs. 17.7, 17.8, 17.9, w_1 (resp w_2, w_3, the coordinates of plastic terms) are plotted versus time again with or without control.

One can see that the effect of control on the vibrations of the first mass m_1 is almost negligible. That is not surprising since we introduced the active control procedure in order to decrease the vibration amplitudes of the mass m_2 and m_3.

In Figs. 17.10 and 17.11, displacement x_2 and acceleration $\ddot{x}_2$ are plotted versus time.

One can see in these last two figures that active control procedure is efficient: Curves for active control (thick lines) are plotted for r equal to $2e-10$ that permits this efficient diminution of oscillations amplitudes and of the amplitudes of accelerations. The choice of this value for r has been made after the procedure explained here after.

In order to chose the best value of r we can study J versus r (see in the Fig. 17.12) for example. General study of J versus q_1, q_2, q_3 and r can be done. We can also study the random mean square (rms) values of $\ddot{x}_2$ (see in the Fig. 17.13), $x_1 - f$ (see in the Fig. 17.15) and $x_2 - x_1$ (see in the Fig. 17.16). The best choice of r is the value which provides the smallest values of rms values of $\ddot{x}_2$, $x_1 - f$ and $x_2 - x_1$. That leads here to $r = 2e - 10$ where rms values of displacements of both masses are almost equal to case without control but where rms values of acceleration of second mass is improved.

The presented graphs shows that active control improves the dynamical behaviour of second masses in comparison to the passive behaviour.

17.9 Extension to Nonlinear Cases

We simply provide an example to explain how to extend the previous control procedure to a class of nonlinear cases.

If we add nonlinear nonsmooth terms between two degrees of freedom initial exhibiting a linear coupling as one can see in the Fig. 17.18, the control is calculated for an associated linear system involving only stiffness similar to k_0 of Fig. 17.3 or k_3 of Fig. 17.18 (such stiffness can be identified for real systems from dynamical tests at very low level of solicitation since in this case only linear behaviour of the system is involved [Schatzman *et. al.*

(1999)]). So we do not build an optimal control for the complete nonlinear system. Then the command is applied to the full nonlinear system.

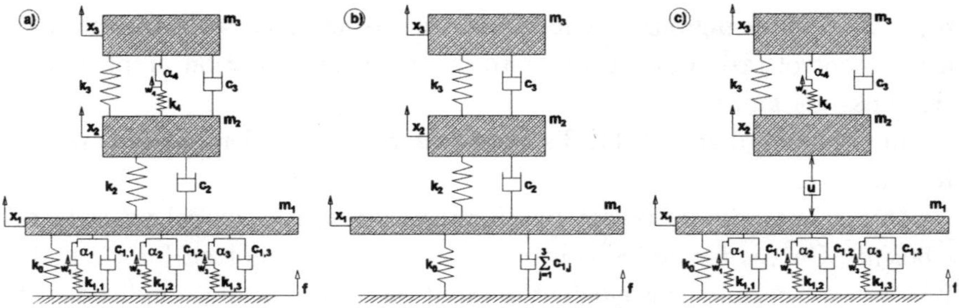

Fig. 17.18 Building-soil-foundation model with 3 DoF. Initial system a), modified system for calculation of command b) and full system with control c).

In this chapter, we consider only linear smooth parts for the calculation of the control force. Introducing nonsmooth terms (elasto plasticity) between masses 2 and 3 (e.g., in order to extend classical Riccati equation) creates hard theoretical problems : If Riccati equation corresponding to such a case has to be extended in the form of differential inclusion, then it can not be solved backward because of the plastic terms.

Such a procedure can be used because the initial studied system exhibits convex hysteretic cycles so that this initial system is more dissipative than the linear associated system.

Existence and uniqueness results are valid for the mathematical problem involving the control procedure, since the calculation of control force is simply the result of a linear procedure and the obtained control force becomes external solicitation for the nonsmooth nonlinear initial problem.

In order to illustrate the control of a structure, which consists of a building and its foundation coupled to the soil, a three degrees of freedom approximation will be used. The initial system is described in Fig. 17.18a). In order to control the buildings floors (modelled as the material points and represented by m_2 and m_3) a force generator situated between two first masses is applied (Fig. 17.18c). In order to compute the command the linear associated system of Fig. 17.18b) is introduced. The controlled system is governed by the following ODEs:

$$\begin{cases} m_1\ddot{x}_1 = -\sum_{j=1}^{3} k_{1,j}w_j + \sum_{j=1}^{3} c_{1,j}(\dot{f} - \dot{x}_1) + k_0(f - x_1) - u, \\ m_2\ddot{x}_2 = k_3(x_3 - x_2) + c_3(\dot{x}_3 - \dot{x}_2) + u + k_4 w_4, \\ m_3\ddot{x}_3 = -k_3(x_3 - x_2) - c_3(\dot{x}_3 - \dot{x}_2) - k_4 w_4, \end{cases} \tag{17.11}$$

and

$$\dot{w}_4 + \beta(\frac{w_4}{\eta_4}) \ni \dot{x}_3 - \dot{x}_2.$$

Let the control law will be governed by the equation:

$$u(t) = -R^{-1}(t)B^T(t)\left[K_x(t) + K_z(t)\right] = F_x(t)x(t) + F_z(t)z(t).$$

The matrices F_x and F_z are called the amplifications matrices.
For t_f large enough ($t_f \longrightarrow \infty$), the matrices A, B and D are the constant ones of the previous Section.

We consider also the same equilibrium matrices Q, S and R and the same cost function than in the Section 17.8.

Therefore also the matrices K_x and K_z (solutions of system of Eqs. (17.7.2)) are approached constant values $\overline{K}_x$ and $\overline{K}_z$. The mentioned situation corresponds to the stationary case.
In order to test the system (17.8) we apply the stochastic process x_s.

17.9.1 *System with 3 DOF under stochastic loading*

System parameters are:
$m_1 = 1500$ kg, $m_2 = 11000$ kg, $m_3 = 11000$ kg, $k_0 = 575000$ N/m, $k_{1,1} = 575000$ N/m, $k_{1,2} = 575000$ N/m, $k_{1,3} = 575000$ N/m, $k_2 = 53000$ N/m, $k_3 = 53000$ N/m, $\eta_1 = 0.1$ m, $\eta_2 = 0.15$ m, $\eta_3 = 0.2$m, $c_{1,1} = 5000$ Ns/m, $c_{1,2} = 5000$ Ns/m, $c_{1,3} = 5000$ Ns/m, $c_2 = 15000$ Ns/m, $c_3 = 15000$ Ns/m, $q_1 = 1$, $q_2 = 1$, $q_3 = 10$, $\omega_d = 18.87$ rad/s, $\xi_d = 0.65$, $t_1 = 3$s, $t_2 = 10$ s and $c = 0.76s^{-1}$.
Initial conditions are:
$x_{1,0} = x_{2,0} = 0.$, $w_1 = 0$, $\dot{x}_{1,0} = \dot{x}_{2,0} = 0$, $x_{3,0} = \dot{x}_{3,0} = 0$.

Since the effect of control on the vibrations of the first mass m_1 is almost negligible, results concerning x_1, $\dot{x}_1$, $\ddot{x}_1$, w_1, w_2, w_3 are not presented here.

In Figs. 17.19 and 17.20, x_3 and $\ddot{x}_3$ are again plotted versus time. One can see in these figures that control procedure is efficient: curves for controlled system (thick lines) are plotted for parameter r equal to $2e-10$ that permits this efficient diminution of oscillations amplitudes and of the amplitudes of accelerations. The choice of this value for r has been made according to the following procedure.

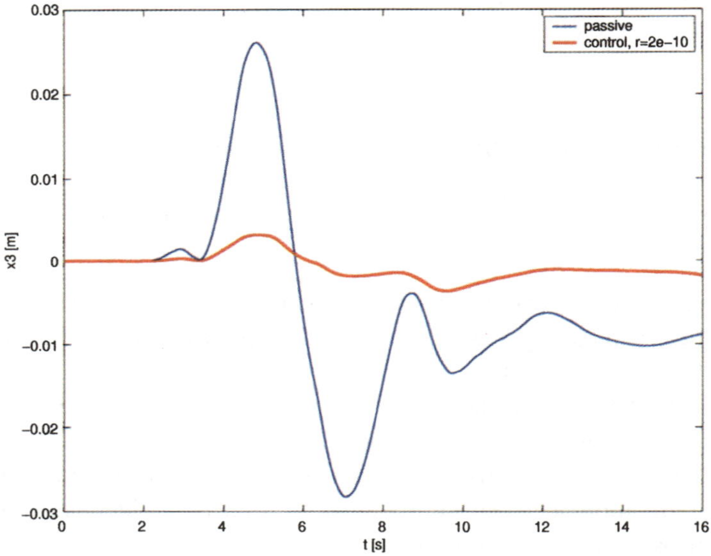

Fig. 17.19 Displacement x_3 versus time t for the passive and controlled system.

In order to chose a good value of r we studied J versus r for example. General study of J versus q_1, q_2, and r could be done. We also studied rms values of $\ddot{x}_2$, $x_1 - f$ and $x_2 - x_1$. The best choice of r is the value which provides the smallest values of rms values of $\ddot{x}_2$, $x_1 - f$ and $x_2 - x_1$. That leads here to $r = 2e - 10$ where rms values of displacements of both masses are almost equal to case without control but where rms values of acceleration of second mass is improved.

Presented graphs shows that the control process improves the dynamical behaviour of second masses in comparison to the passive behaviour.

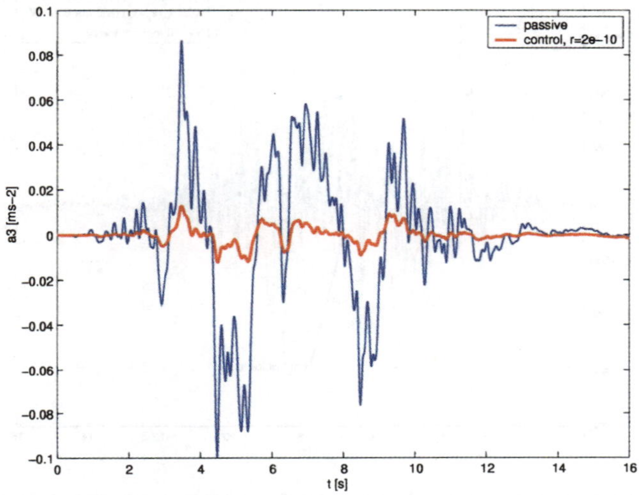

Fig. 17.20 Acceleration $\ddot{x}_3$ (denoted a_3 on y label axis) versus time t for the passive and controlled system.

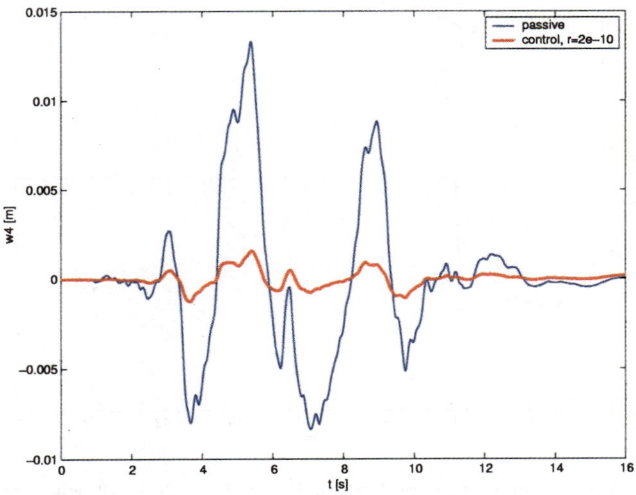

Fig. 17.21 Displacement w_4 versus time t for the passive and the controlled system.

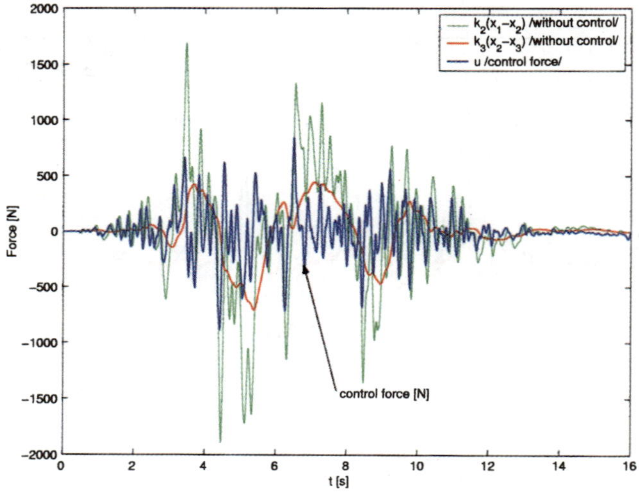

Fig. 17.22 Comparison of control force $u(t)$ for $r = 2\text{e} - 10$ with $k_2(x_1 - x_2)$ and $k_3(x_2 - x_3)$ computed in for the system without control.

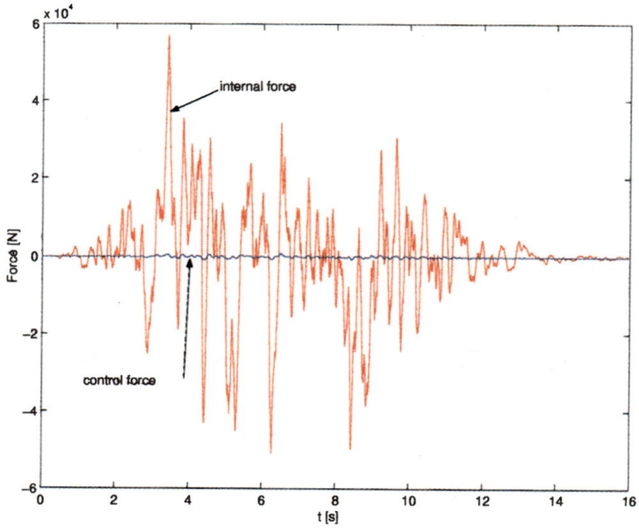

Fig. 17.23 Comparison of control force $u(t)$ for $r = 2\text{e} - 10$ with internal force $\sum_{j=1}^{3} c_{1,j}\dot{f} + k_0 f$ computed for the system without control.

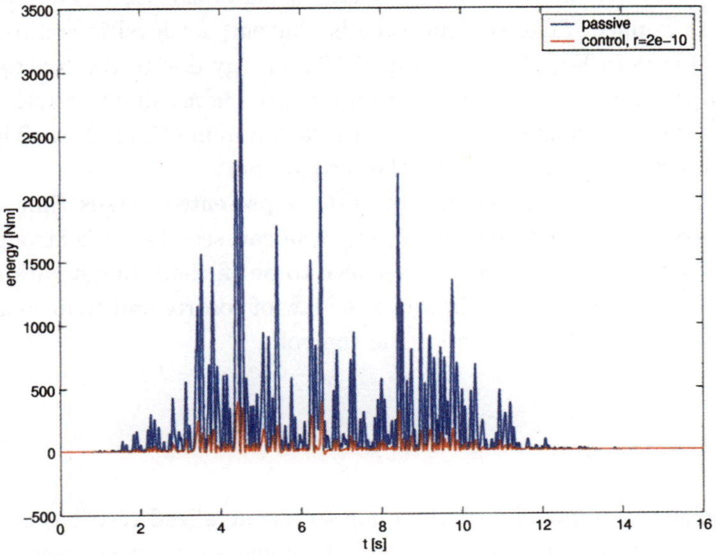

Fig. 17.24 Comparison of energies due to elastic forces between m_1 and m_2 for system of Figs. 3(a) and 3(c).

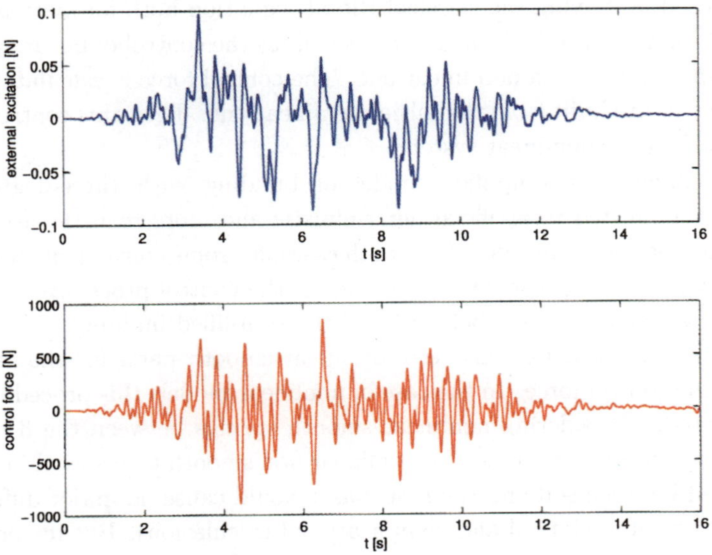

Fig. 17.25 Comparison of control force $u(t)$ and external excitation $f(t)$ in the case of stochastic excitation.

In Fig. 17.21, one can see that plastic displacement w_4 is reduced when control is applied. The control force is efficient; its level is compared to internal forces in Fig. 17.22. In Fig. 17.24, energy due to the control force between the masses m_1 and m_2 is compared to energy due to elastic forces between the same masses in the uncontrolled system of Fig. 17.4. The first one is (much more) smaller than the second one.

In Fig. 17.25, the control force $u(t)$ is presented versus time and it is compared to the external force $f(t)$: one can see that u is strong with respect to f. Even if the control is shown to be efficient, investigations will have to be made in order to improve the law of control and to decrease the quantity of energy to provide for the control.

17.10 Conclusion

In this chapter, a model corresponding to a generalized translational cone including elastoplastic terms coupled with linear spring mass elements has been presented. It is based on discrete rheological models for the expansion of elastoplasticity. The mathematical background has been recalled. Theoretical control procedure has been explained and calculation of control force is proposed by solving the classical Riccati equation both for stationary or non stationary control. This method useful for the control of linear systems has been adapted for a non linear one. The control force is calculated for a linear structure built from the initial nonlinear one. Then this control force is applied to the nonlinear system.

An example of simplified model of building with three-degrees-of-freedom submitted to artificial deterministic and more realistic stochastic excitation has been considered. In both cases, for some convenient choices of parameters defining cost function (energy), the control procedure is shown to improve the dynamical behaviour of the simplified building.

In this Chapter, we consider only linear smooth parts for the calculation of the control force. In the case of nonlinear models this procedure can be adapted. Considering nonlinear smooth springs between the 3 masses (without non smooth terms: the nonlinear non smooth terms would be concentrated between soil and the first mass) would cause no major difference (except size of analytical and computational calculation). But introducing nonlinear smooth and non smooth terms (elastoplasticity) between masses 2 and 3 e.g., creates problems to adapt the previous procedure: the nonlinear smooth part should be identified and a smooth nonlinear associated system would have to be build in order to compute the control force. This has not still been done. The attempt to calculate the control force from a Ricatti equation derived directly from the nonlinear non smooth problems copes with hard theoretical problems: if Riccati equation corresponding to such a case has to be extended in the form of differential inclusion, then it can not be solved backward in the present frame because of the plastic terms. Another difficulty has to be pointed out but it is inherent to the control procedure : if the full solution of the Riccati equation is used, some knowledge is required on the excitation in order to calculate the control force; in practice, it is not always easy to get such a knowledge via measurements. The second one is linked to the great amount of energy (see [Inman (1989)] e.g.) that is necessary to obtain an efficient control.

Bibliography

Abraham, O. N. L. (1993). Dynamic Modelling of Cracked Timoshenko Beams, Division of Mechanical Engineering and Energy Studies, School of Engineering, University of Wales College of Cardiff, Wales, UK.

Acheson, D. J. and Mullin, T. (1993). Upside–down pendulums, *Nature* **366**, pp. 215–216.

Aichara, K. and Matsumoto, G. (1986). Chaotic oscillations and bifurcations in giant squid axons, *Chaos*, Manchester University Press, A. V. Holden, pp. 257-269.

Aizerman, M. A. and Gantmacher, F. R. (1958). On the stability of periodic motions, *J. Appl. Math. Mech.* (translated from Russian), pp. 1065–1078.

Aizerman, M. A. and Gantmacher, F. R. (1963). Absolute Stability of Control Systems, Academy of Sciences of USSR, 1963.

Aizerman, M. A. and Piatnitskij, E. S.(1974). Fundamentals of theory of discontinuous systems, *Aut. and Telemech.* **7, 8**.

Alber, H. D. (1998). *Materials with Memory*, Springer Verlag, Berlin Heidelberg, Lecture Notes in Mathematics 1682.

Alexandrov, V. and Annakulova, H. (1990). A Contact problem of thermoelasticity with wear and heat generation taken into account, *Friction and Wear* **11**, 1, pp. 24–28.

Alexandrov, V. and Annakulova, H. (1992). An Interaction of body's coverings with taking into account of deformability, heat generation and wear due to friction, *Friction and Wear* **13**, 1, pp. 154–160 (in Russian).

Altpeter, F., Ghorbel, F. and Longchamp, R. (1998). Relationship between two friction models: A singular perturbation approach, *Proc.* 37th *IEEE Conf. on Decision*, Tampa, Florida, IEEE.

Anand, G. V. (1972). Natural modes of a coupled nonlinear system, *Int. J. Nonlin. Mech.* **7**, pp. 81.

Anderson, J. R. and Ferri, A. A. (1990). Behaviour of a single degree of freedom system with a generalized friction law, *J. Sound Vibr.* **140**, pp. 287–304.

Andronov, A. A. and Witt, A. A. and Khaikin, S. E. (1966). Theory of Oscillations, Pergamon Press.

Antman, S. S. (1995). Nonlinear Problems of Elasticity, AMS 107, Springer Verlag, New-York.

Archard, J. F. (1959). The temperature of rubbing surfaces, *ZAMP* **2**, 6, pp. 438–455.

Armstrong-Helouvry, B. (1991). *Control of Machines with Friction*, Kluwer Academic Publishers, Massachussets.

Armstrong-Helouvry, B., Dupont, P. and Canudas de Witt, C. (1994). Survey of models, analysis tools and compensation methods for the control of machines with friction, *Automatica* **30**, 7, pp. 1083–1138.

Arnold, V. I. (1976a). Méthodes Mathématiques de la Mécanique Classique, Editions de Moscou, Moscou.

Arnold, V. I. (1976b). Local normal forms of functions, *Inventiones Mathematicae* **35**, pp. 87–109.

Arnold, V. I. (1980). Chapitres Supplémentaires à la Théorie des Equations Différentielles Ordinaires, Editions de Moscou, Moscou.

Arnold, V. I. (1983). Geometrical Models in the Theory of Ordinary Differential Equations, Springer-Verlag, New York, Heidelberg, Berlin.

Arnold, V. I. (1984). Catastroph Theory, Springer, Berlin.

Arnold, V. I., Varchenko, A. and Goussein-Zade, S. (1986). Singularités des Applications Différentiables. I. Classification des Points Critiques, des Caustiques, des Fronts D'ondes, Editions de Moscou, Moscou.

Arnold, V. I., Varchenko, A. and Goussein-Zade, S. (1986). Singularités des Applications Différentiables. II. Monodromie et Comportement Asymptotique des Intégrales, Editions de Moscou, Moscou.

Arrowsmith, D. K. and Taha, K. I. (1983). Bifurcations of a particular forced van der Pol oscillator, *Meccanica* **18**, pp. 195–204.

Awrejcewicz, J. (1981). Analysis of self-excited oscillations caused by friction, Ph.D. Thesis, Technical University of Łódź (in Polish).

Awrejcewicz, J. (1983). Analysis of self-exicted vibration due to nonlinear friction in a one-degree-of-freedom system, *Sci. Bull. Lodz Tech. Univ.* **66**, pp. 21–28.

Awrejcewicz, J. (1985). Self-excited vibrations modelling between two bodies with the relative motion taking into account the elasticity of contacts, *Proc. 24th Symp. Model. in Mech.*, pp. 27-34 (in Polish).

Awrejcewicz, J. (1986). Chaos in simple mechanical systems with friction, *J. of Sound and Vibr.* **109**, pp. 178–180.

Awrejcewicz, J. (1987). Analysis of self-exicted vibration in the system with four-degrees-of-freedom, *Sci. Bull. Lodz Tech. Univ.* **72**, pp. 5–27.

Awrejcewicz, J. (1988). Chaotic motion in a nonlinear oscillator with friction, *KSME J.* **2**, 22, pp. 104-109.

Awrejcewicz, J. (1989a). Bifurcation and Chaos in Simple Dynamical Systems, World Scientific, Singapore.

Awrejcewicz, J. (1989b). A route to chaos in a nonlinear oscillator with delay, *Acta Mech.* **77**, pp. 111-120.

Awrejcewicz, J. (1990a). Bifurcation and Chaos in Dynamical Systems, Technical University of Łódź, Habilitation Thesis (in Polish).

Awrejcewicz, J. (1990b). Parametric and self-exicted vibrations induced by friction in a system with three-degrees-of-freedom, *KSME Journal* **4**, 2, pp. 156–166.

Awrejcewicz, J. (1990c). Routes to chaos in complex physical systems, *J. Tech. Phys.* **31**, pp. 355–372.

Awrejcewicz, J. (1991a). Bifurcation and Chaos in Coupled Oscillators, World Scientific, Singapore.

Awrejcewicz, J. (1991b). Chaos in a system of coupled oscillators, *Acta Phys. Slovaca* **41**, 4, pp. 217–223.

Awrejcewicz, J. (1991c). Nonlinear dynamics of a two-body nonlinear mechanical system, *Comp. Meth. Appl. Mech. and Eng.* **91**, 1–3, pp. 1093–1108.

Awrejcewicz, J. (1991d). Numerical analysis of the oscillations of human vocal cords, *Nonlin. Dyn.* **2**, pp. 35–52.

Awrejcewicz, J. (1991e). Three routes to chaos in simple sinusoidally driven oscillators, *Appl. Math. Mech. ZAMM* **71**, 2, pp. 71–79.

Awrejcewicz, J. (1991f). Chaos in a sinusoidally (parametrically and externally) driven system with three-degrees-of-freedom, *J. Tech. Phys.* **32**, 2, pp. 247–265.

Awrejcewicz, J. (1993). Analysis of self-exicted vibration due to dry friction in a system with two-degrees-of-freedom, *Sci. Bull. Lodz Tech. Univ.* **82**, 635, pp. 23–24.

Awrejcewicz, J. (1996b). Oscillations of the Discrete Deterministic Systems, Warsaw (in Polish).

Awrejcewicz J. (1999). Control of simple vibro-impact systems using time delay loop, *Proc. 2nd Sci. Conf. 'Impact Resistance of Constructions'*, pp. 5-16, Warsaw, November '99, WAT, invited lecture.

Awrejcewicz, J., Andrianov, I. V. and Manevich, L. I. (1998). Asymptotic Approaches in Nonlinear Dynamics: New Trends and Applications, Springer Verlag, Berlin.

Awrejcewicz, J. and Calvisi, M. L. (2002). Mechanical models of Chua's circuit, *Int. J. Bif. and Chaos* **12**, 4, pp. 671–686.

Awrejcewicz, J. and Delfs, J. (1990a). Dynamics of a self-excited stick-slip oscillator with two degrees of freedom. Part I. Investigation of equilibria, *Eur. J. of Mech. A/Solids* **9**, 4, pp. 269–282.

Awrejcewicz, J. and Delfs, J.(1990b). Dynamics of a self-excited stick-slip oscillator with two-degrees-of-freedom. Part II. Slip-stick, slip-slip, stick-slip transitions, periodic and chaotic orbits, *Eur. J. Mech. A/Solids* **9**, 5, pp. 397–418.

Awrejcewicz, J. and Holicke, M. (1996). Detection of stick-slip chaotic oscillations using Melnikov method, *Proc. ICNBC96*, pp. 65-68, Łódź - Dobieszków, September 16 - 18, TU Łódź and ENTPE.

Awrejcewicz, J. and Holicke, M. (1997a). Detection of chaotic orbits in the Duffing oscillator with dry friction using Melnikov's method, *Proc. 1st Nat. Conf. 'Methods and Computer System in Scientific Research Engineering Design'* (ed. T. Szmuc, R. Tadeusiewicz and M. Szymkat and T. Uhl), pp. 369-374, Kraków, November 25-26 (in Polish).

Awrejcewicz, J. and Holicke, M. (1997b). Influence of dry friction on homoclinic chaos, *Proc. 4th Conf. 'Dynamical Systems - Theory and Applications'*, (ed. J. Awrejcewicz, J. Grabski and J. Mrozowski), pp. 59-62, Łódź, December 8-9, LODART S.A.

Awrejcewicz, J. and Holicke, M. M. (1999). Melnikov's method and stick-slip chaotic oscillations in very weekly forced mechanical systems, *Int. J. Bif. Chaos* **9**, 3, pp. 505–518.

Awrejcewicz, J. and Krodkiewski, J. (1984). Analysis of self-excited vibrations due nonlinear friction in two-degrees-of-freedom system, *Sci. Bull. Łódź Tech. Univ.* **67**, pp. 13–30 (in Polish).

Awrejcewicz, J., Kudra, G. and Lamarque, C. -H. (2001). Analysis of bifurcations and chaos in three coupled physical pendulums with impacts, *Proc. Design Eng. Tech. Conf.*, Pittsburgh, 12p., CD-ROM.

Awrejcewicz, J., Kudra, G. and Lamarque, C.-H. (2002). Nonlinear dynamics of triple pendulum with impacts, *J. Tech. Phys.* **43**, 2, pp. 97–112.

Awrejcewicz, J., Kudra, G. and Lamarque, C.-H. Investigation of triple pendulum with impacts, *Int. J. Bif. and Chaos* (to appear).

Awrejcewicz, J. and Mrozowski, J. (1989). Bifurcation and chaos in a particular van der Pol-Duffing oscillator, *J. Sound and Vibr.* **132**, 1, pp. 89-100.

Awrejcewicz, J. and Olejnik, P. (2003). Regular and chaotic stick-slip dynamics in a self-excited two-degree-of-freedom system with friction, *Int. J. Bif. and Chaos* **13**, 4 (to appear),

Awrejcewicz, J. and Pyryev, J. (2002). Thermoelastic contact of a rotating shaft with a rigid bush in conditions of bush wear and stick-slip movements, *Int. J. Eng. Sci.* **40**, pp. 1113–1130.

Awrejcewicz, J. and Reinhardt, W.-D. (1990). Quasiperiodicity, strange non-chaotic and chaotic attractors in the forced system with two-degrees-of-freedom, *J. Appl. Math. Phys. ZAMP* **41**, pp. 713–727.

Awrejcewicz, J. and Reinhardt, W.-D. (1991). Observation of chaos in the nonautonomous system with two-degrees-of-freedom, *J. Appl. Math. Mech. ZAMM* **71**, 9, pp. 357–360.

Awrejcewicz, J. and Someya, T. (1991). Quasi-periodic and chaotic orbits and their bifurcation in a system of coupled oscillators, *J. Sound Vibr.* **146**, 3, pp. 527–532.

Awrejcewicz, J. and Someya, T. (1992). A twisted horseshose in the roll–slide oscillator, *J. Phys. Soc. of Japan* **61**, 5, pp. 1556–1559.

Awrejcewicz, J. and Tomczak, K. (1995). Vibrating systems with impacts - exact and equivalent models, *Proc. Third Conf. on 'Dynamical Systems: Theory and Applications'*, pp. 63-68, Łódź, Technical University Press.

Awrejcewicz, J. and Tomczak, K. (1998). Analysis and control of discrete systems vibration with impacts, *Computer Supported Computations Examples of Applications of the MATLAB and Maple V Programs*, (ed. M. Szymkat), Centre for Advanced Training in Information Engineering, Cracow.

Awrejcewicz, J. and Tomczak, K. (1999a). One- and Two-Degree-of-Freedom Dynamics Control, book A. I. Manevitch(ed.), Problems of Nonlinear Mechanics and Physics of Materials, NGAU (in Russian and in English).

Awrejcewicz, J. and Tomczak, K. (1999b). Analytical stability improvement of the vibro-impact processes, *Proc. ASME Design Eng. Tech. Conf.*, **DETC99/VIB-8002**, Las Vegas, Nevada, September 12-15, pp. 8, published on CD-Rom.

Awrejcewicz, J. and Tomczak, K. (1999c). Stability improvement of the impact dynamical systems - analytical and numerical methods, March 2–5, Hanoi, *EUROMECH Colloquium 386, 15–18 September 1998*, Loughborough University. In: V. I. Babitsky (ed.), *Dynamics of Vibro-Impact Systems*, Springer Verlag, Berlin, *IUTAM Symposion on RECENT DEVELOPMENTS IN NON-LINEAR OSCILLATIONS OF MECHANICAL SYSTEMS*, Vietnam National University.

Awrejcewicz, J. and Tomczak, K. (1999d). Feedback delay loop control of the vibro-impact dynamical systems, *Proc. of the Tenth World Congress on the Theory of Machines and Mechanisms*, pp. 1459-1464, Oulu, Finland, June 20-24.

Awrejcewicz, J. and Tomczak, K. (1999e). Stability improvement of the vibro-impact discrete systems, *Dynamics of Vibro-Impact Systems*, (ed. V. I. Babitsky), Loughborough University, EUROMECH Colloquium 386, pp. 109-118.

Awrejcewicz, J. and Tomczak, K. (2000). Analytical and numerical investigation of vibro-impact dynamics control, *Proc. of the IUTAM/IFToMM Symposium on SYNTHESIS OF NONLINEAR DYNAMICAL SYSTEMS*, (ed. E. Lavendalis and M. Zakrzhevsky), Dordrecht, August 24-28, Kluwer Academic Publisher, pp. 67-76.

Awrejcewicz, J., Tomczak, K. and Lamarque, C. -H. (1996b). Multi-body vibro-impact dynamics, *Proc. EUROMECH - 2nd Eur. Nonlin. Oscill. Conf.*, pp. 9-12, Prague, September 9-13.

Awrejcewicz, J., Tomczak, K. and Lamarque C. -H. (1996a). Controlling systems with impacts, *Proc. Int. Conf. Nonlin., Bif. and Chaos, ICNBC96*, pp. 79-82, Łódź - Dobieszków, 16-18 September, TU Łódź and ENTPE, ISBN 2-86834-110-1.

Awrejcewicz, J., Tomczak, K. and Lamarque, C. -H. (1999). Controlling system with impacts, *Int. J. Bif. Chaos* **9**, 3, pp. 547–553.

Azeez, M. A. F., Vakakis, A. F. and Manevich, L. I. (1999). Exact solutions for the problem of the vibro-impact oscillations of a discrete systems wiyh two-degrees-of-freedom, *PMM - J. Appl. Math. Mech. (Prikladnaya Matematika i Mekhanika)* **63**, 4, pp. 527–530.

Azeez, M. A. F. and Vakakis A. F. (1999). Numerical and experimental analysis of a continuously overhang rotor undergoing vibro-impacts, *Int. J. Nonlin. Mech.* **34**, pp. 415–435.

Azeez, M. A. F. and Vakakis A. F. (2001). Proper orthogonal decomposition (POD) of a class of vibro-impact oscillations, *J. Sound and Vibr.* **240**, 5, pp. 859–889.

Babakov, I. M. (1968). Theory of Vibrations, Nauka, Moskva (in Russian).

Babitsky, V. I. (1998a). Theory of Vibro-Impact Systems and Applications, Springer-Verlag, Berlin, Revised translation from Russian, 1978, Nauka, Moscow.

Babitsky, V. I. (1998b). Dynamics of Vibro-Impact Systems, Springer Verlag, London, Proc. Euromech Colloquium 386-Loughborough-England-September 1998.

Babitsky, V. I. (1999). Hand-held percussion machine as an optimal converter, *Proc. ASME Design Eng. Tech. Conf.*, **DETC99/VIB-8179**, Las Vegas, Nevada, September 12-15, ASME, published on CD-ROM.

Babitsky, V. I. and Krupenin, V. L. (2001). Vibration of Strongly Nonlinear Discontinuous Systems, Springer-Verlag.

Baker, G. L. and Gollub, J. P. (1990). Chaotic Dynamics, Cambridge University Press, Cambridge.

Bargmann, V. (1961). On a Hilbert space of analytic functions and an associated integral transform. Part I, *Communications on Pure and Appl. Math.***XIV**, pp. 187–214.

Bastien, J. (2000). Etude théorique et numérique d'inclusions différentielles maximales monotones. Application à des modèles élasto-plastiques, Université Lyon I, 30 mai 2000, numéro d'ordre 96-2000.

Bastien, J. and Schatzman, M. (1999). Précision de schémas numériques en évolution multivoque, *preprint 307*, www.numerix.univ-lyon1.fr /publis /publiv /1999 /publis.html.

Bastien, J. and Schatzman, M. (2000). Schéma numérique pour des inclusions différentielles avec terme maximal monotone, *C. R. Acad. Sci. Paris Série I Math.* **330**, pp. 611–615.

Bastien, J., Schatzman, M. and Lamarque, C. -H. (2000). Study of some rheological models with a finite number of degrees of freedom, *Eur. J. Mech. A/Solids* **19**, 2, pp. 277-307.

Baumberger, T., Caroli, C., Perrin, B. and Ronsin, O. (1995). Nonlinear analysis of the stick-slip bifurcation in the creep-controlled regime of dry friction, *Phys. Rev. E*, **51**, 5, pp. 4005–4010.

Bayly, P. V. and Virgin, L. N. (1992). Experimental evidence of diffusive dynamics and random walking in a single deterministic mechanical system: the shaken pendulum, *Int. J. Bif. and Chaos*, **2**, 4.

Benedettini, F., Capecchi, D. and Vestroni, F. (1995). Identification of hysteretic oscillators under earthquake loading by nonparametric models, *J. Eng. Mech.*, May, pp. 606–612.

Bensoussan, A. and Rascanu, A. (1997). Stochastic variational inequalities in infinite dimensional spaces, *Numer. Funct. Anal. and Optimiz.* **18**, pp. 19-54.

Bernardin, F. (2000). Equations différentielles stochastiques multivoques et modélisation de bâtiments sous sollicitations sismiques, Université Claude–Bernard Lyon I and Ecole Nationale des Travaux Publics de l'Etat, *Rapport de DEA de Mathématiques*, Lyon.

Birkhoff, G. D. (1927). Dynamical Systems, 9, AMS coll. Publication, New-York.

Bishop, S. R. and Clifford, M. J. (1996). Zones of chaotic behaviour in the parametrically excited pendulum, *J. Sound Vib.* **189**, 1, pp. 142–147.

Blackburn, J. A., Zhou-jing, Y., Vik, S., Smith, H. J. T. and Nerenberg, M. A. H. (1987). Experimental study of chaos in a driven pendulum, *Physica* **D26**.

Bogaevski, V. N. and Povzner, A. (1991). Algebraic Methods in Nonlinear Perturbation Theory, Springer Verlag, New-York.

Boivin, N., Pierre, C. and Shaw, S. W. (1995). Non-linear normal modes, invariance, and modal dynamics approximations of nonlinear systems, *Nonlin. Dyn.* **8**, 3, pp. 315–46.

Bothe, D. (1999). Periodic solutions of non-smooth friction oscillators, *Zeitschrift für angewandte Mathematik und Physik* **50**, 5, pp. 779–808.

Bouleau, N. (1988). Processus Stochastiques et Applications, Hermann, Paris.

Bourquin, F. (1991). Synthèse modale et analyse numérique desmulti-structures 'elastiques, Université Pierre et Marie Curie, PhD Thesis, Paris.

Boussa, D. and Labbé, P. (1989). Seismically induced ratcheting: a probabilistic analysis of a simple case, *Recent Advances in Earthquake Eng. Struct. Dyn.*, pp. 661–678.

Boxler, P. (1991). How to Construct Stochastic Center Manifolds on the level of Vector Fields, *Lyapunov Exponents - Lecture Notes in Mathematics 1486*, Springer Verlag, L. Arnold and H. Crauel and J. -P. Eckmann.

Brezis, H. (1973). *Opérateurs maximaux monotones et semi-groupes de contractions dans les espaces de Hilbert*, North-Holland, Amsterdam, North-Holland Mathematics Studies, No. 5. Notas de Matemática (50).

Brezis, H. (1992). Analyse Fonctionnelle. Théorie et Applications, Masson, Paris.

Brjuno, A. D. (1989). Local Methods in Nonlinear Differential Equations, Springer-Verlag, New-York.

Broer, H. W., Hoveijn, G. A., Lunter, G. A. and Vegter, G. (1998). Resonances in a spring-pendulum: algorithms for equivariant singularity theory, *Nonlinearity* **11**, pp. 1569–1605.

Brogliato, B. (1996). Nonsmooth Impact Mechanics, Models, Dynamics and Control, Springer-Verlag,, London.

Brogliato, B. (1999). Nonsmooth Mechanics, Springer Verlag, London, Chapter 6, pp. 263-282.

Brommundt, E. (1975). On the numerical investigation of nonlinear periodic rotor vibrations. Dynamics of rotors, *I.U.T.A.M. Symp. in Lyngby, Denmark*, Berlin-Heidelberg-New York, Springer-Verlag.

Brommundt, E. (1977). Bifurcation of self-excited rotor vibrations, *VII Internationale Konferenz über nichtlineare Schwingungen*, pp. 113-137, Berlin, Akademie-Verlag.

Budd, C. and Dux, F. (1994). Chattering and related behaviour in impact oscillators, *Phil. Trans. Royal Soc. of London* Series A, **347**, pp. 365–389.

Budd, C. and Dux, F. (1996). Intermittency in impact oscillators close to resonance, *Nonlinearity*, **7**, pp. 1191–1224.

Budd, C. J., Dux, F. and Cliffe, A. (1995). The effect of frequency and clearance variations on single-degree-of-freedom impact oscillators, *J. Sound Vibr.* **184**, 3, pp. 475–502.

Cépa, E. (1995a). Equations Différentielles Stochastiques Multivoques, Université d'Orléans.

Cépa, E. (1995b). *Lecture Notes in Math. Prob., Semin. XXIX*, Springer-Verlag, Berlin, pp. 86-107.

Capecchi, D. and Vestroni, F. (1995). Asymptotic response of a two-degrees-of-freedom elastoplastic system under harmonic excitation. Internal resonance Case, *Nonlin. Dyn.* **7**, pp.317–333.

Carr, J. (1981). Applications of Center Manifold Theory, Springer Verlag, New-York.

Carslaw, H. S. and Jaeger, J. C. (1959). Conduction of Heat in Solids, Clarendon Press, Oxford.

Caughey, T. K. (1965). Classical normal modes in damped linear dynamic systems, *J. Appl. Mech.*, pp. 583-594.

Chatelin, F. (1989). Valeurs Propres de Matrices, Masson, Paris.

Chatterjee, S. and Mallik, A. K. (1996). Bifurcations and chaos in autonomous self excited oscillators with impact damping, *J. Sound Vibr.* **191**, 4, pp. 539–562.

Chen, Q., Manson, G. and Worden, K. (1997). Higher-order FRFs for multi-input nonlinear systems: a case study, *New Advances in Modal Synthesis of Large Structures*, (ed. L. Jézéquel), pp. 549-561, Rotterdam, Balkema.

Chen, S.-L. and Shaw, S. W. (1996). Normal modes for piecewise linear vibratory systems, *Nonlin. Dyn.* **10**, 2, pp. 135–163.

Chichinadze, A. V., Brown, E. D., Hinsburg, A. H. and Ihnatyeva, Z. V. (1979). Calculation, Trial and Selection of Friction Couples, Nauka, Moscow (in Russia).

Chow, S. W. and Hale, J. K. (1982). Methods of Bifurcation Theory, Springer-Verlag, New York, Heidelberg and Berlin.

Christopher, P. A. T. (1979). A modified stroboscopic method for non linear vibration equations, *J. Sound Vibr.* **66**, 1, pp. 91–97.

Chua, L. O. (1992). The genesis of Chua's circuit, *Archive für Elektronik und Übertragungstechnik* **46**, 4, pp. 250–257.

Chua, L. O. (1993). Global unfolding of Chua's circuit, *IEICE Trans. Fundamentals*, **E16–A**, 5, pp. 704–734.

Chua, L. O., Komuro, M. and Matsumoto, T. (1986a). The double scroll family, Parts I and II, *IEEE Trans. Circuits ans Systems*, **CAS-33**, pp. 1072–1118.

Chua, L. O., Komuro, M. and Matsumoto, T. (1986b). The double scroll family. II. Rigorous analysis of bifurcation phenomena, *IEEE Trans. Circuits and Systems* **33**, 11, pp. 1097–1118.

Chua, L. O., Komuro, M. and Matsumoto, T. (1986c). The double scroll family. I. Rigorous proof of chaos, *IEEE Trans. Circuits and Systems* **33**, 11, pp. 1072–1097.

Cipera S., Cerná, R. and Peterka, F. (1996). Characteristics of the motion of mechanical impact systems with dry friction, *Proc. Second Eur. Nonlin. Oscill. Conf.*, Prague, September 9–13, pp. 129–132.

Claude, D. (1986). Découplage des systèmes non linéaires, séries génératrices non commutatives et algèbres de Lie, *SIAM J. Control and Optimization* **24**, 3, pp. 562–578.

Colette, F. S. (1998). A combined tuned absorber and pendulum impact damper under random excitation, *J. Sound Vibr.* **216**, 2, pp. 199–213.

Collet, P. and Eckmann, J. (1980). Iterated Maps on the Interval as Dynamical

Systems, Birkhauser, Basel.

Cone, K. M. and Zadoks, R. I. (1995). A numerical study of an impact oscillator with the addition of dry friction, *J. Sound Vibr.* **188**, pp. 659–683.

Cooke, C. H. and Struble, R. A. (1966). On the existence of periodic solutions and normal mode vibrations of non linear systems, *Quaterly of Appl. Math.* **XXIV**, 3, pp. 177–193.

Coullet, P. H. and Spiegel, E. A. (1985). Amplitude equation for systems with competing instabilities, *SIAM J. Appl. Math.* **43**, 4, pp. 776-821.

Coullet, P. and Tresser, C. (1984). Un scénario typique de transitionvers le chaos: la cascade de dédoublement de période, *J. de mécanique théorique et appliquée*, **Numéro spécial**, pp. 217–240.

Crandall, M.G. and Evans, L. C. (1975). On the relation of the operator $\partial/\partial s + \partial/\partial\tau$ to evolution governed by accretive operators, *Israel J. Math.* **21**, 4, pp. 261–278.

Crouzeix, M. and Mignot, A. L. (1984). Analyse Numérique des Équations Différentielles, Masson, Paris.

Curry, J. and Yorke, J. (1978). A transition from Hopf bifurcation to chaos: computer experiments on maps in R^2; the structure of attractors in dynamical systems, *Lect. Not. Math.* **668**, pp. 48–68.

Deimling, K. (1992). Multivalued Differential Equations, W. de Gruyter, Berlin.

Deimling, K. and Szilágyi, P. (1994). Periodic solutions of dry friction problems, *Z. Angew. Math. Phys.* **45**, pp. 53–60.

Dekker, K. and Verwer, J. G. (1984). Stability of Runge-Kutta Methods for Stiff Nonlinear Differential Equations, North-Holland, Amsterdam.

Den Hartog, J. P. (1956). Mechanical Vibrations, McGraw-Hill, New York.

Di Lieto, A., Fenicia, S. and Mancini, P. (1991). A computer assisted pendulum for didactics, *Eur. J. Phys.* **12**, 51.

Dmitriev, A. S., Kislov, V. I. and Spiro, A. G. (1983). Khaoticeskije kolebania w neavtonomnom generatore s reaktivnoi nelineinostyu (Chaotic vibrations in non–autonomous generator with nonlinear reactance), *Radiotekhnika i elektronika* **12**, pp. 2430–2439 (in Russian).

Doerner, R., Hübinger, B. and Heng, H. and Martienssen, W. (1994). Approaching nonlinear dynamics by studying the motion of a pendulum. II Analyzing chaotic motion, *Int. J. Bif. Chaos* **4**, pp. 761–771.

Dold, A. and Eckmann, B. (1984). Bifurcations Theory and Applications: Lecture Notes in Mathematics, Springer-Verlag, New York, Heidelberg and Berlin.

Dowell, E. H. and Schwartz, H. B. (1983a). Forced response of a cantilever beam with a dry friction damper attached, Part I: Theory, *J. Sound Vibr.* **91**, 2, pp. 255–267.

Dowell, E. H. and Schwartz, H. B. (1983b). Forced response of a cantilever beam with a dry friction damper attached, Part II: Experiment, *J. Sound Vibr.* **91**, 2, pp. 269–291.

Duvaut, G. and Lions, J. L. (1972). *Les inéquations en mécanique et en physique*, Dunod, Paris, Travaux et Recherches Mathématiques, No. 21, pp. 366-387.

Emaci, E., Nayfeh, T. A. and Vakakis A. F. (1997). Numerical and experimantal study of nonlinear localization in a flexible structure with vibro-impacts,

Zeitschrift fur Angewandte Mathematik und Mechanik - ZAMM (J. Appl. Math. Mech.) **77**, 7, pp. 527–541.

Eringen, C. et al.(1975). Continuum Physics, Academic Press, I and II, New-York.

Falconer, K. J. (1990). The Geometry of Fractal Sets, Cambridge University Press, Cambridge, Cambridge Tracts in Mathematics, 85.

Farahanchi, F. and Shaw, S. H. (1994). Chaotic and periodic dynamics of a slider–crank mechanism with sleader clearance, *J. Sound Vibr.* **177**, 3, pp. 307–324.

Fečkan, M. (1997). Bifurcation of periodic solutions in differential inclusions, *Appl. Math.* **42**, pp. 369–393.

Feder, H. J. S. and Feder, J. (1991). Self-organised criticality in a stick-slip process, *Phys. Rev. Lett.* **66**, 20, pp. 2669-2672.

Feely, O. and Chua, L. O. (1992). Nonlinear dynamics of a class of analog to digital converters, *Int. J. Bif. Chaos* **2**, 2, pp. 325-340.

Feeny, B. F. and Moon, F. C. (1993). Bifurcation sequences of a Coulomb friction oscillator, *Nonlin. Dyn.* **4**, pp. 25-37.

Feigenbaum, M. J. (1978). Quantitative universality for a class of nonlinear transformations, *J. Stat. Phys.* **19**, pp. 25–52.

Ferri, A. A. and Bindemann, A. C. (1995). Large amplitude vibration of a beam restrained by a non–linear sleeve joint, *J. Sound Vibr.* **184**, 1, pp. 19–34.

Filippov, A. F. (1988). Differential Equations with Discontinuous Right-Hand Sides, Kluwer Academic Publishers.

Filliatrault, A. (1996). *Eléments de Génie Parasismique et de calcul dynamique des structures*, Editions de l'Ecole Polytechnique de Montréal, Montréal, 1996, pp. 478.

Foale, S. and Bishop, S. R. (1994). Bifurcations in impact oscillations, *Nonlin. Dyn.* **6**, pp. 285–299.

Fougeres, R. and Sidoroff, F. (1989). The evolutive massing model and its application to cyclic plasticity and ageing, *Nuclear Eng. and Design* **114**, pp. 273-284.

Frederiksson, M. H. and Nordmark, A. B. (1997). Bifurcations caused by grazing incidence in many degrees of freedom impact oscillators, *Proc. R. Soc. Lond. A* **453**, pp. 1261–1276.

Fung, Y. C. and Kaplan, A. (1952). Buckling of low arches or curved beams of small curvature, *Technical Note 2840*, U.S. National Advisor Committee for Aeronautics.

Głowiński, R., Lions, J. L. and Trémolières, R. (1976a). *Analyse numérique des inéquations variationnelles. Tome 1*, Dunod, Paris, Théorie générale premières applications, Méthodes Mathématiques de l'Informatique, 5, pp. xii+268.

Głowiński, R., Lions, J. L. and Trémolières, R. (1976b). *Analyse numérique des inéquations variationnelles. Tome 2*, Dunod, Paris, Applications aux phénomènes stationnaires et d'évolution, Méthodes Mathématiques de l'Informatique, 5, pp. xiv+290.

Galvanetto, U., Bishhop, R. S. and Briseghella, L. (1993). Some remarks on the stick-slip vibrations of a two degree-of-freedom mechanical model, *Mech.*

Rev. Commun **20**, 6, pp. 459-466.

Galvanetto, U., Bishop, R. S. and Briseghella, L. (1995). Mechanical stick-slip vibration, *Int. J. Bif. Chaos* **5**, 3, pp. 637-651.

Gessing, R., Latarnik, M. and Skrzywan-Kosek, A. (1981). Exercises in Theory of Nonlinear Control, Wydawnictwa Naukowo-Techniczne, Warsaw (in Polish).

Giacaglia, G. E. (1972). Perturbation Methods in Nonlinear Systems, AMS 8, Springer Verlag, Berlin.

Gifford, S. J. and Tomlinson, G. R. (1989). Recent developments in the application of functional series to non-linear structures, *J. Sound Vibr.* **135**, pp. 289–317.

Gilmore, R. (1981). Catastrophe Theory for Scientists and Engineers, Wiley Interscience, New York.

Glocker, C. (1999). Velocity jumps induced by C^0-constraints, *Proc. ASME Design Eng. Tech. Conf.*, **DETC99/VIB-8344**, Las Vegas, Nevada, September 12-15, ASME, published on CD-ROM.

Glocker, C. and Pfeiffer, F. (1992). Dynamical systems with unilateral contacts, *Nonlin. Dyn.* **3**, pp. 245-259.

Gradshteyn, I. S. and Ryzhik, I. W. R. (1965). *Tables of integrals series and products.* Academic Press, New-York, pp. 933-942.

Grebogi, C. and Ott, E. (1983). Crises, sudden changes in chaotic attractors and transient chaos, *Physica* **7D**, pp. 181–200.

Grüner, G., Zawadowski, A. and Chaikin, P. M. (1981). Nonlinear conductivity and noise due to charge-density-Wave depinning in $NbSe_3$, *Phys. Rev. Lett.* **46**, pp. 511-516.

Guckenheimer, J. and Holmes, P. J. (1983). Nonlinear Oscillations, Dynamical Systems and Bifurcations of Vector Fields, Springer-Verlag, New York.

Gumowski, I. and Mira, C. (1980). Dynamique Chaotique, Cepadues éditions, Toulouse.

Gustavson, F. G. (1966). On constructing formal integrals of a Hamiltonian system near an equilibrium point, *The Astronomical J.* **71**, 8, pp. 670-675.

Hairer, E., Norsett, S. P. and Wanner, G. (1987). Solving Ordinary Differential Equations I, Springer-Verlag, Berlin.

Hairer, E. and Wanner, G. (1991). Solving Ordinary Differential Equations II. Springer-Verlag, Berlin.

Hale, J. K. (1969). Ordinary Differential Equations, Wiley Interscience, New York.

Hale, J. K. and Chow, S. N. (1982). Methods of Bifurcation, Springer Verlag, Berlin.

Hall, G. and Watt, J. M. (1976). Modern Numerical Methods for Ordinary Differential Equations, Clarendon Press, Oxford.

Han, R. P. S. and Luo, A. C. (1995). Chaotic motion of a horizontal impact pair, *J. Sound Vibr.* **181**, 2, pp. 231–250.

Hassard, B. D., Kazarinoff, N. D. and Wan, V. H. (1981). Theory and Applications of Hopf Bifurcation, Cambridge University Press.

Heng, H., Doerner, R., Hübinger, B. and Martienssen, W. (1994). Approaching nonlinear dynamics by studying the motion of a pendulum. I. Observing trajectories in state space, *Int. J. Bif. Chaos* **4**, 4, pp. 751–760.

Hess, D. P. and Soom, A. (1990). Friction at a lubricated linear contacts operating at oscillating sliding velocities, *Transactions of ASME. J. of Tribology* **112**, January, pp. 147–152.

Hinrichs, N., Oestreich, M. and Popp, K. (1998). On the modelling of friction oscillators, *J. Sound Vibr.* **216**, 3, pp. 435–459.

Holicke, M. and Awrejcewicz, J. (1997). Homoclinic chaos in the system of pendulum with friction, *Proc. 4th Conf. 'Dynamical Systems–Theory and Applications'*, (ed. J. Awrejcewicz, J. Grabski, and J. Mrozowski), pp. 121-124, Łódź, December 8-9, LODART S.A.

Holmes, P. J. (1981). Center manifolds, normal forms and bifurcations of vector fields with application to coupling between periodic and steady motions, *Physica D*, pp. 449–481.

Holmes, P. J. and Moon, F. C. (1983). Strange attractors and chaos in nonlinear mechanics, *J. Appl. Mech.* **50**, pp. 1021–1032.

Holmes, P. J. and Rand, R. H. (1980). Bifurcation of periodic motions in two weakly coupled Van der Pol oscillators, *Int. J. Nonlin. Mech.* **15**, pp. 387–399.

Horyacheva, I. H. and Dobychin, M. (1988). Contact Problems in Tribology, Mashynostroyeniye, Moscov (in Russian).

Housner, G. W., Bergman, L. A., Caughey, T. K., Chassiakos, A. G., Claus, R. O., Masri, S. F., Skelton, R. E., Soong, T. T., Spencer, B. F. and Yao, J. T. P. (1997). Structural control: past, present and future, *J. Eng. Mech.* **ASCE 123**, pp. 897–971.

Hsu, L. (1977). On the analysis of fourth order systems near a bifurcation, *Mech. Res. Comm.* **4**, 4, pp. 219–227.

Hsu, L. (1983a). Analysis of critical and post-critical behavior of non linear dynamical systems by the normal form method - Part I. Normalization formulae, *J. Sound Vibr.* **89**, 2, pp. 169–182.

Hsu, L. (1983b). Analysis of critical and post-critical behavior of non linear dynamical systems by the normal form method – Part II, *J. Sound Vibr.* **89**, 2, pp. 183–194.

Hsu, C. S. (1987). Cell-to-Cell Mapping, Springer Verlag, New-York.

Hyams, M. R. and Month, L. A. (1984). The origin of stability indeterminacy in a symmetric Hamiltonian, *J. Appl. Mech.* **51**, pp. 399-409.

Ibrahim, R. A. (1994). Friction-induced vibration, chatter, squel and chaos Part 1–2: Mechanics of contact and friction, *Appl. Mech. Rev.* **47**, 7, pp. 209–227.

Inman, D. J. (1989). Vibrations: with Control, Measurement and Stability, Prentice Hall, Englewood Cliffs, New Jersey, 1989.

Ionescu, I. and Paumier, J. C. (1993). Friction dynamique avec coefficient dépendant de la vitesse de glissement, *C. R. Acad. Sci. Paris Sér. I Math.* **316**, 1, pp. 121–125, 0764–4442, CASMEI.

Iooss, G. (1979). Bifurcation of Maps and Applications, North-Holland, Amsterdam, Math. Studies 36.

Iooss, G. (1988). Local Techniques in Bifurcation Theory and Nonlinear Dynamical Systems, *Chaotic Motions in Nonlinear Dynamical Systems*, Springer Verlag, (ed. W. Szemplińska and G. Iooss and F. Moon), CISM Courses and Lectures, No. 298.

Iooss, G. and Adelmayer, M. (1992). Topics in Bifurcation Theory and Applications, World Scientific, Singapore.

Iooss, G. and Joseph, D. D. (1980). Elementary Stability and Bifurcation Theory, Springer-Verlag, New York.

Irretier, H. (1983). The natural and forced vibrations in a wheel disk, *J. Sound Vibr.* **87**, pp. 161-177.

Itô, K. (1951). On stochastic differential equations, *Mem. Amer. Math. Soc.* **4**, pp. 1–51.

Ivanov, A. P. (1994). Impact oscillations: linear theory of stability and bifurcations, *J. Sound Vibr.* **178**, 3, pp. 361–378.

Iyengar, R. N. and Roy, D. (1996). Nonlinear dynamics of a rigid block on a rigid base, *J. Appl. Mech.* **63**, pp. 55-61.

Jézéquel, L. (1985). Synthèse Modale, Théorie et Extensions, Université Claude Bernard, Doctorat d'Etat No 8539, Lyon I.

Jézéquel, L. and Lamarque, C. -H. (1991). Nonlinear modal synthesis based on normal form theory, *Proc. IX-th Int. Modal Anal. Conf.*, pp. 101-108, Florence - Italy, September 10-12.

Jézéquel, L. and Lamarque, C. -H. (1992). Analysis of nonlinear structural vibrations by normal form theory, *J. Sound Vibr.* **149**, 3, pp. 429–459.

Jakubovich, V. A. and Starzinskii V. M. (1972). Linear Differential Equations with Periodic Coefficients and Their Applications, Nauka, Moscow.

Janin, O. (1996). Etude de Circuits de Chua Avec Discontinuités, Université Claude Bernard Lyon I, DEA Thesis, Lyon.

Janin, O. (2001). Etude de Systèmes Mécaniques à Non–Linéarités Irrégulières de Type Impact, Université Claude Bernard Lyon I, PhD Thesis, Lyon.

Jean, M. and Pratt, E. (1985). A system of rigid bodies with dry friction, *Int. J. Eng. Sci.* **23**, 5, pp. 497–513.

Johnson, T. L. and Rand, R. H. (1979). On the existence and bifurcation of minimal normal modes, *Int. J. Nonlin. Mech.* **14**, pp. 1–12.

Kahan, M. (1996). Approches Stochastiques Pour le Calcul des Ponts aux Séismes, Ecole Nationale des Ponts et Chaussées, Thèse de Doctorat, Marne-la-Vallée.

Karatzas, I. and Tseng, W. S. (1991). Brownian Motion and Stochastic Calculus, Springer Verlag, New-York.

Kauderer, H. (1958). Nonlinear Mechanics, Springer-Verlag, Berlin (in German).

Kennedy, P. (1992). Robust op amp realization of Chua's circuit, *Frequenz* **46**, pp. 66–80.

Kloeden, P. and Platen, E. (1992). Numerical Solutions of Stochastic Differential Equations, Springer Verlag, Berlin.

Komuro, M., Tokunaga, R., Matsumoto, T., Chua, L. O. and Hotta, A. (1991). Global bifurcation analysis of the double scroll circuit, *Int. J. Bif. Chaos* **1**, pp. 139–182.

Kovalenko, A. (1975). Thermoelasticity, Kiev (in Russian).

Krée, P. and Soize, C. (1982). Mécanique Aléatoire, Dunod, Paris.

Krahelskyy, I. and Hittis, N. (1987). Frictional Self-Oscillations, Moscow (in Russian).

Kree, P. (1982). Diffusion equation for multivalued stochastic differential equations, *J. Funct. Anal.* **49**, pp. 73–90.

Kreuzer, E. (1987). Numerische Untersuchung Nichtlinearer Dynamischer Systeme, Springer-Verlag, Berlin-Heidelberg.

Kriukov, B. I. (1984). Vynuzhdennyie Kolebania Sushchestvenno Nelineinykh Sistem (Forced Vibration of Strongly Non-Linear Systems), Mashinostroenie, Moskva (in Russian).

Kunert, A. and Pfeiffer, F. (1989a). Stochastic model for ratling in gear-boxes, *Symp. Nonlin. Dyn. Eng. Syst.*, pp. 425-439, University of Stuttgart, IUTAM.

Kunert, A. and Pfeiffer, F. (1989b). Stochastic model for ratling in gear-boxes, *IUTAM Symp. Nonlin. Dyn. Eng. Sys.*, (ed. W. Schiehlen), pp. 173-180, University of Stuttgart FRG, 21-25 August 1989, IUTAM, Springer Verlag.

Kunze, M. (2000). Non-Smooth Dynamical Systems, Springer Verlag, Berlin, Lecture Notes in Mathematics 1744.

Kuznetsov, A. P., Kuznetsov, S. P., Sataev, I. R. and Chua, L. O. (1996). Multi–parameter criticality in Chua's circuit at period doubling transition to chaos, *Int. J. Bif. Chaos* **6**, 1, pp. 119–148.

Lépingle, D. and Marois, C. (1987). Equations différentielles stochastiques multivoques unidimensionnelles, *Sémin. Proba.*, **21**, pp. 520–533.

Laghdir, M. and Monteiro Marques, M. D. P. (1995). Dynamics of a particle with damping, friction, and percussional effects, *J. Math. Anal. Appl.* **196**, 3, pp. 902–920, 0022–247X, JMANAK.

Lamarque, C. -H. (1992). Contribution à la Modélisation et àl'Identification des Systèmes Mécaniques Non-Linéaires, Ecole Centrale de Lyon, Phd Thesis No 92-32, Ecully.

Lamarque, C. -H. and Bastien, J. (2000). Numerical study of a forced pendulum with dry friction, *Nonlin. Dyn.* **23**, pp. 335–352.

Lamarque, C. -H., Janin, O. (2001). Modal analysis of mechanical systems with impact non-linearities: limitations to a modal superposition, *J. Sound Vibr.* **235**, 4, pp. 567–2000.

Lamarque, C. -H. and Jézéquel, L. (1992). Introduction of nonlinear complex modes by using normal form theory, *Proc. Conf. Exper. Num. Meth. in Struc. Dyn.*, pp. 937-946, Louvain - Belgium, 23-25 September, Katholieke Universiteit Leuven, Prof. dr. ir. P. Sas.

Lamarque, C. -H., Janin, O. and Awrejcewicz, J. (1999). Chua systems with discontinouties, *Int. J. Bif. Chaos* **9**, 4, pp. 591–616.

Lamarque, C. -H. and Malasoma, J. -M. (1992). Computation of basin of attraction for three coexisting attractors, *Eur. J. Mech. A/ Solids* **11**, 6, pp. 781–790.

Lamarque, C. -H. and Malasoma, J. -M. (1995). Nonlinear behaviour of a nonlinear mechanical systems, *Nonlinear Dynamics: New Theoretical and Applied Results*,(ed. J. Awrejcewicz), Akademie Verlag.

Lamarque, C. -H., Malasoma, J. -M. and Ouarzazi, M. N. (1998). Chaos in the convective flow of a fluid mixture in a porous medium, *Nonlin. Dyn.* **15**, pp. 83–102.

Lamba, H. and Budd, C. J. (1994). Scaling of Lyapunov exponents at nonsmooth bifurcations, *Phys. Rev.* **50**, pp. 84–90.

Landa, P. S. and Zaikin, A. A. (1996). Noise-induced phase transitions in a pendulum with a randomly vibrating suspension axis, *Phys. Rev.* **E54**, 4, pp. 1–10.

Leine, R. I., Van der Vrande, B. L. and Van Campen, D. H. (1999). Bifurcations in Nonlinear Discontinuous Systems, Eindhoven University of Technology.

Lenci, S. and Rega, G. (1998a). A procedure for reducing the chaotic response region in an impact mechanical system, *Nonlin. Dyn.* **15**, 4, pp. 391–409.

Lenci, S. and Rega, G. (1998b). Controlling nonlinear dynamics in a two–well impact system. I. Attractors and bifurcation scenario under symmetric excitations, *Int. J. Bif. Chaos* **8**, 12, pp. 2387–2407.

Lenci, S. and Rega, G. (1998c). Controlling nonlinear dynamics in a two-well impact system. II. Attractors and bifurcation scenario under unsymmetric optimal excitation, *Int. J. Bif. Chaos* **8**, 12, pp. 2409–2424.

Lenci, S. and Rega, G. (1999). Non-smooth dynamics and non-classical bifurcations in impulse-impact oscillators, *Proc. ASME Design Eng. Tech. Conf.*, **DETC99/VIB-8054**, Las Vegas, Nevada, September 12-15, ASME, published on CD-ROM.

Li, G. X. and Paidoussis, M. P. (1994). Impact phenomena of rotor-casing dynamical systems, *Nonlin. Dyn.*, **5**, pp. 53–70.

Li, T. Y. and Yorke, J. A. (1975). Period three implies chaos, *Am. Math. Mon.* **82**, pp. 985–992.

Ling, F. F. (1959). A quasi - iterative method for computing interface temperature distribution, *ZAMP*, No. 10, pp. 461–474.

Lorenz, E. N. (1963). Deterministic nonperiodic flow, *J. Atmos. Sci.* **20**, pp. 130–141.

Lozi, R. (1978). Un attracteur étrange du type attracteur de Hénon, *J. Phys.*, pp. 9–10.

Müller, P. C. (1995). Calculation of Lyapunov exponents for dynamic systems with discontinuities, *Chaos, Solitons and Fractals* **5**, pp. 1671–1681.

Madan, R. N. (1993). Chua's Circuit: A Paradigm for Chaos, World Scientific, Singapore, World Scientific Series on Nonlinear Science, Series B, vol. 1.

Mahla, A. I. and Badan Palhares, A. G. (1993). Chua's circuit with a discontinuous nonlinearity, *J. Circuits, Syst. Comp.* **3**, 1, pp. 231–237.

Malasoma, J. M., Lamarque, C.-H. and Jézéquel, L. (1994). Numerical observation of an intermittent transition from an equilibrium point to a chaotic attractor, *Phys. Letters A* **186**, pp. 126–132.

Marsden, J. E. and McCracken, M. (1976). The Hopf bifurcation and its applications, *Appl. Math. Sci.* **19**.

Martins, J. A. C., Oden, J. T. and Simoes, F. M. F. (1990). A study of static and kinetic friction, *Int. J. Eng. Sci.* **28**, 1, pp. 29–92.

Masri, S. F. and Caughey, T. K. (1966). On the stability of the impact damper,

J. Appl. Mech. **33**, pp. 586–592.

Matrosov, V. M. and Finogenko, I. A. (1995). Right-hand solutions of the differential equations of dynamics for mechanical systems with sliding friction, *J. Appl. Math. Mech.* **59**, 6, pp. 837–844.

Matrosov, V. M. and Finogenko, I. A. (1996a). To the theory of differential equations arising in the dynamics of systems with friction. I, *Differ. Equat.* **32**, 5, pp. 610–618.

Matrosov, V. M. and Finogenko, I. A. (1996b). On the existence of right–side solutions to differential equations of dynamics of mechanical systems with dry friction, *Differ. Equat.* **32**, 2, pp. 186–194.

McIntyre, M. E., Schumacher, R. T. and Woodhouse, J. (1981). Aperiodicity in bowered string motion, *Acoustica* **43**, 2, pp. 93-108.

Meirovitch, L. (1967). Analytical Methods in Vibrations, McMillan, London.

Melnikov, V. K. (1963). On the stability of the center for time periodic perturbations, *Trans. Moscow Math. Soc.* **12**, 1, pp. 56-72.

Mesnier, V., Lamarque, C.-H., Malasoma, J.-M. and Awrejcewicz, J. (1999). Comparison of the behaviour of exact and approximate modelling of an nonlinear oscillator with one degree-of-freedom, *J. Tech. Phys.* **39**, 1, pp. 101–108.

Mikhlin, Yu. V. (1995). Matching of local expansions in the theory of nonlinear vibrations, *J. Sound Vibr.* **182**, 2, pp. 577–588.

Mikhlin, Yu. V. (1996). Normal vibrations of a general class of conservative oscillators, *Nonlin. Dyn.* **11**, 1, pp. 1–16.

Mikhlin, Yu. V., A. F., Vakakis, A. F. and Salenger, G. (1998). Direct and inverse problems encountered in vibro-impact oscillations of a discrete systems, *J. Sound Vibr.* **216**, 2, pp. 227–250.

Minorsky, N. (1962). Nonlinear Oscillations, Van Nostrand, Princeton.

Molenaar, J., de Weger, J. G. and v.d. Water, W. (1999). Near-grazing behaviour of the harmonic oscillator, *Proc. ASME Design Eng. Tech. Conf.*, **DETC99/VIB-8036**, Las Vegas, Nevada, September 12-15, ASME, published on CD-ROM.

Monteiro Marques, M. D. P. (1994). An existence, uniqueness and regularity study of the dynamics of systems with one-dimensional friction, *Eur. J. Mech. A/Solids* **13**, pp. 277-306.

Month, L. A. and Rand, R. H. (1977). The stability of bifurcating periodic solutions in a two-degree-of-freedom non linear systems, *J. Appl. Mech.* **44**, pp. 782–784.

Month, L. A. and Rand, R. H. (1980). An application of the Poincaré map to the stability of nonlinear normal modes, *J. Appl. Mech.* **47**, September, pp. 645.

Moon, F., Holmes, W. and Khoury, S. (1991). Symbol dynamics maps of spatial-temporal chaotic vibrationsin a string of impact oscillators, *Chaos* **1**, 1, pp. 65–68.

Moreau, J. J. (1988). Unilateral contact and dry friction in finite freedom dynamics, *Non-smooth Mechanics and Applications*, Springer Verlag, (ed. J. J. Moreau and P. Panagiotopoulos), C.I.S.M. Courses and Lectures, pp.

1-81.

Nayfeh, A. H. (1973). Perturbation Methods, Wiley Interscience, New-York.

Nayfeh, A. H. (1993). Method of Normal Forms, Wiley, New-York.

Nayfeh, A. H. and Mook, D. T. (1979). Nonlinear Oscillations, Wiley, New-York.

Nordmark, A. B. (1991). Non-periodic motion caused by grazing incidence in an impact oscillator, *J. Sound Vibr.* **145**, 2, pp. 279–297.

Nordmark, A. B. (1992). Effects due to low velocity impact in mechanical oscillators, *Int. J. Bif. Chaos* **2**, 3, pp. 597–605.

Nordmark, A. B. (1997). Universal limit mapping in grazing bifurcations, *Phys. Rev. E* **55**, 1, pp. 266–270.

Nqi, F. and Schatzman, M. (1997). Computation of Lyapunov Exponents for a Dynamical System with Impact, *Publications du Laboratoire d'Analyse Numérique Lyon-Saint Etienne CNRS-UMR 5585*, (ed. LAN UMR CNRS 5585), LAN.

Oden, J. T. and Martins, J. C. A. (1985). Models and computational methods for dynamic friction phenomena, *Comput. Meth. Appl. Mech. Eng.* **52**, pp. 527-634.

Oksendal, B. (1992). Stochastic Differential Equations. An Introduction with Applications, Springer Verlag, Berlin.

Olesiak, Z. S. and Pyryev, Yu. (1999). On dynamical thermoelastic contact of two solids with heat and friction generation and wear, *Proc. Third Int. Cong. Ther. Stresses*, pp. 599-602, Cracow.

Olesiak, Z. S. and Pyryev, Yu. (2000). A nonlinear, nonstationary problem of frictional contact with inertia and heat generation taken into account, *Acta Mech.*, **143**, pp. 67–78.

Olesiak, Z. S., Pyryev, Yu. and Yevtushenko, A. (1997). Determination of temperature and wear during braking, *Wear* **210**, pp. 120–126.

Pak, C. H. (1989). On the stability behaviour of bifurcated normal modes in coupled nonlinear systems, *J. Appl. Mech.* **56**, March, pp. 155–161.

Pak, C. H. and Rosenberg, R. M. (1968). On the existence of normal mode vibrations in non linear systems, *Quaterly Appl. Math.* **XXVI**, 3, pp. 403–416.

Palmov, V. (1998). Vibrations of Elasto-Plastic Bodies, Springer-Verlag, Berlin.

Panet, M., Paoli, L. and Schatzman, M. (1995). Theoretical and numerical study for a model of vibrations with unilateral constraints, *Proc. Contact Mech. Int. Symp. 2*, (ed. M. Raous and M. Jean and J. J. Moreau), pp. 457-464, 19-23 September 1994, Plenum Press New York.

Paoli, L. (1993). Analyse Numérique de Vibrations Avec Contraintes Unilatérales, University of Lyon I, PhD. Thesis, Lyon, France.

Paoli, L. and Schatzman, M. (1993a). Vibrations avec contraintes unilatérales et perte d'énergie aux impacts, en dimension finie, *C. R. Acad. Sci. Paris Sér. I Math.* **317**, 1, pp. 97–101.

Paoli, L. and Schatzman, M. (1993b). Schéma numérique pour un modèle de vibrations avec contraintes unilatérales et perte d'énergie aux impacts, en dimension finie, *C. R. Acad. Sci. Paris Sér. I Math.* **317**, 2, pp. 211–215.

Paoli, L. and Schatzman, M. (1993c). Mouvement à nombre fini de degrés de liberté avec contraintes unilatérales: cas avec perte d'énergie, *Modèl. Math.*

Anal. Numér. **27**, pp. 673–717.

Paoli, L. and Schatzman, M. (1994). A numerical scheme for a dynamical impact problem with loss of energy in finite dimension, *Rapport interne No 167 - Février*, (ed. L. Paoli and M. Schatzman), Equipe d'Analyse Numérique LyonSaint-Etienne.

Paoli, L., Schatzman, M. and Panet, M. (1992). Vibrations with an obstacle and finite number of degrees of freedom, *Proc. EUROMECH 280*, pp. 159-164, Rotterdam, October 29-31, 1991, Ecole Centrale de Lyon, Balkema.

Pardoux, E. and Talay, D. (1985). Discretization and simulation of stochastic differential equations, *Acta Appl. Math.* **3**, pp. 23–47.

Peterka, F. (1996). Dynamics of mechanical systems with impacts and dry friction, *Proc. Second Eur. Nonlin. Oscill. Conf.*, Prague, September 9–13, pp. 33–44.

Peterka, F. (1999a). Global analysis of the double impact oscillator dynamics, *Proc. 1999 ASME Design Tech. Conf.*, September 1999, Las-Vegas, Nevada, **DETC99**, VIB-8039.

Peterka, F. (1999b). Analysis of motion of the impact-dry-friction pair of bodies and its application to the investigation of the impact dampers dynamics, *Proc. 1999 ASME Design Tech. Conf.*, September 1999, Las-Vegas, Nevada, **DETC99**, VIB-8350.

Peterka, F. and Szöllös, O. (1996). The stability of a symmetric two-degree-of-freedom system with impacts, *Proc. Second Eur. Nonlin. Oscill. Conf.*, Prague, September 9-13, pp. 345–348.

Peterka, F. and Vacik, J. (1992). Transition to chaotic motion in mechanical systems with impacts, *J. Sound Vibr.* **154**, 1, pp. 95–115.

Pfeiffer, F. (1988a). Seltsame Attraktoren in Zahnradgetrieben, *Ingenieur–Archiv* **58**, Springer Verlag, pp. 113–125.

Pfeiffer, F. (1988b). Über unstetige, insbesondere stoerregte Schwingungen, *Z. Flugwiss. Weltraumforsch* **12**, pp. 358-367.

Pfeiffer, F. (1992). Introduction to Dynamics, B. G. Teubner Verlagsgesselschaft, Stuttgart (in German).

Pfeiffer, F. (1994). Unsteady processes in machines, *Chaos* **4**, 4, pp. 693-705.

Pfeiffer, F. (1999). Impacts with friction. Theory and experiments, *Proc. ASME Design Eng. Tech. Conf.*, **DETC99/VIB-8346**, Las Vegas, Nevada, September 12-15, ASME, published on CD-ROM.

Pfeiffer, F. and Glocker, C. (1996). Multibody Dynamics with Unilateral Contacts, Wiley, New-York.

Pfeiffer, F. and Hajek, M. (1992). Stick-slip motions of turbine blade dampers, *Phil. Trans. R. Soc. London* **338**, pp. 503-517.

Pfeiffer, F. and Prestl, W. (1994). Hammering in diesel-engine driveline systems, *Nonlin. Dyn.* **5**,pp. 477–492.

Pilipchuk, V. N., Vakakis, A. F. and Azeez, M. A. F. (1997). Study of a class of subharmonic motions using a non-smooth temporal transformation (NSTT), *Physica D* **100**, pp. 145–164.

Pierre, C. and Shaw, S. W. (1991). Non-linear normal modes and invariant manifolds, *J. Sound Vibr.* **150**, 1, pp. 170-173.

Poincaré, H. (1987). Les Méthodes Nouvelles de la Mécanique Céleste, Librairie A. Blanchard, Paris.

Popp, K. and Stelter, P. (1990). Stick-slip vibrations and chaos, *Phil. Trans. R. Soc. Lond. A* **332**, pp. 89-105.

Pratt, T. K. and Williams, R. (1981). Non-linear analysis of stick-slip motion, *J. Sound Vibr.* **74**, 4, pp. 531-542.

Pyryev, Yu. (1994). Dynamical model of thermoelastic contact in conditions of frictional heating and limiting of heat expansion, *Friction and Wear* **15**, 6, pp. 941-948 (in Russian).

Pyryev, Yu. (2000). Frictional contact of the cylinder and the bush with inertia, heat generation and wear taking into account, *Phys.-Chem. Mech. Mater.* **36**, 3, pp. 53-58 (in Russian).

Pyryev, Yu. and Hrylitskyy, D. (1996). Transient problem of frictional contact for the cylinder with heat generation and wear, *Appl. Mech. Tech. Phys.* **37**, 6, pp. 99-104.

Rand, R. H. (1974). A direct method for non-linear normal modes, *Int. J. Nonlin. Mech.* **9**, pp. 363-368.

Rand, R. H., Pak, C. and Vakakis, A. F. (1992). Bifurcation of nonlinear normal modes in a class of two degree-of-freedom systems, *Acta Mech.* **3**, pp. 129-145.

Roberti, V. (1994). Contrôle de Structures: Théories et Applications, Ecole Centrale de Lyon.

Roberti, V., Lamarque, C. -H. and Jézéquel, L. (1995). Control strategies for a slender structure, *J. Struc. Contr.***1**, 2, pp. 25-48.

Roseau, M. (1986). Vibrations des Systèmes Mécaniques, Méthodes Analytiqueset Applications, Masson, Paris.

Rosenberg, R. M. (1961). On normal vibrations of a general class of non linear dual mode systems, *J. Appl. Mech.*, **June**, pp. 275-283.

Rosenberg, R. M. (1962). The normal modes of non linear n-degree-of-freedom systems, *J. Appl. Mech.*, **March**, pp. 7-20.

Rosenberg, R. M. (1964). On the existence of normal mode vibrations of non linear systems with two-degree-of-freedom, *Quaterly of Appl. Math.* **XXII**, 3, pp. 217-234.

Rosenberg, R. M. and Atkinson, C. P. (1959). On the natural modes and their stability in non linear two-degree-of-freedom systems, *J. Appl. Math.*, **September**, pp. 377-385.

Ruelle, D. and Takens, F. (1971). On the nature of turbulence, *Comm. Math. Phys.* **20**, pp. 167–192.

Rumpel, R. J. (1996). On the qualitative behaviour of nonlinear oscillators with dry friction, *Zeitschrift für Ang. Math. Mech.* **76**, pp. 1-2.

Schatzman, M. (1978). A class of nonlinear differential equations of second order in time, *Nonlin. Anal.* **2**, 3, pp. 355-373.

Schatzman, M., Bastien, J. and Lamarque, C. -H. (1999). An ill-posed mechanical problem with friction, *Eur. J. Mech. A/Solids* **18**, 3, pp. 415-420.

Schelleng, J. C. (1971). The bowed string and the player, *J. Acou. Soc. Am.* **53**, 1, pp. 26-41.

Sedov, L. I. (1971). A Course in Continuum Mechanics, Wolters-Noordhoff Publishing, Groningen.

Sekar, P. (1996). Chaotic Vibrations in Systems with Contacts and Impact Nonlinearities, Indian Institute of Technology, PhD Thesis, Madras, India.

Seydel, R. (1988). From Equilibrium to Chaos, Practical Bifurcation and Stability Analysis, Elsevier, New York.

Sharkovski, A. N. (1964). Coexistence of the cycles of a continuous mapping of the line into itself, *Ukr. Mat. Zh.* **16**, 1, pp. 61-71 (in Russian).

Shaw, S. W. (1985). The dynamics of a harmonically excited system having rigid amplitude constraints, Part I: Subharmonic motions and local bifurcations, and Part II: Chaotic motions and global bifurcations, *J. Appl. Mech.* **52**, pp. 453-464.

Shaw, S. W. (1986). On the dynamic response of a system with dry friction, *J. Sound Vibr.* **108**, 2, pp. 305-325.

Shaw, S. W. (1994). An invariant manifold approach to nonlinear normal modes of oscillation, *J. Nonlin. Sci.* **4**, pp. 419-448.

Shaw, S. W. and Holmes, P. J. (1983). A periodically forced piecewise linear oscillator, *J. Sound Vibr.* **90**, 1, pp. 129-155.

Shaw, S. W. and Pierre, C. (1992). Normal modes for non-linear vibratory systems, *J. Sound Vibr.* **164**, 1, pp. 85-124.

Shaw, S. W. and Pierre, C. (1994). Normal modes of vibration for non-linear continuous systems, *J. Sound Vibr.* **169**, 3, pp. 319-347.

Shaw, S. W. and Rand, R. H. (1989). The transition to chaos in a simple mechanical system, *Int. J. Nonlin. Mech.* **24**, 1, pp. 41-56.

Shaw, J. and Shaw, S. W. (1989). The onset of chaos in a two-degree-of-freedom impacting system, *J. Appl. Mech.* **56**, pp. 168-174.

Skeledon, A. C. (1994). Dynamics of a parametrically excited double pendulum, *Physica* **D75**, pp. 541-558.

Skeledon, A. C. and Mullin, T. (1992). Mode interaction in a double pendulum, *Phys. Lett.* **A166**, pp. 224-229.

Smith, H. L. (1986a). Periodic solutions of periodic competitive and cooperative systems, *SIAM J. Math. Anal.* **17**, 6, pp. 1289–1318.

Smith, H. L. (1986b). Normal forms for periodic systems, *J. Math. Anal. Applicat.* **113**, pp. 578–590.

Sparrow, C. (1982). Bifurcations in the Lorenz Equations, Springer-Verlag, New-York, Lecture Notes in Appl. Math.

Sri Namachchivaya, N. and Lin, Y. K. (1991). Method of stochastic normal forms, *Int. J. Nonlin. Mech.* **26**, 6, pp. 931-943.

Stelter, P. (1992). Nonlinear vibrations of structures induced by dry friction, *Nonlin. Dyn.* **3**, pp. 329-345.

Steward, W. C. (1968). Current voltage characteristics of Josephson junctions, *Appl. Phys. Lett.* **12**, pp. 277-286.

Stewart, D. E. (1996). A numerical method for friction problems with multiple contacts, J. Austral. Math. Soc. Ser. B **37**, pp. 288-308.

Stewart, D. E. and Trinkle, J. C. (1996). An implicit time-stepping scheme for rigid body dynamics with inelastic collisions and Coulomb friction, Inter.

J. Numer. Meth. Eng. **39**, 15, pp. 2673-2691.

Stewart, D. E. and Trinkle, J. C. (1997). Dynamics, friction, and complementarity problems, *Complementarity and Variational Problems (Baltimore, MD, 1995)*, SIAM, pp. 425-439, Philadelphia, PA.

Stiegelmeyr, A. and Pfeiffer, F. (1999). A time stepping algorithm for mechanical systems with unilateral contacts, *Proc. ASME Design Eng. Tech. Conf.*, **DETC99/VIB-8348**, Las Vegas, Nevada, September 12-15, ASME, published on CD-ROM.

Stoker, J. J. (1950). Nonlinear Vibrations, Interscience Publishers, New York.

Stuart, J. T. (1971). Non linear stability theory, *Annual Rev. Fluid Mech.* **3**, pp. 347-370.

Sung, C. K. and Yu, W. S. (1992). Dynamics of a harmonically excited impact damper: bifurcations and chaotic motion, *J. Sound Vibr.* **158**, 2, pp. 317-329.

Sygniewicz, J. (1991). Modelling of cooperation of the piston with piston rings and barrel, 615/149, Łódź Technical University (in Polish).

Szemplińska-Stupnicka, W., Plaut, R. H. and Hsieh, J. C. (1989). Period doubling and chaos in unsymmetric structures under parametric excitation, *J. Appl. Mech.* **56**, pp. 947-952.

Szemplińska-Stupnicka, W., Tyrkiel, E. and Zubrzycki, A. (2000). The global bifurcations that lead to transient tumbling chaos in a parametrically driven pendulum, *Int. J. Bif. Chaos* **10**, 9, pp. 2161-2175.

Szemplińska-Stupnicka, W. (1969). On the normal coordinates in an analysis of steady - state forced vibrations of a nonlinear multiple-degree-of freedom system, *Arch. Appl. Mech.* **5**, 21, pp. 603-621.

Szemplińska-Stupnicka, W. (1983). Nonlinear normal modes and the generalized Ritz method in the problems of vibrations of non linear elastic continuous systems, *J. Nonlin. Mech.* **18**, 2, pp. 149-165.

Szemplińska-Stupnicka, W. (1990). The Behaviour of Nonlinear Vibrating Systems; vol. 1 - Fundamental Concepts and Methods: Applications to Single-Degree-of-Freedom Systems, Kluwer Academic Publishers, Dordrecht.

Szemplińska-Stupnicka, W. (1990). The behaviour of Non-Linear Vibrating Systems, vol. 2, Kluwer, Dordrecht.

Szemplińska-Stupnicka, W.,Tyrkiel, E. and Zubrzycki, A. (2001). On the stability 'in the large' and unsafe initial disturbances in nonlinear oscillator, *Comp. Assisted Mech. Engng. Sci.* **8**, pp. 155-168.

The Numerical Algorithms Group (1997). NAG Fortran Library Manual, Mark 18, N.A.G., Oxford, Scientific Library and Software.

Thompson, I. M. T. and Ghaffari, R. (1982). Chaos after period doubling bifurcations in the resonance of an impact oscillator, *Phys. Lett.* **91A**, pp. 5–8.

Thompson, J. M. T. and Stewart, H. B. (1986). Nonlinear Dynamics and Chaos, Geometrical Methods for Engineers and Scientists, John Wiley and Sons, Chichester-New York-Brisbane-Toronto-Singapore.

Timoshenko, S. and Goodier, J. N. (1951). Theory of Elasticity, McGraw-Hill Book Company, New York.

Tomita, K. (1986). Periodically forced nonlinear oscillators, *Chaos*, Manchester University Press, A. V. Holden, pp. 211-236.

Tomlinson, G. R. and Chen, Q. (1996). Parametric identification of systems with dry friction and nonlinear stiffness using a time series model, *J. Vibr. Acous.* **118**, 2, pp. 252-263.

Toulemonde, C. and Gontier, C. (1998). Sticking motions of impact oscillators, *Eur. J. Mech. A/Solids* **17**, 2, pp. 339-366.

Trinkle, J. C., Pang, J. S., Sudarsky, S. and Lo, G. (1997). On dynamic multi-rigid-body contact problems with Coulomb friction, *Z. Angew. Math. Mech.* **77**, 4, pp. 267-279.

Tworzydlo, J., Becker, V. B. and Olden, J. T. (1992). Numerical modelling of friction induced vibrations and dynamic instabilities, *Friction-Induced Vibr., Chatter, Squeal and Chaos, ASME DE* **49**, pp. 13-32.

Tyrkiel, E., Szemplińska-Stupnicka, W. and Zubrzycki, A. (2000). On the boundary crises of chaotic attractors in nonlinear oscillators, *Comp. Assisted Mech. Eng. Sci.* **7**, pp. 743-755.

Ueda, Y. (1979). Randomly transitional phenomena in the system governed by Duffing's equation, *J. Stat. Phys.* **20**, pp. 181–196.

Ueda, Y. and Akamatsu, N. (1981). Chaotically transitional phenomena in the forced negative resistance oscillator, *Inst. Electr. Electro. Eng. Trans. Circ. Sys. CAS* **28**, pp. 217–223.

Ueda, Y., Duomoto, H. and Suehiro, Y. (1978). Randomly transitional Phenomena in the system governed by Mathieu's equation with nonlinear restoring term, *Report of Electric and Electronic Communication in Japan NLP-24*.

Ueda, Y. and Nanahara, T. (1978). Computer experiments on the solutions of nonlinear differential-difference equations for the phase locked loop with delay, *Report of Electric and Electronic Communication in Japan NLP-20*.

Ulitko, A. F. (1990). Three-dimensional motion of elastic solids, *Mech. Solid Body* **6**, pp. 55 (in Russian).

Urabe, M. (1967). Nonlinear Autonomous Oscillations, Analytical Theory, Academic Press, New York.

Urabe, M. and Reiter, A. (1966). Numerical Computation of nonlinear forced oscillations by Galerkin's procedure, *J. Math. Anal. Appl.* **14**, pp. 107-140.

Vakakis, A. F. (1992). Nonlinear normal oscillations in a strongly nonlinear discrete system, *J. Sound Vibr.* **158**, 2, pp. 341-361.

Vakakis, A. F., Manevitch, L. I., Mikhlin, Y. V., Pilipchuk, V. N. and Zevin, A. A. (1996). Normal Modes and Localization in Nonlinear Systems, Wiley, New York.

Valkering, T. P. (1994). Non-trivial dynamics in a driven string with impact nonlinearity, *J. Sound Vibr.* **175**, 3, pp. 397-422.

Van Campen, D. H., Van De Horst, E. L. B., Van Der Spek, J. A. W. and De Kraker, A. (1995). Dynamics of a multi-DOF beam system with discontinuous support, *Nonlin. Dyn.* **8**, pp. 453-466.

Van De Vorst, E. L. B., Van Campen, D. H. and De Kraker, A. (1996). Periodic solutions of a multi-DOF beam system with impact, *J. Sound Vibr.* **192**, pp. 913-925.

Van Der Spek, J. A. W., De Hoon, C. A. L., De Kraker, A. and Van Campen, D.H. (1994). Application of cell mapping methods to a discontinuous dynamic system, *Nonlin. Dyn.* **6**, pp. 87-99.

Van Horssen, W. T. (1999). A perturbation method based on integrating vectors and multiple scales, *SIAM J. Appl. Math.* **59**, 4, pp. 1444-1467.

Verhulst, F. (1996). Nonlinear Differential Equations and Dynamical Systems, Springer Verlag, Berlin.

Weinstein, A. (1973). Normal modes for non linear Hamiltonian systems, *Invertiones Mathematicae* **20**, pp. 47-57.

Whiston, G. S. (1987). Global dynamics of a vibro-impacting linear oscillator, *J. Sound Vibr.* **118**, pp. 395-429.

Whiston, G. S. (1992). Singularities in vibro-impact dynamics, *J. Sound Vibr.* **152**, 3, pp. 427-460.

Whiteman, W. E. and Ferri, A. A. (1996). Displacement-dependant dry friction damping of a beam-like structure, *J. Sound Vibr.* **198**, 3, pp. 313-329.

Wiggins, S. (1988). Global Bifurcation and Chaos, Analytical Methods, AMS 73, Springer Verlag, New York.

Wiggins, S. (1990). Introduction to Applied Nonlinear Dynamical Systems and Chaos, Springer, New York.

Wojtkiewicz, S. F. and Bergman, L. A. (2001). A moment specification algorithm for control of nonlinear systems driven by Gaussian white noise, *Nonlin. Dyn.* **24**, 1, pp. 17-30.

Wolf, J. P. (1994). Foundation vibration analysis using simple physical models, Prentice Hall, Englewood Cliffs, New Jersey.

Wolf, A., Swift, B., Swinney, H. and Vastano, J. (1985). Determining Lyapunov exponents from a time series, *Physica* **D16**, pp. 285-317.

Yagasaki, K. (1992). Chaotic dynamics od quasiperiodically forced oscillators detected by Melnikov's method, *SIAM J. Math. Anal.* **23**, pp. 1230-1254.

Yagasaki, K. (1994). Chaos in a pendulum with feedback control, *Nonlin. Dyn.* **6**, pp. 125-142.

Yagasaki, K. (1995). Melnikov's method and chaos in quasiperiodically forced systems, *Nonlinear Dynamics: New Theoretical Applied Results*, (ed. J. Awrejcewicz), Akademie Verlag.

Yagasaki, K. (1996). A simple feedback control system: bifurcations of periodic orbits and chaos, *Nonlin. Dyn.* **9**, pp. 391-417.

Yamada, K. and Kobori, T. (2001). Fundamental dynamics and control strategies for aseismic structural control, *Int. J. Solids Struc.* **38**, pp. 6079–6121.

Yamada, T. and Watanabe, S. (1971). On the uniqueness of solutions of stochastic differential equations, *J. Math. Kyoto Univ.* **11**, pp. 155-167.

Yim, S. C. S. and Lin, H. (1991). Nonlinear impact and chaotic response of slender rocking objects, *J. Eng. Mech.* **117**, 9, pp. 2079-2100.

Yosida, K. (1980). Functional Analysis, Springer Verlag, Berlin.

Zeidler, E. (1986). Nonlinear Functional Analysis and its Applications, vol.I, IIA, IIB, III, IV, Springer Verlag, New York.

Zeidler, E. (1995). Applied Functional Analysis, AMS 108, Springer Verlag, New York.

Zevin, A. A. (1984). Some conditions for the existence and stability of periodic oscillations in nonlinear non–autonomous Hamiltonian systems, *J. Appl. Math. Mech.* **48**, 4, pp. 451-458.

Zhu, W. Q., Ying, Z. G. and Soong, T. T. (2001). An optimal nonlinear feedback control strategy for randomly excited structural systems, *Nonlin. Dyn.* **24**, 1, pp. 31–51.

Zuo, L. and Curnier, A. (1992). Dynamics of piecewise linear systems, *Proc. Euromech 280*, (ed. L. Jézéquel and C. -H. Lamarque), pp. 141-148, Rotterdam, October 29 - 31 1991, Ecole Centrale de Lyon, Balkema.

Index